Edited by
Kikuo Ujihara

Output Coupling in Optical Cavities and Lasers

Related Titles

W. Vogel, D.-G. Welsch

Quantum Optics

2006

ISBN: 978-3-527-40507-7

S.M. Dutra

Cavity Quantum Electrodynamics
The Strange Theory of Light in a Box

2004

ISBN: 978-0-471-44338-4

H.-A. Bachor, T.C. Ralph

A Guide to Experiments in Quantum Optics

2004

ISBN: 978-3-527-40393-6

Kikuo Ujihara

Output Coupling in Optical Cavities and Lasers

A Quantum Theoretical Approach

WILEY-VCH

WILEY-VCH Verlag GmbH & Co. KGaA

The Author

Prof. Kikuo Ujihara
3-52-1-601 Chofugaoka, Chofu
Tokyo 182-0021
Japan

Library of Congress Card No.: applied for

British Library Cataloguing-in-Publication Data
A catalogue record for this book is available from the British Library.

Bibliographic information published by the Deutsche Nationalbibliothek
The Deutsche Nationalbibliothek lists this publication in the Deutsche Nationalbibliografie; detailed bibliographic data are available on the Internet at http://dnb.d-nb.de

Printed in the Federal Republic of Germany
Printed on acid-free paper

Cover Design Adam-Design, Weinheim
Typesetting Macmillan Publishing Solutions, Bangalore, India
Printing and Binding Strauss GmbH, Mörlenbach

ISBN: 978-3-527-40763-7

For Mieko

Contents

Output Coupling in Optical Cavities and Lasers: A Quantum Theoretical Approach
Kikuo Ujihara

ISBN: 978-3-527-40763-7

Preface

After a half-century from the birth of the laser, we now see lasers in a variety of locations in academic institutions and industrial settings, as well as in everyday life. The species of laser are diverse. The core of quantum-mechanical laser theory was established in the 1960s by the Haken school and Scully. Semiclassical gas laser theory was also established in the 1960s by Lamb. Subsequently, many theoretical works on lasers have appeared for specific types of lasers or for specific operation modes. So, laser science is now mature and seems to leave little to be elucidated. Laser science has evolved into many branches of quantum-optical science, including coherent interaction, nonlinear optics, optical communications, quantum-optical information, quantum computation, laser-cooled atoms, and Bose–Einstein condensation, as well as gravitational wave detection by laser interferometer. Laser light is typical classical light, in that it closely simulates the coherent state of light, while in recent years light with non-classical quality has claimed more and more attention.

The role of laser theory is to clarify the character and quality of laser light and to show how it arises. The Haken school considered the laser linewidth and the amplitude distribution, while Scully considered the number distribution of laser photons. Laser linewidth and photon number distribution are complementary aspects of the same laser phenomenon viewed from wave phase or corpuscular viewpoints. Analysis of a laser from these viewpoints is involved because of the interaction of many atoms and the optical field as well as the pumping and damping processes. Thus, a common recipe for treating the laser field is to assume a single-mode field and reduce the number of degrees of freedom of the field to one. Then one has a single time-dependent variable for the field or a photon distribution for a single mode. The cost of reducing the number of degrees of freedom for the field to one is to lose information regarding the spatial field distribution, especially the relation between the fields inside and outside the laser cavity.

The theme of this book is to discuss how to deal with this defect of standard laser theories. To fully incorporate the field degrees of freedom in a laser is to treat the output coupling of the laser cavity rigorously. When the output coupling loss of the cavity is incorporated, cavity mode quantization becomes a difficult task because of the associated losses. Usual field quantization, relying on the field expansion in terms of orthogonal modes, becomes impossible because decaying

Output Coupling in Optical Cavities and Lasers: A Quantum Theoretical Approach
Kikuo Ujihara

ISBN: 978-3-527-40763-7

cavity modes are non-orthogonal. A direct approach to this problem is to set the laser cavity in a much larger cavity that simulates the "universe." Quantization is accomplished using the normal modes of the larger cavity that includes the laser cavity. The cost of this procedure is to have an infinite number of field modes instead of the single mode in conventional theories.

The burden of the infinite number of field modes can be relaxed if we go to a collective field variable expressing the total electric field. Then, the laser equation of motion can be solved for the total electric field. Thermal or vacuum fluctuation affecting the laser field is incorporated automatically in this procedure. Quantum noise is introduced as the fluctuating force associated with the decay of the atomic dipole. The resulting expressions for the laser linewidth both below and above threshold have a common correction factor compared with the formula resulting from the theory assuming a single mode. This factor, called the excess noise factor, attracted the attention of many scientists, who discussed the origin of the factor. Various approaches to derive the factor have been published. In particular, Siegman proposed that the excess noise factor is the result of non-orthogonality or bi-orthogonality of cavity modes. The non-orthogonality is, in turn, a consequence of the open character of the laser cavity as compared to the closed structure of a fictitious "single"-mode cavity.

Using the orthogonal modes of the "universe," it can be shown that the relation between the field inside and outside the cavity is not determined simply by the transmission coefficient of the cavity mirror, because the thermal or vacuum field exists everywhere. Outside the cavity, the total field is the sum of the transmitted field and the ambient thermal or vacuum field.

In this book, we present a laser theory that takes into account the output coupling of the cavity and uses the orthogonal modes of the "universe." We analyze the wave aspect of the laser field in both a semiclassical and quantum-mechanical manner. In the quantum-mechanical analysis, we obtain the excess noise factor. We also present a simplified method to avoid the use of the modes of the "universe" where again the excess noise factor is derived. We analyze the spontaneous emission process in a cavity with output coupling to show that the respective spontaneous emission process in a cavity is *not* enhanced by the excess noise factor. In order to consider the physical origin of the excess noise factor, the theories of the excess noise factor are surveyed. Also, to compare the method taken in this book with other methods to treat output coupling, quantum theories on cavity output coupling or the input–output relation are surveyed.

We begin in Chapter 1 with a classical analysis of one-dimensional optical cavities with output coupling. Chapter 2 gives a quantum-mechanical analysis of the same cavities embedded in a larger cavity. Chapter 3 describes the necessary preliminaries for a quantum-mechanical laser analysis. This includes the Langevin force for the field in the case of the single-mode approximation, and those for atomic polarization and atomic inversion. As a reference for a full laser analysis that incorporates cavity output coupling, a laser theory assuming a single mode, which we call a quasimode, is presented in Chapter 4. Standard, conventional results on laser operation, especially on laser linewidth, are derived. Chapter 5 displays, for a laser with output coupling, the complete equations of motion for the field modes, atomic polarizations, and atomic inversions. The atomic variables are eliminated to obtain an equation for the total field. In Chapter 6 is shown the contour integral method of solution for the field equation utilizing the poles in the normalization constant in the field mode functions.

Chapter 7 gives semiclassical, linear gain analysis. Ignoring the Langevin forces and the gain saturation effect, it solves the field equation of motion using the Fourier expansion of the normalization constant of the field mode functions. The space-time structure of the linear build-up of the field is clarified. Chapter 8, giving a semiclassical, nonlinear gain analysis, improves Chapter 7 by incorporating the saturation effect in atomic inversion, but still ignoring the Langevin forces. The steady-state field distribution is determined. Chapter 9 improves Chapter 7 by incorporating the Langevin forces, but ignoring the saturation effect. This amounts to a quantum, linear gain analysis. The laser linewidth below threshold is determined. Chapter 10 gives a quantum, nonlinear gain analysis. This includes both the Langevin forces and the saturation effect, summarizing the results of the previous three chapters. The expression for the linewidth is shown to have two corrections compared to that in Chapter 4, one of which is the excess noise factor. Chapter 11 presents a simplified method of laser analysis that combines the effects of the Langevin forces and optical boundary conditions for traveling waves. Chapter 12 summarizes the results obtained in Chapters 7–11 and discusses various physical aspects of laser oscillation. The spontaneous emission process in a cavity with output coupling is analyzed in Chapter 13. Chapter 14 surveys theories of excess noise factor and finally Chapter 15 surveys quantum theories of output coupling.

The book is structured so that the reader can begin with basic quantum-mechanical knowledge and step up to rather complicated laser wave analyses. Knowledge of preliminary quantum mechanics, some preliminary operator algebra, simple contour integrals, Fourier transforms, and differential equations is assumed. Knowledge of elementary laser theory is also assumed. Knowledge of basic semiclassical laser analysis is preferable. Leaps in transforming one equation into the next are avoided as often as possible. Wherever the description of a topic is short and poor, the relevant literature is cited for the reader's reference. Problems are provided in Chapters 1–5.

Fully quantum-mechanical theories of excess noise or output coupling exist, but most of them are in a sophisticated form. Unfortunately, to treat a realistic cavity is involved, as we will see in this book. But theories to be compared easily with experiments will be of particular importance in view of the developing field of nonclassical light. We hope some of the published papers cited in this book meet this demand. There is no doubt that future papers will appear to improve the situation.

Kikuo Ujihara
Tokyo, August 2009

Acknowledgments

Thanks are due to Ichiro Takahashi for reading the manuscript.

The author is grateful to the authors of various papers who readily granted permission to cite portions of their work or to reproduce their figures. He also acknowledges the publishers of the journals or books, The American Physical Society (Ref. [30] of Chapter 14, Refs [6, 7, 14] of Chapter 15, Figure 13.1–13.4, and Figures 14.2–14.4), The Optical Society of America (Ref. [15] of Chapter 15, and Figure 15.1), The IEEE (Refs [6, 8] of Chapter 14), and Springer-Verlag (Ref. [25] of Chapter 14, and Ref. [2] of Appendices), for giving permission to cite or reproduce figures from their journals or books. The origins of the citations and figures are individually given in the book.

Output Coupling in Optical Cavities and Lasers: A Quantum Theoretical Approach
Kikuo Ujihara

ISBN: 978-3-527-40763-7

1
A One-Dimensional Optical Cavity with Output Coupling: Classical Analysis

In this chapter, a one-dimensional optical cavity with output coupling is considered. The optical cavity has transmission loss at one or both of the end surfaces. The classical, natural cavity mode is defined, and decaying or growing mode functions are derived using the cavity boundary conditions. A series of resonant modes appears. But these modes are not orthogonal to each other and are not suitable for quantum-mechanical analysis of the optical field inside or outside of the cavity. Hypothetical boundaries are added at infinity in order to obtain orthogonal wave mode functions that satisfy the cavity and infinity boundary conditions. These new mode functions are suitable for field quantization, where each mode is quantized separately and the electric field of an optical wave is made up of contributions from each mode. Some results of quantization are described in the next chapter. Chapter 3 deals with the usual quasimode model: a perfect cavity with distributed internal loss or with a fictitious loss reservoir.

1.1
Boundary Conditions at Perfect Conductor and Dielectric Surfaces

In a source-free space, the electric field **E** and the magnetic field **H** described using a vector potential **A** satisfy the following equations:

$$\nabla^2 \mathbf{A}(\mathbf{r}) - \frac{1}{c^2}\left(\frac{\partial}{\partial t}\right)^2 \mathbf{A}(\mathbf{r}) = 0 \tag{1.1}$$

$$\mathbf{E}(\mathbf{r}) = -\frac{\partial}{\partial t}\mathbf{A}(\mathbf{r}) \tag{1.2}$$

$$\mathbf{H}(\mathbf{r}) = \frac{1}{\mu}\nabla \times \mathbf{A}(\mathbf{r}) \tag{1.3}$$

Output Coupling in Optical Cavities and Lasers: A Quantum Theoretical Approach
Kikuo Ujihara

ISBN: 978-3-527-40763-7

where c is the velocity of light and μ is the magnetic permeability of the medium. We work in a Coulomb gauge where

$$\mathrm{div}\,\mathbf{A}(\mathbf{r}) = 0 \tag{1.4}$$

In this chapter we consider one-dimensional, plane vector waves that are polarized in the x-direction and propagated to the z-direction. Therefore we write

$$\mathbf{A}(\mathbf{r}) = A(z,t)\mathbf{x} \tag{1.5}$$

where $\mathbf{x}$ is the unit vector in the x-direction. At the surface of a perfect conductor that is vertical to the z-axis, the tangential component of the electric vector vanishes. The tangential component of the magnetic field should be proportional to the surface current. In the absence of a forced current, this condition is automatically satisfied: the magnetic field that is consistent with the electric field induces the necessary surface current. At the interface between two dielectric media, or at the interface between a dielectric medium and vacuum, the tangential components of both the electric and magnetic fields must be continuous. Thus, at the surface z_c of a perfect conductor,

$$\frac{\partial}{\partial t}A(z_c,t) = 0 \tag{1.6}$$

and at the interface z_i of dielectrics 1 and 2,

$$\frac{\partial}{\partial t}A_1(z_i,t) = \frac{\partial}{\partial t}A_2(z_i,t) \tag{1.7}$$

$$\left.\frac{\partial}{\partial z}A_1(z,t)\right|_{z=z_i} = \left.\frac{\partial}{\partial z}A_2(z,t)\right|_{z=z_i} \tag{1.8}$$

In Equation 1.8 we have dropped the magnetic permeability μ_1 and μ_2, assuming that both of them are equal to that in vacuum, μ_0, which is usually valid in the optical region of the frequency spectrum.

1.2
Classical Cavity Analysis

1.2.1
One-Sided Cavity

Consider a one-sided cavity depicted in Figure 1.1. This cavity consists of a lossless non-dispersive dielectric of dielectric constant ε_1, which is bounded by a perfect conductor at $z = -d$ and vacuum at $z = 0$. The outer space $0 < z$ is a vacuum of dielectric constant ε_0. Subscripts 1 and 0 will be used for the regions $-d < z < 0$ and $0 < z$, respectively. The velocity of light in the regions 1 and 0 are c_1 and c_0, respectively.

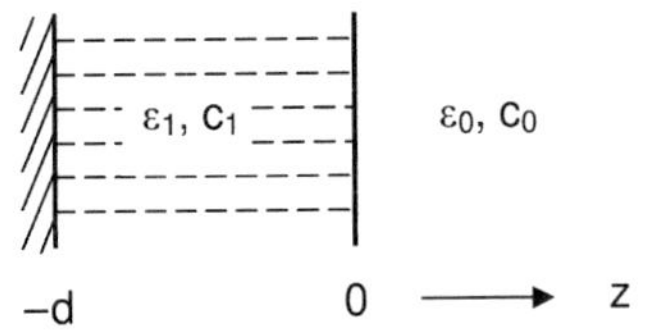

Figure 1.1 The one-sided cavity model.

The natural oscillating field mode of the cavity, the cavity resonant mode, is defined as the mode that has only an outgoing wave in the outer space $0 < z$. For reasons that will be described in Chapter 14, we also derive a mode that has only an incoming wave outside. For simplicity, let us call these the outgoing mode and incoming modes, respectively. Let the mode functions be

$$\begin{aligned} A(z,t) &= u(z)e^{-i\omega t}, \qquad -d < z < 0 \\ &= ve^{-i(\omega t \mp k_0 z)}, \qquad 0 < z \end{aligned} \tag{1.9}$$

where v is a constant. We define the wavenumber k by

$$k_i = \omega / c_i, \qquad i = 0, 1 \tag{1.10}$$

The upper and lower signs in the second line in Equation 1.9 are for the outgoing mode and the incoming mode, respectively. Substituting Equation 1.9 into Equation 1.1 via Equation 1.5 we obtain

$$\begin{aligned} -\frac{\omega^2}{c_1^2}u &= \left(\frac{d}{dz}\right)^2 u, \qquad -d < z < 0 \\ k_0 &= \frac{\omega}{c_0}, \qquad 0 < z \end{aligned} \tag{1.11}$$

Thus we can set

$$\begin{aligned} u(z) &= Ae^{ik_1 z} + Be^{-ik_1 z} \\ v &= C \end{aligned} \tag{1.12}$$

where $k_1 = \omega_k / c_1$. Putting this into Equation 1.6 for $z = -d$ and into Equations 1.7 and 1.8 for $z = 0$, we obtain

$$\begin{aligned} Ae^{-ik_1 d} + Be^{ik_1 d} &= 0 \\ A + B &= C \\ ik_1(A - B) &= \pm ik_0 C \end{aligned} \tag{1.13}$$

We then have

$$e^{2ik_1 d} = \frac{\mp k_0 - k_1}{k_1 \mp k_0} = \frac{\mp c_1 - c_0}{c_0 \mp c_1} \tag{1.14}$$

For the outgoing mode (upper sign) we have

$$e^{2ik_1 d} = \frac{-c_1 - c_0}{c_0 - c_1} = -\frac{c_0 + c_1}{c_0 - c_1} \tag{1.15}$$

Because we are assuming that both c_1 and c_0 are real and that the velocity of light in the dielectric is smaller than that in vacuum ($c_1 < c_0$), k_1 is a complex number K_{1out}. We reserve k_1 for the real part of K_{1out}. Then we obtain

$$\begin{aligned} K_{1out,m} &= k_{1m} - i\gamma \\ k_{1m} &= \frac{1}{2d}(2m+1)\pi, \qquad m = 0, 1, 2, 3, \ldots \\ \gamma &= \frac{1}{2d}\ln\left(\frac{c_0 + c_1}{c_0 - c_1}\right) = \frac{1}{2d}\ln\left(\frac{1}{r}\right) \end{aligned} \tag{1.16}$$

There is an eigenmode every π/d in the wavenumber. Note that the imaginary part is independent of the mode number. The coefficient

$$r = \frac{c_0 - c_1}{c_0 + c_1} \tag{1.17}$$

is the amplitude reflectivity of the coupling surface, $z = 0$, for the wave incident from the left, that is, from inside the cavity. The corresponding eigenfrequency of the mode is

$$\begin{aligned} \Omega_m &\equiv \Omega_{kout,m} = \omega_{cm} - i\gamma_c \\ \omega_{cm} &= \frac{c_1}{2d}(2m+1)\pi, \qquad m = 0, 1, 2, 3, \ldots \\ \gamma_c &= \frac{c_1}{2d}\ln\left(\frac{c_0 + c_1}{c_0 - c_1}\right) = \frac{c_1}{2d}\ln\left(\frac{1}{r}\right) \end{aligned} \tag{1.18a}$$

where we have defined the complex angular frequency Ω_m. In subsequent chapters, a typical cavity eigenfrequency, with a certain large number m, will be denoted as

$$\Omega_c = \omega_c - i\gamma_c \tag{1.18b}$$

The separation of the mode frequencies is $\Delta\omega_c = c_1\pi/d$.

Likewise, for the incoming mode (lower sign) we have

$$e^{2ik_1 d} = \frac{c_1 - c_0}{c_0 + c_1} = -\frac{c_0 - c_1}{c_0 + c_1} \tag{1.19}$$

from which we obtain

$$K_{1in,m} = K^*_{1out,m} = k_{1m} + i\gamma \tag{1.20a}$$

and

$$\Omega_{kin,m} = \Omega^*_{kout,m} = \omega_{cm} + i\gamma_c \equiv \Omega^*_m \tag{1.20b}$$

Going back to Equation 1.13 we now get the ratios of A, B, and C. Thus, except for an undetermined constant factor, for the outgoing mode we have

$$A(z,t) = u_m(z)e^{-i\Omega_m t} \tag{1.21a}$$

$$u_m(z) = \begin{cases} \sin\{\Omega_m(z+d)/c_1\}, & -d < z < 0 \\ \sin\{\Omega_m d/c_1\}e^{i\Omega_m(z/c_0)}, & 0 < z \end{cases} \tag{1.21b}$$

and for the incoming mode we have

$$A(z,t) = \tilde{u}_m(z)e^{-i\Omega_m^* t} \tag{1.22a}$$

$$\tilde{u}_m(z) = \begin{cases} \sin\{\Omega_m^*(z+d)/c_1\}, & -d < z < 0 \\ \sin\{\Omega_m^* d/c_1\}e^{-i\Omega_m^*(z/c_0)}, & 0 < z \end{cases} \tag{1.22b}$$

where the suffix m signifies the cavity mode. We note that the outgoing mode is temporally decaying whereas the incoming mode is growing. Inside the cavity, the field is a superposition of a pair of right-going and left-going waves with decaying or growing amplitudes. We note that $\tilde{u}_m(z) = u_m^*(z)$, meaning that the complex conjugate of the incoming mode function is the time-reversed outgoing mode function.

We also note that different members of the outgoing mode are non-orthogonal in the sense that

$$\int_{-d}^{0} u_m^*(z)\, u_{m'}(z)dz \neq 0, \qquad m \neq m' \tag{1.23}$$

Similarly, members of the incoming mode are mutually non-orthogonal. However, a member of the outgoing mode and a member of the incoming mode are approximately orthogonal. That is, if normalized properly, it can be shown that

$$\int_{-d}^{0} \tilde{u}_{out,m}^*(z)\, u_{in,m'}(z)dz \cong \delta_{m,m'} \tag{1.24}$$

The approximation here neglects the integrals of spatially rapidly oscillating terms. This is justified when the cavity length d is much larger than the optical wavelength $\lambda_k = 2\pi c_1/\omega_k$ or when $m \gg 1$ in Equation 1.16. These relationships among the outgoing and incoming mode functions will be discussed in Chapter 14 in relation to the quantum excess noise or the excess noise factor of a laser.

1.2.2
Symmetric Two-Sided Cavity

Consider a symmetrical, two-sided cavity depicted in Figure 1.2. This cavity consists of a lossless non-dispersive dielectric of dielectric constant ε_1, which is

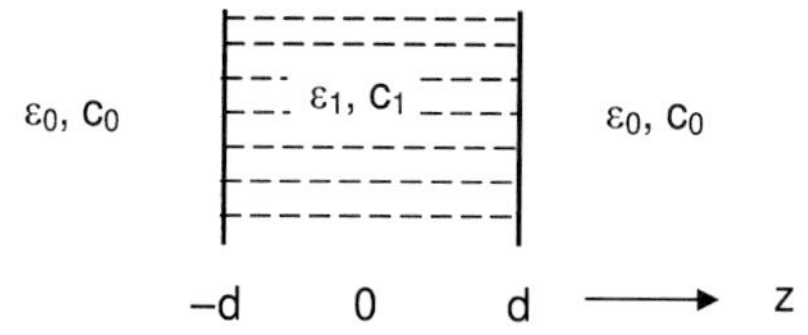

Figure 1.2 The symmetrical two-sided cavity model.

bounded by external vacuum at both $z = -d$ and $z = d$. Subscripts 1 and 0 will be used for the internal region $-d < z < d$ and external region $d < z$ and $z < -d$, respectively. The velocity of light in the regions 1 and 0 are c_1 and c_0, respectively.

Let the mode functions be

$$\begin{aligned} A(z,t) &= u(z)e^{-i\omega t}, && -d < z < d \\ &= ve^{-i(\omega t \mp k_0 z)}, && d < z \\ &= we^{-i(\omega t \pm k_0 z)}, && z < -d \end{aligned} \tag{1.25}$$

where again the upper signs are for the outgoing mode and the lower ones are for the incoming mode, and both v and w are constants. Following a similar procedure as above, this time we get symmetric and antisymmetric mode functions for both outgoing and incoming modes.

The symmetric outgoing mode function is (problem 1-1)

$$A(z,t) = \begin{cases} \cos(\Omega z/c_1)e^{-i\Omega t}, & -d < z < d \\ \cos(\Omega d/c_1)e^{-i\Omega\{t-(z-d)/c_0\}}, & d < z \\ \cos(\Omega d/c_1)e^{-i\Omega\{t+(z+d)/c_0\}}, & z < -d \end{cases} \tag{1.26}$$

where

$$\begin{aligned} \Omega &= \Omega_m = \omega_m - i\gamma_c \\ \omega_m &= \frac{c_1}{d}m\pi, \qquad m = 0, 1, 2, 3, \ldots \\ \gamma_c &= \frac{c_1}{2d}\ln\left(\frac{c_0 + c_1}{c_0 - c_1}\right) = \frac{c_1}{2d}\ln\left(\frac{1}{r}\right) \end{aligned} \tag{1.27}$$

The antisymmetric outgoing mode function is

$$A(z,t) = \begin{cases} \sin(\Omega z/c_1)e^{-i\Omega t}, & -d < z < d \\ \sin(\Omega d/c_1)e^{-i\Omega\{t-(z-d)/c_0\}}, & d < z \\ -\sin(\Omega d/c_1)e^{-i\Omega\{t+(z+d)/c_0\}}, & z < -d \end{cases} \tag{1.28}$$

where

$$
\begin{aligned}
\Omega &= \Omega_m = \omega_m - i\gamma_c \\
\omega_m &= \frac{c_1}{2d}(2m+1)\pi, \qquad m = 0, 1, 2, 3, \ldots \\
\gamma_c &= \frac{c_1}{2d}\ln\left(\frac{c_0 + c_1}{c_0 - c_1}\right) = \frac{c_1}{2d}\ln\left(\frac{1}{r}\right)
\end{aligned}
\tag{1.29}
$$

The symmetric and antisymmetric incoming mode functions are given by Equations 1.26 and 1.28, respectively, with Ω_m replaced by Ω_m^*. Note that the antisymmetric mode functions for $0 < z$, if shifted to the left by d ($z \to z+d$), coincide with the mode functions for the one-sided cavity in Equations 1.21a and 1.22a, as is expected from the mirror symmetry of the two-sided cavity. The relations 1.23 and 1.24 also hold in this cavity model.

1.3 Normal Mode Analysis: Orthogonal Modes

As we have seen in the previous section, the natural resonant modes (outgoing mode) of the cavity, as well as the associated incoming modes, are non-orthogonal and associated with time-decaying or growing factors. This feature is not suitable for straightforward quantization. For straightforward quantization, we need orthogonal, stationary modes describing the cavity. For this purpose, we introduce artificial boundaries at large distances so as to get such field modes.

1.3.1 One-Sided Cavity

1.3.1.1 Mode Functions of the "Universe"

For the one-sided cavity, we add a perfectly reflective boundary of a perfect conductor at $z = L$ as in Figure 1.3. Then we have three boundaries: at $z = -d$ and $z = L$ the boundary condition 1.6 applies, whereas at $z = 0$ the conditions 1.7 and 1.8 apply. The region $-d < z < L$ is our "universe," within which the region $-d < z < 0$ is the cavity and the region $0 < z < L$ is the outside space.

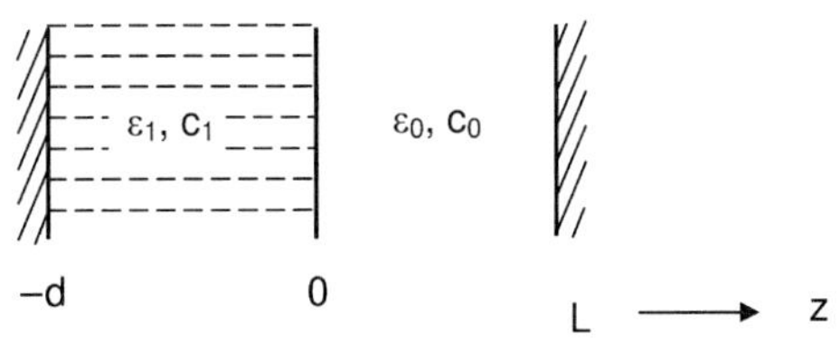

Figure 1.3 The one-sided cavity embedded in a large cavity.

Here, again, subscripts 1 and 0 will be used for the regions $-d < z < 0$ and $0 < z < L$, respectively. Assuming, again, the form of Equation 1.5 for the field, we assume the following form of the field:

$$A(z,t) = Q(t)U(z) \tag{1.30}$$

We try solutions of the form:

$$A_1(z,t) = Q(t)U_1(z,t), \quad -d < z < 0 \tag{1.31a}$$

$$A_0(z,t) = Q(t)U_0(z,t), \quad 0 < z < L \tag{1.31b}$$

Then Equation 1.1 gives

$$\left(\frac{d}{dt}\right)^2 Q(t) + \omega^2 Q(t) = 0 \tag{1.32a}$$

and

$$\begin{aligned} \left(\frac{d}{dz}\right)^2 U_1(z) + (k_1)^2 U_1(z) = 0 \\ \left(\frac{d}{dz}\right)^2 U_0(z) + (k_0)^2 U_0(z) = 0 \end{aligned} \tag{1.32b}$$

where

$$k_i = \omega/c_i = \omega(\varepsilon_i\mu_0)^{1/2}, \quad i = 0, 1 \tag{1.33}$$

Thus we assume the following spatial form:

$$U(z) = \begin{cases} U_1(z) = a_1 e^{ik_1 z} + b_1 e^{-ik_1 z}, & -d < z < 0 \\ U_0(z) = a_0 e^{ik_0 z} + b_0 e^{-ik_0 z}, & 0 < z < L \end{cases} \tag{1.34}$$

Applying the boundary conditions yields

$$a_1 e^{-ik_1 d} + b_1 e^{ik_1 d} = 0 \tag{1.35a}$$

$$a_1 + b_1 = a_0 + b_0 \tag{1.35b}$$

$$a_1 k_1 - b_1 k_1 = a_0 k_0 - b_0 k_0 \tag{1.35c}$$

$$a_0 e^{ik_0 L} + b_0 e^{-ik_0 L} = 0 \tag{1.35d}$$

For non-vanishing coefficients, we need the determinantal equation (problem 1-2)

$$\tan(k_0 L) = -(k_0/k_1)\tan(k_1 d) \tag{1.36}$$

or

$$c_1 \tan\frac{\omega d}{c_1} + c_0 \tan\frac{\omega L}{c_0} = 0 \tag{1.37}$$

Under this condition, the function A can be determined except for a constant factor as

$$\begin{aligned} A_1(z,t) &= f \sin k_1(z+d)\cos(\omega t+\phi), \qquad -d < z < 0 \\ A_0(z,t) &= f\frac{k_1 \cos k_1 d}{k_0 \cos k_0 L}\sin k_0(z-L)\cos(\omega t+\phi) \\ &= f\left(\frac{k_1}{k_0}\cos k_1 d \sin k_0 z + \sin k_1 d \cos k_0 z\right)\cos(\omega t+\phi), \quad 0 < z < L \end{aligned} \tag{1.38}$$

where ϕ is an arbitrary phase and f is an arbitrary constant. Equation 1.37 has been used in the last line.

1.3.1.2 Orthogonal Spatial Modes of the "Universe"

Now the allowed values of $k_{0,1}$ or ω are determined by Equation 1.37. If we choose a large L, $L \gg d$, it can be seen that the solution is distributed rather uniformly with approximate frequency, in k_0, of π/L, and that there is no degeneracy in k_0 and thus in ω. It can be shown that the space part of the jth mode functions in Equation 1.38, that is,

$$U_j(z) = \begin{cases} \sin k_{1j}(z+d), & -d < z < 0 \\ \left(\dfrac{k_{1j}}{k_{0j}}\cos k_{1j} d \sin k_{0j} z + \sin k_{1j} d \cos k_{0j} z\right), & 0 < z < L \end{cases} \tag{1.39}$$

form an orthogonal set in the sense that

$$\int_{-d}^{L} \varepsilon(z) U_i(z) U_j(z) dz = 0, \qquad i \neq j \tag{1.40a}$$

To show this relation, let us consider the integral

$$\begin{aligned} I &= \int_{-d}^{L} \frac{1}{\mu_0}\frac{\partial}{\partial z}U_i(z)\frac{\partial}{\partial z}U_j(z)dz \\ &= \frac{1}{\mu_0}U_i(z)\frac{\partial}{\partial z}U_j(z)\bigg|_{-d}^{0} + \frac{1}{\mu_0}U_i(z)\frac{\partial}{\partial z}U_j(z)\bigg|_{0}^{L} \\ &\quad - \frac{1}{\mu_0}\int_{-d}^{0} U_i(z)\left(\frac{\partial}{\partial z}\right)^2 U_j(z)dz - \frac{1}{\mu_0}\int_{0}^{L} U_i(z)\left(\frac{\partial}{\partial z}\right)^2 U_j(z)dz \end{aligned}$$

$$= \frac{k_{1j}^2}{\mu_0}\int_{-d}^{0} U_i(z)\,U_j(z)dz \;+\; \frac{k_{0j}^2}{\mu_0}\int_{0}^{L} U_i(z)\,U_j(z)dz$$

$$= \omega_j^2\left(\int_{-d}^{0} \varepsilon_1\,U_i(z)\,U_j(z)dz \;+\int_{0}^{L} \varepsilon_0\,U_i(z)\,U_j(z)dz\right) \tag{1.40b}$$

$$= \omega_j^2\int_{-d}^{L} \varepsilon(z)\,U_i(z)\,U_j(z)dz$$

In the second line, the values at $z = -d$ and $z = L$ vanish because of the condition on the perfect boundary, while the values at $z = 0$ cancel because of the continuity of both the function and its derivative. The Helmholtz equation 1.32a and 1.32b was used on going from the third to the fourth line. Finally, Equation 1.33 was used to go to the fifth line. Because we can interchange $U_i(z)$ and $U_j(z)$ in the first line, we also have

$$I = \omega_i^2\int_{-d}^{L} \varepsilon(z)\,U_i(z)\,U_j(z)dz \tag{1.40c}$$

Thus we have

$$0 = \left(\omega_j^2 - \omega_i^2\right)\int_{-d}^{L} \varepsilon(z)\,U_i(z)\,U_j(z)dz \tag{1.40d}$$

Since the modes are non-degenerate, the integral must vanish, which proves Equation 1.40a.

1.3.1.3 Normalization of the Mode Functions of the "Universe"

For later convenience, we normalize the mode function 1.39 as

$$U_j(z) = N_j u_j(z) \tag{1.41a}$$

$$u_j(z) = \begin{cases} \sin k_{1j}(z+d), & -d<z<0 \\ \left(\dfrac{k_{1j}}{k_{0j}}\cos k_{1j}d\sin k_{0j}z + \sin k_{1j}d\cos k_{0j}z\right), & 0<z<L \end{cases} \tag{1.41b}$$

with the orthonormality property

$$\int_{-d}^{L} \varepsilon(z)\,U_i(z)\,U_j(z)dz = \delta_{i,j} \tag{1.42a}$$

where the Kronecker delta symbol

$$\delta_{i,j} = \begin{cases} 1, & i=j \\ 0, & i\neq j \end{cases} \tag{1.42b}$$

It will be left for the reader to derive the normalization constant (problem 1-3):

$$
\begin{aligned}
N_j &= \sqrt{\frac{2}{\varepsilon_1\{d + (\cos k_{1j}d/\cos k_{0j}L)^2 L\}}} \\
&= \sqrt{\frac{2}{\varepsilon_1\{d + (1 - K\sin^2 k_{1j}d)L\}}} \qquad (1.43) \\
K &= 1 - \left(\frac{k_{0j}}{k_{1j}}\right)^2 = 1 - \left(\frac{c_1}{c_0}\right)^2
\end{aligned}
$$

The condition 1.37 has been used in the second line. Note that K is a constant for a given cavity. As will be discussed in Section 1.4, we will take the limit $L \to \infty$ and ignore the quantity d in Equation 1.43 in later applications of the one-sided cavity model.

1.3.1.4 Expansion of the Field in Terms of Orthonormal Mode Functions and the Field Hamiltonian

If the mode functions in Equation 1.41a form a complete set, which will be discussed in the last part of this section, a vector potential of any spatio-temporal distribution in the entire space $-d \le z \le L$, which vanishes at both ends, may be expanded in terms of these functions in the form

$$A(z,t) = \sum_k Q_k(t)\, U_k(z) \qquad (1.44)$$

where $Q_k(t)$ is the time-varying expansion coefficient. The corresponding electric and magnetic fields are found from Equations 1.2 and 1.3. In the following, we want to calculate the total Hamiltonian associated with the waves in Equations 1.39:

$$
\begin{aligned}
H &= \int_{-d}^{L}\left[\frac{\varepsilon}{2}E(z,t)^2 + \frac{\mu}{2}H(z,t)^2\right]dz \\
&= \int_{-d}^{L}\left[\frac{\varepsilon}{2}\left(\frac{\partial}{\partial t}A(z,t)\right)^2 + \frac{1}{2\mu}\left(\frac{\partial}{\partial z}A(z,t)\right)^2\right]dz
\end{aligned}
\qquad (1.45)
$$

Writing

$$\frac{d}{dt}Q_k = P_k \qquad (1.46)$$

we perform the integrations in Equation 1.45, which include, for the regions both inside and outside the cavity, the squared electric and magnetic fields for every member k and cross-terms of electric fields coming from different members k and k', and similar cross-terms for the magnetic field. The integration is done in

Appendix A. The resultant expression is very simple due to the orthogonality of the mode functions:

$$H = \frac{1}{2}\sum_k \left(P_k^2 + \omega_k^2 Q_k^2\right) \tag{1.47}$$

1.3.2
Symmetric Two-Sided Cavity

1.3.2.1 Mode Functions of the "Universe"

For the symmetric two-sided cavity, we impose a periodic boundary condition instead of perfect boundary conditions. Figure 1.4 depicts a two-sided cavity of a lossless non-dispersive dielectric of dielectric constant ε_1 extending from $z = -d$ to $z = d$. The exterior regions are vacuum with dielectric constant ε_0. We assume a periodicity with period $L + 2d$ and set another dielectric from $z = L + d$ to $z = L + 3d$. The region $-d < z < L + d$ is one period of our "universe" within which the region $-d < z < d$ is the cavity. The "universe" may alternatively be thought to exist in the symmetric region $-L/2 - d < z < L/2 + d$.

Here, again, subscripts 1 and 0 will be used for the regions $-d < z < d$ and $d < z < L + d$, respectively. Assuming again the form of Equation 1.5 for the field, we assume a solution of the form

$$A(z, t) = Q_k(t)\, U_k(z) \tag{1.48}$$

Equation 1.1 then yields

$$\left(\frac{d}{dt}\right)^2 Q_j(t) = -\omega_j^2 Q_j(t) \tag{1.49}$$

$$\left(\frac{d}{dz}\right)^2 U_j(z) = -k_j^2 U_j(z) \tag{1.50}$$

where $k_j = \omega_j / c$. A general solution of Equation 1.50 in the one period may be written as

$$U_{0j}(z) = A_j e^{ik_{0j}z} + B_j e^{-ik_{0j}z} \quad (d < z < L + d) \tag{1.51}$$

$$U_{1j}(z) = C_j e^{ik_{1j}z} + D_j e^{-ik_{1j}z} \quad (-d < z < d) \tag{1.52}$$

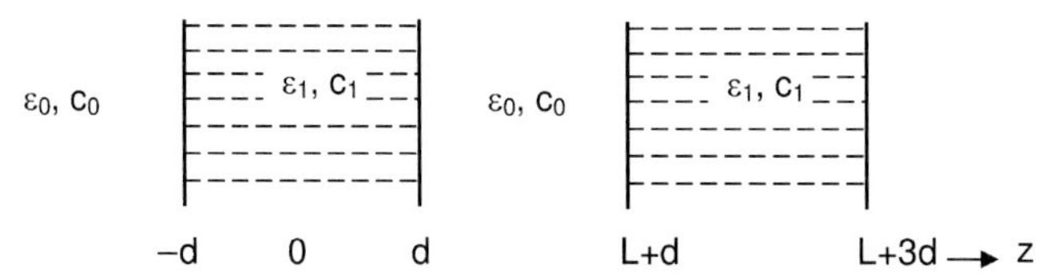

Figure 1.4 The two-sided cavity with the cyclic boundary condition.

where

$$k_{0,1j} = \omega_j / c_{0,1} \tag{1.53}$$

Applying the continuity boundary conditions at $z = d$ and the periodic boundary conditions at $z = -d$ and $Z = L + d$, one has

$$\begin{aligned} U_{1j}(d) &= U_{0j}(d) \\ U'_{1j}(d) &= U'_{0j}(d) \\ U_{1j}(-d) &= U_{0j}(L+d) \\ U'_{1j}(-d) &= U'_{0j}(L+d) \end{aligned} \tag{1.54}$$

The last two equations are obtained by combining the continuous conditions at $z = -d$ with the cyclic boundary conditions. With Equations 1.51 and 1.52, the coefficients A_j, B_j, C_j, and D_j must satisfy

$$\begin{aligned} C_j e^{ik_{1j}d} + D_j e^{-ik_{1j}d} &= A_j e^{ik_{0j}d} + B_j e^{-ik_{0j}d} \\ C_j k_{1j} e^{ik_{1j}d} - D_j k_{1j} e^{-ik_{1j}d} &= A_j k_{0j} e^{ik_{0j}d} - B_j k_{0j} e^{-ik_{0j}d} \\ C_j e^{-ik_{1j}d} + D_j e^{ik_{1j}d} &= A_j e^{ik_{0j}(L+d)} + B_j e^{-ik_{0j}(L+d)} \\ C_j k_{1j} e^{-ik_{1j}d} - D_j k_{1j} e^{ik_{1j}d} &= A_j k_{0j} e^{ik_{0j}(L+d)} - B_j k_{0j} e^{-ik_{0j}(L+d)} \end{aligned} \tag{1.55}$$

It is left to the reader to show that the determinantal equation for non-zero values of the coefficients is

$$\left(1 - \frac{k_{1j}}{k_{0j}}\right)^2 \sin^2\left(k_{1j}d - \frac{k_{0j}L}{2}\right) = \left(1 + \frac{k_{1j}}{k_{0j}}\right)^2 \sin^2\left(k_{1j}d + \frac{k_{0j}L}{2}\right) \tag{1.56}$$

which reduces to two equations:

$$\tan(k_{1j}d) = -\frac{c_0}{c_1}\tan\left(\frac{k_{0j}L}{2}\right) \qquad (a \text{ mode}) \tag{1.57a}$$

$$\tan(k_{1j}d) = -\frac{c_1}{c_0}\tan\left(\frac{k_{0j}L}{2}\right) \qquad (b \text{ mode}) \tag{1.57b}$$

Thus we have two sets of eigenvalues of wavenumber k_j or eigenfrequency ω_j. We refer to the modes determined by Equation 1.57a as a modes and those determined by Equation 1.57b as b modes. Graphical examination shows that the a mode and b mode solutions appear alternately on the angular frequency axis. Then we derive mode functions from Equations 1.55 and 1.57a and 1.57b as (problem 1-4):

$$U_j^a(z) = \alpha_j \times \begin{cases} \sin(k_{1j}z) & (-d < z < d) \\ \sin(k_{1j}d)\cos\{k_{0j}(z-d)\} & \\ + \dfrac{c_0}{c_1}\cos(k_{1j}d)\sin\{k_{0j}(z-d)\} & (d < z < L + d) \end{cases} \tag{1.58a}$$

$$U_j^b(z) = \beta_j \times \begin{cases} \cos(k_{1j}z) & (-d < z < d) \\ \cos(k_{1j}l)\cos\{k_{0j}(z-d)\} & \\ - \dfrac{c_0}{c_1}\sin(k_{1j}l)\sin\{k_{0j}(z-d)\} & (d < z < L + d) \end{cases} \tag{1.58b}$$

1.3.2.2 Orthonormal Spatial Modes of the "Universe"

It can be shown that the two different members, each from either *a* mode or *b* mode, are orthogonal in the sense of Equation 1.40a. They are normalized in the sense of Equation 1.42a if the constants α_j and β_j are given by

$$\alpha_j = \sqrt{\frac{2}{\varepsilon_1\{2d + (1 - K\sin^2 k_{1j}d)L\}}} \tag{1.59a}$$

$$\beta_j = \sqrt{\frac{2}{\varepsilon_1\{2d + (1 - K\cos^2 k_{1j}d)L\}}} \tag{1.59b}$$

where K was defined in Equation 1.43. This can be derived by repeated use of the determinantal equations 1.57a and 1.57b. As will be discussed in Section 1.4, we will take the limit $L \to \infty$ and ignore the quantity $2d$ in Equations 1.59a and 1.59b in later applications of the two-sided cavity model.

1.3.2.3 Expansion of the Field in Terms of Orthonormal Mode Functions and the Field Hamiltonian

If the mode functions in Equations 1.58a and 1.58b form a complete set, which will be discussed in Section 1.6, a vector potential of any spatio-temporal distribution in the entire space $-d < z < L + d$ or $-L/2 - d < z < L/2 + d$ may be expanded in terms of these functions in the same form as in Equation 1.44,

$$A(z,t) = \sum_k Q_k(t)\, U_k(z) \tag{1.60}$$

where $Q_k(t)$ is the time-varying expansion coefficient. The total Hamiltonian defined as in Equation 1.45, with the upper limit of integration replaced by $L + d$, can be evaluated again defining the "momentum" P_k associated with the "amplitude" Q_k as in Equation 1.46. Using Equations 1.2 and 1.3, we perform the integrations as in Equation 1.45, which include, for the regions both inside and outside the cavity, the squared electric and magnetic field for every member k from both the *a* mode and *b* mode functions, and cross-terms of electric fields coming from different members k and k' and similar cross-terms for the magnetic field.

All the cross-terms vanish on integration due to the orthogonality of the mode functions. The resultant expression is the same as Equation 1.47:

$$H = \frac{1}{2}\sum_k (P_k^2 + \omega_k^2 Q_k^2) \tag{1.61}$$

Note that the mode index k here includes both a mode and b mode functions.

1.4
Discrete versus Continuous Mode Distribution

The length L, expressing the extent of the outside region, was introduced for mathematical convenience. As we have seen, this allowed us to obtain discrete, orthogonal mode functions, which are stationary. We eventually normalized them. Because the physical content of the outside region is the free space outside the cavity, there is no reason to have a finite value of L. On the contrary, if L is finite (comparable to d), various artifacts may arise due to reflections at the perfect boundary at $z = L$ in the case of one-sided cavity or at the neighboring cavity surface in the case of the two-sided cavity. For this reason, we take the limit $L \to \infty$ in what follows.

In this limit, in the case of the one-sided cavity, the normalization constant reduces to

$$N_j = \sqrt{\frac{2}{\varepsilon_1 L(1 - K\sin^2 k_{1j}d)}} \tag{1.62a}$$

and the normalized mode function is

$$U_j(z) = \sqrt{\frac{2}{\varepsilon_1 L(1 - K\sin^2 k_{1j}^2 d)}} \times \begin{cases} \sin k_{1j}(z+d), & -d < z < 0 \\ \left(\dfrac{k_{1j}}{k_{0j}}\cos k_{1j}d \sin k_{0j}z + \sin k_{1j}d \cos k_{0j}z\right), & 0 < z < L \end{cases} \tag{1.62b}$$

The mode distribution in the frequency domain is determined by Equation 1.37. For $L \to \infty$, the spacing $\Delta\omega$ of the two eigenfrequencies is

$$\Delta\omega = (c_0/L)\pi \tag{1.63}$$

which is infinitely small and the modes distribute continuously. The density of modes (the number of modes per unit angular frequency) is

$$\rho(\omega) = \frac{L}{\pi c_0} \tag{1.64}$$

We note that the maxima of the normalization constant N_j occur at the cavity resonant frequencies given by Equation 1.18a.

In the case of the two-sided cavity, the normalization constants become

$$\alpha_j = \sqrt{\frac{2}{\varepsilon_1 L(1 - K \sin^2 k_{1j} d)}} \tag{1.65a}$$

$$\beta_j = \sqrt{\frac{2}{\varepsilon_1 L(1 - K \cos^2 k_{1j} d)}} \tag{1.65b}$$

It is easy to see from Equations 1.57a and 1.57b that the a mode and b mode appear in pairs along the frequency axis and, in the limit $L \to \infty$, every pair is degenerate. The separation of the pairs is now

$$\Delta\omega = (2c_0/L)\pi \tag{1.66}$$

so that the density of modes is

$$\rho_a(\omega) = \rho_b(\omega) = \frac{1}{2}\rho(\omega) = \frac{L}{2\pi\, c_0} \tag{1.67}$$

For both the one-sided and the two-sided cavities, the overall density of modes becomes independent of the cavity size and is equal to $L/\pi\, c_0$.

In what follows we sometimes encounter the summation of some mode-dependent quantity B_k over modes of the "universe." Such a summation is converted to an integral as follows:

$$\sum_k B_k \to \int_0^\infty B_{\omega_k} \rho(\omega_k) d\omega_k \tag{1.68a}$$

for the case of a one-sided cavity, and

$$\begin{aligned} \sum_k B_k &\to \int_0^\infty \left\{ B^a_{\omega_k} \rho_a(\omega_k) + B^b_{\omega_k} \rho_b(\omega_k) \right\} d\omega_k \\ &= \frac{1}{2} \int_0^\infty (B^a_{\omega_k} + B^b_{\omega_k}) \rho(\omega_k) d\omega_k \end{aligned} \tag{1.68b}$$

for the case of a two-sided cavity. Correspondingly, the Kronecker delta symbol becomes a Dirac delta function by the rule

$$\rho(\omega_k)\delta_{k,k'} \to \delta(k - k') \tag{1.69}$$

because for a k-dependent variable f_k we should have $\sum_k f_k \delta_{k,k'} = \int dk f(k) \delta(k - k')$. We note that the maxima of the a (b) mode occur at the cavity resonant frequencies of the antisymmetric (symmetric) modes given by Equation 1.29 (Equation 1.27).

1.5
Expansions of the Normalization Factor

The squared normalization constant for the one-sided cavity, Equation 1.62a, divided by $2/(\varepsilon_1 L)$ has two expansions that are frequently used in subsequent sections and chapters (problems 1-6 and 1-7):

$$\begin{aligned}\frac{1}{1-K\sin^2 k_{1j}d} &= \frac{2c_0}{c_1}\left\{\sum_{n=0}^{\infty}\frac{1}{1+\delta_{0,n}}(-r)^n\cos 2nk_{1j}d\right\}\\ &= \sum_{m=-\infty}^{\infty}\frac{c_0\gamma_c/d}{\gamma_c^2+(\omega_j-\omega_{cm})^2}\end{aligned} \tag{1.70a}$$

where the coefficient r was defined in Equation 1.17 and $\omega_j = c_1 k_{1j}$. The coefficients ω_{cm} and γ_c were defined in Equation 1.18a. The first expansion is a Fourier series expansion and the second one in terms of cavity resonant modes comes from the Mittag–Leffler theorem [1], which states a partial fraction expansion based on the residue theory. Similar expansions exist for the normalization constants for the two-sided cavity in Equations 1.65a and 1.65b [2]. The expansion for Equation 1.65a is the same as in Equation 1.70a with $\omega_{cm} \to \omega_{cm}^a$:

$$\begin{aligned}\frac{1}{1-K\sin^2 k_{1j}d} &= \frac{2c_0}{c_1}\left\{\sum_{n=0}^{\infty}\frac{1}{1+\delta_{0,n}}(-r)^n\cos 2nk_{1j}d\right\}\\ &= \sum_{m=-\infty}^{\infty}\frac{c_0\gamma_c/d}{\gamma_c^2+(\omega_j-\omega_{cm}^a)^2}\\ \frac{1}{1-K\cos^2 k_{1j}d} &= \frac{2c_0}{c_1}\left\{\sum_{n=0}^{\infty}\frac{1}{1+\delta_{0,n}}(r)^n\cos 2nk_{1j}d\right\}\\ &= \sum_{m=-\infty}^{\infty}\frac{c_0\gamma_c/d}{\gamma_c^2+(\omega_j-\omega_{cm}^b)^2}\end{aligned} \tag{1.70b}$$

where $\omega_{cm}^a = (2m+1)(\pi c_1/2d)$ and $\omega_{cm}^b = 2m(\pi c_1/2d)$ (m is an integer); ω_{cm}^a (ω_{cm}^b) is the resonant frequency of the antisymmetric (symmetric) mode function defined in Equation 1.29 (Equation 1.27).

1.6
Completeness of the Modes of the "Universe"

Concerning the expansion of the field in terms of the mode functions of the "universe," it was mentioned above Equation 1.44 that the latter mode functions must form a complete set. Completeness of a set of functions means the possibility of expanding an arbitrary function, in a defined region of the variable(s), in terms of them. The set of orthogonal functions in Equations 1.41a and 1.41b fulfills this property. Assume that an arbitrary function $\Psi(z)$ defined in the region $-d < z < L$ is expanded as

$$\Psi(z) = \sum_i A_i U_i(z) \tag{1.71}$$

where A_i is a constant. Multiplying both sides by $\varepsilon(z) U_j(z)$ and integrating, we have

$$\int_{-d}^{L} \varepsilon(z) U_j(z) \Psi(z) dz = \int_{-d}^{L} \sum_i A_i \varepsilon(z) U_j(z) U_i(z) dz = \sum_i A_i \delta_{ji} = A_j \tag{1.72}$$

where we have used Equation 1.42a in the second equality. Substituting this result in Equation 1.71 we have

$$\begin{aligned}\Psi(z) &= \sum_i \int_{-d}^{L} \varepsilon(z') U_i(z') \Psi(z') dz' U_i(z) \\ &= \int_{-d}^{L} \left\{ \sum_i \varepsilon(z') U_i(z') U_i(z) \right\} \Psi(z') dz'\end{aligned} \tag{1.73}$$

Because $\Psi(z)$ is arbitrary, the quantity in the curly bracket should be a delta function:

$$\sum_i \varepsilon(z') U_i(z') U_i(z) = \delta(z' - z) \tag{1.74}$$

In integral form it reads

$$\int_0^\infty \varepsilon(z') U_i(z') U_i(z) \rho(\omega_i) d\omega_i = \delta(z' - z) \tag{1.75}$$

This is a necessary condition for completeness. Conversely, if Equation 1.75 holds, we can use Equation 1.73 to find the expansion coefficient in the form of Equation 1.72. Thus Equation 1.75 is also sufficient for completeness.

Whether the mode functions in Equations 1.41a and 1.41b really fulfill this condition is another problem. For example, for the case $-d < z < 0$ and $-d < z' < 0$, using Equation 1.62b, we need to show that

$$\int_0^\infty d\omega \frac{L}{\pi c_0} \varepsilon_1 \frac{2}{\varepsilon_1 L} \frac{1}{1 - K \sin^2 k_1 d} \sin k_1(z + d) \sin k_1(z' + d) = \delta(z' - z) \tag{1.76}$$

The squared normalization constant N_ω^2 is expanded in terms of cos $2nk_1 d$, $n = 0$, 1, 2, 3,..., as in Equation 1.70a. So, except for constant factors, the integrand becomes a sum of integrals of the form

$$\int_0^\infty \cos\{2nd \pm (z - z')\} k_1 dk_1$$

or

$$\int_0^{\infty} \cos\{2nd \pm (z + z' + 2d)\} k_1 dk_1$$

We apply the formula [3]

$$\int_0^{\infty} \cos zk\, dk = \pi\delta(z) \tag{1.77}$$

Noting that $\delta(z \neq 0) = 0$, we find for the above combination of z and z' that

$$\begin{aligned} &\int_0^{\infty} d\omega \frac{L}{\pi c_0} \varepsilon_1 \frac{2}{\varepsilon_1 L} \frac{1}{1 - K\sin^2 k_1 d} \sin k_1(z + d) \sin k_1(z' + d) = \delta(z' - z) \\ &-d < z < 0, \quad -d < z' < 0 \end{aligned} \tag{1.78}$$

where we have discarded the term $-\delta(z + z' + 2d)$ because it is meaningful only at the perfect boundary, $z = z' = -d$, where all the fields vanish physically. Other combinations of the regions for z and z' can be examined in the same way. We have

$$\begin{aligned} \sum_k \varepsilon(z') U_k(z') U_k(z) &= \int_0^{\infty} \varepsilon(z') U_k(z') U_k(z) \rho(\omega_k) d\omega_k \\ &= \delta(z' - z) \end{aligned} \tag{1.79}$$

for $-d < z < L$, $-d < z' < L$, except $z = z' = 0$. The exception at $z = z' = 0$ occurs because at $z = 0$ the dielectric constant is unspecified. Also, the boundary conditions demand that the fields should be continuous across this boundary, so that a delta function at $z = 0$ is prohibited. The completeness of the mode functions in the case of two-sided cavities can similarly be examined.

▶ Exercises

1.1 For the symmetrical, two-sided cavity model, derive the resonant frequencies for the outgoing modes. Also derive the resonant frequencies of the incoming modes.
1-1. Set $u(z) = A\exp(ik_1 z) + B\exp(-ik_1 z)$. Then the boundary conditions at $z = d$ and $z = -d$ give, respectively,

$$\frac{A}{B} = \frac{k_1 \pm k_0}{k_1 \mp k_0} e^{-2ik_1 d}, \qquad \frac{A}{B} = \frac{k_1 \mp k_0}{k_1 \pm k_0} e^{2ik_1 d}$$

Therefore we have

$$e^{2ik_1 d} = +\frac{k_1 \pm k_0}{k_1 \mp k_0} \quad \text{or} \quad -\frac{k_1 \pm k_0}{k_1 \mp k_0}$$

For $e^{2ik_1 d} = +(k_1 \pm k_0)/(k_1 \mp k_0)$ we have $A = B$ and have symmetric modes. With the upper signs, a symmetric outgoing mode is obtained; and with the lower signs, an incoming symmetric mode is obtained. For $e^{2ik_1 d} = -(k_1 \pm k_0)/(k_1 \mp k_0)$ we have $A = -B$ and have antisymmetric modes. With the upper signs, an antisymmetric outgoing mode is obtained; and with the lower signs, an antisymmetric incoming mode is obtained.

1.2 Derive the determinantal equation 1.37 and the mode function in Equation 1.38.
1-2. Delete b_1 and b_0 from Equations 1.35b and 1.35c using Equations 1.35a and 1.35d and divide side by side to obtain Equation 1.36. Next express U_1 and U_0 in terms of a_1 and a_0. Determine a_0/a_1 by the modified version of Equation 1.35c to eliminate a_0. Finally, set $2ia_1 \exp(-ik_1 d) = \frac{1}{2} f \exp(-i\phi)$ to obtain Equation 1.38.

1.3 Derive the normalization constant in Equation 1.43 for the one-sided cavity model.
1-3. Use the form in the first line of Equation 1.38 for $0 < z < L$ and use the determinantal equation 1.37.

1.4 Derive the mode functions for the symmetrical two-sided cavity model given in Equations 1.58a and 1.58b.
1-4. See the solution to 1-2.

1.5 Show the orthogonality of mode functions in Equations 1.58a and 1.58b for the symmetric cavity under the cyclic boundary conditions following the example in Equations 1.40b–1.40d. In the limit $L \to \infty$, an a mode and a b mode can be degenerate. Are they orthogonal?
1-5. An a mode is antisymmetric and a b mode is symmetric with respect to the center of the cavity $z = 0$. So, if we have the symmetric region $-L/2 - d < z < L/2 + d$ as a cycle under the cyclic boundary condition, the a mode and b mode are easily seen to be orthogonal even if they are degenerate.

1.6 Show that the Fourier series expansion in Equation 1.70a for the squared normalization constant is valid.
1-6. Multiply both sides by the denominator on the left and compare the coefficients of $\cos 2nk_{1j}d$, $n = 0, 1, 2, 3, \ldots$, on both sides. Note that $K = 1 - (c_1/c_0)^2$ and $r = (c_0 - c_1)/(c_0 + c_1)$.

1.7 Show that the denominator in the squared normalization constant in Equation 1.70a vanishes at $\omega_j = \omega_{cm} \mp i\gamma_c$. That is, these ω_j are simple poles.
1-7. Rewrite the $\sin^2$ term as follows:

$$\sin^2 k_{1j}d \to \left(\frac{e^{ik_{1j}d} - e^{-ik_{1j}d}}{2i}\right)^2 = \left(\frac{e^{2i(k_{1m}-i\gamma)d} + e^{-2i(k_{1m}-i\gamma)d} - 2}{-4}\right)$$

$$e^{2i(k_{1m}-i\gamma)d} = -(1/r), \qquad e^{-2i(k_{1m}-i\gamma)d} = -r$$

Therefore

$$\sin^2 k_{1j}d = \frac{\{(1+r^2)/r\}+2}{4} = \frac{(1+r)^2}{4r} = \frac{1}{1-\{(1-r)/(1+r)\}^2} = \frac{1}{K}$$

References

1 Carrier, G.F., Krook, M., and Pearson, C.E. (**1966**) *Functions of a Complex Variable*, McGraw-Hill, New York.

2 Feng, X.P. and Ujihara, K. (**1990**) *Phys. Rev. A*, *41*, 2668–2676.

3 Heitler, W. (**1954**) *The Quantum Theory of Radiation*, 3rd edn, Clarendon, Oxford.

2
A One-Dimensional Optical Cavity with Output Coupling: Quantum Analysis

2.1
Quantization

Now we have a complete, orthonormal set of mode functions of the universe. One method of quantization of the field, then, is to separately quantize the respective field modes and then to add them up to form the total quantized field. Thus we quantize the system represented by the Hamiltonian in Equation 1.47 or 1.61 by imposing the following commutation relations on the variables Q_k and P_k:

$$\left[\hat{Q}_i, \hat{Q}_j\right] = \left[\hat{P}_i, \hat{P}_j\right] = 0, \qquad \left[\hat{Q}_i, \hat{P}_j\right] = i\hbar\delta_{ij} \tag{2.1}$$

where a hat symbol is attached to a quantum-mechanical operator, and the commutator is defined for two operators as

$$\left[\hat{A}, \hat{B}\right] \equiv \hat{A}\hat{B} - \hat{B}\hat{A} \tag{2.2}$$

Now $\hat{Q}_k$ and $\hat{P}_k$ are operators acting on the kth mode. Imposing the above commutation relation is equivalent to imposing an uncertainty relation between the variables. The uncertainty relation is one of the fundamental postulates of quantum mechanics. The "position" and "momentum" operators (the field amplitude and its time derivative operators) of the same mode do not commute with each other. The variances of these two variables are related by an uncertainty relation. Although the inter-mode spacing is infinitely small, the variables belonging to different modes are assumed to be independent variables and to commute with each other.

The Hamiltonian becomes the Hamiltonian operator

$$\hat{H} = \frac{1}{2}\sum_k \left(\hat{P}_k^2 + \omega_k^2\hat{Q}_k^2\right) \tag{2.3}$$

We define the annihilation and creation operators for the kth mode by

Output Coupling in Optical Cavities and Lasers: A Quantum Theoretical Approach
Kikuo Ujihara

ISBN: 978-3-527-40763-7

$$\hat{a}_k = (2\hbar\omega_k)^{-1/2}(\omega_k\hat{Q}_k + i\hat{P}_k) \tag{2.4}$$

$$\hat{a}_k^\dagger = (2\hbar\omega_k)^{-1/2}(\omega_k\hat{Q}_k - i\hat{P}_k) \tag{2.5}$$

The inverse relation is

$$\hat{Q}_k = (\hbar/2\omega_k)^{1/2}(\hat{a}_k + \hat{a}_k^\dagger) \tag{2.6}$$

$$\hat{P}_k = -i(\hbar\omega_k/2)^{1/2}(\hat{a}_k - \hat{a}_k^\dagger) \tag{2.7}$$

The commutation relation 2.1 can be rewritten in terms of the new operators as

$$\left[\hat{a}_i, \hat{a}_j^\dagger\right] = \delta_{ij}, \qquad \left[\hat{a}_i, \hat{a}_j\right] = 0, \qquad \left[\hat{a}_i^\dagger, \hat{a}_j^\dagger\right] = 0 \tag{2.8}$$

Substituting Equations 2.6 and 2.7 into Equation 2.3 and applying Equation 2.8, we have

$$\hat{H} = \sum_k \hbar\omega_k\left(\hat{a}_k^+\hat{a}_k + \frac{1}{2}\right) = \sum_k \hat{H}_k \tag{2.9}$$

The product $\hat{a}_k^\dagger\hat{a}_k$ is called the photon number operator of the mode k, and the term $\frac{1}{2}$ represents the zero-point energy of the mode.

2.2 Energy Eigenstates

The state $|\Psi\rangle$ of the free radiation field obeys the Schrödinger equation

$$i\hbar\frac{\partial}{\partial t}|\Psi\rangle = H|\Psi\rangle \tag{2.10}$$

a solution of which is

$$|\Psi\rangle = \prod_k |\varphi_k\rangle = \prod_k \exp(-iE_{k,n}t/\hbar)|n_k\rangle \tag{2.11}$$

Here the solution for mode k is

$$|\varphi_k\rangle = \exp(-iE_{k,n}t/\hbar)|n_k\rangle \tag{2.12}$$

where $|n_k\rangle$ and $E_{k,n}$ are the nth energy eigenstate and the corresponding energy of the kth mode, respectively. The eigenstate satisfies the eigenvalue equation (see Appendix B)

$$\hat{H}_k|n_k\rangle = E_{k,n}|n_k\rangle = \left(n_k + \frac{1}{2}\right)\hbar\omega_k|n_k\rangle, \qquad n_k = 0, 1, 2, 3, \ldots \tag{2.13}$$

The integer number n_k is the eigenvalue for the photon number operator $\hat{a}_k^\dagger\hat{a}_k$, and represents the photon number in the mode. The general solution to Equation

2.10 is a superposition of the pure states of Equation 2.11. The annihilation and creation operators have the following effects when operated on the energy eigenstate (see Appendix B):

$$\begin{aligned} \hat{a}_k|n_k\rangle &= \sqrt{n_k}|n_k - 1\rangle, \qquad n_k = 1, 2, 3, \ldots \\ \hat{a}_k|0_k\rangle &= 0 \end{aligned} \tag{2.14}$$

and

$$\hat{a}_k^\dagger|n_k\rangle = \sqrt{n_k + 1}|n_k + 1\rangle, \qquad n_k = 0, 1, 2, 3, \ldots \tag{2.15}$$

The non-vanishing matrix elements of these are therefore

$$a_{k,n_k-1,n_k} = \langle n_k - 1|\hat{a}_k|n_k\rangle = \sqrt{n_k} \tag{2.16a}$$

and

$$a^\dagger_{k,n_k+1,n_k} = \langle n_k + 1|\hat{a}_k^\dagger|n_k\rangle = \sqrt{n_k + 1} \tag{2.16b}$$

Going back to the definition of the electric field (Equation 1.2), the operator form of the vector potential and the electric field are then given by

$$\hat{A}(z,t) = \sum_j \hat{Q}_j U_j(z) = \sum_j (\hbar/2\omega_j)^{1/2}(\hat{a}_j + \hat{a}_j^\dagger) U_j(z) \tag{2.17}$$

and

$$\hat{E}(z,t) = -\sum_j \hat{P}_j U_j(z) = \sum_j i(\hbar\omega_j/2)^{1/2}(\hat{a}_j - \hat{a}_j^\dagger) U_j(z) \tag{2.18}$$

Present-day detectors of the optical field cannot follow the very high frequency of oscillation and detect only some time-averaged intensity as discussed by Glauber [1]. In this case, the physically meaningful quantity is not the total oscillating field but the product of so-called positive and negative frequency parts of the fields defined, respectively, as

$$\hat{E}^{(+)}(z,t) = \sum_j i(\hbar\omega_j/2)^{1/2}\hat{a}_j U_j(z) \tag{2.19a}$$

and

$$\hat{E}^{(-)}(z,t) = -\sum_j i(\hbar\omega_j/2)^{1/2}\hat{a}_j^\dagger U_j(z) \tag{2.19b}$$

with

$$\hat{E}(z,t) = \hat{E}^{(+)}(z,t) + \hat{E}^{(-)}(z,t) \tag{2.19c}$$

2.3 Field Commutation Relation

Here we derive a commutation relation for the electric fields at two different space-time points for the one-sided cavity. We arbitrarily choose one space point z_A from inside the cavity and another z_B from outside. The corresponding time variables are t_A and t_B, respectively. In order to include the time variables in the commutation relation, we go to the Heisenberg picture. According to the general rule to go to the Heisenberg picture, we have

$$\begin{aligned}\hat{a}_{kH} &= e^{i\hat{H}_k t/\hbar}\hat{a}_k e^{-i\hat{H}_k t/\hbar}\\ \hat{a}^{\dagger}_{kH} &= e^{i\hat{H}_k t/\hbar}\hat{a}^{\dagger}_k e^{-i\hat{H}_k t/\hbar}\end{aligned}\tag{2.20}$$

where the subscript H indicates the operator in the Heisenberg picture. Taking the non-vanishing matrix elements, we have

$$a_{kH,n_k-1,n_k} = \langle n_k - 1|e^{i\hat{H}_k t/\hbar}\hat{a}_k e^{-i\hat{H}_k t/\hbar}|n_k\rangle = \sqrt{n_k}\, e^{-i\omega_k t}\tag{2.21a}$$

$$a^{+}_{kH,n_k+1,n_k} = \langle n_k + 1|e^{i\hat{H}_k t/\hbar}\hat{a}^{+}_k e^{-i\hat{H}_k t/\hbar}|n_k\rangle = \sqrt{(n_k + 1)}\, e^{i\omega_k t}\tag{2.21b}$$

where we have used the fact that $\hat{H}|n_k\rangle = \hbar\omega(n_k + \frac{1}{2})|n_k\rangle$ and $\langle n_k|\hat{H} = \hbar\omega (n_k + \frac{1}{2})\ \langle n_k|$. The commutator for the creation and annihilation operators is unchanged:

$$\begin{aligned}\left[\hat{a}_{kH}, \hat{a}^{\dagger}_{kH}\right] &= \left[e^{i\hat{H}_k t/\hbar}\hat{a}_k e^{-i\hat{H}_k t/\hbar}, e^{i\hat{H}_k t/\hbar}\hat{a}^{\dagger}_k e^{-i\hat{H}_k t/\hbar}\right]\\ &= e^{i\hat{H}_k t/\hbar}\left[\hat{a}_k, \hat{a}^{\dagger}_k\right]e^{-i\hat{H}_k t/\hbar} = e^{i\hat{H}_k t/\hbar}e^{-i\hat{H}_k t/\hbar} = 1\end{aligned}\tag{2.22}$$

Using Equation 1.62b we have

$$\begin{aligned}\hat{E}_H(z,t) &= i\sum_k\left[\frac{\hbar\omega_k}{\varepsilon_1 L(1 - K\sin^2 k_1 d)}\right]^{1/2}\sin k_1(z + d)(\hat{a}_k e^{-i\omega_k t} - \hat{a}^{+}_k e^{i\omega_k t}), \quad -d<z<0\\ &= i\sum_k\left[\frac{\hbar\omega_k}{\varepsilon_1 L(1 - K\sin^2 k_1 d)}\right]^{1/2}\left(\frac{k_1}{k_0}\cos k_1 d\sin k_0 z_B + \sin k_1 d\cos k_0 z_B\right)\\ &\times \left(\hat{a}_k e^{-i\omega_k t} - \hat{a}^{+}_k e^{i\omega_k t}\right),\ \ 0 < z < L\end{aligned}\tag{2.23}$$

Therefore, the commutator is

$$\begin{aligned}&\left[\hat{E}_H(z_A, t_A), \hat{E}_H(z_B, t_B)\right]\\ &= 2i\sum_k\frac{\hbar\omega_k}{\varepsilon_1 L(1 - K\sin^2 k_1 d)}\sin k_1(z_A + d)\\ &\times\left(\frac{k_1}{k_0}\cos k_1 d\sin k_0 z_B + \sin k_1 d\cos k_0 z_B\right)\sin\omega_k(t_B - t_A)\end{aligned}\tag{2.24}$$

Using the series expansion (Equation 1.70a) and the density of modes (Equation 1.64) we have

$$\left[\hat{E}_H(z_A,t_A),\hat{E}_H(z_B,t_B)\right]$$
$$= 2i\int_0^\infty \frac{\hbar\omega_k}{\pi c_0\varepsilon_1}\frac{2k_1}{k_0}\left[\sum_{n=0}^{\infty}\frac{1}{1+\delta_{0,n}}(-r)^n\cos 2nk_1 d\right]\sin k_1(z_A+d) \tag{2.25}$$
$$\times\left(\frac{k_1}{k_0}\cos k_1 d\sin k_0 z_B+\sin k_1 d\cos k_0 z_B\right)\sin\omega_k(t_B-t_A)d\omega_k$$

Evaluating the products of the sinusoidal functions we obtain

$$\left[\hat{E}_H(z_A,t_A),\hat{E}_H(z_B,t_B)\right]=\frac{i\hbar}{\pi c_0\varepsilon_1}\frac{k_1^2}{k_0(k_1+k_0)}\int_0^\infty\omega_k\sum_{n=0}^{\infty}(-r)^n$$
$$\times\left[\sin\omega_k\left(\frac{z_B}{c_0}-\frac{z_A}{c_1}+\frac{2nd}{c_1}+t_B-t_A\right)\right.$$
$$-\sin\omega_k\left(\frac{z_B}{c_0}+\frac{z_A+2d}{c_1}+\frac{2nd}{c_1}+t_B-t_A\right) \tag{2.26}$$
$$-\sin\omega_k\left(\frac{z_B}{c_0}-\frac{z_A}{c_1}+\frac{2nd}{c_1}-t_B+t_A\right)$$
$$\left.+\sin\omega_k\left(\frac{z_B}{c_0}+\frac{z_A+2d}{c_1}+\frac{2nd}{c_1}-t_B+t_A\right)\right]d\omega_k$$

Now, we remember that the Dirac delta function is given by the integral [2]

$$\delta(t)=(1/\pi)\lim_{K\to\infty}\int_0^K\cos\omega t\,d\omega \tag{2.27}$$

Differentiation with respect to t yields

$$\delta'(t)=(-1/\pi)\lim_{K\to\infty}\int_0^K\omega\sin\omega t\,d\omega \tag{2.28}$$

Therefore, we finally obtain

$$\left[\hat{E}_H(z_A,t_A),\hat{E}_H(z_B,t_B)\right]=\frac{i\hbar}{\varepsilon_1 c_1}\frac{c_0}{c_0+c_1}\sum_{n=0}^{\infty}(-r)^n$$
$$\times\left[-\delta'\left(\frac{z_B}{c_0}-\frac{z_A}{c_1}+\frac{2nd}{c_1}+t_B-t_A\right)+\delta'\left(\frac{z_B}{c_0}+\frac{z_A+2d}{c_1}+\frac{2nd}{c_1}+t_B-t_A\right)\right. \tag{2.29}$$
$$\left.+\delta'\left(\frac{z_B}{c_0}-\frac{z_A}{c_1}+\frac{2nd}{c_1}-t_B+t_A\right)-\delta'\left(\frac{z_B}{c_0}+\frac{z_A+2d}{c_1}+\frac{2nd}{c_1}-t_B+t_A\right)\right],$$
$$-d<z_A<0,\qquad 0<z_B$$

The resultant expression for the commutator consists of an infinite number of terms. They have non-zero values when their arguments are zero. The electric fields at these two space-time points cannot be determined independently. The physical meanings of these terms are obvious. For example, the third term in the sum gives the intensity and the time of arrival at z_B of the disturbance when a flash of light is emitted instantaneously at z_A at time t_A. The first of these terms ($n=0$) is the disturbance transmitted directly from z_A to z_B. Subsequent terms are those reflected at the output interface, at $z=0$, n times with n round trips in the cavity, before reaching z_B. The coefficients of these terms are in powers of the product of the reflection coefficients at $z=0$ and at $z=-d$, which are $r=(c_0-c_1)/(c_0+c_1)$ and –1, respectively. Solving for n for the third term, we have

$$(-r)^n=(-1)^n\exp\left\{-\gamma_c\left[(t_B-t_A)-\left(\frac{z_B}{c_0}-\frac{z_A}{c_1}\right)\right]\right\} \tag{2.30}$$

where γ_c is given in Equation 1.18. Note that the decay with time is similar to that in the outgoing mode function in Equation 1.21. The three other terms in Equation 2.29 correspond to respective different propagation sequences.

2.4 Thermal Radiation and the Fluctuation–Dissipation Theorem

An important theoretical issue concerning the output coupling of an optical cavity is the treatment of the thermal radiation noise, although the physical magnitude of the thermal noise in a laser is usually negligible compared with the so-called quantum noise. When an optical cavity has a loss, statistical mechanics tell us that there should be a noise associated with the loss, the so-called fluctuation–dissipation theorem. When the cavity loss comes from the output coupling, where does the noise come from? Is it thermal noise? The answer is "yes," as will be shown explicitly in Chapters 9 and 10. Especially, Chapter 10 will show that the relevant thermal noise penetrates into the cavity from outside. Also, existing theories related to this issue will be reviewed in Chapter 15. In this section, we develop the noise theory using the cavity models with output coupling discussed so far.

2.4.1 The Density Operator of the Thermal Radiation Field

For the ith mode of the "universe," the thermal radiation field of the mode is described by the density operator $\hat{\rho}_i$ of the mode, where

$$\hat{\rho}_i=(1-e^{-\beta\hbar\omega_i})\sum_{m_i=0}^{\infty}e^{-m_i\beta\hbar\omega_i}|m_i\rangle\langle m_i| \tag{2.31}$$

Here, $\beta = (kT)^{-1}$ with k the Boltzmann constant and T the absolute temperature. The factor $(1 - e^{-\beta\hbar\omega_i})e^{-m_i\beta\hbar\omega_i} \equiv p_{m_i}$ is the probability that m_i photons appear in the mode at temperature T. The ensemble average of an operator $\hat{O}_i(t)$ acting on the ith mode is given by

$$\begin{aligned}\langle \hat{O}_i(t) \rangle &= \mathrm{Tr}\{\hat{\rho}_i\, \hat{O}_i(t)\} = \sum_{n_i} \langle n_i | \sum_{m_i} p_{m_i} |m_i\rangle\langle m_i|\hat{O}_i(t)|n_i\rangle \\ &= \sum_{m_i} p_{m_i} \langle m_i|\hat{O}_i(t)|m_i\rangle\end{aligned} \tag{2.32}$$

Because the modes of the "universe" are mutually independent, the density operator for the total field is the direct product of the respective density operators:

$$\hat{\rho} = \prod_i \hat{\rho}_i \tag{2.33}$$

We are assuming that the thermal radiation field described by Equation 2.33 is prepared at $t=0$ and the field oscillates freely afterwards. Here we are working in the Heisenberg picture, so that the density operator does not change with time, but the field operators do change, in general, with time. The ensemble average of operator $\hat{O}(t)$ is given by

$$\langle \hat{O}(t) \rangle = \mathrm{Tr}\{\rho\, \hat{O}(t)\} \tag{2.34}$$

The ensemble average of $\hat{O}_i(t)$ under Equation 2.34 reduces to that in Equation 2.32 because a trace over mode $j \neq i$, for example, simply creates a factor $\sum_{m_j} p_{m_j} = 1$.

2.4.2
The Correlation Function and the Power Spectrum

Now we want to discuss a theorem that connects system loss and the correlation function of the associated noise and then derive the correlation function for our model optical cavity. Assume that the thermal radiation field described by the density operators in Equations 2.31 and 2.33 is prepared at $t=0$. The positive and negative frequency parts of the electric field are now written as (see Equation 2.19a and 2.19b)

$$\hat{E}_T^{(+)}(z,t) = \sum_j i(\hbar\omega_j/2)^{1/2}\hat{a}_j(0)\, U_j(z)e^{-i\omega_j t} \tag{2.35a}$$

and

$$\hat{E}_T^{(-)}(z,t) = -\sum_j i(\hbar\omega_j/2)^{1/2}\hat{a}_j^\dagger(0)\, U_j(z)e^{i\omega_j t} \tag{2.35b}$$

The suffix T denotes the thermal field. Here the motion of the annihilation operator for $t \geq 0$ is obtained from the Heisenberg equation

$$
\begin{aligned}
i\hbar \frac{d}{dt}\hat{a}_j &= [\hat{a}_j, \hat{H}] = [\hat{a}_j, \hat{H}_j] = \left[\hat{a}_j, \hbar\omega_j\left(\hat{a}_j^\dagger \hat{a}_j + \frac{1}{2}\right)\right] \\
&= \hbar\omega_j\left(\hat{a}_j\hat{a}_j^\dagger \hat{a}_j - \hat{a}_j^\dagger \hat{a}_j \hat{a}_j\right) = \hbar\omega_j(\hat{a}_j\hat{a}_j^\dagger - \hat{a}_j^\dagger \hat{a}_j)\hat{a}_j \\
&= \hbar\omega_j\hat{a}_j
\end{aligned}
\tag{2.36}
$$

where the commutation relation of Equation 2.8 has been used. Motion of the creation operator can be obtained similarly. From the matrix elements in Equations 2.21a and 2.21b we see that

$$
\left\langle \hat{E}_T^{(+)}(z,t) \right\rangle = \left\langle \hat{E}_T^{(-)}(z,t) \right\rangle = 0 \tag{2.37}
$$

The normally ordered correlation function or the coherence function of second order is defined as the ensemble average of the product $\hat{E}_T^{(-)}(z',t')$ $\hat{E}_T^{(+)}(z,t)$:

$$
G(z',t',z,t) = \left\langle \hat{E}_T^{(-)}(z',t')\hat{E}_T^{(+)}(z,t) \right\rangle \tag{2.38}
$$

In a normally ordered product, all the annihilation operators come to the right of the creation operators. The anti-normally ordered correlation function is similarly defined as

$$
G_a(z',t',z,t) = \left\langle \hat{E}_T^{(+)}(z',t')\hat{E}_T^{(-)}(z,t) \right\rangle \tag{2.39}
$$

Calculating the average using Equation 2.34 we have

$$
G(z',t',z,t) = \sum_j \frac{1}{2}\hbar\omega_j \langle n_j \rangle U_j(z') U_j(z) e^{-i\omega_j(t-t')} \tag{2.40}
$$

and

$$
G_a(z',t',z,t) = \sum_j \frac{1}{2}\hbar\omega_j (\langle n_j \rangle + 1) U_j(z') U_j(z) e^{i\omega_j(t-t')} \tag{2.41}
$$

Here we have used

$$
\begin{aligned}
\left\langle \hat{a}_i^\dagger(0)\hat{a}_j(0) \right\rangle &= \langle n_j \rangle \delta_{ij} \\
\left\langle \hat{a}_i(0)\hat{a}_j^\dagger(0) \right\rangle &= (\langle n_j \rangle + 1)\delta_{ij}
\end{aligned}
\tag{2.42a}
$$

where

$$
\langle n_j \rangle = \left\langle \hat{a}_j^\dagger(0)\hat{a}_j(0) \right\rangle = (e^{\beta\hbar\omega_j} - 1)^{-1} \tag{2.42b}
$$

The constant β is defined below Equation 2.31.

The correlation function in Equation 2.40 with $t=t'$ can be rewritten by use of Equation 1.64 as

$$G(z',t,z,t) = \int_0^\infty \rho(\omega)\frac{1}{2}\hbar\omega\langle n_\omega\rangle U_\omega(z')U_\omega(z)d\omega \tag{2.43}$$

We define the power spectrum $I(z',z,\omega)$ of the thermal radiation field as the integrand of this expression:

$$I(z',z,\omega) = \rho(\omega)\frac{1}{2}\hbar\omega\langle n_\omega\rangle U_\omega(z')U_\omega(z)H(\omega) \tag{2.44}$$

Here $H(\omega)$ is the unit step function. Using the formula $\int_{-\infty}^{\infty} e^{ixt}dt = 2\pi\delta(x)$, it is easy to show that (problem 2-5)

$$I(z',z,\omega) = \frac{1}{2\pi}\int_{-\infty}^{\infty} G(z',0,z,t)e^{i\omega t}dt \tag{2.45}$$

That is, the power spectrum is the Fourier transform of the correlation function. The factor $N_k^2 = (1 - K\sin^2 k_1 d)^{-1}$ included in $U_\omega(z')U_\omega(z)$ has simple poles at $\omega = \omega_{cm} \pm i\gamma_c$, which appear in Equations 1.18 and 1.20, but here m also takes negative integer values. Thus from the Mittag–Leffler theorem [3] in Equation 1.70 we have

$$(1 - K\sin^2 k_{1\omega}d)^{-1} = \sum_{m=-\infty}^{\infty} \frac{c_0\gamma_c/d}{\gamma_c^2 + (\omega - \omega_{cm})^2} \tag{2.46}$$

Substituting this into Equation 2.44 we have

$$I(z',z,\omega) = \sum_{m=0}^{\infty} \frac{\hbar\omega\langle n_\omega\rangle}{\pi\,\varepsilon_1 d}\frac{\gamma_c}{\gamma_c^2 + (\omega - \omega_{cm})^2} u_\omega(z')u_\omega(z)H(\omega) \tag{2.47}$$

where $u_\omega(z) = \sin k_{1\omega}(z+d)$, if $-d < z < 0$, as was defined in Equation 1.41.

Here we have omitted unphysical negative m values. We see that each cavity resonant mode contributes a term with a Lorentzian profile of width $2\gamma_c$.

2.4.3 The Response Function and the Fluctuation–Dissipation Theorem

Let us consider a classical current source at $z=z_0$ with sinusoidal time dependence $J\exp(-i\omega t)\delta(z - z_0)$ and coupled linearly to the field at $t=0$. We define the response function $Y(z, z_0, \omega)$ as the asymptotic ratio of the induced field at z to the current as the time goes to infinity:

$$Je^{-i\omega t}Y(z,z_0,\omega) = \lim_{t\to\infty}\left\langle \hat{E}^{(+)}(z,t)\right\rangle \tag{2.48}$$

where the thermal average is taken in order to extract systematic motions. Then, if the current source has density $\Im(z,\omega)$ in the space and angular frequency domain, the electric field may be written as

$$\hat{E}^{(+)}(z,t) = \int dz_0 \int_{-\infty}^{\infty} d\omega\, Y(z,z_0,\omega)\Im(z_0,\omega)e^{-i\omega t} + \hat{E}_T^{(+)}(z,t) \tag{2.49}$$

In order to obtain an expression for the response function, we solve the equation of motion of the field operator $\hat{a}_k$ in the presence of the driving current $J\exp(-i\omega t)\delta(z-z_0)$. The interaction is [4]

$$\begin{aligned}\hat{H}_{int} &= -\int_{-d}^{L} \hat{A}(z,t)\Big[\{J\exp(-i\omega t) + \text{C.C.}\}\delta(z-z_0)\Big]dz \\ &= -\sum_k \left(\frac{\hbar}{2\omega_k}\right)^{1/2} (\hat{a}_k + \hat{a}_k^+)\,U_k(z_0)\left(Je^{-i\omega t} + \text{C.C.}\right)\end{aligned} \tag{2.50}$$

where we have used Equation 2.17 in the second line. Here we have written the current in the form $J\exp(-i\omega t) + \text{C.C.}$ rather than in the form $\text{Re}[J\exp(-i\omega t)]$, in accordance with the definition of the positive electric field in the form of Equation 2.19c. We solve

$$i\hbar\frac{d}{dt}\hat{a}_k = [\hat{a}_k,\, \hat{H} + \hat{H}_{int}] = \hbar\omega_k\hat{a}_k - \left(\frac{\hbar}{2\omega_k}\right)^{1/2} U_k(z_0)\left(Je^{-i\omega t} + \text{C.C.}\right) \tag{2.51}$$

Solving for $\hat{a}_k$ and substituting into Equation 2.19a we have

$$\begin{aligned}\hat{E}^{(+)}(z,t) = &\sum_k \left(-\frac{i}{2}\right) U_k(z)\,U_k(z_0)\left(\frac{1-e^{i(\omega-\omega_k)t}}{\omega-\omega_k}Je^{-i\omega t} - \frac{1-e^{-i(\omega+\omega_k)t}}{\omega+\omega_k}J^*e^{i\omega t}\right) \\ &+ \hat{E}_T^{(+)}(z,t)\end{aligned} \tag{2.52}$$

where the initial values of the $\hat{a}_k$ constitute the thermal field given by Equation 2.35a. Since we are interested in the response to $J\exp(-i\omega t)\delta(z-z_0)$, we discard the second term in parentheses. Thus, with $t \to \infty$ as in the definition (Equation 2.48), the response function is obtained as

$$Y(z,z_0,\omega) = -\frac{i}{2}\sum_k U_k(z)\,U_k(z_0)\zeta(\omega-\omega_k) \tag{2.53a}$$

where the zeta function [2] is

$$\zeta(\omega) = \frac{P}{\omega} - i\pi\delta(\omega) = -i\int_0^{\infty} e^{i\omega t}dt \tag{2.53b}$$

From Equations 2.53a and 2.53b we see that

$$\begin{aligned}\text{Re}\,Y(z,z_0,\omega) &= -\frac{\pi}{2}\sum_k U_k(z)\,U_k(z_0)\delta(\omega-\omega_k) \\ &= -\frac{\pi}{2}\rho_\omega U_\omega(z)\,U_\omega(z_0)\end{aligned} \tag{2.54}$$

Thus from Equations 2.44 and 2.45 we see that

$$\hbar\omega\langle n_k\rangle \mathrm{Re}Y(z,z',\omega) = -\pi I(z',z,\omega) = -\frac{1}{2}\int_{-\infty}^{\infty} G(z',0,z,t)e^{i\omega t}dt \qquad (2.55)$$

This is a fluctuation–dissipation theorem connecting the response function and the correlation function. Knowledge of either one is sufficient to know the other. The term "dissipation" here is related to the radiation loss of the field energy stored inside the cavity into the outside space through the coupling surface. The reader can check the equality of the stored energy lost per second and the magnitude of the pointing vector outside the cavity for a source-free field [5].

2.4.4
Derivation of the Langevin Noise for a Single Cavity Resonant Mode

Here we derive the Langevin noise force widely assumed in laser theories. We assume that only a single cavity mode is involved and that other cavity modes are spectrally distant from the one in question. Then the power spectrum in Equation 2.47 may be replaced by

$$\begin{aligned} I(z,z',\omega) &= \frac{\hbar\omega_c\langle n_{\omega_c}\rangle}{\pi\varepsilon_1 d}\frac{\gamma_c}{\gamma_c^2+(\omega-\omega_c)^2}u^*_{\omega_c}(z')u_{\omega_c}(z)H(\omega), \\ & \qquad |\omega-\omega_c| \ll \frac{1}{2}\Delta\omega_c \end{aligned} \qquad (2.56)$$

The inequality describes the large departure of the cavity mode ω_c from other modes; and $\Delta\omega_c = \pi c_1/d$ is the cavity mode separation. Since the bandwidth concerned is narrow compared with the cavity mode separation, that is, $2\gamma_c \ll \Delta\omega_c$, we have replaced $u_\omega(z')u_\omega(z)$ by $u_{\omega_c}(z')u_{\omega_c}(z)$. Here ω_c is the resonant frequency, which was defined in Equation 1.18b. Also, we have replaced $\hbar\omega\langle n_\omega\rangle$ by $\hbar\omega_c\langle n_{\omega_c}\rangle$, which is valid if $|\omega-\omega_c| \ll kT/\hbar$ holds. The real part of the response function Y (z, z_0, ω) for the cavity is given by Equations 2.55 and 2.56 as

$$\mathrm{Re}Y(z,z',\omega) = -\frac{1}{\varepsilon_1 d}\frac{\gamma_c}{\gamma_c^2+(\omega-\omega_c)^2}u_{\omega_c}(z')u_{\omega_c}(z)H(\omega) \qquad (2.57)$$

where we have used the approximation $\hbar\omega\langle n_\omega\rangle = \hbar\omega_c\langle n_{\omega_c}\rangle$. Because of the form of the zeta function in Equations 2.53a and 2.53b, the imaginary part of the response function is given by

$$\begin{aligned} \mathrm{Im}Y(z,z_0,\omega) &= -\frac{1}{2}P\int_0^\infty \frac{\rho_{\omega'}U_{\omega'}(z)U_{\omega'}(z_0)}{\omega-\omega'}d\omega' \\ &= \frac{P}{\pi}\int_0^\infty \frac{\mathrm{Re}Y(z,z_0,\omega')}{\omega-\omega'}d\omega' \end{aligned} \qquad (2.58a)$$

where we have used Equation 2.54 in the second line. Thus, the imaginary part becomes

$$\mathrm{Im}Y(z,z_0,\omega) = -\frac{\gamma_c u_{\omega_c}(z_0)u_{\omega_c}(z)}{\pi\varepsilon_1 d} P\int_0^\infty \frac{d\omega'}{(\omega-\omega')\left\{\gamma_c^2+(\omega-\omega_c)^2\right\}} \tag{2.58b}$$

The integral can be evaluated as follows. First, we replace the lower limit of integration by $-\infty$ on the grounds that this does not affect the results because the important region of the integrand is compressed around the frequency ω or ω_c. Next, we choose, for example, a contour along a large semicircle in the upper half-plane of the frequency ω' and examine the integral along the small circle around the pole $\omega' = \omega_c + i\gamma_c$ in the upper half-plane and that along the small semicircle above the pole $\omega' = \omega$. The integral along the large semicircle vanishes because the integrand vanishes for large radius of the semicircle. Then the integral is equal to the sum of integrals around the poles. We have

$$\begin{aligned} &P\int_0^\infty \frac{d\omega'}{(\omega-\omega')\left\{\gamma_c^2+(\omega'-\omega_c)^2\right\}} \\ &= 2\pi i\frac{1}{\omega-\omega_c-i\gamma_c}\frac{1}{2i\gamma_c} - \pi i\frac{1}{\gamma_c^2+(\omega-\omega_c)^2} \\ &= \pi\frac{\omega-\omega_c}{\left\{\gamma_c^2+(\omega-\omega_c)^2\right\}\gamma_c} \end{aligned} \tag{2.58c}$$

Therefore we have

$$\mathrm{Im}Y(z,z_0,\omega) \simeq -\frac{u_{\omega_c}(z_0)u_{\omega_c}(z)}{\varepsilon_1 d}\frac{\omega-\omega_c}{\gamma_c^2+(\omega-\omega_c)^2} \tag{2.58d}$$

Adding the real and imaginary parts in Equations 2.57 and 2.58d we have

$$Y(z,z',\omega) = -\frac{1}{\varepsilon_1 d}\frac{1}{\gamma_c - i(\omega-\omega_c)}u_{\omega_c}(z')u_{\omega_c}(z) \tag{2.59}$$

Then, inverse Fourier transforming Equation 2.55 using Equation 2.57 we have the correlation function

$$G(z',t',z,t) = \frac{\hbar\omega_c\langle n_{\omega_c}\rangle}{\varepsilon_1 d}u_{\omega_c}(z')u_{\omega_c}(z)e^{-\gamma_c|t-t'|-i\omega_c(t-t')} \tag{2.60}$$

where the contour integral in the lower (upper) half-plane of ω has been performed for $t > t'$ $(t<t')$ with the relevant pole at $\omega = \omega_c - i\gamma_c$ $(\omega = \omega_c + i\gamma_c)$. Below, we use this function to describe the correlation of the so-called Langevin noise.

When a current source distribution $\Im(z,\omega)$ exists, Equations 2.49 and 2.37 give

$$\left\langle \hat{E}^{(+)}(z,t)\right\rangle = \int_{-d}^{0} dz_0 \int_{-\infty}^{\infty} d\omega \left(-\frac{1}{\varepsilon_1 d}\right) \frac{1}{\gamma_c - i(\omega-\omega_c)} u_{\omega_c}(z_0) u_{\omega_c}(z) \times H(\omega) \Im(z_0,\omega) e^{-i\omega t} \tag{2.61}$$

where we have used Equation 2.59 for the response function. Differentiation with respect to time t changes the integrand by a factor $-i\omega = -(\gamma_c + i\omega_c) + \{\gamma_c - i(\omega - \omega_c)\}$. Thus we have

$$\frac{d}{dt}\left\langle \hat{E}^{(+)}(z,t)\right\rangle = -(\gamma_c + i\omega_c)\left\langle \hat{E}^{(+)}(z,t)\right\rangle - \int_{-d}^{0} \frac{dz_0}{\varepsilon_1 d} u_{\omega_c}(z_0) u_{\omega_c}(z) \Im(z_0,t) \tag{2.62}$$

where we have Fourier transformed the source density in the second term. If we remove from this equation the ensemble averaging, then we may have the Langevin force $f(z,t)$ that drives the electric field, a fact expected from Equation 2.49 as a result of the presence of the thermal radiation field. Thus, in the absence of the current source, we should have

$$\frac{d}{dt}\hat{E}^{(+)}(z,t) = -(\gamma_c + i\omega_c)\hat{E}^{(+)}(z,t) + \hat{f}(z,t) \tag{2.63}$$

where the Langevin noise operator $\hat{f}(z,t)$ has the mean

$$\left\langle \hat{f}(z,t)\right\rangle = 0 \tag{2.64}$$

Now, for simplicity, we truncate the rapid oscillation of the cavity field by writing

$$\hat{E}^{(+)}(z,t) = \tilde{\hat{E}}^{(+)}(z,t)e^{-i\omega_c t}, \qquad \hat{f}(z,t) = \tilde{\hat{f}}(z,t)e^{-i\omega_c t} \tag{2.65}$$

Thus we have

$$\frac{d}{dt}\tilde{\hat{E}}^{(+)}(z,t) = -\gamma_c \tilde{\hat{E}}^{(+)}(z,t) + \tilde{\hat{f}}(z,t) \tag{2.66}$$

At the same time, the sinusoidal oscillation in the correlation function in Equation 2.60 drops out: we write

$$\tilde{G}(z',t',z,t) = \left\langle \tilde{\hat{E}}_T^{(-)}(z',t')\tilde{\hat{E}}_T^{(+)}(z,t)\right\rangle = \frac{\hbar\omega_c\langle n_{\omega_c}\rangle}{\varepsilon_1 d} u_{\omega_c}(z') u_{\omega_c}(z) e^{-\gamma_c|t-t'|} \tag{2.67}$$

Then, it follows that

$$\begin{aligned}\left\langle \tilde{\tilde{f}}^{\dagger}(z',t')\tilde{\tilde{f}}(z,t)\right\rangle &= \left\langle \left[\left(\frac{d}{dt'}+\gamma_c\right)\tilde{\tilde{E}}^{(-)}(z',t')\right]\left[\left(\frac{d}{dt}+\gamma_c\right)\tilde{\tilde{E}}^{(+)}(z,t)\right]\right\rangle \\ &= \left[\frac{\partial}{\partial t'}\frac{\partial}{\partial t}+\gamma_c\left(\frac{\partial}{\partial t'}+\frac{\partial}{\partial t}\right)+\gamma_c^2\right]\tilde{G}(z',t',z,t)\end{aligned} \tag{2.68}$$

Let us consider

$$x(t,t') = \left[\frac{\partial}{\partial t'}\frac{\partial}{\partial t} + \gamma_c\left(\frac{\partial}{\partial t'}+\frac{\partial}{\partial t}\right) + \gamma_c^2\right]e^{-\gamma_c|t-t'|} \tag{2.69a}$$

We see that $x(t, t')$ always vanishes for $t \neq t'$. Let us examine the region $t-\varepsilon < t' < t+\varepsilon$ with $\varepsilon \to 0$. For $t' < t-\varepsilon$, we have $(\partial/\partial t')\exp\{-\gamma_c(t-t')\} = \gamma_c\exp\{-\gamma_c(t-t')\}$. In the limit $t' \to t-\varepsilon$, this becomes $\gamma_c\exp\{-\gamma_c\varepsilon\}$. Similarly, for $t' > t+\varepsilon$ and for $t' \to t+\varepsilon$, we have $(\partial/\partial t')\exp\{-\gamma_c(t'-t)\} \to -\gamma_c\exp(-\gamma_c\varepsilon)$. Thus, in the region $t-\varepsilon < t' < t+\varepsilon$, the derivative in terms of t and t', $(\partial/\partial t)(\partial/\partial t')\exp\{-\gamma_c|t-t'|\}$, which is equal to $-(\partial/\partial t')^2\exp\{-\gamma_c|t-t'|\}$, becomes $2\gamma_c\exp(-\gamma_c\varepsilon)/(2\varepsilon)$. Therefore, this derivative yields a narrow square region around $t' = t$ on the t'-axis, the area of which tends to $2\gamma_c$ as $\varepsilon \to 0$ (*see Figure 2.1*).

The term $(\partial/\partial t') + (\partial/\partial t)$ makes at most $2\gamma_c \times 2\varepsilon \to 0$, and the term of γ_c^2 yields $\gamma_c^2 \times 2\varepsilon \to 0$. Therefore, we have

$$\begin{aligned}\int_{-\infty}^{\infty} x(t,t')dt' &= \lim_{\varepsilon\to 0}\left\{\int_{-\infty}^{t-\varepsilon} x(t,t')dt' + \int_{t+\varepsilon}^{\infty} x(t,t')dt' + \int_{t-\varepsilon}^{t+\varepsilon} x(t,t')dt'\right\} \\ &\to 0 + 0 + 2\gamma_c\end{aligned} \tag{2.69b}$$

Thus, we have $x(t,t') \to 2\gamma_c\delta(t-t')$. Then Equation 2.68 becomes

$$\left\langle \tilde{\tilde{f}}^{\dagger}(z',t')\tilde{\tilde{f}}(z,t)\right\rangle = \frac{2\gamma_c\hbar\omega_c\langle n_{\omega_c}\rangle}{\varepsilon_1 d}u_{\omega_c}(z')u_{\omega_c}(z)\delta(t-t') \tag{2.70a}$$

We can eliminate the tildes on the noise operators because of the delta correlated nature:

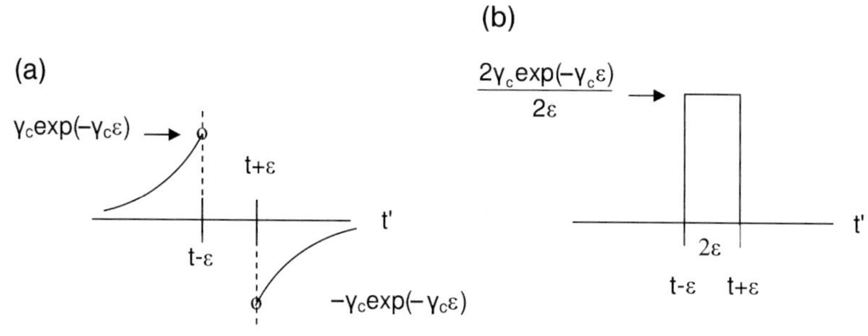

Figure 2.1 The function $x(t,t')$: (a) $(\partial/\partial t')x(t,t')$; and (b) $(\partial/\partial t)(\partial/\partial t')x(t,t')$.

$$\left\langle \hat{f}^{\dagger}(z',t')\hat{f}(z,t)\right\rangle = \frac{2\gamma_c \hbar\omega_c \langle n_{\omega_c}\rangle}{\varepsilon_1 d} u_{\omega_c}(z')u_{\omega_c}(z)\delta(t-t') \qquad (2.70b)$$

So, we have a Markovian noise for the field mode ω_c. The set of Equations 2.63 or 2.66, 2.64, 2.70a and 2.70b gives the usual description of the thermal Langevin force used in quasimode laser theory. The reader may show that the noise function $\hat{f}(z,t)$ can be simulated by $\{(d/dt)+i\Omega_c\}\hat{E}_T(z,t)$ around a cavity resonant mode. This is to be expected since Equation 2.63 should hold for the thermal field $\hat{E}_T(z,t)$ around the frequency ω_c.

The Langevin force derived here is valid only for the case of narrower field bandwidth than the cavity mode spacing. In Chapters 9 and 10 the thermal noise is treated more rigorously, taking the cavity output coupling into account. For a Langevin equation applicable to a cavity with two-side output coupling, see Equation 15.6a.

2.4.5
Excitation of the Cavity Resonant Mode by a Current Impulse

Looking back at Equation 2.61, if the driving current is an impulse of the form $J\delta(t-t_0)\delta(z-z_A)$, then the current density is

$$\begin{aligned}\Im(z,\omega) &= \frac{1}{2\pi}\int_{-\infty}^{+\infty} J\delta(t-t_0)\delta(z-z_A)e^{i\omega t}dt \\ &= \frac{1}{2\pi}J\delta(z-z_A)e^{i\omega t_0}\end{aligned} \qquad (2.71)$$

Then, the induced electric field (the net field minus the thermal field) is

$$\begin{aligned}\left\langle \hat{E}^{(+)}(z,t)\right\rangle &= \int_{-d}^{0} dz_0 \int_{-\infty}^{\infty} d\omega \left(-\frac{1}{\varepsilon_1 d}\right)\frac{1}{\gamma_c - i(\omega-\omega_c)} u_\omega(z_0)u_\omega(z) \\ &\quad \times H(\omega)\frac{1}{2\pi}J\delta(z_0-z_A)e^{-i\omega(t-t_0)}\end{aligned} \qquad (2.72)$$

where we have retrieved the ω dependences of the functions u. The integration over ω can be done on the complex ω-plane by noting that the pole is at $\omega=\omega_c-i\gamma_c\equiv\Omega_c$ in the lower half-plane. If we expand the numerator in exponential functions, we will have exponents with $-i\omega[(t-t_0)\pm\{(z+d)\pm(z_m+d)\}/c_1]$ (see Equation 1.41). For simplicity, we assume that we are concerned with phenomena that are slow in a time scale of order $|(z+d)\pm(z_m+d)|/c_1 \le 2d/c_1$, that is, we examine the changes on a time scale that is greater than the round-trip time in the cavity. Since this assumption expects an optical spectrum that is narrower than $c_1/(2d)$, this is consistent with the choice of only one cavity mode, which anticipates an optical spectrum that is narrower than the cavity mode spacing $\Delta\omega_c = c_1\pi/d$. Then the contour of integration can be determined by the sign of $t-t_0$. The result is

$$R_n = (c_{n-1} - c_n)/(c_{n-1} + c_n) \tag{2.82b}$$

$$J_n = (1 + R_n)^{-1} = \frac{c_{n-1} + c_n}{2c_{n-1}} \tag{2.82c}$$

The parameter R_n is the amplitude reflection coefficient for a wave incident from the left to the boundary of the nth and $(n-1)$th layers. Later we will use the relation

$$\det(M_n) = 1 - R_n^2 \tag{2.83}$$

The field values at the ends of the region comprising the mth, $(m+1)$th,. . ., and the nth layers are related by

$$\begin{pmatrix} X_n \\ Y_n \end{pmatrix} = J_{n,m} M_{n,m} \begin{pmatrix} X_{m-1} \\ Y_{m-1} \end{pmatrix}, \qquad N \geq n \geq m \geq 1 \tag{2.84a}$$

where

$$M_{n,m} = M_n \times M_{n-1} \times \cdots \times M_{m+1} \times M_m \tag{2.84b}$$

$$J_{n,m} = J_n \times J_{n-1} \times \cdots \times J_{m+1} \times J_m \tag{2.84c}$$

The matrix elements of $M_{n,m}$ are polynomials of exponential functions of $i\omega$ with real exponents and real coefficients. The elements have the symmetry property

$$(M_{n,m})_{11} = (M_{n,m})_{22}^* \tag{2.85a}$$

$$(M_{n,m})_{12} = (M_{n,m})_{21}^* \tag{2.85b}$$

2.5.2
The Modes of the "Universe"

2.5.2.1 The Eigenmode Frequency

As described above, the cavity extends over the region $-D \leq z \leq 0$, while the zeroth region, $0 \leq z \leq L$, is the outside, vacuum region. The mode function of the "universe" should satisfy all the relevant boundary conditions. The field values at both ends of the zeroth region are related by the phase of the propagation as

$$\begin{pmatrix} X_0 \\ Y_0 \end{pmatrix} = \begin{pmatrix} e^{-ik_0 L}, & 0 \\ 0, & e^{ik_0 L} \end{pmatrix} \begin{pmatrix} X_L \\ Y_L \end{pmatrix} \tag{2.86}$$

where the suffix L denotes the location $z = L$. The boundary conditions that the electric field vanishes at the perfect conductors, $u_0(L) = 0$ and $u_N(-D) = 0$, read, respectively,

$$X_L + Y_L = 0 \tag{2.87}$$

and

$$X_N + Y_N = 0 \tag{2.88}$$

Then we have

$$Y_0 = -X_0 e^{2ik_0 L} \tag{2.89}$$

Thus Equation 2.84a for $n = N$ and $m = 1$ reads

$$\begin{pmatrix} X_N \\ -X_N \end{pmatrix} = J_{N,1} M_{N,1} \begin{pmatrix} X_0 \\ -X_0 e^{2ik_0 L} \end{pmatrix} \tag{2.90}$$

Therefore, for a non-trivial solution to exist, we should have

$$\begin{aligned} e^{2ik_0 L} &= \frac{(M_{N,1})_{11} + (M_{N,1})_{21}}{(M_{N,1})_{22} + (M_{N,1})_{12}} \\ &= e^{2i\phi(\omega)} \end{aligned} \tag{2.91}$$

The eigenmode frequencies are those satisfying this equation. Here

$$\phi(\omega) = \arg\{(M_{N,1})_{11} + (M_{N,1})_{21}\}, \qquad (0 \le \phi \le 2\pi) \tag{2.92}$$

The second form in Equation 2.91 results from the symmetry properties, Equations 2.85a and 2.85b. Equation 2.91 yields

$$k_0 L = \phi(\omega) + p\pi \tag{2.93}$$

where p is an integer. Since the zeroth region simulates free space, L is assumed to be very large. Since it may be argued that the phase angle $\phi(\omega)$ as a function of ω varies much more slowly than $k_0 L = \omega(L/c_0)$, the eigenmode frequencies are non-degenerate. In the limit of large L/c_0, the separation between the neighboring solutions for ω is

$$\Delta\omega = c_0 \pi / L \tag{2.94}$$

The density of modes is thus

$$\rho = 1/\Delta\omega = L/c_0\pi \tag{2.95}$$

2.5.2.2 The Mode Functions

The orthogonality of the mode functions can be proved, just as we did in Equations 1.40a–1.40d, by performing the integration:

$$I = \int_{-D}^{L} \frac{1}{\mu_0} \left(\frac{\partial}{\partial z} u_i(z)\right) \left(\frac{\partial}{\partial z} u_j(z)\right) dz \tag{2.96}$$

Note that we are assuming that the magnetic permeability is μ_0 for all the layers. Using the boundary conditions and the Helmholtz equation 2.77 we can show that

$$I = \omega_i^2 \int_{-D}^{L} \varepsilon(z) u_i(z) u_j(z)\, dz = \omega_j^2 \int_{-D}^{L} \varepsilon(z) u_i(z) u_j(z)\, dz \tag{2.97}$$

Therefore, we have the orthogonality relation

$$\int_{-D}^{L} \varepsilon(z) u_i(z) u_j(z)\, dz = 0, \qquad i \neq j \tag{2.98}$$

For the jth mode we can show that

$$\int_{-D}^{L} \varepsilon(z) \{u_j(z)\}^2 dz = 2 \sum_{n=0}^{n} \varepsilon_n \left|\alpha_{jn}\right|^2 l_n \tag{2.99}$$

To show this formula, let us consider the integral in the nth region:

$$\begin{aligned} I_n &= \int_{-d_n}^{-d_{n-1}} \varepsilon_n \{u_j(z)\}^2 dz \\ &= \int_{-d_n}^{-d_{n-1}} \varepsilon_n \left\{ \alpha_{jn}^2 e^{2ik_{jn}z} + \beta_{jn}^2 e^{-2ik_{jn}z} + 2\alpha_{jn}\beta_{jn} \right\} dz \\ &= \frac{\varepsilon_n \alpha_{jn}^2}{2ik_{jn}} \left(e^{-2ik_{jn}d_{n-1}} - e^{-2ik_{jn}d_n}\right) + \frac{\varepsilon_n \beta_{jn}^2}{-2ik_{jn}} \left(e^{2ik_{jn}d_{n-1}} - e^{2ik_{jn}d_n}\right) + 2\varepsilon_n \alpha_{jn}\beta_{jn} l_n \end{aligned} \tag{2.100}$$

Using the matrix relation 2.81 for the nth layer, we have

$$\begin{aligned} \alpha_{jn}^2 e^{-2ik_{jn}d_{n-1}} &= (1+R_n)^{-2} \left(\alpha_{j(n-1)} e^{-ik_{j(n-1)}d_{n-1}} + R_n \beta_{j(n-1)} e^{ik_{j(n-1)}d_{n-1}}\right)^2 \\ \beta_n^2 e^{2ik_{jn}d_{n-1}} &= (1+R_n)^{-2} \left(R_n \alpha_{j(n-1)} e^{-ik_{j(n-1)}d_{n-1}} + \beta_{j(n-1)} e^{ik_{j(n-1)}d_{n-1}}\right)^2 \end{aligned} \tag{2.101}$$

Thus

$$\begin{aligned} &\frac{\varepsilon_n \left(\alpha_{jn}^2 e^{-2ik_{jn}d_{n-1}} - \beta_n^2 e^{2ik_{jn}d_{n-1}}\right)}{2ik_{jn}} \\ &\quad = \frac{\varepsilon_n (1 - R_n^2)}{2ik_{jn}(1+R_n)^2} \left(\alpha_{j(n-1)}^2 e^{-2ik_{j(n-1)}d_{n-1}} - \beta_{j(n-1)}^2 e^{2ik_{j(n-1)}d_{n-1}}\right) \end{aligned} \tag{2.102}$$

Now the factor $\varepsilon_n(1-R_n^2)/\left\{2ik_{jn}(1+R_n)^2\right\} = \varepsilon_{n-1}/2ik_{j(n-1)}$, as can be shown by use of Equation 2.82b and the fact that $\varepsilon_n = \varepsilon_0 (c_0/c_n)^2$. Therefore, the integral in Equation 2.100 becomes

$$\begin{aligned} I_n &= \int_{-d_n}^{-d_{n-1}} \varepsilon_n \{u_j(z)\}^2 dz \\ &= \varepsilon_n \frac{\beta_{jn}^2 e^{2ik_{jn}d_n} - \alpha_{jn}^2 e^{-2ik_{jn}d_n}}{2ik_{jn}} - \varepsilon_{n-1} \frac{\beta_{j(n-1)}^2 e^{2ik_{j(n-1)}d_{n-1}} - \alpha_{j(n-1)}^2 e^{-2ik_{j(n-1)}d_{n-1}}}{2ik_{j(n-1)}} \\ &\quad + 2\varepsilon_n \alpha_{jn}\beta_{jn} l_n \end{aligned} \tag{2.103}$$

Thus the first term of the integral I_n cancels with the second term of I_{n+1}. For the total integral (Equation 2.99) the remaining terms are

$$\int_{-D}^{L} \varepsilon(z)\{u_j(z)\}^2 dz$$
$$= \varepsilon_N \frac{\beta_{jN}^2 e^{2ik_{jN}D} - \alpha_{jN}^2 e^{-2ik_{jN}D}}{2ik_{jN}} - \varepsilon_0 \frac{\beta_{j0}^2 e^{2ik_{j0}L} - \alpha_{j0}^2 e^{-2ik_{j0}L}}{2ik_{j0}} \tag{2.104}$$
$$+ 2\sum_{n=0}^{N} \varepsilon_n |\alpha_{jn}|^2 l_n$$

The first and second terms vanish because the electric field should vanish at the surface of a perfect conductor (see Equation 2.78a). Thus using Equation 2.75b we arrive at Equation 2.99.

The inverse of the square root of the quantity in Equation 2.99 gives the normalization constant for the jth mode function. In the limit of large L we have

$$N_j^2 = \left(2\varepsilon_0 |\alpha_{j0}|^2 L\right)^{-1} \tag{2.105}$$

Thus a formal expression of the normalized mode function is

$$U_{jn}(z) = N_j u_{jn}(z) = \frac{\alpha_{jn} e^{ik_{jn}z} + \text{C.C.}}{(2\varepsilon_0 L)^{1/2} |\alpha_{j0}|}, \qquad N \geq n \geq 0 \tag{2.106}$$

These satisfy the orthonormality relation

$$\int_{-D}^{L} \varepsilon(z) U_i(z) U_j(z)\, dz = \delta_{ij} \tag{2.107}$$

For a complete expression, we still need the expression for the ratio $\alpha_{jn}/|\alpha_{j0}|$ in terms of the cavity parameters. We derive the expression as a product of $\alpha_{j0}/|\alpha_{j0}|$ and α_{jn}/α_{j0}. Substituting Equations 2.80a and 2.80b into Equation 2.84a with $m = 1$ we have

$$\begin{pmatrix} \alpha_{jn} e^{-ik_{jn}d_n} \\ \text{C.C.} \end{pmatrix} = J_{n,1} M_{n,1} \begin{pmatrix} \alpha_{j0} \\ \text{C.C.} \end{pmatrix}, \qquad N \geq n \geq 0 \tag{2.108}$$

Note that for a real standing wave mode we need $\beta = \alpha^*$ as in Equation 2.75b. Now Equation 2.89 reads

$$\beta_{j0} = \alpha_{j0}^* = -\alpha_{j0} e^{2ik_{j0}L} \tag{2.109a}$$

or

$$\frac{\alpha_{j0}}{\alpha_{j0}^*} = -e^{-2ik_{j0}L} \tag{2.109b}$$

Thus, using Equation 2.91 we have for the phase factor of α_{j0}

$$\frac{\alpha_{j0}}{|\alpha_{j0}|} = \pm i \left\{ \frac{(M_{N,1})_{22} + (M_{N,1})_{12}}{(M_{N,1})_{11} + (M_{N,1})_{21}} \right\}^{1/2} \tag{2.110}$$

Next, from Equations 2.108 and 2.109a or 2.109b we have

$$\alpha_{jn} e^{-ik_{jn}d_n} = J_{n,1}\Big((M_{n,1})_{11}\alpha_{j0} - (M_{n,1})_{12}\alpha^*_{j0} e^{2ik_{j0}L} \Big) \tag{2.111}$$

Substituting Equation 2.91 we have

$$\frac{\alpha_{jn}}{\alpha_{j0}} = J_{n,1} \left(\frac{(M_{n,1})_{11}\{(M_{N,1})_{22} + (M_{N,1})_{12}\} - (M_{n,1})_{12}\{(M_{N,1})_{11} + (M_{N,1})_{21}\}}{(M_{N,1})_{22} + (M_{N,1})_{12}} \right) e^{ik_{jn}d_n} \tag{2.112}$$

The numerator of the fraction can be shown to be equal to

$$\{(M_{N,n+1})_{22} + (M_{N,n+1})_{12}\} \det(M_{n,1}) \tag{2.113}$$

Thus using Equations 2.83, 2.82c, and 2.84c we have

$$\frac{\alpha_{jn}}{\alpha_{j0}} = I_{n,1} \frac{(M_{N,n+1})_{22} + (M_{N,n+1})_{12}}{(M_{N,1})_{22} + (M_{N,1})_{12}} e^{ik_{jn}d_n}, \qquad N \geq n \geq 0 \tag{2.114}$$

where

$$I_{n,1} = \prod_{i=1}^{n} (1 - R_i) \tag{2.115}$$

Thus we have the ratio $\alpha_{jn}/|\alpha_{j0}|$ from Equations 2.110 and 2.114. Using the result and noting the symmetry property in Equation 2.85a and 2.85b, we have for the mode function in Equation 2.106

$$\begin{aligned} U_{jn}(z) = & \frac{-i}{(2\varepsilon_0 L)^{1/2}} \frac{I_{n,1}}{|(M_{N,1})_{11} + (M_{N,1})_{21}|} \\ & \times \Big[\{(M_{N,n+1})_{22} + (M_{N,n+1})_{12}\} e^{ik_{jn}(z+d_n)} - \text{C.C.} \Big], \\ & \qquad N \geq n \geq 0 \end{aligned} \tag{2.116}$$

For $N = 1$, in particular, if we use Equation 2.82a with the conventions

$$M_{1,1} = \begin{pmatrix} e^{-ik_1 d}, & re^{-ik_1 d} \\ re^{ik_1 d}, & e^{ik_1 d} \end{pmatrix}, \qquad M_{1,1+1} = \begin{pmatrix} 1, & 0 \\ 0, & 1 \end{pmatrix}, \tag{2.117}$$

$$I_{0,1} = 1, \qquad d_0 = 0$$

we can show that Equation 2.116 reproduces the normalized mode function given by Equation 1.62b. We have chosen the minus sign in Equation 2.116 so that the formula fits with the function in Equation 1.62b.

Thus we have obtained the orthonormal mode functions of the "universe" for an arbitrarily stratified one-sided cavity. Thus the quantization can be carried out as in Section 2.1.

▶ Exercises

2.1 Prove the commutation rules in Equation 2.8.
2-1. Substitute Equations 2.4 and 2.5 into Equation 2.8 and use Equation 2.1 or substitute Equations 2.6 and 2.7 into Equation 2.1.

2.2 Derive the Hamiltonian in Equation 2.9.
2-2. Substitute Equations 2.6 and 2.7 into Equation 2.3 and use Equation 2.8.

2.3 Derive the expressions for the matrix elements in Equations 2.21a and 2.21b.
2-3. Note Equations 2.13 and use the relation $e^{\hat{A}} = \sum_{n=0}^{\infty} \hat{A}^n/n!$ for an operator $\hat{A}$.

2.4 For the density operator in Equation 2.31 show that $\hat{\rho}_i = 1$.
2-4. $\mathrm{Tr}\,\rho_i = \sum_{m_i} \langle m_i|\rho_i|m_i\rangle = (1 - e^{-\beta\hbar\omega_i}) \sum_{m_i=0}^{\infty} e^{-m_i\beta\hbar\omega_i} = 1.$

2.5 Prove the relation between the power spectrum and the correlation function described in Equation 2.45. Use Equations 2.40 and 2.44.
2-5. From Equation 2.40

$$G(z', 0, z, t) = \sum_j \frac{1}{2}\hbar\omega_j\langle n_j\rangle U_j(z')U_j(z)e^{-i\omega_j t}$$

Fourier transforming this quantity we have

$$\begin{aligned}
\int_{-\infty}^{\infty} G(z', 0, z, t)e^{i\omega t}dt &= \sum_j \frac{1}{2}\hbar\omega_j\langle n_j\rangle U_j(z')U_j(z)\int_{-\infty}^{\infty} e^{-i(\omega_j-\omega)t}dt \\
&= \sum_j \frac{1}{2}\hbar\omega_j\langle n_j\rangle U_j(z')U_j(z)2\pi\delta(\omega_j - \omega) \\
&= \int_0^{\infty} d\omega_j\rho_{\omega_j}\frac{1}{2}\hbar\omega_j\langle n_j\rangle U_j(z')U_j(z)2\pi\delta(\omega_j - \omega) \\
&= 2\pi\rho_\omega\frac{1}{2}\hbar\omega\langle n_\omega\rangle U_\omega(z')U_\omega(z)H(\omega)
\end{aligned}$$

which on rearranging and using Equation 2.44 gives the required result.

2.6 Derive the equation of motion 2.51 for the annihilation operator a_k under the presence of a sinusoidal current source described in Equation 2.50.
2-6. Start with

$$i\hbar\frac{d}{dt}\hat{a}_k = [\hat{a}_k,\ \hat{H} + \hat{H}_{int}] = [\hat{a}_k,\ \hat{H}] + [\hat{a}_k,\ \hat{H}_{int}]$$

The first term is given by Equation 2.36. For the second term, use the commutation rules in Equation 2.8.

2.7 Derive the form of the correlation function in Equation 2.60.

2-7. Equation 2.55 gives

$$\pi I(z', z, \omega) = \frac{1}{2}\int_{-\infty}^{\infty} G(z', 0, z, t')e^{i\omega t'}dt'$$

Inverse Fourier transforming both sides (without the factor $1/(2\pi)$) we have

$$\int_{-\infty}^{\infty} \pi I(z', z, \omega)e^{-i\omega t}d\omega = \int_{-\infty}^{\infty}\frac{1}{2}\int_{-\infty}^{\infty} G(z', 0, z, t')e^{i\omega t'}e^{-i\omega t}dt'd\omega$$

The right-hand side yields $\pi G(z', 0, z, t)$, and the left-hand side yields, by Equation 2.56,

$$\int_{-\infty}^{\infty}\frac{\hbar\omega_c\langle n_{\omega_c}\rangle}{\varepsilon_1 d}\frac{\gamma_c}{\gamma_c^2 + (\omega - \omega_c)^2}u^*_{\omega_c}(z')u_{\omega_c}(z)e^{-i\omega t}d\omega$$

where we have dropped $H(\omega)$ considering the narrow region of importance given by the Lorentzian function. For correlation between t' and t we may change the time t in the above integral to t–t'. The contour integral in the lower (upper) half-plane of ω is appropriate for $t > t'$ $(t < t')$ with the relevant pole at $\omega = \omega_c - i\gamma_c$ $(\omega = \omega_c + i\gamma_c)$. For $t > t'$, for example, the factor $1/(\omega - \omega_c + i\gamma_c)$ yields an integral $-2\pi i$, while the factor $1/(\omega - \omega_c - i\gamma_c)$ gives $1/(-2i\gamma_c)$, and the exponential becomes $e^{-i(\omega_c - i\gamma_c)(t-t')}$.

2.8 Derive the expectation value of the electric field in Equation 2.73 under the presence of an impulsive driving force in Equation 2.71.

2-8. See the solution to 2-7. In this case the only pole is $\omega = \omega_c - i\gamma_c$ in the lower half-plane.

References

1 Glauber, R.J. (**1963**) *Phys. Rev.*, *130*, 2529–2539.

2 Heitler, W. (**1954**) *The Quantum Theory of Radiation*, 3rd edn, Clarendon, Oxford.

3 Carrier, G.F., Krook, M., and Pearson, C.E. (**1966**) *Functions of a Complex Variable*, McGraw-Hill, New York.

4 Kibble, T.W.B. (**1970**) *Quantum electrodynamics*, in *Quantum Optics* (eds S.M. Kay and A. Maitland), Academic Press, London.

5 Ujihara, K. (**1978**) *Phys. Rev. A*, *18*, 659–670.

3
A One-Dimensional Quasimode Laser: General Formulation

In Chapters 3 and 4 we consider, as an introduction to laser theory, a simplified one-dimensional laser model, where the laser cavity is made of perfectly reflecting mirrors, or perfect conductors, at $z = -d$ and $z = 0$. The mirror transmission loss is replaced by a fictitious decay mechanism expressed by a single decay constant. This replacement simplifies the spatial aspect of laser analysis: the spatial distribution of the laser field is always fixed to a cavity resonant mode of the perfect cavity. Only the field amplitude changes with time. Therefore, the analysis can be made mostly in the time domain. We will call this fictitious perfect cavity mode a quasimode to distinguish it from the more natural, resonant modes of the cavity with finite transmission loss. The introduction of atoms as the amplifying medium and a description of the associated quantum noise sources are made in Chapter 3. Chapter 4 includes the semiclassical and the quantum analysis of the laser with the assumed perfect cavity with additional loss mechanism.

3.1
Cavity Resonant Modes

Here we consider one-dimensional plane vector waves that are polarized in the x-direction and propagated to the z-direction as before. Consider the perfect cavity depicted in Figure 3.1. The cavity consists of a lossless non-dispersive dielectric of dielectric constant ε_1, which is bounded by perfect conductors at $z = -d$ and at $z = 0$.

The natural oscillating field mode of the cavity, the cavity resonant mode, is defined as one that satisfies the perfect boundary condition of Equation 1.6 at $z = -d$ and at $z = 0$. It is easy to show that the normalized mode function is (problem 3-1)

$$U_k(z) = \sqrt{\frac{2}{\varepsilon_1 d}} \sin\frac{\omega_k}{c_1}(z + d) \tag{3.1}$$

with

$$\omega_k = k\frac{c_1 \pi}{d}, \qquad k = 1, 2, 3, \ldots \tag{3.2}$$

Output Coupling in Optical Cavities and Lasers: A Quantum Theoretical Approach
Kikuo Ujihara

ISBN: 978-3-527-40763-7

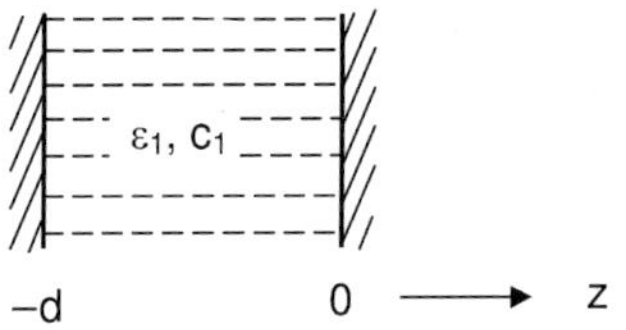

Figure 3.1 The perfect cavity.

The form of the function in Equation 3.1 has been chosen for easy comparison with that in Equation 1.21b for the one-sided cavity. These resonant modes are orthogonal to each other and make up a complete set for the region $-d < z < 0$. Writing the vector potential again as in Equation 1.44,

$$A(z,t) = \sum_k Q_k(t)\, U_k(z) \tag{3.3}$$

and defining the time derivative of the mode amplitude as in Equation 1.46,

$$\frac{d}{dt} Q_k = P_k \tag{3.4}$$

the total Hamiltonian of the field is now

$$\begin{aligned} H_f &= \int_{-d}^{0} \left[\frac{\varepsilon_1}{2} E(z,t)^2 + \frac{\mu}{2} H(z,t)^2\right] dz \\ &= \int_{-d}^{0} \left[\frac{\varepsilon_1}{2}\left(\frac{\partial}{\partial t} A(z,t)\right)^2 + \frac{1}{2\mu}\left(\frac{\partial}{\partial z} A(z,t)\right)^2\right] dz \end{aligned} \tag{3.5}$$

Here we have added a subscript f to indicate the optical field. This is easily evaluated to obtain (problem 3-2)

$$H_f = \frac{1}{2}\sum_k (P_k^2 + \omega_k^2 Q_k^2) \tag{3.6}$$

The quantization procedure goes just as in Equations 2.1–2.19: we have the creation and annihilation operators satisfying the commutation relations

$$\left[\hat{a}_i, \hat{a}_j^\dagger\right] = \delta_{ij}, \qquad \left[\hat{a}_i, \hat{a}_j\right] = 0, \qquad \left[\hat{a}_i^\dagger, \hat{a}_j^\dagger\right] = 0 \tag{3.7}$$

and the Hamiltonian

$$\hat{H}_f = \sum_k \hbar\omega_k (\hat{a}_k^\dagger \hat{a}_k + \tfrac{1}{2}) = \sum_k \hat{H}_k \tag{3.8}$$

3.2 The Atoms

We assume a two-level atom having upper laser level 2 with normalized wavefunction $\phi_2(\mathbf{r})$ and lower laser level 1 with wavefunction $\phi_1(\mathbf{r})$. We describe the atoms in the second quantized form, where the electron field amplitude is an operator [1,2]. The electronic wavefunction for the mth atom, now an operator, is

$$\psi_m(\mathbf{r}) = \hat{b}_{m1}\phi_1(\mathbf{r}) + \hat{b}_{m2}\phi_2(\mathbf{r}) \tag{3.9}$$

where $\hat{b}_{mi}$ is the annihilation operator for the ith level. The atomic Hamiltonian is written as

$$\hat{H}_a = \sum_m \hbar\nu_m \hat{b}^\dagger_{m2}\hat{b}_{m2} \tag{3.10}$$

where $\hat{b}^\dagger_{mi}$ is the creation operator for the ith level of the mth atom. The angular frequency ν_m is the transition frequency of the mth atom. The product $\hat{b}^\dagger_{m2}\hat{b}_{m2}$ is the number operator for level 2 of the mth atom. The Hamiltonian is evaluated with the lower atomic level as the origin. The product $\hat{b}^\dagger_{m2}\hat{b}_{m1}$ is the flipping operator from level 1 to level 2, and $\hat{b}^\dagger_{m1}\hat{b}_{m2}$ is that for the reverse. The atomic operators obey the anticommutation relations

$$\begin{aligned}
&\hat{b}_{mi}\hat{b}^\dagger_{m'i'} + \hat{b}^\dagger_{m'i'}\hat{b}_{mi} = \delta_{mm'}\delta_{ii'},\\
&\hat{b}_{mi}\hat{b}_{m'i'} + \hat{b}_{m'i'}\hat{b}_{mi} = 0,\\
&\hat{b}^\dagger_{mi}\hat{b}^\dagger_{m'i'} + \hat{b}^\dagger_{m'i'}\hat{b}^\dagger_{mi} = 0
\end{aligned} \tag{3.11}$$

3.3 The Atom–Field Interaction

When an atom is put in a field that is described by a vector potential, the atomic energy changes by an amount [3]

$$\hat{H}_{int} = -\frac{e}{m}\hat{\mathbf{A}}(\mathbf{r}) \cdot \hat{\mathbf{p}} \tag{3.12}$$

if we ignore a small quantity that is proportional to $\hat{\mathbf{A}}^2$. Here, e is the electron charge, m the electron mass, and $\hat{\mathbf{p}}$ the electron momentum. Here we are assuming that the vector potential has only an x-component and that only the x-component of the electron momentum $\hat{p}_x$ is effective. In the second quantized form it reads

$$\begin{aligned}
\hat{p}_x &= \int \psi^\dagger(\mathbf{r})\hat{p}_x\psi(\mathbf{r})d\mathbf{r} = \sum_{i,j=1,2} p_{ij}\hat{b}^\dagger_i\hat{b}_j\\
p_{ij} &= \int \phi^\dagger_i\hat{p}_x\phi_j d\mathbf{r}
\end{aligned} \tag{3.13}$$

Because in an atom the electron momentum follows the Heisenberg equation

$$\hat{p}_x = m\frac{d}{dt}\hat{x} = m[\hat{x}, \hat{H}_a]/(i\hbar) \tag{3.14}$$

the (i,j) matrix element is

$$\begin{aligned} p_{ij} &= \int \phi_i^* \hat{p}_x \phi_j d\mathbf{r} = m\int \phi_i^* [\hat{x}, \hat{H}_a]\phi_j d\mathbf{r}/(i\hbar) \\ &= m(E_j - E_i)x_{ij}/(i\hbar) = -im\omega_{ji}x_{ij} \end{aligned} \tag{3.15}$$

where $\omega_{ji} = (E_j - E_i)/\hbar$. Thus, using Equation 2.17 for the vector potential, we have

$$\begin{aligned} \hat{H}_{int} &= -\frac{e}{m}\sum_k (\hbar/2\omega_k)^{1/2}(\hat{a}_k + \hat{a}_k^\dagger)U_k(z)\sum_{i,j=1,2}(-im\omega_{ji}x_{ij})\hat{b}_i^\dagger \hat{b}_j \\ &= i\sum_k (\hbar/2\omega_k)^{1/2}(\hat{a}_k + \hat{a}_k^\dagger)U_k(z)\sum_{i,j=1,2}\omega_{ji}ex_{ij}\hat{b}_i^\dagger \hat{b}_j \end{aligned} \tag{3.16}$$

Here, z is interpreted as the location of the atom, which contains an approximation, called the electric dipole approximation, valid only when the spatial spread of the atom is much smaller than the optical wavelength.

Instead of the interaction described by Equation 3.12, some textbooks, for example, that by Loudon [4], consider the electric dipole interaction of one atom,

$$\hat{H}_{int} = -e\hat{\mathbf{r}}\cdot\hat{\mathbf{E}}(z) \tag{3.17}$$

where e is the electron charge, $\hat{\mathbf{r}}$ the effective displacement, and $\hat{\mathbf{E}}(z)$ the electric field at the position of the atom. This contains also the electric dipole approximation. Here we assume again that $\hat{\mathbf{E}}$ has only an x-component. The inner product makes only the x-component $\hat{x}$ in $\hat{\mathbf{r}}$ effective. The quantized form of $\hat{x}$ is

$$\begin{aligned} \hat{x} &= \int \psi^\dagger(\mathbf{r})\hat{x}\psi(\mathbf{r})d\mathbf{r} = \sum_{i,j=1,2} x_{ij}\hat{b}_i^\dagger \hat{b}_j \\ x_{ij} &= \int \phi_i^* \hat{x}\phi_j \end{aligned} \tag{3.18}$$

For $\hat{\mathbf{E}}$ we use Equation 2.18,

$$\hat{E}(z,t) = -\sum_k \hat{P}_k U_k(z) = \sum_k i(\hbar\omega_k/2)^{1/2}(\hat{a}_k - \hat{a}_k^\dagger)U_k(z) \tag{3.19}$$

Thus the interaction Hamiltonian of one atom is

$$\hat{H}_{int} = -e\sum_{i,j} x_{ij}\hat{b}_i^\dagger \hat{b}_j \sum_k i(\hbar\omega_k/2)^{1/2}(\hat{a}_k - \hat{a}_k^\dagger)U_k(z) \tag{3.20}$$

Here z is the location of the atom. Usually, an atom lacks a permanent electric dipole and thus $x_{ii} = 0$ in Equations 3.16 and 3.20. Then, there occur four kinds of terms in these equations, with operators $\hat{b}_1^\dagger \hat{b}_2 \hat{a}_j$, $\hat{b}_1^\dagger \hat{b}_2 \hat{a}_j^\dagger$, $\hat{b}_2^\dagger \hat{b}_1 \hat{a}_j$, and $\hat{b}_2^\dagger \hat{b}_1 \hat{a}_j^\dagger$. The physical content of the first kind of term is the annihilation of a photon with downward atomic flip, which contravenes energy conservation. Similarly, the fourth term means the creation of a photon with upward atomic transition, which also is incompatible with energy conservation. The second and third terms, respectively, imply the creation of one photon with downward atomic transition and the annihilation of one photon with upward atomic transition, which are both energy conserving. In this book we assume that the energy non-conserving terms may be ignored. This approximation is called the rotating-wave approximation.

Thus, when there are a number of atoms, each labeled by m, the interaction Hamiltonian is

$$\hat{H}_{int} = \sum_{k,m} \hbar(\kappa_{km}\hat{a}_k^\dagger \hat{b}_{m1}^\dagger \hat{b}_{m2} + \kappa_{km}^* \hat{a}_k \hat{b}_{m2}^\dagger \hat{b}_{m1}) \tag{3.21}$$

$$\kappa_{km} = i\nu_m(1/2\hbar\omega_k)^{1/2} U_k(z_m) p_m, \qquad p_m = ex_{m12}, \qquad \nu_m = \omega_{21} \tag{3.22a}$$

from Equation 3.16 and

$$\kappa_{km} = i(\omega_k/2\hbar)^{1/2} U_k(z_m) p_m, \qquad p_m = ex_{m12} \tag{3.22b}$$

from Equation 3.20. The difference between these two expressions for the atom–field coupling coefficient is negligibly small for phenomena in the optical frequency region.

In Equation 3.21 the sequence of the operators in the products is written in a mixed order. The order is normal if the photon annihilation operator is set to the rightmost position and the creation operator to the leftmost position, and it is anti-normal if the order is reversed. Any order is allowed as long as the product sequence is not changed during the calculation [5].

3.4
Equations Governing the Atom–Field Interaction

We examine the motion of the field using the Hamiltonians obtained so far. Because the cavity field modes are orthogonal, it is natural to assume single-mode operation, as is usually done in laser theories. Now the total Hamiltonian, including the optical field, the laser active atoms, and their interaction through dipole interaction, is

$$\hat{H}_t = \hat{H}_f + \hat{H}_a + \hat{H}_{int} \tag{3.23}$$

Here we assume that the field Hamiltonian is written, from Equation 3.8, as

$$\hat{H}_f = \hbar\omega_c\left(\hat{a}^\dagger \hat{a} + \frac{1}{2}\right) \tag{3.24}$$

dropping the mode suffix from the operators. The cavity resonance angular frequency is written as ω_c. The atomic Hamiltonian is given by Equation 3.10. The interaction Hamiltonian in Equation 3.21 is rewritten as

$$\hat{H}_{int} = \sum_m \hbar(\kappa_m \hat{a}^\dagger \hat{b}^\dagger_{m1}\hat{b}_{m2} + \kappa^*_m \hat{a}\hat{b}^\dagger_{m2}\hat{b}_{m1}) \tag{3.25}$$

with the coupling coefficient in Equation 3.22b,

$$\kappa_m = i(\omega_c/2\hbar)^{1/2} U_c(z_m) p_m \tag{3.26}$$

Because we are most interested in the motion of the laser field, we first examine the motion of the annihilation operator of the field in the Heisenberg picture (problem 3-3):

$$\frac{d}{dt}\hat{a} = \frac{1}{i\hbar}[\hat{a}, \hat{H}_t] = \frac{1}{i\hbar}\left(\hat{a}\hat{H}_t - \hat{H}_t\hat{a}\right) = -i\omega_c\hat{a} - i\sum_m \kappa_m(\hat{b}^\dagger_{m1}\hat{b}_{m2}) \tag{3.27}$$

where we have used the commutators in Equation 2.8 with $i = j$. The first term is the free motion originating from the field Hamiltonian, and the second term stems from the interaction Hamiltonian. From this result we know that the product $\hat{b}^\dagger_{m1}\hat{b}_{m2}$ drives the optical field; thus $\hat{b}^\dagger_{m1}\hat{b}_{m2}$ is the quantum counterpart of the classical atomic dipole. Thus we wish next to know the motion of the atomic dipole operator $\hat{b}^\dagger_{m1}\hat{b}_{m2}$, which flips the atom from the upper level 2 to the lower level 1. Again using the Heisenberg equation, we have (problem 3-4)

$$(d/dt)(\hat{b}^\dagger_{m1}\hat{b}_{m2})(t) = -i\nu_m(\hat{b}^\dagger_{m1}\hat{b}_{m2})(t) + i\kappa^*_m\hat{a}(t)\hat{\sigma}_m(t) \tag{3.28}$$

where we have used the anticommutators in Equation 3.11,

$$\begin{aligned}[\hat{b}^\dagger_{m1}\hat{b}_{m2}, \hat{b}^\dagger_{m2}\hat{b}_{m2}] &= \hat{b}^\dagger_{m1}\hat{b}_{m2}\hat{b}^\dagger_{m2}\hat{b}_{m2} - \hat{b}^\dagger_{m2}\hat{b}_{m2}\hat{b}^\dagger_{m1}\hat{b}_{m2} \\ &= \hat{b}^\dagger_{m1}(1 - \hat{b}^\dagger_{m2}\hat{b}_{m2})\hat{b}_{m2} - \hat{b}^\dagger_{m2}(-\hat{b}^\dagger_{m1}\hat{b}_{m2})\hat{b}_{m2} = \hat{b}^\dagger_{m1}\hat{b}_{m2}\end{aligned} \tag{3.29}$$

because $\hat{b}_{m2}\hat{b}_{m2} = 0$ due to the second member in Equation 3.11, and

$$\begin{aligned}[\hat{b}^\dagger_{m1}\hat{b}_{m2}, \hat{b}^\dagger_{m2}\hat{b}_{m1}] &= \hat{b}^\dagger_{m1}\hat{b}_{m2}\hat{b}^\dagger_{m2}\hat{b}_{m1} - \hat{b}^\dagger_{m2}\hat{b}_{m1}\hat{b}^\dagger_{m1}\hat{b}_{m2} \\ &= \hat{b}^\dagger_{m1}(1 - \hat{b}^\dagger_{m2}\hat{b}_{m2})\hat{b}_{m1} - \hat{b}^\dagger_{m2}(1 - \hat{b}^\dagger_{m1}\hat{b}_{m1})\hat{b}_{m2} \\ &= \hat{b}^\dagger_{m1}\hat{b}_{m1} - \hat{b}^\dagger_{m2}\hat{b}_{m2} \equiv -\hat{\sigma}_m\end{aligned} \tag{3.30}$$

because $\hat{b}_{m2}\hat{b}_{m1} = -\hat{b}_{m1}\hat{b}_{m2}$ and $\hat{b}^\dagger_{m1}\hat{b}^\dagger_{m2} = -\hat{b}^\dagger_{m2}\hat{b}^\dagger_{m1}$. Here, the operator $\hat{\sigma}_m = \hat{b}^\dagger_{m2}\hat{b}_{m2} - \hat{b}^\dagger_{m1}\hat{b}_{m1}$ is the atomic inversion operator, which probes the population difference between the upper and lower atomic levels. Note that an atom operator commutes with a field operator at this first stage of calculation.

Now that a new operator appears in the equation, we further examine the motion of the new operator $\hat{\sigma}_m$ (problem 3-5):

$$(d/dt)\hat{\sigma}_m(t) = 2i\{\kappa_m\hat{a}^\dagger(t)(\hat{b}^\dagger_{m1}\hat{b}_{m2})(t) - \kappa^*_m\hat{a}(t)(\hat{b}^\dagger_{m2}\hat{b}_{m1})(t)\} \tag{3.31}$$

Here a new operator $\hat{b}^\dagger_{m2}\hat{b}_{m1}$ appears. But this is just the Hermitian adjoint of $\hat{b}^\dagger_{m1}\hat{b}_{m2}$. So, Equations 3.27, 3.28, and 3.31 and their adjoints form a closed set of equations. These equations describe the coherent interaction between the field and the atoms, that is, there are no dissipation or random forces that tend to hinder coherent motions of the field and atoms. Now that the three coupled equations for the three operators are derived, these operators are generally mutually mixed and cannot be interchanged during the calculations, unlike at the first stage where we interchanged the atom and field operators in the appropriate Heisenberg equations under the given interaction Hamiltonian.

By inspection of the details in the calculations in Equations 3.29 and 3.30, one may notice that the general rule of reduction of a product of four atomic operators is [1]

$$\hat{b}^{\dagger}_{mi}\hat{b}_{mj}\hat{b}^{\dagger}_{mk}\hat{b}_{ml} = \hat{b}^{\dagger}_{mi}\hat{b}_{ml}\delta_{jk} \tag{3.32}$$

This is a modified form of the commutation relations in Equation 3.11.

3.5 Laser Equation of Motion: Introducing the Langevin Forces

Up to now we have considered the interaction of the cavity field mode and the atoms. In order to consider a laser, we need to take into account (i) the pumping process to make population inversion in the atoms, (ii) the cavity loss, which has up to now been ignored, and (iii) the effects of atomic environment on the atoms. These three processes are random processes if seen microscopically and introduce randomness in the motion of the field or the atoms: thus these are incoherent processes. The pumping process induces a relaxation of the inversion to a certain value depending on the strength of the pumping. The cavity loss causes the field amplitude to damp or relax to zero. And the atomic environment, for example, collisions or vacuum fluctuations, causes the atomic dipole oscillation to lose its phase or to relax to zero amplitude.

3.5.1 The Field Decay

The difficult point here is that, if we introduce appropriate relaxation terms into the equations obtained above, the quantum-mechanical consistency is ruined. For example, if, in the equation for the field (Equation 3.27), we add a decay term such as

$$\frac{d}{dt}\hat{a} = -i\omega_c\hat{a} - \gamma_c\hat{a} \tag{3.33}$$

where we have assumed the absence of atoms, the solution to this equation and the adjoint yield

$$[\hat{a}, \hat{a}^{\dagger}] = [\hat{a}(0)e^{-i\omega_c t-\gamma_c t}, \hat{a}^{\dagger}(0)e^{+i\omega_c t-\gamma_c t}] = [\hat{a}(0), \hat{a}^{\dagger}(0)]e^{-2\gamma_c t} = e^{-2\gamma_c t} \tag{3.34}$$

This means that, although the commutation relation (Equation 3.7) holds at $t = 0$, it is violated for $t > 0$. Thus, consistent quantum-mechanical analysis becomes impossible.

For the remedy, it is known that, if we are to use the decay term, we must add, at the same time, a fluctuating noise term, or the Langevin force term $\hat{\Gamma}_f(t)$, which makes the commutator revive on average. Thus we write

$$\frac{d}{dt}\hat{a} = -i\omega_c\hat{a} - \gamma_c\hat{a} + \hat{\Gamma}_f(t) \tag{3.35}$$

We assume that the ensemble average of the noise with respect to the damping mechanism, the so-called damping reservoir, satisfies

$$\begin{aligned} &\langle\hat{\Gamma}_f(t)\rangle = 0, && \left\langle\hat{\Gamma}_f^\dagger(t)\right\rangle = 0, \\ &\left\langle\hat{\Gamma}_f^\dagger(t)\hat{\Gamma}_f(t')\right\rangle = C_1\delta(t-t'), && \left\langle\hat{\Gamma}_f(t)\hat{\Gamma}_f^\dagger(t')\right\rangle = C_2\delta(t-t') \end{aligned} \tag{3.36}$$

Here the angle bracket signifies the quantum-mechanical expectation value averaged over the reservoir. Below, we will show that the diffusion coefficients satisfy

$$C_1 = 2\gamma_c\langle n_c\rangle, \qquad C_2 = 2\gamma_c(\langle n_c\rangle + 1) \tag{3.37}$$

where $\langle n_c\rangle$ is the reservoir average of the number of thermal photons belonging to the cavity mode. From Equation 3.35 we have

$$\hat{a}(t) = \int_0^t e^{(-i\omega_c-\gamma_c)(t-t')}\hat{\Gamma}_f(t')dt' + \hat{a}(0)e^{(-i\omega_c-\gamma_c)t} \tag{3.38}$$

Thus, taking note that $\hat{\Gamma}(t)$ and $\hat{a}(0)$ are mutually independent, we have

$$\begin{aligned} &\langle[\hat{a}(t),\hat{a}^\dagger(t)]\rangle \\ &= \left\langle\left[\int_0^t e^{(-i\omega_c-\gamma_c)(t-t')}\hat{\Gamma}_f(t')dt' + \hat{a}(0)e^{-i\omega_c t-\gamma_c t},\right.\right. \\ &\qquad \left.\left.\int_0^t e^{(+i\omega_c-\gamma_c)(t-t')}\hat{\Gamma}_f^\dagger(t')dt' + \hat{a}^\dagger(0)e^{+i\omega_c t-\gamma_c t}\right]\right\rangle \\ &= \left\langle\left[\int_0^t e^{(-i\omega_c-\gamma_c)(t-t')}\hat{\Gamma}_f(t')dt', \int_0^t e^{(+i\omega_c-\gamma_c)(t-t')}\hat{\Gamma}_f^\dagger(t')dt'\right]\right\rangle \\ &\quad + \langle[\hat{a}(0),\hat{a}^\dagger(0)]e^{-2\gamma_c t}\rangle \\ &= \int_0^t dt' \int_0^t dt' \, e^{(-i\omega_c-\gamma_c)(t-t')+(+i\omega_c-\gamma_c)(t-t')}\left\langle[\hat{\Gamma}_f(t'),\hat{\Gamma}_f^\dagger(t')]\right\rangle + e^{-2\gamma_c t} \end{aligned} \tag{3.39}$$

We require that this reservoir average be equal to unity. Using Equation 3.36 we have

$$\langle[\hat{a}(t),\hat{a}^\dagger(t)]\rangle = (C_2 - C_1)\frac{1-e^{-2\gamma_c t}}{2\gamma_c} + e^{-2\gamma_c t} = 1 \tag{3.40}$$

The last equality holds if

$$C_2 - C_1 = 2\gamma_c \tag{3.41}$$

Thus we can retain on average the commutation relation with the above condition described by Equation 3.41. Now, in thermal equilibrium, we should have $\langle \hat{a}^\dagger(t)\hat{a}(t)\rangle = \langle n_c\rangle$, where the thermal photon number of the cavity mode is given by Equation 2.42b. Then, following the above calculation, we see that for $t \to \infty$ the thermal photon number is $\langle \hat{a}^\dagger(t)\hat{a}(t)\rangle = C_1/2\gamma_c$ (problem 3-6). Therefore we obtain Equation 3.37: $C_1 = 2\gamma_c\langle n_c\rangle$ and $C_2 = 2\gamma_c(\langle n_c\rangle + 1)$.

We have derived the Langevin force associated with the cavity damping assuming the Markovian (delta-correlated) nature of the noise and requiring the preservation of the field commutation relation. The argument does not depend on the origin or nature of the decay constant γ_c. In a semiclassical analysis of a laser, the loss is typically introduced through assumed finite conduction loss distributed uniformly in the "perfect" cavity. In subsequent laser analyses in this and in the next chapters, we assume the form of Equation 3.35 for the cavity loss without arguing the precise origin of the decay constant.

The quantum-mechanical version of the conduction loss is to add a number of absorbing atoms, loss atoms, in the cavity. The number of atoms is so large that their absorbing power does not saturate, and these atoms constitute a reservoir. Then the total cavity field plus loss atom system conserves energy. In this model of the cavity, the cavity loss rate is determined by the atom–field coupling strength and the spatial and spectral number density of the atoms. This reservoir model is outlined in Appendix C.

Equations 3.35 and 3.36 are another form of the fluctuation–dissipation theorem stated in Equation 2.55. To see this, we construct the correlation function $\langle \hat{a}^\dagger(t)\hat{a}(t')\rangle$ using Equation 3.38. It is easy to show that (problem 3-7)

$$\langle \hat{a}^\dagger(t)\hat{a}(t')\rangle = \langle n_c\rangle e^{-i\omega_c(t'-t)-\gamma_c|t'-t|} \tag{3.42}$$

and that its Fourier transform from the domain of $t' - t$ to ω is

$$\int_{-\infty}^{+\infty} \langle \hat{a}^\dagger(t)\hat{a}(t')\rangle e^{i\omega(t'-t)} d(t'-t) = \langle n_c\rangle \frac{2\gamma_c}{\gamma_c^2 + (\omega - \omega_c)^2} \tag{3.43}$$

Comparing Equation 3.43 with Equation 2.55 together with Equation 2.56, the similarity is obvious. If we had constructed the correlation function $G(z', t', z, t)$ using the single-mode cases of Equations 2.18 and 2.19 together with Equation 3.1, a more precise comparison might have been possible.

We note that the cavity field decaying in the form of Equation 3.35 has a Lorentzian spectrum of full width at half-maximum (FWHM)$2\gamma_c$. We will call this width the cavity width.

3.5.2 Relaxation in Atomic Dipole and Atomic Inversion

For the motion of the dipole operator in Equation 3.28 and for the atomic inversion in Equation 3.31, we respectively add a relaxation or damping term and the associated Langevin force term as

$$\frac{d}{dt}\hat{a} = -i\omega_c\hat{a} - \gamma_c\hat{a} - i\sum_m \kappa_m(\hat{b}^\dagger_{m1}\hat{b}_{m2}) + \hat{\Gamma}_f(t) \tag{3.44}$$

$$\begin{aligned}(d/dt)(\hat{b}^\dagger_{m1}\hat{b}_{m2})(t) = &-i\nu_m(\hat{b}^\dagger_{m1}\hat{b}_{m2})(t) - \gamma_m(\hat{b}^\dagger_{m1}\hat{b}_{m2})(t)\\ &+ i\kappa^*_m\hat{a}(t)\hat{\sigma}_m(t) + \hat{\Gamma}_m(t)\end{aligned} \tag{3.45}$$

$$\begin{aligned}(d/dt)\hat{\sigma}_m(t) = &-\Gamma_{mp}\{\hat{\sigma}_m(t) - \sigma^0_m\} + 2i\{\kappa_m\hat{a}^\dagger(t)(\hat{b}^\dagger_{m1}\hat{b}_{m2})(t)\\ &- \kappa^*_m\hat{a}(t)(\hat{b}^\dagger_{m2}\hat{b}_{m1})(t)\} + \hat{\Gamma}_{m\|}(t)\end{aligned} \tag{3.46}$$

where we have added the reformed equation for the field for later convenience. The newly added terms represent incoherent motions of respective operators. The damping constant γ_m for the atomic dipole comes from collision of other particles with the atom and is ultimately determined by the spontaneous emission due to the vacuum fluctuation of the field. For vacuum fluctuation as the cause of the spontaneous emission, all the existing three-dimensional field modes should be taken into account even though we are considering the motion of a single cavity mode. Except in the case of a microcavity laser, where the cavity volume is nearly the wavelength cubed, the three-dimensional field is, roughly, that of a free field. The constant Γ_{mp} is the relaxation constant of the population inversion and σ^0_m is the equilibrium atomic inversion under the pumping but in the absence of the field. The latter constant is a measure of the strength of the pumping. We assume that each atom has its own dipole reservoir and pumping reservoir.

Here we insert a brief note on the atomic bandwidth. The atomic dipole, in the absence of the field described by Equation 3.45 without the third term, will have a power spectrum similar in form to the one for the field described in Equation 3.43. There will appear another Lorentzian profile with FWHM $2\gamma_m$. We will call this the atomic width or the natural width of the mth atom.

The discussion on the nature of the Langevin forces, or the noise terms, $\hat{\Gamma}_m$ and $\hat{\Gamma}_{m\|}$, are rather involved. To derive their characteristics, one also assumes the respective Markovian nature of the forces and the conservation of quantum-mechanical consistency described by Equation 3.32. For details, the reader is referred to the book by Haken [1] for example. Here we cite the results. We write the time rates of change of the incoherent part of the populations and dipoles in a multi-level atom as

$$\frac{d}{dt}\left(\hat{b}^\dagger_j\hat{b}_j\right) = \sum_k w_{kj}\left(\hat{b}^\dagger_k\hat{b}_k\right) - \sum_k w_{jk}\left(\hat{b}^\dagger_j\hat{b}_j\right) + \hat{\Gamma}_{jj}(t) \tag{3.47a}$$

$$\frac{d}{dt}\left(\hat{b}^\dagger_j\hat{b}_k\right) = -\gamma_{jk}\left(\hat{b}^\dagger_j\hat{b}_k\right) + \hat{\Gamma}_{jk}(t) \tag{3.47b}$$

where w_{kj} is the transition rate from level k to level j due to pumping and incoherent damping of atoms. Then the reservoir averages of the noise sources and of the product of the Langevin noise terms are given as

$$\langle\hat{\Gamma}_{ij}(t)\rangle = 0, \qquad \left\langle\hat{\Gamma}_{ij}^{\dagger}(t)\right\rangle = 0 \tag{3.48}$$

$$\langle\hat{\Gamma}_{ij}(t)\hat{\Gamma}_{kl}(t')\rangle = G_{ij,kl}\delta(t-t') \tag{3.49}$$

with the diffusion coefficient

$$\begin{aligned}
G_{ii,jj} &= \delta_{ij}\left\{\sum_k w_{ki}\left\langle\hat{b}_k^{\dagger}\hat{b}_k\right\rangle + \sum_k w_{ik}\left\langle\hat{b}_i^{\dagger}\hat{b}_i\right\rangle\right\} - w_{ij}\left\langle\hat{b}_i^{\dagger}\hat{b}_i\right\rangle - w_{ji}\left\langle\hat{b}_j^{\dagger}\hat{b}_j\right\rangle \\
G_{ij,ij} &= 0, \qquad\qquad i \neq j \\
G_{ij,ji} &= \sum_k w_{ki}\left\langle\hat{b}_k^{\dagger}\hat{b}_k\right\rangle - \sum_k w_{ik}\left\langle\hat{b}_i^{\dagger}\hat{b}_i\right\rangle + (\gamma_{ij}+\gamma_{ji})\left\langle\hat{b}_i^{\dagger}\hat{b}_i\right\rangle, \quad i \neq j
\end{aligned} \tag{3.50}$$

where the angle bracket signifies the quantum-mechanical expectation value averaged over the atomic reservoirs. The constants in Equation 3.46 concerning the pumping process in a two-level atom are related to the transition rates, if we note that $\left\langle\hat{b}_1^{\dagger}\hat{b}_1\right\rangle + \left\langle\hat{b}_2^{\dagger}\hat{b}_2\right\rangle = 1$, as

$$\Gamma_{mp} = w_{m12} + w_{m21}, \qquad \sigma_m^0 = \frac{w_{m12} - w_{m21}}{w_{m12} + w_{m21}} \tag{3.51}$$

or

$$w_{m12} = \frac{1}{2}\Gamma_{mp}(1+\sigma_m^0), \qquad w_{m21} = \frac{1}{2}\Gamma_{mp}(1-\sigma_m^0) \tag{3.52}$$

There is an equation, known as the Einstein relation, connecting the diffusion coefficient and the drift coefficients (the constants C_1 and C_2 in Equations 3.36 and the decay constant in Equation 3.35, for example), which, in the case of the atoms, leads ultimately to Equation 3.50 – see, for example, the book by Sargent *et al.* [6].

▶ Exercises

3.1 Show that the normalized mode functions for the cavity described in Figure 3.1 are given by Equations 3.1 and 3.2. Show that they are mutually orthogonal.
3-1. Assume a solution of the form that satisfies the vanishing boundary condition at $z=-d$: $U(z) = C\sin\kappa(z+d)$. *The vanishing boundary condition at* $z=0$ yields $\kappa d = k\pi$, $k=1, 2, 3, \ldots$, so that we have $\omega_k = c_1\kappa = kc_1\pi/d$, $k=1, 2, 3, \ldots$. For orthonormality, calculate $\int_{-d}^{0}\varepsilon_1 C^2 \sin(k\pi/d)(z+d)\sin(k'\pi/d)(z+d)dz$, which is $\varepsilon_1 C^2 d/2$ for $k=k'$ and vanishes for $k \neq k'$. By the normalization condition in Equation 1.42a, we have $C = \sqrt{2/(\varepsilon_1 d)}$.

3.2 Derive the Hamiltonian in Equation 3.6 from Equations 3.3–3.5.
3-2. We have

$$H_f = \int_{-d}^{0} \left[\frac{\varepsilon_1}{2} \left(\frac{\partial}{\partial t} A(z,t) \right)^2 + \frac{1}{2\mu} \left(\frac{\partial}{\partial z} A(z,t) \right)^2 \right] dz$$
$$= \int_{-d}^{0} \left[\frac{\varepsilon_1}{2} \left\{ \sum_k P_k U_k(z) \right\}^2 + \frac{1}{2\mu} \left\{ \sum_k Q_k U'_k(z) \right\} \right]^2$$

The first term yields, due to problem 3-1, $\frac{1}{2}\sum_k P_k^2$. The second term reads

$$\int_{-d}^{0} \frac{1}{2\mu} \left\{ \sum_k Q_k U'_k(z) \right\}^2 dz = \frac{1}{2\varepsilon_1 \mu} \sum_k \sum_{k'} Q_k Q_{k'} \int_{-d}^{0} \varepsilon_1 U'_k(z) U'_{k'}(z) dz$$
$$= \frac{1}{2\varepsilon_1 \mu} \sum_k \sum_{k'} Q_k Q_{k'} \left[\varepsilon_1 U_k(z) U'_{k'} |_{-d}^{0} - \int_{-d}^{0} \varepsilon_1 U_k(z) U''_{k'}(z) dz \right]$$
$$= \frac{1}{2\varepsilon_1 \mu} \sum_k \sum_{k'} Q_k Q_{k'} \times \int_{-d}^{0} \varepsilon_1 U_k(z) k_{k'}^2 U_{k'}(z) dz = \sum_k \frac{k_k^2}{2\varepsilon_1 \mu} Q_k^2$$

where a prime on a function indicates differentiation with respect to z. On going from the second to the third line, the Helmholtz equation for the U has been used. Note that $k_k^2/(\varepsilon_1 \mu) = k_k^2 c_1^2 = \omega_k^2$.

3.3 Derive the equation of motion for the annihilation operator described in Equation 3.27.
3-3. For the first term, see Equation 2.36. For the second term, use the commutator in Equation 2.8.

3.4 Derive the equation of motion for the atomic dipole described in Equation 3.28.
3-4. See Equations 3.29 and 3.30.

3.5 Derive the equation of motion for the atomic inversion described in Equation 3.31.
3-5. We have $[\sigma_m, H_f] = 0$.
That $[\sigma_m, H_a] = 0$ can be shown by repeated use of Equation 3.11. Then,

$$[\hat{b}^\dagger_{m2}\hat{b}_{m2}, \hat{b}^\dagger_{m1}\hat{b}_{m2}] = \hat{b}^\dagger_{m2}\hat{b}_{m2}\hat{b}^\dagger_{m1}\hat{b}_{m2} - \hat{b}^\dagger_{m1}\hat{b}_{m2}\hat{b}^\dagger_{m2}\hat{b}_{m2}$$
$$= \hat{b}^\dagger_{m2}(-\hat{b}^\dagger_{m1}\hat{b}_{m2})\hat{b}_{m2} - \hat{b}^\dagger_{m1}(1 - \hat{b}^\dagger_{m2}\hat{b}_{m2})\hat{b}_{m2} = -\hat{b}^\dagger_{m1}\hat{b}_{m2}$$

$$[\hat{b}^\dagger_{m1}\hat{b}_{m1}, \hat{b}^\dagger_{m1}\hat{b}_{m2}] = \hat{b}^\dagger_{m1}\hat{b}_{m1}\hat{b}^\dagger_{m1}\hat{b}_{m2} - \hat{b}^\dagger_{m1}\hat{b}_{m2}\hat{b}^\dagger_{m1}\hat{b}_{m1}$$
$$= \hat{b}^\dagger_{m1}(1 - \hat{b}^\dagger_{m1}\hat{b}_{m1})\hat{b}_{m2} - \hat{b}^\dagger_{m1}(-\hat{b}^\dagger_{m1}\hat{b}_{m2})\hat{b}_{m1} = \hat{b}^\dagger_{m1}\hat{b}_{m2}$$

because $\hat{b}_{m2}\hat{b}_{m2} = 0$ and $\hat{b}^\dagger_{m1}\hat{b}^\dagger_{m1} = 0$. Thus $[\sigma_m, \hat{b}^\dagger_{m1}\hat{b}_{m2}] = -2\hat{b}^\dagger_{m1}\hat{b}_{m2}$. Similarly,

$$[\hat{b}^\dagger_{m2}\hat{b}_{m2}, \hat{b}^\dagger_{m2}\hat{b}_{m1}] = \hat{b}^\dagger_{m2}\hat{b}_{m2}\hat{b}^\dagger_{m2}\hat{b}_{m1} - \hat{b}^\dagger_{m2}\hat{b}_{m1}\hat{b}^\dagger_{m2}\hat{b}_{m2}$$
$$= \hat{b}^\dagger_{m2}(1 - \hat{b}^\dagger_{m2}\hat{b}_{m2})\hat{b}_{m1} - \hat{b}^\dagger_{m2}(-\hat{b}^\dagger_{m2}\hat{b}_{m1})\hat{b}_{m2} = \hat{b}^\dagger_{m2}\hat{b}_{m1}$$

$$[\hat{b}^\dagger_{m1}\hat{b}_{m1}, \hat{b}^\dagger_{m2}\hat{b}_{m1}] = \hat{b}^\dagger_{m1}\hat{b}_{m1}\hat{b}^\dagger_{m2}\hat{b}_{m1} - \hat{b}^\dagger_{m2}\hat{b}_{m1}\hat{b}^\dagger_{m1}\hat{b}_{m1}$$
$$= \hat{b}^\dagger_{m1}(-\hat{b}^\dagger_{m2}\hat{b}_{m1})\hat{b}_{m1} - \hat{b}^\dagger_{m2}(1 - \hat{b}^\dagger_{m1}\hat{b}_{m1})\hat{b}_{m1} = -\hat{b}^\dagger_{m2}\hat{b}_{m1}$$

and $[\sigma_m, \hat{b}^\dagger_{m2}\hat{b}_{m1}] = 2\hat{b}^\dagger_{m2}\hat{b}_{m1}$. Note that the rule in Equation 3.32 yields these results more quickly.

3.6 Using the Langevin equation (Equation 3.35), show that $\lim_{t\to\infty}\langle\hat{a}^\dagger(t)\hat{a}(t)\rangle = C_1/(2\gamma_c)$.

3-6. Equation 3.35 leads to Equation 3.38. Thus, noting that $\hat{\Gamma}(t)$ and $\hat{a}(0)$ are mutually independent, we get

$$\langle\hat{a}^\dagger(t)\hat{a}(t)\rangle = \left\langle\left\{\int_0^t e^{(+i\omega_c-\gamma_c)(t-t')}\hat{\Gamma}^\dagger_f(t')dt' + \hat{a}^\dagger(0)e^{+i\omega_c t-\gamma_c t}\right\}\right.$$
$$\left.\times\left\{\int_0^t e^{(-i\omega_c-\gamma_c)(t-t')}\hat{\Gamma}_f(t')dt' + \hat{a}(0)e^{-i\omega_c t-\gamma_c t}\right\}\right\rangle$$
$$= \left\langle\int_0^t e^{(+i\omega_c-\gamma_c)(t-t'')}\hat{\Gamma}^\dagger_f(t'')dt''\int_0^t e^{(-i\omega_c-\gamma_c)(t-t')}\hat{\Gamma}_f(t')dt'\right\rangle$$
$$+ \langle\hat{a}^\dagger(0)\hat{a}(0)e^{-2\gamma_c t}\rangle$$
$$= \int_0^t dt''\int_0^t dt'\, e^{(-i\omega_c-\gamma_c)(t-t')+(+i\omega_c-\gamma_c)(t-t'')}\left\langle\hat{\Gamma}^\dagger_f(t'')\hat{\Gamma}_f(t')\right\rangle$$
$$+ \langle\hat{a}^\dagger(0)\hat{a}(0)\rangle e^{-2\gamma_c t}$$
$$= \int_0^t dt''\int_0^t dt'\, e^{(-i\omega_c-\gamma_c)(t-t')+(+i\omega_c-\gamma_c)(t-t'')}C_1\delta(t''-t')$$
$$+ \langle\hat{a}^\dagger(0)\hat{a}(0)\rangle e^{-2\gamma_c t}$$
$$= C_1\frac{1-e^{-2\gamma_c t}}{2\gamma_c} + \langle\hat{a}^\dagger(0)\hat{a}(0)\rangle e^{-2\gamma_c t}$$

3.7 Derive the correlation function in Equation 3.42.

3-7. We have

$$
\begin{aligned}
\langle \hat{a}^{\dagger}(t)\hat{a}(t')\rangle &= \left\langle \left\{ \int_0^t e^{(i\omega_c-\gamma_c)(t-t'')}\hat{\Gamma}_f^{\dagger}(t'')dt'' + \hat{a}^{\dagger}(0)e^{(i\omega_c-\gamma_c)t} \right\} \right. \\
&\quad \times \left. \left\{ \int_0^{t'} e^{(-i\omega_c-\gamma_c)(t'-t''')}\hat{\Gamma}_f(t''')dt''' + \hat{a}(0)e^{(-i\omega_c-\gamma_c)t'} \right\} \right\rangle \\
&= \int_0^t dt'' \int_0^{t'} dt'''\, e^{(i\omega_c-\gamma_c)(t-t'')}e^{(-i\omega_c-\gamma_c)(t'-t''')}\left\langle \hat{\Gamma}_f^{\dagger}(t'')\hat{\Gamma}_f(t''')\right\rangle \\
&\quad + \langle \hat{a}^{\dagger}(0)\hat{a}(0)\rangle e^{(i\omega_c-\gamma_c)t}e^{(-i\omega_c-\gamma_c)t'} \\
&= \int_0^t dt'' \int_0^{t'} dt'''\, e^{(i\omega_c-\gamma_c)(t-t'')}e^{(-i\omega_c-\gamma_c)(t'-t''')}C_1\delta(t''-t''') \\
&\quad + \langle \hat{a}^{\dagger}(0)\hat{a}(0)\rangle e^{(i\omega_c-\gamma_c)t}e^{(-i\omega_c-\gamma_c)t'} \\
&= \begin{cases} C_1 \int_0^{t'} dt''\, e^{i\omega_c(t-t')}e^{-\gamma_c(t+t'-2t'')}, & t>t' \\ C_1 \int_0^{t} dt''\, e^{i\omega_c(t-t')}e^{-\gamma_c(t+t'-2t'')}, & t<t' \end{cases} \\
&\quad + \langle \hat{a}^{\dagger}(0)\hat{a}(0)\rangle e^{(i\omega_c-\gamma_c)t}e^{(-i\omega_c-\gamma_c)t'} \\
&= \begin{cases} C_1 e^{i\omega_c(t-t')}\frac{e^{-\gamma_c(t-t')}-e^{-\gamma_c(t+t')}}{2\gamma_c}, & t>t' \\ C_1 e^{i\omega_c(t-t')}\frac{e^{-\gamma_c(t'-t)}-e^{-\gamma_c(t+t')}}{2\gamma_c}, & t<t' \end{cases} \\
&\quad + \langle \hat{a}^{\dagger}(0)\hat{a}(0)\rangle e^{(i\omega_c-\gamma_c)t}e^{(-i\omega_c-\gamma_c)t'}
\end{aligned}
$$

If we note that $C_1/(2\gamma_c) = \langle n_c \rangle$ we obtain Equation 3.42 for $t, t' \to \infty$.

References

1 Haken, H. (**1970**) *Laser theory*, in *Licht und Materie, IC*, Handbuch der Physik, vol. XXV/2c (eds. S. Flügge and L. Genzel), Springer, Berlin.

2 Heitler, W. (**1954**) *The Quantum Theory of Radiation*, 3rd edn, Clarendon, Oxford.

3 Kibble, T.W.B. (**1970**) Quantum electrodynamics, in *Quantum Optics* (eds S.M. Kay and A. Maitland), Academic Press, London.

4 Loudon, R. (**1973**) *Quantum Theory of Light*, Clarendon, Oxford.

5 Allen, L. and Eberly, J.H. (**1975**) *Optical Resonance and Two-Level Atoms*, John Wiley and Sons, Inc., New York.

6 Sargent, M., III, Scully, M.O., and Lamb, W.E., Jr. (**1974**) *Laser Physics*, Addison-Wesley, Reading, MA.

4
A One-Dimensional Quasimode Laser: Semiclassical and Quantum Analysis

Now that we have a complete set of equations (Equations 3.44–3.46) with known properties of the Langevin forces (Equations 3.37 and 3.50), we can in principle analyze the laser in a quantum mechanically consistent manner. However, the problem to be solved is rather hard because of the presence of random force terms. So, before solving the complete equations, we first consider the "average" equations, leaving the Langevin force terms ignored. Such an approach is called a semiclassical theory because the quantum-mechanical consistency is then not fully preserved. After solving the average equations, we can introduce the random forces to take fully into account the quantum effects. The semiclassical analysis has two main steps. In the first step, we assume that the population inversion is determined by the pumping process only and it is not affected by the presence of the laser field. This is a linear gain analysis in that the amplification by the atoms is linear with respect to the laser field amplitude. This theory applies to the sub-threshold region of laser operation. In the second step, we allow for a nonlinear behavior of the amplifying atoms. The amplifying capability of atoms then decreases because of the consumption of inverted atoms by the presence of the laser field because of the stimulated emission. This is called the saturation effect. This nonlinear gain analysis yields the steady-state laser amplitude, which keeps the inversion level or the field gain just balancing the field loss due to the cavity loss. In this book we limit ourselves to a steady-state oscillation and do not consider pulsed laser operation or temporal variation of the field intensity. The quantum analysis is also divided into two main steps: the linear gain analysis and the nonlinear gain analysis.

4.1
Semiclassical Linear Gain Analysis

Here, we assume a steady state with constant population inversion, and set $\hat{\sigma}_m(t) = \sigma_m$ in Equation 3.45. We consider the following equations, ignoring the noise terms:

Output Coupling in Optical Cavities and Lasers: A Quantum Theoretical Approach
Kikuo Ujihara

ISBN: 978-3-527-40763-7

$$\frac{d}{dt}\hat{a}(t) = -i\omega_c\hat{a}(t) - \gamma_c\hat{a}(t) - i\sum_m \kappa_m(\hat{b}^\dagger_{m1}\hat{b}_{m2})(t) \tag{4.1}$$

$$(d/dt)(\hat{b}^\dagger_{m1}\hat{b}_{m2})(t) = -i\nu_m(\hat{b}^\dagger_{m1}\hat{b}_{m2})(t) - \gamma_m(\hat{b}^\dagger_{m1}\hat{b}_{m2})(t) + i\kappa^*_m\hat{a}(t)\sigma_m \tag{4.2}$$

From Equation 3.19 we have the positive frequency part of the electric field for a single mode in question:

$$\hat{E}^{(+)}(z,t) = i(\hbar\omega_c/2)^{1/2}U_c(z)\hat{a}(t) \tag{4.3}$$

Thus multiplying both sides of Equation 4.1 by $i(\hbar\omega_c/2)^{1/2}U_c(z)$ and integrating we have

$$\begin{aligned}\hat{E}^{(+)}(z,t) = {} & \sqrt{\frac{\hbar\omega_c}{2}}U_c(z)e^{(-i\omega_c-\gamma_c)t}\int_0^t e^{(i\omega_c+\gamma_c)t'}\sum_m \kappa_m\left(\hat{b}^\dagger_{m1}\hat{b}_{m2}\right)(t')dt' \\ & + \hat{E}^{(+)}(z,0)e^{(-i\omega_c-\gamma_c)t}\end{aligned} \tag{4.4}$$

The second term comes from the initial field and vanishes for large t. So we will ignore this term. Since $\kappa^*_m = -i\nu_m(1/2\hbar\omega_c)^{1/2}U_c(z_m)p^*_m$ according to Equation 3.22a, we replace $\kappa^*_m\hat{a}(t)$ in Equation 4.2 by $-\{\nu_m p^*_m/(\hbar\omega_c)\}\hat{E}^{(+)}(z_m,t)$. Integrating Equation 4.2 (problem 4-1) and substituting the result into Equation 4.4, we have

$$\begin{aligned}\hat{E}^{(+)}(z,t) = \sum_m \Bigg[& \frac{|p_m|^2\nu_m^2\sigma_m}{2\hbar\omega_c}U_c(z)U_c(z_m) \\ & \times\int_0^t e^{(-i\omega_c-\gamma_c)(t-t')}\int_0^{t'} e^{(-i\nu_m-\gamma_m)(t'-t'')}\hat{E}^{(+)}(z_m,t'')dt''dt'\Bigg]\end{aligned} \tag{4.5}$$

where we have ignored the term coming from the initial value of $\hat{b}^\dagger_{m1}\hat{b}_{m2}$, which represents a switching-on effect. This form of equation, describing $\hat{E}^{(+)}(z,t)$ rather than $\hat{a}(t)$, has been derived for later comparison with the situation where the field inside the cavity is non-uniform because of the coupling loss at the cavity ends. Anyway, this form expresses the contribution of the mth atom at z_m to the field at (z,t) through the product of the mode functions at z and z_m, which we saw in the expression of the response function in Equation 2.53a (see Equation 5.34 below). We go back to $\hat{a}(t)$ by dividing Equation 4.5 by $i(\hbar\omega_c/2)^{1/2}U_c(z)$ to obtain

$$\begin{aligned}\hat{a}(t) = \sum_m \Bigg[& \frac{|p_m|^2\nu_m^2\sigma_m}{2\hbar\omega_c}U_c(z_m)U_c(z_m)\int_0^t e^{(-i\omega_c-\gamma_c)(t-t')} \\ & \times\int_0^{t'} e^{(-i\nu_m-\gamma_m)(t'-t'')}\hat{a}(t'')dt''dt'\Bigg]\end{aligned} \tag{4.6}$$

It is not easy to go further if the atoms are of different nature. For simplicity, we assume here equally pumped, homogeneously broadened atoms and set

$$\nu_m = \nu_0, \qquad p_m = p_a, \qquad \gamma_m = \gamma, \qquad \sigma_m = \sigma \tag{4.7}$$

Then we can write

$$\hat{a}(t) = k^2 N\sigma \int_0^t e^{(-i\omega_c-\gamma_c)(t-t')} \int_0^{t'} e^{(-i\nu_0-\gamma)(t'-t'')}\hat{a}(t'')dt''dt' \tag{4.8}$$

with

$$k^2 N\sigma = \frac{|p_a|^2 \nu_0^2 \sigma}{2\hbar\omega_c} \int_{-d}^{0} N\, dz_m\, U_c^2(z_m) = \frac{|p_a|^2 \nu_0}{2\varepsilon_1 \hbar} N\sigma = \sum_m |\kappa_m|^2 \sigma \tag{4.9}$$

where N is the density of atoms in the z-direction and we have set $\nu_0 = \omega_c$ in the second equality, which is usually highly accurate. Differentiating Equation 4.8 twice we have

$$\ddot{\hat{a}} + \{i(\omega_c + \nu_0) + \gamma_c + \gamma\}\dot{\hat{a}} - \{k^2 N\sigma - (i\nu_0 + \gamma)(i\omega_c + \gamma_c)\}\hat{a} = 0 \tag{4.10a}$$

Because the field is oscillating at a high frequency ω, which is still to be determined, we write $\hat{a}(t) = \tilde{a}(t)e^{-i\omega t}$ and rewrite Equation 4.10a as (problem 4-2)

$$\begin{aligned} &\ddot{\tilde{a}} + \{i(\omega_c + \nu_0 - 2\omega) + \gamma_c + \gamma\}\dot{\tilde{a}} \\ &\quad - [k^2 N\sigma - \{i(\nu_0 - \omega) + \gamma\}\{i(\omega_c - \omega) + \gamma_c\}]\tilde{a} = 0 \end{aligned} \tag{4.10b}$$

Now that $\tilde{a}(t)$ is slowly varying, we ignore the second derivative. Then the amplitude $\tilde{a}(t)$ simply grows or decays exponentially depending on the coefficients of the second and third terms. Since we are anticipating operation below threshold, the amplitude should decay as

$$\tilde{a} \sim \exp\left[\frac{k^2 N\sigma - \{i(\nu_0 - \omega) + \gamma\}\{i(\omega_c - \omega) + \gamma_c\}}{i(\omega_c + \nu_0 - 2\omega) + \gamma_c + \gamma} t\right] \tag{4.11}$$

The laser threshold conditions are obtained if we set the exponent to zero. Thus we have the threshold oscillation frequency and the threshold atomic inversion as

$$\omega_{th} = \frac{\gamma\omega_c + \gamma_c \nu_0}{\gamma + \gamma_c} \tag{4.12}$$

$$k^2 N\sigma_{th} = \gamma\gamma_c - (\nu_0 - \omega_{th})(\omega_c - \omega_{th}) = \gamma\gamma_c \left\{1 + \frac{(\omega_c - \nu_0)^2}{(\gamma + \gamma_c)^2}\right\} \tag{4.13a}$$

or using Equation 4.9

$$\sigma_{th} = \frac{2\hbar\varepsilon_1 \gamma\gamma_c}{|p_a|^2 \nu_0 N}(1 + \delta^2) \tag{4.13b}$$

where the squared relative detuning is

$$\delta^2 = \frac{(\nu_0 - \omega_c)^2}{(\gamma + \gamma_c)^2} \tag{4.13c}$$

Equation 4.12 shows the linear pulling effect between the cavity resonance and the atomic resonance frequencies. The laser frequency tends to be pulled to the narrower resonance. The threshold inversion is smaller for larger atomic density and larger electric dipole matrix element, and is larger for larger atomic width, larger cavity loss, and larger detuning between the cavity and the atomic resonances. We rewrite Equation 4.13a as

$$\mathcal{g} N\sigma_{th} = \gamma_c \tag{4.13d}$$

where the coefficient

$$\mathcal{g} = \frac{k^2}{\gamma(1+\delta^2)} = \frac{|p_a|^2 \nu_0}{2\varepsilon_1 \hbar\gamma(1+\delta^2)} \tag{4.14}$$

is interpreted as the amplitude gain per unit density of inverted atoms per unit time. This is equal to half the stimulated transition rate per atom per unit density of photons.

4.2
Semiclassical Nonlinear Gain Analysis

Now we take the saturation of the atomic inversion due to stimulated emission into account in the context of semiclassical analysis. We start with Equations 3.44 and 3.45, with the noise terms discarded, and obtain, as in Equation 4.5,

$$\begin{aligned}\hat{E}^{(+)}(z,t) = \sum_m \Bigg[& \frac{|p_m|^2 \nu_m}{2\hbar} U_c(z) U_c(z_m) \\ & \times \int_0^t e^{(-i\omega_c-\gamma_c)(t-t')} \int_0^{t'} e^{(-i\nu_m-\gamma_m)(t'-t'')} \hat{E}^{(+)}(z_m,t'')\sigma_m(t'')dt''dt' \Bigg]\end{aligned} \tag{4.15}$$

where we have assumed that $\omega_c \sim \nu_m$. We want to find the time variation of the field amplitude $\tilde{E}^{(+)}(z,t)$, where

$$\hat{E}^{(+)}(z,t) = \tilde{E}^{(+)}(z,t)e^{-i\omega t} \tag{4.16}$$

and the angular frequency ω is the center frequency of oscillation to be determined. Then we have

$$\begin{aligned}\tilde{E}^{(+)}(z,t) = \sum_m \Bigg[& \frac{|p_m|^2 \nu_m}{2\hbar} U_c(z) U_c(z_m) \times \int_0^t e^{(i\omega-i\omega_c-\gamma_c)(t-t')} \\ & \times \int_0^{t'} e^{(i\omega-i\nu_m-\gamma_m)(t'-t'')} \sigma_m(t'')\tilde{E}^{(+)}(z_m,t'')dt''dt' \Bigg]\end{aligned} \tag{4.17}$$

Here, again, we go to the homogeneously broadened case for simplicity:

$$\nu_m = \nu_0, \qquad p_m = p_a, \qquad \gamma_m = \gamma \tag{4.18}$$

Unlike in Equation 4.7 we do not assume $\sigma_m = \sigma$ here, since the atomic saturation may differ for different atoms. Differentiating twice with respect to time, we have

$$\begin{aligned}&\left(\frac{\partial}{\partial t}\right)^2 \tilde{E}^{(+)}(z,t) + [\{\gamma_c + i(\omega_c - \omega)\} + \{\gamma + i(\nu_0 - \omega)\}]\left(\frac{\partial}{\partial t}\right)\tilde{E}^{(+)}(z,t)\\ &\quad + \{\gamma_c + i(\omega_c - \omega)\}\{\gamma + i(\nu_0 - \omega)\}\tilde{E}^{(+)}(z,t)\\ &\quad - \sum_m \frac{|p_a|^2 \nu_0}{2\hbar} U_c(z) U_c(z_m)\sigma_m(t)\tilde{E}^{(+)}(z_m,t) = 0\end{aligned} \tag{4.19}$$

We go to the steady state and have

$$\begin{aligned}\tilde{E}^{(+)}(z) &= \frac{1}{\{\gamma_c + i(\omega_c - \omega)\}\{\gamma + i(\nu_0 - \omega)\}}\\ &\quad \times \sum_m \frac{|p_a|^2 \nu_0}{2\hbar} U_c(z) U_c(z_m)\sigma_m \tilde{E}^{(+)}(z_m)\end{aligned} \tag{4.20}$$

Dividing both sides by

$$\tilde{E}^{(+)}(z) = i(\hbar\omega_c/2)^{1/2} U_c(z)\tilde{a} \tag{4.21}$$

we obtain the steady-state condition:

$$1 = \frac{1}{\{\gamma_c + i(\omega_c - \omega)\}\{\gamma + i(\nu_0 - \omega)\}} \sum_m \frac{|p_a|^2 \nu_0}{2\hbar} U_c^2(z_m)\sigma_m \tag{4.22}$$

Now Equation 3.46 for steady state without the noise term reads

$$\Gamma_p\{\sigma_m - \sigma^0\} = 2i\{\kappa_m \tilde{a}^\dagger (b_{m1}^\dagger b_{m2}) - \kappa_m^* \tilde{a}(b_{m2}^\dagger b_{m1})\} \tag{4.23}$$

where we have written, according to Equation 4.16,

$$\hat{a}(t) = \tilde{a}e^{-i\omega t}, \qquad \hat{b}_{m1}^\dagger \hat{b}_{m2}(t) = b_{m1}^\dagger b_{m2} e^{-i\omega t} \tag{4.24}$$

Also, we have assumed a uniform pumping $\Gamma_{mp} = \Gamma_p$ and $\sigma_m^0 = \sigma^0$. Equation 4.2 for steady state then gives (problem 4-3)

$$b_{m1}^\dagger b_{m2} = \frac{i\kappa_m^* \tilde{a}\sigma_m}{\gamma + i(\nu_0 - \omega)} \tag{4.25}$$

Substituting Equation 4.25 into Equation 4.23 we have

$$\Gamma_p\{\sigma_m - \sigma^0\} = -4\frac{\gamma|\kappa_m|^2|\tilde{a}|^2\sigma_m}{\gamma^2 + (\nu_0 - \omega)^2} \tag{4.26}$$

where we have set $\tilde{a}^\dagger\tilde{a} = \tilde{a}\tilde{a}^\dagger = |\tilde{a}|^2$, ignoring the operator aspect of the field amplitude. This is permissible for a semiclassical analysis. This equation shows the balance with respect to the atomic inversion. The left-hand side describes the supply rate of inversion by pumping, while the right-hand side describes the net consumption rate of the inversion by stimulated emission and absorption. The quantity $2\gamma|\kappa_m|^2/\left\{\gamma^2+(\nu_0-\omega)^2\right\}$, which is nearly equal to twice g in Equation 4.14, is the stimulated transition probability per photon for the field at frequency ω. Thus we have

$$\sigma_m = \frac{\sigma^0}{1+4\{[\gamma|\kappa_m|^2|\tilde{a}|^2]/[\gamma^2+(\nu_0-\omega)^2]\}/\Gamma_p} \tag{4.27}$$

This equation represents the saturation of the atomic inversion, or of the atomic gain, that is, decrease of inversion due to stimulated emission. Saturation becomes appreciable when the rate of decrease of the inversion due to stimulated emission is comparable to its rate of increase due to pumping. If the field amplitude $\tilde{a}$ is small, the inversion is equal to the unsaturated value σ^0. However, it decreases with increasing field amplitude. Now, since we can write (see above Equation 4.5)

$$\kappa_m\tilde{a} = -\frac{p_a}{\hbar}\tilde{E}^{(+)}(z_m) \tag{4.28}$$

we rewrite Equation 4.27 as

$$\sigma_m = \frac{\sigma^0}{1+\left|\tilde{E}^{(+)}(z_m)\right|^2/|E_s|^2} \tag{4.29}$$

where the saturation parameter

$$|E_s|^2 = \frac{\Gamma_p\hbar^2}{4\gamma p_a^2}\{\gamma^2+(\nu_0-\omega)^2\} \tag{4.30}$$

If we substitute Equation 4.29 into the steady-state condition Equation 4.22, we can in principle determine the field amplitude $\left|\tilde{E}^{(+)}(z_m)\right|$ and the oscillation frequency ω.

However, a difficulty arises because the summation over atoms m contains $\sin^2(z_m+d)$ in the denominator due to the form of the electric field $\tilde{E}^{(+)}(z_m)$ given by Equations 4.3 and 3.1. Thus the inversion σ_m is locally depleted where the mode amplitude is large. This is the so-called spatial hole in the laser gain. For simplicity, we ignore the spatial holes and assume that we can replace $\sin^2(z_m+d)$by its space average $\frac{1}{2}$ and write

$$\left|\tilde{E}^{(+)}\right|^2 = \frac{1}{d}\int_{-d}^{0}\left|\tilde{E}^{(+)}(z_m)\right|^2 dz_m \tag{4.31}$$

We replace $|\tilde{E}^{(+)}(z_m)|^2$ in Equation 4.29 by $|\tilde{E}^{(+)}|^2$ and the corresponding steady-state inversion σ_m by σ_{ss}, which is independent of atom index m,

$$\sigma_{ss} = \frac{\sigma^0}{1+\left|\tilde{E}^{(+)}\right|^2/|E_s|^2} \tag{4.32}$$

We then substitute Equation 4.32 into Equation 4.22 and evaluate the summation over m similarly to obtain

$$\omega = \frac{\gamma\omega_c + \gamma_c\nu_0}{\gamma + \gamma_c} \tag{4.33}$$

and

$$\begin{aligned}\left|\tilde{E}^{(+)}\right|^2 &= |E_s|^2 \left[\frac{|p_a|^2 \nu_0 N\sigma^0}{2\hbar\varepsilon_1} \frac{1}{\gamma_c\gamma - (\omega_c - \omega)(\nu_0 - \omega)} - 1\right] \\ &= |E_s|^2 \left[\frac{|p_a|^2 \nu_0 N\sigma^0}{2\hbar\varepsilon_1} \frac{1}{\gamma_c\gamma\left\{1 + (\nu_0 - \omega_c)^2/(\gamma + \gamma_c)^2\right\}} - 1\right]\end{aligned} \tag{4.34}$$

Note that the squared field amplitude obtained is the spatial average value. The oscillation frequency in Equation 4.33 is the same as the threshold oscillation frequency for the linear gain analysis in Equation 4.12. The saturation parameter in Equation 4.30 reduces to

$$|E_s|^2 = \frac{\Gamma_p \hbar^2 \gamma}{4p_a^2}(1 + \delta^2) \tag{4.35}$$

where δ is given by Equation 4.13c. The threshold atomic inversion is obtained by setting $\left|\tilde{E}^{(+)}\right| = 0$ in Equation 4.34 as

$$\sigma_{th}^0 = \frac{2\hbar\varepsilon_1\gamma\gamma_c}{|p_a|^2 \nu_0 N}(1 + \delta^2) \tag{4.36}$$

which is the same as in Equation 4.13b for the linear gain analysis. Then, Equation 4.34 can be rewritten as

$$\frac{\sigma^0}{1 + \left|\tilde{E}^{(+)}\right|^2/|E_s|^2} = \sigma_{th}^0 \tag{4.37}$$

Comparison with Equation 4.32 shows that the steady-state atomic inversion is the same as the threshold inversion:

$$\sigma_{ss} = \sigma_{th}^0 \tag{4.38}$$

This is an example of the well-known fact that the gain of an oscillator above threshold is clamped at the threshold value.

4.3 Quantum Linear Gain Analysis

We now return to the case of fixed atomic inversion, a linear gain regime, but regain the noise terms. We use Equations 3.44 and 3.45 with the inversion operator replaced by a constant $\hat{\sigma}_m(t) = \sigma_m$:

$$\frac{d}{dt}\hat{a} = -i\omega_c\hat{a} - \gamma_c\hat{a} - i\sum_m \kappa_m(\hat{b}^\dagger_{m1}\hat{b}_{m2}) + \hat{\Gamma}_f(t) \tag{4.39a}$$

$$\begin{aligned}(d/dt)(\hat{b}^\dagger_{m1}\hat{b}_{m2})(t) &= -i\nu_m(\hat{b}^\dagger_{m1}\hat{b}_{m2})(t) - \gamma_m(\hat{b}^\dagger_{m1}\hat{b}_{m2})(t)\\ &\quad + i\kappa^*_m\hat{a}(t)\sigma_m + \hat{\Gamma}_m(t)\end{aligned} \tag{4.39b}$$

We follow the procedure used above to go from Equations 4.1 and 4.2 to Equation 4.5. Multiplying both sides of Equation 4.39a by $i(\hbar\omega_c/2)^{1/2}U_c(z)$ and integrating, we have

$$\begin{aligned}\hat{E}^{(+)}(z,t) &= \sqrt{\frac{\hbar\omega_c}{2}}U_c(z)e^{(-i\omega_c-\gamma_c)t}\int_0^t e^{(i\omega_c+\gamma_c)t'}\\ &\left\{\sum_m \kappa_m\left(\hat{b}^\dagger_{m1}\hat{b}_{m2}\right)(t') + i\hat{\Gamma}_f(t')\right\}dt' + \hat{E}^{(+)}(z,0)e^{(-i\omega_c-\gamma_c)t}\end{aligned} \tag{4.40}$$

We replace $\kappa^*_m\hat{a}(t)$ in Equation 4.39b by $-\{\nu_m p^*_m/(\hbar\omega_c)\}\hat{E}^{(+)}(z_m,t)$. Integrating Equation 4.39b (problem 4-4) and substituting the result in Equation 4.40, we obtain, noting that $\kappa_m = i\nu_m(1/2\hbar\omega_c)^{1/2}U_c(z_m)p_m$,

$$\begin{aligned}\hat{E}^{(+)}(z,t) &= \sum_m\left[\frac{|p_m|^2\nu_m^2\sigma_m}{2\hbar\omega_c}U_c(z)U_c(z_m)\int_0^t e^{(-i\omega_c-\gamma_c)(t-t')}\right.\\ &\quad\left.\times\int_0^{t'} e^{(-i\nu_m-\gamma_m)(t'-t'')}\hat{E}^{(+)}(z_m,t'')dt''dt'\right]\\ &\quad + i\sum_m\left[\frac{\nu_m p_m}{2}U_c(z)U_c(z_m)\int_0^t e^{(-i\omega_c-\gamma_c)(t-t')}\right.\\ &\quad\left.\times\int_0^{t'} e^{(-i\nu_m-\gamma_m)(t'-t'')}\hat{\Gamma}_m(t'')dt''dt'\right]\\ &\quad + i\sqrt{\frac{\hbar\omega_c}{2}}U_c(z)\int_0^t e^{(-i\omega_c-\gamma_c)(t-t')}\hat{\Gamma}_f(t')dt'\end{aligned} \tag{4.41}$$

Here, we have again ignored the terms coming from the initial values $\hat{a}(0)$ and $(\hat{b}^\dagger_{m1}\hat{b}_{m2})(0)$. This form of integral equation for the field has been derived for comparison with those in Chapter 9, where we take into account the output coupling. Going again to the homogeneously broadened atoms and to the slowly varying part of the field annihilation operator $\tilde{a}(t)$ as in Equation 4.10b, we obtain

$$\begin{aligned}&\ddot{\tilde{a}} + \{i(\omega_c + \nu_0 - 2\omega) + \gamma_c + \gamma\}\dot{\tilde{a}} - [k^2N\sigma - \{i(\nu_0 - \omega) + \gamma\}\\ &\times\{i(\omega_c - \omega) + \gamma_c\}]\tilde{a} = \dot{\tilde{\Gamma}}_f + \{i((\nu_0 - \omega) + \gamma)\tilde{\Gamma}_f\} - i\sum_m \kappa_m\tilde{\Gamma}_m\end{aligned} \tag{4.42}$$

where we have written $\hat{\Gamma}_f(t) = \tilde{\Gamma}_f(t)e^{-i\omega t}$ and $\hat{\Gamma}_m(t) = \tilde{\Gamma}_m(t)e^{-i\omega t}$. The factor $k^2N\sigma$ was defined in Equation 4.9. We again ignore the second derivative of the slowly varying amplitude $\ddot{\tilde{a}}$. Also, we ignore the term $\dot{\tilde{\Gamma}}_f$, assuming that the time variation of $\tilde{\Gamma}_f$ is slower than the dipole relaxation rate γ. Then, we have

$$\tilde{a}(t) = \tilde{a}(0)e^{s_0 t} + \frac{1}{i(\omega_c + \nu_0 - 2\omega) + \gamma_c + \gamma} \times \int_0^t e^{s_0(t-t')}\left[\{i(\nu_0 - \omega) + \gamma\}\tilde{\Gamma}_f(t') - i\sum_m \kappa_m \tilde{\Gamma}_m(t')\right] dt' \tag{4.43}$$

where

$$s_0 = \frac{k^2 N\sigma - \{i(\nu_0 - \omega) + \gamma\}\{i(\omega_c - \omega) + \gamma_c\}}{i(\omega_c + \nu_0 - 2\omega) + \gamma_c + \gamma} \tag{4.44}$$

Going back to $\hat{a}(t)$ we have

$$\hat{a}(t) = \hat{a}(0)e^{(s_0 - i\omega)t} + \frac{1}{i(\omega_c + \nu_0 - 2\omega) + \gamma_c + \gamma} \times \int_0^t e^{(s_0 - i\omega)(t-t')}\left[\{i(\nu_0 - \omega) + \gamma\}\hat{\Gamma}_f(t') - i\sum_m \kappa_m \hat{\Gamma}_m\right] dt' \tag{4.45}$$

We remember that the same exponential constant was obtained in Equation 4.11 for the semiclassical linear gain analysis. If the atomic inversion is below threshold, the first term represents an exponential decay of the initial field value. The second integral term shows the same exponential decay but incessantly excited by the lasting noise terms. Since the exponent s_0 is the same as that in Equation 4.11 for the semiclassical linear gain analysis, we obtain the same threshold frequency and threshold population inversion as in Equations 4.12 and 4.13b by setting $s_0 = 0$:

$$\omega_{th} = \frac{\gamma\omega_c + \gamma_c\nu_0}{\gamma + \gamma_c} \tag{4.46}$$

$$\sigma_{th} = \frac{\gamma\gamma_c}{k^2 N}\left\{1 + \frac{(\nu_0 - \omega_c)^2}{(\gamma + \gamma_c)^2}\right\} = \frac{2\hbar\omega_{th}\varepsilon_1\gamma\gamma_c}{|p_a|^2\nu_0^2 N}(1 + \delta^2) = \frac{\gamma_c}{Ng} \tag{4.47}$$

In order to calculate the steady-state field spectrum, we calculate the correlation function of the field using the correlation functions of the noise sources given in Equations 3.36 and 3.49. It is well known that the power spectrum of a field is given by the Fourier transform of the correlation function of the field. We ignore the first term in Equation 4.45, which can be ignored in the steady state, that is, for $t \to \infty$. Then we construct

$$\langle \hat{a}^\dagger(t)\hat{a}(t')\rangle = \frac{1}{(\omega_c + \nu_0 - 2\omega)^2 + (\gamma_c + \gamma)^2}$$
$$\times \left\langle \int_0^t e^{(s_0^* + i\omega)(t-t'')} \left[\{-i(\nu_0 - \omega) + \gamma\} \hat{\Gamma}_f^\dagger(t'') + i \sum_m \kappa_m^* \hat{\Gamma}_m^\dagger(t'') \right] dt'' \right. \tag{4.48}$$
$$\left. \times \int_0^{t'} e^{(s_0 - i\omega)(t'-t''')} \left[\{i(\nu_0 - \omega) + \gamma\} \hat{\Gamma}_f(t''') - i \sum_{m'} \kappa_{m'} \hat{\Gamma}_{m'}(t''') \right] dt''' \right\rangle$$

Note that the average sign here refers to the field reservoir as well as atomic reservoirs in addition to the quantum-mechanical expectation value. Using Equations 3.36 and 3.48, assuming the independence of reservoirs for different atoms as well as the independence of the field reservoir from the atomic reservoirs, we rewrite it as

$$\langle \hat{a}^\dagger(t)\hat{a}(t')\rangle = \frac{1}{(\omega_c + \nu_0 - 2\omega)^2 + (\gamma_c + \gamma)^2}$$
$$\times \left\{ \int_0^t \int_0^{t'} e^{(s_0^* + i\omega)(t-t'')} e^{(s_0 - i\omega)(t'-t''')} \{(\nu_0 - \omega)^2 + \gamma^2\} \left\langle \hat{\Gamma}_f^\dagger(t'')\hat{\Gamma}_f(t''') \right\rangle dt''dt''' \right. \tag{4.49}$$
$$\left. + \int_0^t \int_0^{t'} e^{(s_0^* + i\omega)(t-t'')} e^{(s_0 - i\omega)(t'-t''')} \sum_m |\kappa_m|^2 \langle \hat{\Gamma}_m^\dagger(t'')\hat{\Gamma}_m(t''')\rangle dt'dt''' \right\}$$

The correlation function for the field reservoir is given in Equations 3.36 and 3.37. That for the atomic reservoir is obtained from Equations 3.49 and 3.50 as

$$\langle \hat{\Gamma}_m^\dagger(t'')\hat{\Gamma}_m(t''')\rangle = \langle \hat{\Gamma}_{21}(t'')\hat{\Gamma}_{12}(t''')\rangle = G_{21,12}\delta(t'' - t''')$$
$$G_{21,12} = w_{12}\left\langle \frac{1}{2}(1-\sigma)\right\rangle - w_{21}\left\langle \frac{1}{2}(1+\sigma)\right\rangle + 2\gamma\left\langle \frac{1}{2}(1+\sigma)\right\rangle$$
$$= \frac{1}{2}\Gamma_p(1+\sigma^0)\left\langle \frac{1}{2}(1-\sigma)\right\rangle - \frac{1}{2}\Gamma_p(1-\sigma^0)\left\langle \frac{1}{2}(1+\sigma)\right\rangle \tag{4.50}$$
$$+ 2\gamma\left\langle \frac{1}{2}(1+\sigma)\right\rangle = \gamma(1+\sigma)$$

for all m. The last equality holds since $\langle\sigma\rangle = \sigma^0 = \sigma$ in this linear gain analysis. Carrying out the integration we have

$$\langle \hat{a}^\dagger(t)\hat{a}(t')\rangle = \frac{2\{(\nu_0 - \omega)^2 + \gamma^2\}\gamma_c\langle n_c\rangle + \sum_m |\kappa_m|^2 \gamma(1+\sigma)}{(\omega_c + \nu_0 - 2\omega)^2 + (\gamma_c + \gamma)^2}$$
$$\times \begin{cases} \dfrac{e^{(s_0^* + i\omega)t + (s_0 - i\omega)t'} - e^{(s_0^* + i\omega)(t-t')}}{s_0 + s_0^*}, & t > t' \\[2ex] \dfrac{e^{(s_0^* + i\omega)t + (s_0 - i\omega)t'} - e^{(s_0 - i\omega)(t'-t)}}{s_0 + s_0^*}, & t < t' \end{cases} \tag{4.51}$$

Below threshold, the real part of the exponent s_0 is negative. Then the first terms after the curly bracket vanish for long time t and t' and can be ignored for the steady state. In the steady state, the correlation function depends only on the time difference $t-t'$. The summation over m of $|\kappa_m|^2$ can be written as $K^2 N$ by Equation 4.9. So we have

$$\langle \hat{a}^\dagger(t+\tau)\hat{a}(t)\rangle = \frac{2\{(\nu_0-\omega_o)^2+\gamma^2\}\gamma_c\langle n_c\rangle + k^2 N\gamma(1+\sigma)}{(\omega_c+\nu_0-2\omega_o)^2+(\gamma_c+\gamma)^2} \times \begin{cases} \dfrac{e^{(s_0^*+i\omega_o)\tau}}{2|\mathrm{Re}\, s_0|}, & \tau>0 \\ \dfrac{e^{-(s_0-i\omega_o)\tau}}{2|\mathrm{Re}\, s_0|}, & \tau<0 \end{cases} \tag{4.52}$$

where we have rewritten the central frequency ω as ω_o. The explicit expression for $\mathrm{Re}\, s_0$ can be obtained from Equation 4.44 with ω replaced by ω_o, the undetermined center frequency of oscillation. For simplicity, we assume that this frequency is the same as the threshold frequency, that is, we set $\omega \to \omega_o = \omega_{th}$. Then we have

$$\begin{aligned} 2|\mathrm{Re}\ s_0| &= \frac{2(\gamma+\gamma_c)[\gamma\gamma_c(1+\delta^2)-k^2 N\sigma]}{(\gamma+\gamma_c)^2+\delta^2(\gamma-\gamma_c)^2} \\ &= \frac{2(\gamma+\gamma_c)\gamma\gamma_c(1+\delta^2)[1-\sigma/\sigma_{th}]}{(\gamma+\gamma_c)^2+\delta^2(\gamma-\gamma_c)^2} \end{aligned} \tag{4.53}$$

where the relative detuning δ is given by

$$\delta^2 = \left(\frac{\omega_c-\nu_0}{\gamma+\gamma_c}\right)^2 \tag{4.54}$$

In the second line in Equation 4.53 we have used Equation 4.47. Using Equation 4.53, we can rewrite Equation 4.52 as

$$\begin{aligned} &\langle \hat{a}^\dagger(t+\tau)\hat{a}(t)\rangle \\ &= \frac{2\{(\nu_0-\omega_o)^2+\gamma^2\}\gamma_c\langle n_c\rangle + k^2 N\gamma(1+\sigma)}{2(\gamma+\gamma_c)\gamma\gamma_c(1+\delta^2)[1-\sigma/\sigma_{th}]} \times \begin{cases} e^{(s_0^*+i\omega_o)\tau}, & \tau>0 \\ e^{-(s_0-i\omega_o)\tau}, & \tau<0 \end{cases} \\ &= \frac{\gamma\{\langle n_c\rangle + N_2/(N\sigma_{th})\}}{(\gamma+\gamma_c)[1-\sigma/\sigma_{th}]} \times \begin{cases} e^{(s_0^*+i\omega_o)\tau}, & \tau>0 \\ e^{-(s_0-i\omega_o)\tau}, & \tau<0 \end{cases} \end{aligned} \tag{4.55}$$

Here $N_2 = N(1+\sigma)/2$ is the density of atoms in the upper state. In order to obtain the power spectrum, we Fourier transform the τ-dependent part of Equation 4.55 to obtain (problem 4-5)

$$\begin{aligned}
I(\omega) &= \int_{-\infty}^{+\infty} \langle \hat{a}^\dagger(t+\tau)\hat{a}(t)\rangle e^{-i\omega\tau} d\tau \\
&\propto \int_{-\infty}^{0} e^{-(s_0 - i\omega_o)\tau - i\omega\tau} d\tau + \int_{0}^{+\infty} e^{(s_0^* + i\omega_o)\tau - i\omega\tau} d\tau \\
&= \frac{-2\mathrm{Re}\, s_0}{(\omega_o - \omega - \mathrm{Im}\, s_0)^2 + (\mathrm{Re}\, s_0)^2}
\end{aligned} \tag{4.56}$$

Thus the power spectrum is a Lorentzian with the full width at half-maximum (FWHM) $\Delta\omega$ given as

$$\Delta\omega = 2|\mathrm{Re}\, s_0| \tag{4.57}$$

where the right-hand side was given in Equation 4.53. This shows that the linewidth decreases as the atomic inversion σ approaches the threshold value. Schawlow and Townes [1] gave the laser linewidth in terms of measurable quantities including the power output. We can obtain the stored energy W inside the cavity in terms of the expectation value of the intensity, using Equation 4.52,

$$\langle \hat{a}^\dagger(t)\hat{a}(t)\rangle = \frac{2\gamma^2\gamma_c(1+\delta^2)\langle n_c\rangle + k^2 N\gamma(1+\sigma)}{\delta^2(\gamma-\gamma_c)^2 + (\gamma_c+\gamma)^2}\frac{1}{2|\mathrm{Re}\, s_0|} \tag{4.58}$$

Here, again, we have replaced ω_o by ω_{th}. By going back from $\hat{a}(t)$ to the electric field $\hat{E}^{(+)}(z,t)$ by multiplying by the factor $i(\hbar\omega_c/2)^{1/2}U_c(z)$, composing the real electric field operator by adding $\hat{E}^{(-)}(z,t)$, and integrating the electromagnetic energy in the cavity, and using Equation 3.1, we have

$$\begin{aligned}
W &= \left\langle \int_{-d}^{0} \varepsilon_1 \left\{ \hat{E}^{(+)}(z,t) + \hat{E}^{(-)}(z,t) \right\}^2 dz \right\rangle \\
&= \int_{-d}^{0} \varepsilon_1 \left\{ \left\langle \hat{E}^{(+)}(z,t)\hat{E}^{(-)}(z,t) \right\rangle + \left\langle \hat{E}^{(-)}(z,t)\hat{E}^{(+)}(z,t) \right\rangle \right\} dz \\
&= \int_{-d}^{0} \varepsilon_1 |U_c(z)|^2 (\hbar\omega_c/2) \left\{ \langle \hat{a}(t)\hat{a}^\dagger(t)\rangle + \langle \hat{a}^\dagger(t)\hat{a}(t)\rangle \right\} dz \\
&= \varepsilon_1(\hbar\omega_c/2)\left\{ \langle \hat{a}(t)\hat{a}^\dagger(t)\rangle + \langle \hat{a}^\dagger(t)\hat{a}(t)\rangle \right\} \frac{2}{\varepsilon_1 d}\frac{d}{2} \\
&= \hbar\omega_c \left\{ \langle \hat{a}^\dagger(t)\hat{a}(t)\rangle + \frac{1}{2} \right\}
\end{aligned} \tag{4.59}$$

(The reason for the appearance of ω_c instead of ω, the oscillation frequency, is that we are neglecting the presence of the atoms in estimating the stored energy. The difference is usually negligible.)

However, the zero-point energy cannot be measured directly and Glauber [2] showed that the energy available for measurement is

$$W = \hbar\omega_c \langle \hat{a}^\dagger(t)\hat{a}(t) \rangle \tag{4.60}$$

which states that the expectation value of the photon number in the cavity is just $\langle \hat{a}^\dagger(t)\hat{a}(t) \rangle$, as expected. The output power P is, by assumption, equal to $2\gamma_c W$. Thus we have

$$\begin{aligned} P &= 2\gamma_c W = 2\gamma_c \hbar\omega_c \langle \hat{a}^\dagger(t)\hat{a}(t) \rangle \\ &= \frac{2\gamma^2\gamma_c(1+\delta^2)\langle n_c \rangle + k^2 N\gamma(1+\sigma)}{\delta^2(\gamma-\gamma_c)^2 + (\gamma_c+\gamma)^2} \frac{2\gamma_c\hbar\omega_c}{2|\mathrm{Re}\, s_0|} \end{aligned} \tag{4.61}$$

So we have (see Equations 4.57 and 4.47)

$$\begin{aligned} \Delta\omega = 2|\mathrm{Re}\, s_0| &= \frac{2\gamma_c\hbar\omega_c}{P} \frac{2\gamma^2\gamma_c(1+\delta^2)\langle n_c \rangle + k^2 N\gamma(1+\sigma)}{\delta^2(\gamma-\gamma_c)^2 + (\gamma_c+\gamma)^2} \\ &= \frac{4\hbar\omega_c\gamma_c^2}{P} \frac{\gamma^2(1+\delta^2)}{(\gamma_c+\gamma)^2 + \delta^2(\gamma-\gamma_c)^2} \left\{ \langle n_c \rangle + \frac{N_2}{N\sigma_{th}} \right\} \end{aligned} \tag{4.62a}$$

This is the standard form of the expression for the laser linewidth given by Haken [3]. A similar result, without the second factor, was given by Sargent *et al.* [4]. This is inversely proportional to the output power. The major contribution, the term of N_2, comes from the noise associated with the decay of the atomic polarization, usually called quantum noise. The physical content is the spontaneous emission events that occur at random in the atoms in the upper level 2. The spontaneously emitted photons tend to destroy the phase of the oscillating laser field, giving the finite linewidth. A smaller contribution comes from the thermal noise, the term of $\langle n_c \rangle$, associated with the damping of the field. Physically, this is the thermal radiation field mixed into the oscillating field and amplified by the atoms, also disturbing the continuity of the phase of the oscillating field. Note that $\langle n_c \rangle$, the average number of thermal photons in the cavity mode in question, is much smaller than unity for optical frequencies at moderate temperature, $\hbar\omega \gg kT$, where k is the Boltzmann constant and T is the absolute temperature. On the other hand, the coefficient $N_2/(N\sigma_{th})$, the incomplete inversion factor, is of the order of unity or larger. This coefficient is sometimes called the spontaneous emission factor. We see that the smaller the threshold population inversion as compared to N_2, the larger the laser linewidth. This is because the presence of the lower-level atoms causes the number of upper-level atoms N_2 to increase to retain the necessary gain. We also see that any detuning between the atomic and the cavity resonances, appearing in the form of δ, increases the laser linewidth.

The famous Schawlow–Townes linewidth formula was given as [1]

$$\Delta\omega_{ST} = \frac{4\hbar\omega_c\gamma_c^2}{P} \tag{4.62b}$$

We see that here the thermal noise is neglected, a complete inversion ($N_1 = 0$) is assumed, zero detuning ($\delta = 0$) is assumed, and also a large polarization decay rate as compared to the cavity decay rate $\gamma \gg \gamma_c$ is assumed. Note that the Schawlow–Townes linewidth $\Delta\omega_{ST}$ applies for the linear gain regime.

4.4
Quantum Nonlinear Gain Analysis

Now we take the saturation of the atomic inversion into account again in the context of quantum-mechanical analysis. Here we assume a steady state of constant amplitude and concentrate on the phase diffusion of the oscillating field. We start with Equations 3.44 and 3.45, with the noise terms included, and obtain Equation 4.41, but with σ_m moved into the integral over t'' in the form $\hat{\sigma}_m(t'')$:

$$\begin{aligned}\hat{E}^{(+)}(z,t) = \sum_m \Bigg[\frac{|p_m|^2 \nu_m^2}{2\hbar\omega_c} U_c(z) U_c(z_m) \\ \times \int_0^t e^{(-i\omega_c - \gamma_c)(t-t')} \int_0^{t'} e^{(-i\nu_m - \gamma_m)(t'-t'')} \hat{E}^{(+)}(z_m, t'') \hat{\sigma}_m(t'') dt'' dt' \Bigg] \\ + i \sum_m \Bigg[\frac{\nu_m p_m}{2} U_c(z) U_c(z_m) \\ \times \int_0^t e^{(-i\omega_c - \gamma_c)(t-t')} \int_0^{t'} e^{(-i\nu_m - \gamma_m)(t'-t'')} \hat{\Gamma}_m(t'') dt'' dt' \Bigg] \\ + i\sqrt{\frac{\hbar\omega_c}{2}} U_c(z) \int_0^t e^{(-i\omega_c - \gamma_c)(t-t')} \hat{\Gamma}_f(t') dt'\end{aligned} \tag{4.63}$$

Going to homogeneously broadened and uniformly pumped atoms, where $\nu_m = \nu_0$, $p_m = p_a$, $\gamma_m = \gamma$, $\Gamma_{mp} = \Gamma_p$, and $\sigma_m^0 = \sigma^0$, we require the time variation of the field amplitude $\tilde{a}(t)$ as in Equation 4.42:

$$\begin{aligned}\ddot{\tilde{a}} + \{i(\omega_c + \nu_0 - 2\omega) + \gamma_c + \gamma\}\dot{\tilde{a}} \\ - \left[\frac{|p_a|^2 \nu_0^2}{2\hbar\omega_c} \sum_m \hat{\sigma}_m(t) U_c^2(z_m) - \{i(\nu_0 - \omega) + \gamma\}\{i(\omega_c - \omega) + \gamma_c\} \right] \tilde{a} \\ = [\dot{\tilde{\Gamma}}_f + \{i(\nu_0 - \omega) + \gamma\}\tilde{\Gamma}_f] - i\sum_m \kappa_m \tilde{\Gamma}_m\end{aligned} \tag{4.64}$$

We again discard the second time derivative of the slowly varying field amplitude. For the moment we look for the "average" steady-state amplitude and oscillation frequency, ignoring the noise terms and assuming constant (in time) atomic inversions. What we then obtain is Equation 4.22 of the semiclassical nonlinear gain analysis:

$$\frac{|p_a|^2 \nu_0^2}{2\hbar\omega_c} \sum_m \sigma_m U_c^2(z_m) - \{i(\nu_0 - \omega) + \gamma\}\{i(\omega_c - \omega) + \gamma_c\} = 0 \tag{4.65}$$

Here σ_m is a classical quantity instead of an operator. Following the arguments from Equations 4.22 to 4.38, we have the steady-state oscillation frequency, the oscillation amplitude, and the atomic inversion as

$$\omega = \frac{\gamma\omega_c + \gamma_c \nu_0}{\gamma + \gamma_c} \tag{4.66}$$

$$\left|\tilde{E}^{(+)}\right|^2 = |E_s|^2 \left[\frac{|p_a|^2 \nu_0 N \sigma^0}{2\hbar\varepsilon_1} \frac{1}{\gamma_c\gamma\left\{1 + (\nu_0 - \omega_c)^2/(\gamma + \gamma_c)^2\right\}} - 1 \right] \tag{4.67}$$

$$\sigma_{ss} = \sigma^0_{th} = \frac{\gamma\gamma_c}{k^2 N}\left\{1 + \frac{(\nu_0 - \omega_c)^2}{(\gamma + \gamma_c)^2}\right\} = \frac{2\hbar\varepsilon_1\gamma\gamma_c}{|p_a|^2 \nu_0 N}\left\{1 + \frac{(\nu_0 - \omega_c)^2}{(\gamma + \gamma_c)^2}\right\} \tag{4.68}$$

where $|E_s|^2$ is given by Equation 4.35. We have set $\nu_0 = \omega_c$ in the first fraction in Equation 4.67, which is usually highly accurate.

We now return to Equation 4.64, with the noise terms being revived. The coefficient before $\tilde{a}$, or the left-hand member of Equation 4.65, vanishes for this steady state. Thus we have

$$\begin{aligned} \{i(\omega_c + \nu_0 - 2\omega) + \gamma_c + \gamma\}\dot{\tilde{a}} = \dot{\tilde{\Gamma}}_f + \{i(\nu_0 - \omega) + \gamma)\tilde{\Gamma}_f\} \\ - i\sum_m \kappa_m \tilde{\Gamma}_m \end{aligned} \tag{4.69}$$

We ignore the first noise term, as in Equation 4.43. Now, since we are assuming a stable amplitude, we decompose the field amplitude as a product of constant amplitude and time-varying phase, where we assume both the amplitude and the phase to be real:

$$\tilde{a}(t) = \bar{a}e^{i\phi(t)} \tag{4.70}$$

Then we have

$$\frac{d}{dt}\phi(t) = -i\frac{\{i(\nu_0 - \omega) + \gamma\}\tilde{\Gamma}_f - i\sum_m \kappa_m \tilde{\Gamma}_m}{\{i(\omega_c + \nu_0 - 2\omega) + \gamma_c + \gamma\}\bar{a}} e^{-i\phi(t)} \tag{4.71}$$

In order to obtain the real phase, we add the Hermitian conjugate of the right-hand member and divide by 2 to obtain

$$\frac{d}{dt}\phi(t) = -i\frac{\{i(\nu_0 - \omega) + \gamma\}\tilde{\Gamma}_f - i\sum_m \kappa_m \tilde{\Gamma}_m}{2\{i(\omega_c + \nu_0 - 2\omega) + \gamma_c + \gamma\}\bar{a}} e^{-i\phi(t)} + \text{H.C.} \tag{4.72}$$

As in the previous section, we need the field correlation function for determination of the field spectrum and the laser linewidth. Because we have constant amplitude, here we have

$$\left\langle \hat{a}^\dagger(t + \Delta t)\hat{a}(t)\right\rangle = \left\langle \tilde{a}^\dagger(t + \Delta t)\tilde{a}(t)\right\rangle e^{i\omega\Delta t} = \bar{a}^2 e^{i\omega\Delta t}\left\langle e^{-i\phi(t+\Delta t)} e^{i\phi(t)}\right\rangle \tag{4.73}$$

Now we assume that the small phase change can be expanded as

$$\left\langle e^{-i\phi(t+\Delta t)}e^{i\phi(t)}\right\rangle = \left\langle e^{-i\Delta\phi(t)}\right\rangle = 1 - \langle i\Delta\phi(t)\rangle + \frac{1}{2}\left\langle \{i\Delta\phi(t)\}^2\right\rangle$$
$$\simeq e^{-(1/2)\left\langle\{\Delta\phi(t)\}^2\right\rangle} \tag{4.74}$$

So if the last expression reduces to the form $\exp(-s|\Delta t|)$, s gives the half-width at half-maximum (HWHM) of the laser line; or, $\left\langle(\Delta\phi)^2\right\rangle/|\Delta t|$ is the full width at half-maximum (FWHM) (see Section 10.6 for the detail). Thus, what we want to calculate now is the phase diffusion in a time Δt that is large compared to the delta-correlated fluctuation times of the noise sources. For this purpose, we construct

$$\Delta\phi(t) = \phi(t+\Delta t) - \phi(t)$$
$$= \int_t^{t+\Delta t}\left[-i\frac{\{i(\nu_0-\omega)+\gamma\}\tilde{\Gamma}_f - i\sum_m \kappa_m\tilde{\Gamma}_m}{2\{i(\omega_c+\nu_0-2\omega)+\gamma_c+\gamma\}\bar{a}}e^{-i\phi(t)} + \text{H.C.}\right]dt' \tag{4.75}$$

and

$$\left\langle\{\Delta\phi(t)\}^2\right\rangle = \left\langle\int_t^{t+\Delta t}\left[i\frac{\{-i(\nu_0-\omega)+\gamma\}\tilde{\Gamma}_f^\dagger(t')+i\sum_m\kappa_m^*\tilde{\Gamma}_m^\dagger(t')}{2\{-i(\omega_c+\nu_0-2\omega)+\gamma_c+\gamma\}\bar{a}}e^{i\phi(t')}+\text{H.C.}\right]dt'\right.$$
$$\left.\times\int_t^{t+\Delta t}\left[-i\frac{\{i(\nu_0-\omega)+\gamma\}\tilde{\Gamma}_f(t'')-i\sum_m\kappa_m\tilde{\Gamma}_m(t'')}{2\{i(\omega_c+\nu_0-2\omega)+\gamma_c+\gamma\}\bar{a}}e^{-i\phi(t'')}+\text{H.C.}\right]dt''\right\rangle \tag{4.76}$$

In the first integral in Equation 4.76 we have interchanged the sequence of the two mutually conjugate terms, so as to visualize the appearance in the double integral of the normally ordered product such as $\tilde{\Gamma}_f^\dagger(t')\tilde{\Gamma}_f(t'')$. Non-vanishing anti-normally ordered products like $\tilde{\Gamma}_f(t')\tilde{\Gamma}_f^\dagger(t'')$ also appear from the product of the Hermitian conjugate (H.C.) terms. The evaluation of the double integral goes just as in Equations 4.48 and 4.49 using Equations 3.36, 3.49, and 4.50 for the noise correlation functions. Equation 3.49 gives

$$\langle\hat{\Gamma}_m(t'')\hat{\Gamma}_m^\dagger(t''')\rangle = \langle\hat{\Gamma}_{12}(t'')\hat{\Gamma}_{21}(t''')\rangle = G_{12,21}\delta(t''-t''')$$
$$G_{12,21} = w_{21}\left\langle\frac{1}{2}(1+\sigma)\right\rangle - w_{12}\left\langle\frac{1}{2}(1-\sigma)\right\rangle + 2\gamma\left\langle\frac{1}{2}(1-\sigma)\right\rangle$$
$$= \frac{1}{2}\Gamma_p(1-\sigma^0)\left\langle\frac{1}{2}(1+\sigma)\right\rangle - \frac{1}{2}\Gamma_p(1+\sigma^0)\left\langle\frac{1}{2}(1-\sigma)\right\rangle + 2\gamma\left\langle\frac{1}{2}(1-\sigma)\right\rangle \tag{4.77}$$

for all m. Thus we have (problem 4-6)

$$\left\langle\{\Delta\phi(t)\}^2\right\rangle = \frac{|\Delta t|}{4\bar{a}^2}\frac{1}{(\gamma+\gamma_c)^2+(\nu_0+\omega_c-2\omega)^2}$$
$$\times\left[\{(\nu_0-\omega)^2+\gamma^2\}\{2\gamma_c\langle n_c\rangle + 2\gamma_c(\langle n_c\rangle+1)\}\right.$$
$$\left.+\sum_m|\kappa_m|^2(G_{21,12}+G_{12,21})\right] \tag{4.78}$$

Since by Equations 4.50 and 4.77 $G_{21,12} + G_{12,21} = 2\gamma$, treating the summation over m as in Equation 4.9, we have

$$\left\langle \{\Delta\phi(t)\}^2 \right\rangle = \frac{|\Delta t|}{4\bar{a}^2} \frac{1}{(\gamma+\gamma_c)^2 + (\nu_0 + \omega_c - 2\omega)^2} \times \left[\{(\nu_0 - \omega)^2 + \gamma^2\}\{2\gamma_c\langle n_c\rangle + 2\gamma_c(\langle n_c\rangle + 1)\} + 2\gamma k^2 N \right] \tag{4.79}$$

We use Equation 4.66 for ω and Equation 4.68 for k^2N. Also, we apply Equation 4.61 for the stored energy W and the output power P:

$$P = 2\gamma_c W = 2\gamma_c \hbar\omega_c \bar{a}^2 \tag{4.80}$$

Thus we have

$$\frac{\left\langle \{\Delta\phi(t)\}^2 \right\rangle}{|\Delta t|} = \frac{\hbar\omega_c}{P} \frac{2\gamma_c^2\gamma^2(1+\delta^2)}{(\gamma+\gamma_c)^2 + (\gamma-\gamma_c)^2\delta^2} \left[\langle n_c\rangle + \frac{1}{2} + \frac{N}{2N\sigma_{th}^0} \right] \tag{4.81}$$

As stated below Equation 4.74, this gives the FWHM of the laser linewidth:

$$\Delta\omega = \frac{\left\langle \{\Delta\phi(t)\}^2 \right\rangle}{|\Delta t|} = \frac{2\hbar\omega_c\gamma_c^2}{P} \frac{\gamma^2(1+\delta^2)}{(\gamma+\gamma_c)^2 + (\gamma-\gamma_c)^2\delta^2} \left[\langle n_c\rangle + \frac{N_2}{N_2 - N_1} \right] \tag{4.82}$$

Here, $N_2 - N_1 = N\sigma_{ss} = N\sigma_{th}$, the steady-state population inversion. This is just half the linewidth in Equation 4.62a for the laser in the linear gain regime. The reason for the decreased linewidth is that, in the saturated region of the gain, the amplitude fluctuation is suppressed due to the stabilizing effect by the nonlinear gain. There remains only the phase diffusion, which still gives a finite linewidth. In the unsaturated region, the amplitude fluctuation contributes the same amount to the linewidth.

The forms of noises $\langle n_c\rangle$ versus $N_2/(N_2 - N_1)$ in Equation 4.82 are the same as in Equation 4.62a obtained for the linear gain analysis. In the case of Equation 4.62a, these forms appeared directly from the normally ordered correlation functions in Equations 3.36 and 4.50. However, in the case of Equation 4.82, these factors originally appeared in the forms of $\langle n_c\rangle + \frac{1}{2}$ and $N/(2N\sigma_{th})$, respectively, as seen from Equation 4.81. These forms appeared because of the symmetrically ordered correlation functions used for the evaluation of the real phase of the field. In particular, the symmetric ordering appeared in Equation 4.76 because of the Hermitian conjugate terms. So, in this case of nonlinear gain analysis the anti-normally ordered correlation functions in Equations 3.36 and 4.77 were also taken into account. It should be noted that different orderings of the noise operators lead to the same form of the noise contributions.

In this chapter, we have obtained standard results for a laser, such as the oscillation frequency, threshold conditions, output power, and laser linewidth, assuming a completely lossless cavity and introducing a decay term for the laser field. We call this laser theory the *quasimode theory*. In subsequent chapters we will use many of the concepts introduced in this chapter, especially the concept of the Langevin noise forces for the atomic polarization associated with the damping of the polarization oscillation. On the contrary, the concept of the Langevin noise force associated with the cavity loss will be discarded in subsequent chapters, wherein the cavity output coupling is rigorously treated on the basis of the mode functions of the universe introduced in Chapter 1. We will call such a treatment the *continuous mode theory* in contrast to the quasimode theory.

▶ Exercises

4.1 Integrate Equation 4.2.

4-1. We have

$$(\hat{b}_{m1}^{\dagger}\hat{b}_{m2})(t) = \int_0^t e^{(-i\nu_m-\gamma_m)(t-t')} i\kappa_m^* \hat{a}(t')\sigma_m dt' + e^{(-i\nu_m-\gamma_m)t}(\hat{b}_{m1}^{\dagger}\hat{b}_{m2})(0)$$

4.2 Derive Equation 4.10b from Equation 4.10a.

4-2. We have

$$\hat{a}(t) = \tilde{a}(t)e^{-i\omega t}$$

$$\dot{\hat{a}}(t) = \dot{\tilde{a}}(t)e^{-i\omega t} - i\omega\tilde{a}(t)e^{-i\omega t}$$

$$\ddot{\hat{a}}(t) = \ddot{\tilde{a}}(t)e^{-i\omega t} - 2i\omega\dot{\tilde{a}}(t)e^{-i\omega t} - \omega^2\tilde{a}(t)e^{-i\omega t}$$

$$\ddot{\tilde{a}}(t) - 2i\omega\dot{\tilde{a}}(t) - \omega^2\tilde{a}(t) + \{i(\omega_c + \nu_0) + \gamma_c + \gamma\}\{\dot{\tilde{a}}(t) - i\omega\tilde{a}(t)\}$$
$$- \{k^2 N\sigma - (i\nu_0 + \gamma)(i\omega_c + \gamma_c)\}\tilde{a} = 0$$

4.3 Derive Equation 4.25 from Equation 4.2.

4-3. We have

$$(d/dt)(\hat{b}_{m1}^{\dagger}\hat{b}_{m2})(t) = -i\nu_m(\hat{b}_{m1}^{\dagger}\hat{b}_{m2})(t) - \gamma_m(\hat{b}_{m1}^{\dagger}\hat{b}_{m2})(t) + i\kappa_m^*\hat{a}(t)\sigma_m$$

$$\hat{a}(t) = \tilde{a}e^{-i\omega t}, \qquad \hat{b}_{m1}^{\dagger}\hat{b}_{m2}(t) = b_{m1}^{\dagger}b_{m2}e^{-i\omega t}$$

$$\{(d/dt)(b_{m1}^{\dagger}b_{m2}) - i\omega b_{m1}^{\dagger}b_{m2}\}e^{-i\omega t} = (-i\nu_m - \gamma_m)b_{m1}^{\dagger}b_{m2}e^{-i\omega t} + i\kappa_m^*\tilde{a}e^{-i\omega t}\sigma_m$$

$$(d/dt) = 0 \rightarrow \{-i\omega + (i\nu_m + \gamma_m)\}b_{m1}^{\dagger}b_{m2} = i\kappa_m^*\tilde{a}\sigma_m$$

For homogeneous atoms $\nu_m = \nu_0, \quad \gamma_m = \gamma$ and we have Equation 4.25.

4.4 Integrate Equation 4.39b.

4-4. We have

$$(\hat{b}_{m1}^{\dagger}\hat{b}_{m2})(t) = \int_0^t e^{(-i\nu_m-\gamma_m)(t-t')}\{i\kappa_m^*\hat{a}(t')\sigma_m + \hat{\Gamma}_m(t')\}dt'$$
$$+ e^{(-i\nu_m-\gamma_m)t}(\hat{b}_{m1}^{\dagger}\hat{b}_{m2})(0)$$

4.5 Derive the last line in Equation 4.56 from the second.

4-5. Noting that $\mathrm{Re}\, s_0 < 0$ we have

$$\int_{-\infty}^{0} e^{-(s_0-i\omega_o)\tau-i\omega\tau}d\tau + \int_{0}^{+\infty} e^{(s_0^*+i\omega_o)\tau-i\omega\tau}d\tau$$
$$= \frac{1-0}{-\{i(\omega-\omega_0)+s_0\}} + \frac{0-1}{-\{i(\omega-\omega_0)-s_0^*\}}$$
$$= \frac{1}{-\mathrm{Re}\, s_0 - i(\omega-\omega_0+\mathrm{Im}\, s_0)} + \frac{-1}{\mathrm{Re}\, s_0 - i(\omega-\omega_0+\mathrm{Im}\, s_0)}$$
$$= \frac{-2\mathrm{Re}\, s_0}{(\omega_o-\omega-\mathrm{Im}\, s_0)^2+(\mathrm{Re}\, s_0)^2}$$

4.6 Derive Equation 4.78 from Equation 4.76.

4-6. We have

$$\left\langle\{\Delta\phi(t)\}^2\right\rangle = \left\langle \int_t^{t+\Delta t}\left[i\frac{\{-i(\nu_0-\omega)+\gamma\}\tilde{\Gamma}_f^\dagger(t')+i\sum_m \kappa_m^*\tilde{\Gamma}_m^\dagger(t')}{2\{-i(\omega_c+\nu_0-2\omega)+\gamma_c+\gamma\}\bar{a}}e^{i\phi(t')}+\mathrm{H.C.}\right]dt'\right.$$
$$\left.\times\int_t^{t+\Delta t}\left[-i\frac{\{i(\nu_0-\omega)+\gamma\}\tilde{\Gamma}_f(t'')-i\sum_{m'}\kappa_{m'}\tilde{\Gamma}_{m'}(t'')}{2\{i(\omega_c+\nu_0-2\omega)+\gamma_c+\gamma\}\bar{a}}e^{-i\phi(t'')}+\mathrm{H.C.}\right]dt''\right\rangle$$
$$= \int_t^{t+\Delta t}dt'\int_t^{t+\Delta t}dt''\left[\frac{\{\gamma^2+(\nu_0-\omega)^2\}}{4\bar{a}^2\{(\omega_c+\nu_0-2\omega)^2+(\gamma_c+\gamma)^2\}}\right.$$
$$\times\left\{\left\langle\tilde{\Gamma}_f^\dagger(t')\tilde{\Gamma}_f(t'')\right\rangle e^{i\phi(t')-i\phi(t'')}+\left\langle\tilde{\Gamma}_f(t')\tilde{\Gamma}_f^\dagger(t'')\right\rangle e^{-i\phi(t')+i\phi(t'')}\right\}$$
$$+\frac{1}{4\bar{a}^2\{(\omega_c+\nu_0-2\omega)^2+(\gamma_c+\gamma)^2\}}$$
$$\left.\times\sum_m|\kappa_m|^2\left\{\langle\tilde{\Gamma}_m^\dagger(t')\tilde{\Gamma}_m(t'')\rangle e^{i\phi(t')-i\phi(t'')}+\langle\tilde{\Gamma}_m(t')\tilde{\Gamma}_m^\dagger(t'')\rangle e^{-i\phi(t')+i\phi(t'')}\right\}\right]$$

The correlation functions of the Langevin forces are delta-correlated as in Equations 3.36 and 3.37 as well as in Equations 4.50 and 4.77. Thus we arrive at Equation 4.78.

References

1 Schawlow, A.L. and Townes, C.H. (**1958**) *Phys. Rev.*, **112**, 1940–1949.
2 Glauber, R.J. (**1963**) *Phys. Rev.*, **130**, 2529–2539.
3 Haken, H. (**1970**) Laser theory, in *Licht und Materie, IC*, Handbuch der Physik, vol. XXV/2c (eds. S. Flügge and L. Genzel), Springer, Berlin.
4 Sargent, M., III, Scully, M.O., and Lamb, W.E., Jr. (**1974**) *Laser Physics*, Addison-Wesley, Reading, MA.

5
A One-Dimensional Laser with Output Coupling: Derivation of the Laser Equation of Motion

In this chapter, we derive the general form of the laser equation, taking into account the coherent interaction between the atoms and the oscillating laser field, and the incoherent processes, including pumping and damping of the atoms as well as the decay of the atomic dipole oscillation. These incoherent processes are associated with respective Langevin noise sources. The cavity loss, on the other hand, is treated as a natural process of transmission loss without special Langevin forces artificially introduced. Thus the decay of the laser field inside the cavity is caused by the transmission of optical energy to the outside. However, this finite transmission at the coupling surface allows the ambient thermal field to penetrate into the laser cavity, which constitutes the noise source required by the fluctuation–dissipation theorem. The mathematical tool to treat the above natural cavity decay is the continuous multimode description of the field, that is, the modes of the "universe," introduced in Section 1.3. The natural, decaying cavity modes cannot be used for direct quantization because of their non-orthogonality. (The use of such natural modes for a quantum-mechanical description of the field inside and outside the cavity has been tried by Dutra and Nienhuis [1] but without going into the analysis of laser operation.) In this chapter, we derive the laser equation of motion in a one-dimensional, one-sided optical cavity, taking into account the output coupling at the output end of the cavity. This equation gives the basis of the *continuous mode theory* of the laser. The equation will be solved in the following five chapters, where we will find a new correction factor for the laser linewidth both below and above threshold.

5.1
The Field

In Sections 5.1 to 5.3 we consider the coherent interaction between the atoms and the field in an optical cavity having output coupling. As the model of the one-dimensional laser cavity, we use the one-sided cavity model discussed in Section 1.3.1 (see Figure 1.3). The jth mode function of the "universe" is given by

Output Coupling in Optical Cavities and Lasers: A Quantum Theoretical Approach
Kikuo Ujihara

ISBN: 978-3-527-40763-7

$$U_j(z) = N_j u_j(z) \tag{5.1a}$$

$$u_j(z) = \begin{cases} \sin k_{1j}(z+d), & -d<z<0 \\ \dfrac{k_{1j}}{k_{0j}} \cos k_{1j}d \sin k_{0j}z + \sin k_{1j}d \cos k_{0j}z, & 0<z<L \end{cases} \tag{5.1b}$$

The modes of the universe have the orthonormality and completeness properties

$$\int_{-d}^{L} \varepsilon(z) U_i(z) U_j(z) dz = \delta_{i,j} \tag{5.2}$$

$$\begin{aligned} \sum_i \varepsilon(z') U_i(z') U_i(z) &= \int_0^{\infty} \varepsilon(z') U_i(z') U_i(z) \rho(\omega_i) d\omega_i \\ &= \delta(z' - z) \end{aligned} \tag{5.3}$$

Equation 5.3 holds for $-d<z<L$, $-d<z'<L$, except $z = z' = 0$. The density of modes is

$$\rho(\omega) = \frac{L}{\pi c_0} \tag{5.4}$$

The normalization constant is

$$N_j = \sqrt{\frac{2}{\varepsilon_1 L(1 - K \sin^2 k_{1j}d)}} \tag{5.5}$$

with

$$K = 1 - \left(\frac{k_{0j}}{k_{1j}}\right)^2 = 1 - \left(\frac{c_1}{c_0}\right)^2 \tag{5.6}$$

The factor in the denominator in the normalization constant has two kinds of expansions (see Equation 1.70a):

$$\frac{1}{1 - K \sin^2 k_{1j}d} = \frac{2c_0}{c_1} \left\{ \sum_{n=0}^{\infty} \frac{1}{1 + \delta_{0,n}} (-r)^n \cos 2nk_{1j}d \right\} \tag{5.7}$$

where

$$r = (c_0 - c_1)/(c_0 + c_1) \tag{5.8}$$

and

$$\begin{aligned} \frac{1}{1 - K \sin^2 k_{1j}d} &= \sum_{m=-\infty}^{\infty} \frac{c_0 \gamma_c / d}{\gamma_c^2 + (\omega_j - \omega_{cm})^2} \\ &= \sum_{m=-\infty}^{\infty} \frac{c_0}{2d} \left\{ \frac{1}{\gamma_c + i(\omega_j - \omega_{cm})} + \frac{1}{\gamma_c - i(\omega_j - \omega_{cm})} \right\} \end{aligned} \tag{5.9}$$

The field is assumed to be oriented in the x-direction and the vector potential is expanded in terms of the mode functions as

$$\hat{A}(z,t) = \sum_j \hat{Q}_j(t)\,U_j(z) \tag{5.10}$$

and the electric field operator as

$$\hat{E}(z,t) = -\sum_j \hat{P}_j U_j(z) = \sum_j i(\hbar\omega_j/2)^{1/2}(\hat{a}_j - \hat{a}_j^{\dagger})\,U_j(z) \tag{5.11}$$

where $\hat{P}_j = (d/dt)\hat{Q}_j$. The positive and negative frequency parts of the electric field are

$$\hat{E}^{(+)}(z,t) = \sum_j i(\hbar\omega_j/2)^{1/2}\hat{a}_j\,U_j(z) \tag{5.12}$$

$$\hat{E}^{(-)}(z,t) = -\sum_j i(\hbar\omega_j/2)^{1/2}\hat{a}_j^{\dagger}\,U_j(z) \tag{5.13}$$

The annihilation and creation operators obey the commutation relations

$$\left[\hat{a}_i, \hat{a}_j^{\dagger}\right] = \delta_{ij}, \qquad \left[\hat{a}_i, \hat{a}_j\right] = 0, \qquad \left[\hat{a}_i^{\dagger}, \hat{a}_j^{\dagger}\right] = 0 \tag{5.14}$$

The field Hamiltonian is given by

$$\hat{H} = \sum_j \hat{H}_j = \sum_j \hbar\omega_j\left(\hat{a}_j^{\dagger}\hat{a}_j + \frac{1}{2}\right) \tag{5.15}$$

5.2
The Atoms

For the model of laser atoms, we use the same model as was described in Section 3.2. We assume two-level atoms having upper laser level 2 and lower laser level 1. We describe the atoms in the second quantized form [2, 3]. The Hamiltonian of the atoms, evaluated with respect to the lower atomic levels, is written as

$$\hat{H}_a = \sum_m \hbar\omega_m \hat{b}_{m2}^{\dagger}\hat{b}_{m2} \tag{5.16}$$

Here $\hat{b}_{mi}^{\dagger}$ and $\hat{b}_{mi}$ are the creation and annihilation operators, respectively, for the ith level of the mth atom. The product $\hat{b}_{m2}^{\dagger}\hat{b}_{m2}$ is the number operator for the level 2 of the mth atom. The angular frequency ω_m is the transition frequency of the mth atom between the two levels. The product $\hat{b}_{m1}^{\dagger}\hat{b}_{m2}$ is the flipping operator from level 2 to level 1 associated with the emission of a photon. The product $\hat{b}_{m2}^{\dagger}\hat{b}_{m1}$ is the flipping operator for the reverse process. These two flipping operators are mutually Hermitian adjoints, and their classical counterparts are the positive and negative frequency parts, respectively, of the oscillating dipole composed of the two electronic states. The atomic operators obey the anticommutation relations

$$\begin{aligned}&\hat{b}_{mi}\hat{b}^{\dagger}_{m'i'} + \hat{b}^{\dagger}_{m'i'}\hat{b}_{mi} = \delta_{mm'}\delta_{ii'}, \qquad \hat{b}_{mi}\hat{b}_{m'i'} + \hat{b}_{m'i'}\hat{b}_{mi} = 0,\\ &\hat{b}^{\dagger}_{mi}\hat{b}^{\dagger}_{m'i'} + \hat{b}^{\dagger}_{m'i'}\hat{b}^{\dagger}_{mi} = 0\end{aligned} \tag{5.17}$$

The general rule for the reduction of the product of four operators is

$$\hat{b}^{\dagger}_{mi}\hat{b}_{mj}\hat{b}^{\dagger}_{mk}\hat{b}_{ml} = \hat{b}^{\dagger}_{mi}\hat{b}_{ml}\delta_{jk} \tag{5.18}$$

Note that any operator having either two successive annihilation or two successive creation operators for the same atom and for the same state vanish by Equation 5.17.

5.3
The Atom–Field Interaction

The interaction between the field modes and the atoms is formally the same as in Section 3.3 for the quasimode laser. The coherent part of the interaction is described by the interaction Hamiltonian under the rotating-wave approximation and the electric dipole approximation

$$\hat{H}_{int} = \sum_{j,m} \hbar(\kappa_{jm}\hat{a}^{\dagger}_j\hat{b}^{\dagger}_{m1}\hat{b}_{m2} + \kappa^*_{jm}\hat{a}_j\hat{b}^{\dagger}_{m2}\hat{b}_{m1}) \tag{5.19}$$

where the atom–field coupling coefficient, using Equation 3.22a, is

$$\kappa_{jm} = i\nu_m(1/2\hbar\omega_j)^{1/2}U_j(z_m)p_m \tag{5.20}$$

Here $p_m = ex_{m21}$ is the electric dipole matrix element of the mth atom and e is the electron charge.

Under the total Hamiltonian

$$\begin{aligned}\hat{H}_t &= \hat{H}_f + \hat{H}_a + \hat{H}_{int}\\ &= \sum_j \hbar\omega_j\left(\hat{a}^{\dagger}_j\hat{a}_j + \tfrac{1}{2}\right) + \sum_m \hbar\nu_m\hat{b}^{\dagger}_{m2}\hat{b}_{m2} + \sum_{j,m}\hbar\left(\kappa_{jm}\hat{a}^{\dagger}_j\hat{b}^{\dagger}_{m1}\hat{b}_{m2} + \kappa^*_{jm}\hat{a}_j\hat{b}^{\dagger}_{m2}\hat{b}_{m1}\right)\end{aligned} \tag{5.21}$$

the equations of motion for the mode amplitude $\hat{a}_j$, the dipole amplitude $\hat{b}^{\dagger}_{m1}\hat{b}_{m2}$, and the atomic inversion $\hat{\sigma}_m = \hat{b}^{\dagger}_{m2}\hat{b}_{m2} - \hat{b}^{\dagger}_{m1}\hat{b}_{m1}$ are derived by the Heisenberg equation as (problem 5-1)

$$\frac{d}{dt}\hat{a}_j(t) = -i\omega_j\hat{a}_j(t) - i\sum_m \kappa_{jm}(\hat{b}^{\dagger}_{m1}\hat{b}_{m2})(t) \tag{5.22}$$

$$(d/dt)(\hat{b}_{m1}^{\dagger}\hat{b}_{m2})(t) = -i\nu_m(\hat{b}_{m1}^{\dagger}\hat{b}_{m2})(t) + i\sum_j \kappa_{jm}^* \hat{a}_j(t)\hat{\sigma}_m(t) \tag{5.23}$$

$$(d/dt)\hat{\sigma}_m(t) = 2i\sum_j \{\kappa_{jm}\hat{a}_j^{\dagger}(t)(\hat{b}_{m1}^{\dagger}\hat{b}_{m2})(t) - \kappa_{jm}^*\hat{a}_j(t)(\hat{b}_{m2}^{\dagger}\hat{b}_{m1})(t)\} \tag{5.24}$$

These three equations are the basis for the analysis of coherent interaction of the field and the atoms. For the derivation of these equations, see Section 3.4. The major difference from Section 3.4 is the appearance of the multitude of field operators representing the continuous spectrum of the "universal" modes of the field associated with the optical cavity having output coupling. We will call this formalism the *continuous mode theory*.

5.4 Langevin Forces for the Atoms

As was discussed in Section 3.5.2, the laser atoms are in reality surrounded by their respective environment. One factor is the pumping mechanism intentionally added to create the population inversion that is necessary for optical field amplification. The pumping mechanism usually contains unavoidable pumping to the lower laser level, not only to the upper level. Another factor is the environment: collisions with other atoms, phonons, and so on cause relaxation of the laser level populations. Especially, vacuum fluctuation causes the upper level population to decrease through spontaneous emission. All these mechanisms yield resultant pumping rates to upper and lower laser levels and a steady-state population inversion in the absence of the laser field. These mechanisms also disturb the atomic polarization and yield the decay rate of the atomic dipole oscillation. We describe these effects in terms of the pumping and relaxation terms for the atomic inversion, and in terms of the decay term for the atomic polarization. These incoherent random processes inevitably accompany random forces for both the atomic inversion and the atomic polarization, as was discussed in Section 3.5.2. The field decay through the cavity loss, as was discussed in Section 3.5.1, does not appear here because we are not assuming any phenomenological decay of the field energy stored in the cavity. In this continuous mode theory, the transmission at the cavity mirror is automatically incorporated in the "universal" mode functions, which appear in the electric field operators in Equations 5.11–5.13. Thus we have

$$\frac{d}{dt}\hat{a}_j(t) = -i\omega_j\hat{a}_j(t) - i\sum_m \kappa_{jm}(\hat{b}_{m1}^{\dagger}\hat{b}_{m2})(t) \tag{5.25}$$

$$\begin{aligned}(d/dt)(\hat{b}_{m1}^{\dagger}\hat{b}_{m2})(t) = & -i\nu_m(\hat{b}_{m1}^{\dagger}\hat{b}_{m2})(t) - \gamma_m(\hat{b}_{m1}^{\dagger}\hat{b}_{m2})(t) \\ & + i\sum_j \kappa_{jm}^*\hat{a}_j(t)\hat{\sigma}_m(t) + \hat{\Gamma}_m(t)\end{aligned} \tag{5.26}$$

$$(d/dt)\hat{\sigma}_m(t) = -\Gamma_{mp}\{\hat{\sigma}_m(t) - \sigma_m^0\} + 2i\sum_j \{\kappa_{jm}\hat{a}_j^\dagger(t)(\hat{b}_{m1}^\dagger\hat{b}_{m2})(t) \\ -\kappa_{jm}^*\hat{a}_j(t)(\hat{b}_{m2}^\dagger\hat{b}_{m1})(t)\} + \hat{\Gamma}_{m\|}(t) \quad (5.27)$$

These are the basic equations for the quantum-mechanical analysis of laser operation in a cavity having output coupling. As stated above, we have now no decay terms for the field modes, but have a multitude of field modes. The collection of different field mode amplitudes eventually forms the total field amplitude of interest. In contrast to the analysis of the quasimode laser in Chapter 4, where we had only one field mode, the collection of field mode amplitudes that composes the total field is of paramount importance, and the single mode amplitude as in Equation 5.25 is important only as a step towards the calculation of the total field.

5.5
Laser Equation of Motion for a Laser with Output Coupling

In this book, we restrict ourselves to the steady-state operation of a laser with well-stabilized amplitude, and we ignore the fluctuation of the atomic inversion. So, the Langevin force term $\hat{\Gamma}_{m\|}$ will make no contribution within our treatment. On the other hand, the dipolar noise term $\hat{\Gamma}_m$, which is the cause of the quantum noise, will be a major factor in determining the laser linewidth. The thermal noise will be derived from the initial field, which is persistent and does not decay because we have no decay term here, unlike in the quasimode laser analysis in Section 3.5.1. The thermal noise is small quantitatively, but its appearance is an important theoretical result.

As for the atomic dipoles, their collective motion is important, as it constitutes the gain for the field amplitude. The atomic dipole is, in turn, driven by the atomic inversion through interaction with the collective field amplitude.

In the quasimode laser analysis, it was assumed only that the atoms are distributed uniformly in the z-direction with a given density. Except for the neglect of the spatial holes in the nonlinear gain analysis, no explicit discussion of the local effects of the atoms was made. However, in this continuous mode analysis, the local effects of the atomic dipoles become important, because the electric field distribution will not be a mere sinusoidal function but will have a slowly varying spatial envelope function in addition to the local sinusoidal variation. Through this spatial variation of the field, the contributions of the atomic dipoles vary spatially. It will be assumed that enough atoms exist so that, for every local envelope field, there exist a sufficient number of atoms that constitute the local amplifying medium.

In order to convert the single mode equation (Equation 5.25) into that of the total field amplitude expressed in terms of the positive frequency part of the electric field, we use Equation 5.12. For this purpose, we first integrate Equations 5.25 and 5.26 to obtain

$$\hat{a}_j(t) = \hat{a}_j(0)e^{-i\omega_j t} - ie^{-i\omega_j t}\int_0^t e^{i\omega_j t'}\sum_m \kappa_{jm}\left(b^\dagger_{m1}b_{m2}\right)(t')dt' \tag{5.28}$$

$$\begin{aligned}\left(\hat{b}^\dagger_{m1}\hat{b}_{m2}\right)(t) &= \left(\hat{b}^\dagger_{m1}\hat{b}_{m2}\right)(0)e^{-(i\nu_m+\gamma_m)t} + ie^{-(i\nu_m+\gamma_m)t}\\ &\times\int_0^t e^{(i\nu_m+\gamma_m)t'}\sum_i\{\kappa^*_{im}\hat{a}_i(t')\}\hat{\sigma}_m(t')\,dt'\\ &+ e^{-(i\nu_m+\gamma_m)t}\int_0^t e^{(i\nu_m+\gamma_m)t'}\hat{\Gamma}_m(t')\,dt'\end{aligned} \tag{5.29}$$

The first term in Equation 5.28 is persistent because we are not assuming the presence of any decaying term for the field unlike in Equation 4.1 for the quasimode laser. The first term in Equation 5.29 decays over a long time and is unimportant in the steady state. We will ignore this term as we did in the quasimode laser analysis. Multiplying both sides of Equation 5.28 by $i(\hbar\omega_j/2)^{1/2}U_j(z)$ and summing over j we have

$$\begin{aligned}\hat{E}^{(+)}(z,t) &= i\sum_j\sqrt{\frac{\hbar\omega_j}{2}}U_j(z)\hat{a}_j(0)e^{-i\omega_j t}\\ &+\sum_j\sqrt{\frac{\hbar\omega_j}{2}}U_j(z)e^{-i\omega_j t}\int_0^t e^{i\omega_j t'}\sum_m\kappa_{jm}\left(\hat{b}^\dagger_{m1}\hat{b}_{m2}\right)(t')dt'\end{aligned} \tag{5.30}$$

Before we substitute Equation 5.29 into Equation 5.30 we replace the summation over i in Equation 5.29 by the total field amplitude. We have, using Equation 5.20,

$$\begin{aligned}\sum_i\{\kappa^*_{im}\hat{a}_i(t')\} &= \sum_i(-i)\nu_m(1/2\hbar\omega_j)^{1/2}U_j(z_m)p^*_m\hat{a}_i(t')\\ &= -\sum_i\left(p^*_m\nu_m/\hbar\omega_j\right)i(\hbar\omega_j/2)^{1/2}U_j(z_m)\hat{a}_i(t')\end{aligned} \tag{5.31}$$

Since the spectral width of the effective optical field in the atom–field interaction in a laser is much smaller than the central frequency ω of the laser oscillation, the ω_j in the first round brackets in the second line can safely be replaced by ω. Then we have by Equation 5.12

$$\sum_i\{\kappa^*_{im}\hat{a}_i(t')\} = -\left(\frac{p^*_m\nu_m}{\hbar\omega}\right)\hat{E}^{(+)}(z_m,t') \tag{5.32}$$

Note that, if we had used the coupling coefficient in Equation 3.22b instead of that in Equation 3.22a, the factor $p^*_m\nu_m/\hbar\omega$ would have been replaced by $p^*_m/\hbar$. Thus using Equation 5.32 in Equation 5.29, substituting the result into Equation 5.30, and using Equation 5.20 again, we have (problem 5-2)

$$\hat{E}^{(+)}(z,t) = \hat{F}_t(z,t) + \hat{F}_q(z,t) + \sum_m \left[\frac{|p_m|^2 v_m^2}{2\hbar\omega} \int_0^t \sum_j U_j(z) U_j(z_m)\, e^{-i\omega_j(t-t')} \times \int_0^{t'} e^{-(i\nu_m+\gamma_m)(t'-t'')}\, \hat{E}^{(+)}(z_m,t'')\hat{\sigma}_m(t'')dt''dt' \right] \tag{5.33a}$$

where

$$\hat{F}_t(z,t) = i\sum_j \sqrt{\frac{\hbar\omega_j}{2}} U_j(z)\hat{a}_j(0)e^{-i\omega_j t} \tag{5.33b}$$

$$\hat{F}_q(z,t) = \sum_m \left[\frac{ip_m \nu_m}{2} \int_0^t \sum_j U_j(z) U_j(z_m) e^{-i\omega_j(t-t')} \int_0^{t'} e^{-(i\nu_m+\gamma_m)(t'-t'')} \hat{\Gamma}_m(t'')dt''dt' \right] \tag{5.33c}$$

Equation 5.33a is the basic equation for the total (collective) electric field to be solved for analysis of the laser having output coupling.

The meaning of the terms in Equation 5.33a may be given as follows. The first term, or Equation 5.33b, expressing the thermal noise, is the initial electric field (see Equation 5.12). The jth mode excites the jth mode function. The second term, or Equation 5.33c, is the quantum noise term, made up of contributions from each atom. The summand in the square brackets in Equation 5.33c can be rewritten in the form proportional to

$$\int_0^\infty d\omega\; Y(z,z_m,\omega) J_\omega(z_m)e^{-i\omega t} = -\pi \sum_j U_j(z) U_j(z_m) \hat{J}_{\omega_j}(z_m)e^{-i\omega_j t} \tag{5.34}$$

where

$$\begin{aligned} \hat{J}_{\omega_j}(z_m) &= \int_0^t dt'\, e^{i\omega_j t'} \hat{J}_m(t') \\ \hat{J}_m(t') &= \int_0^{t'} e^{-(i\nu_m+\gamma_m)(t'-t'')} \hat{\Gamma}_m(t'')dt'' \end{aligned} \tag{5.35}$$

Here we have used Equation 2.53a for the response function and assumed the absence of any pole in $J_\omega(z_m)$. Thus, regarding the second integral in Equation 5.33c or in Equation 5.35 as an exciting current, we can express the field as a superposition of frequency components excited by each frequency component of the current. The third term in Equation 5.33a represents the field induced by stimulated emission and absorption events that is similarly excited by the effective current

$$\int_0^{t'} e^{-(i\nu_m+\gamma_m)(t'-t'')}\, \hat{E}^{(+)}(z_m,t'')\hat{\sigma}_m(t'')dt''$$

As stated earlier, the time variation of the atomic inversion will not be considered in this book. In the subsequent linear gain analysis, we assume a constant inversion value independent of the field amplitude. In saturated, nonlinear gain analysis, on the other hand, we use an averaged steady-state value for the atomic inversion, the averaging being carried out over the fluctuating forces: that is, we use the steady-state value obtained with the fluctuating noise forces ignored. The inversion will be constant in time but will be dependent on location.

Equation 5.33a is an integral equation with respect to the time variable. This equation also contains integration over z_m, coming from the summation over the atomic index m, as we have assumed a sufficiently dense and uniform distribution of the atoms in the z-direction. This equation also includes a summation over the field modes of a product of two mode functions and one complex, exponential function. Because of the form of the normalization factor for the mode function, which contains the mode index in a form given by Equation 5.5, some idea is required in evaluating the summation. One needs to treat this normalization factor accurately because the information concerning the structure of the optical cavity is contained in this factor.

The Langevin noise forces appear in the discussion of the laser linewidth, and their correlation functions will be needed. For the thermal noise described by Equation 5.33b, the necessary correlation functions are

$$\begin{aligned} \left\langle a_j^\dagger(0)a_j(0)\right\rangle &= \langle n_j\rangle \\ \left\langle a_j(0)a_j^\dagger(0)\right\rangle &= \langle n_j\rangle + 1 \end{aligned} \tag{5.36}$$

where

$$\langle n_j\rangle = \frac{1}{e^{\hbar\omega_j/kT} - 1} \tag{5.37}$$

is the Planck distribution. Here k is the Boltzmann constant and T is the absolute temperature. The angle bracket signifies the ensemble average of the quantum-mechanical expectation value over the thermal field of temperature T. For the quantum noise, we will need the correlation function of the Langevin noise force as in Equations 4.50 and 4.77:

$$\begin{aligned} \langle \hat{\Gamma}_m^\dagger(t'')\hat{\Gamma}_m(t''')\rangle &= G_{21,12}\delta(t''-t''') \\ G_{21,12} &= \frac{1}{2}\Gamma_p(1+\sigma^0)\left\langle\frac{1}{2}(1-\sigma)\right\rangle - \frac{1}{2}\Gamma_p(1-\sigma^0)\left\langle\frac{1}{2}(1+\sigma)\right\rangle \\ &\quad + 2\gamma\left\langle\frac{1}{2}(1+\sigma)\right\rangle \end{aligned} \tag{5.38}$$

$$\begin{aligned} \langle \hat{\Gamma}_m(t'')\hat{\Gamma}_m^\dagger(t''')\rangle &= G_{12,21}\delta(t''-t''') \\ G_{12,21} &= \frac{1}{2}\Gamma_p(1-\sigma^0)\left\langle\frac{1}{2}(1+\sigma)\right\rangle - \frac{1}{2}\Gamma_p(1+\sigma^0)\left\langle\frac{1}{2}(1-\sigma)\right\rangle \\ &\quad + 2\gamma\left\langle\frac{1}{2}(1-\sigma)\right\rangle \end{aligned} \tag{5.39}$$

Here the angle bracket signifies ensemble average of the quantum-mechanical expectation value over the atomic reservoirs. If the pumping and damping or the broadening of the atoms are non-uniform, the parameters in Equations 5.38 and 5.39 should have suffices, say m, indicating the individual atoms.

▶ Exercises

5.1 Derive the equations for the coherent interaction in Equations 5.22–5.24 from the total Hamiltonian in Equation 5.21.
5-1. See Problems 3, 4, and 5 of Chapter 3.

5.2 Derive Equations 5.33a–5.33c
5-2. Using Equation 5.32 in Equation 5.29 and dropping the initial value term we have

$$\left(\hat{b}_{m1}^{\dagger}\hat{b}_{m2}\right)(t) = -i\left(\frac{p_m^* v_m}{\hbar\omega}\right)e^{-(iv_m+\gamma_m)t}\int_0^t e^{(iv_m+\gamma_m)t'}\hat{E}^{(+)}(z_m,t')\hat{\sigma}_m(t')dt' + e^{-(iv_m+\gamma_m)t}\int_0^t e^{(iv_m+\gamma_m)t'}\hat{\Gamma}_m(t')dt'$$

Substituting this equation and Equation 5.20 into Equation 5.30 we have

$$\begin{aligned}\hat{E}^{(+)}(z,t) &= \hat{F}_t(z,t) \\ &+ \sum_j \sqrt{\frac{\hbar\omega_j}{2}}U_j(z)e^{-i\omega_j t}\int_0^t e^{i\omega_j t'}\sum_m iv_m\left(\frac{1}{2\hbar\omega_j}\right)^{1/2}U_j(z_m)p_m \\ &\times e^{-(iv_m+\gamma_m)t''}\int_0^{t'} e^{(iv_m+\gamma_m)t}\hat{\Gamma}_m(t'')dt'dt'' \\ &+ \sum_j \sqrt{\frac{\hbar\omega_j}{2}}U_j(z)e^{-i\omega_j t}\int_0^t e^{i\omega_j t'}\sum_m v_m\left(\frac{1}{2\hbar\omega_j}\right)^{1/2}U_j(z_m)p_m \\ &\times \left(\frac{p_m^* v_m}{\hbar\omega}\right)e^{-(iv_m+\gamma_m)t'}\int_0^{t'} e^{(iv_m+\gamma_m)t''}\hat{E}^{(+)}(z_m,t'')\hat{\sigma}_m(t'')dt'dt''\end{aligned}$$

Rearranging the sequences of the sums and the integrals yields Equations 5.33a–5.33c.

References

1 Dutra, S.M. and Nienhuis, G. (**2000**) *Phys. Rev. A, 62*, 06385.
2 Haken, H. (**1970**) *Laser theory*, in *Licht und Materie, IC*, Handbuch der Physik, vol. XXV/2c (eds. S. Flügge and L. Genzel), Springer, Berlin.
3 Heitler, W. (**1954**) *The Quantum Theory of Radiation*, 3rd edn, Clarendon, Oxford.

6
A One-Dimensional Laser with Output Coupling: Contour Integral Method

The main problem in solving the basic equation (Equation 5.33a) is the treatment of the summation over the continuous mode index j, which also appears in the noise terms in Equations 5.33b and 5.33c. In particular, the summand includes the mode function, which has a j-dependent quantity in its denominator. There are two routes to get around this difficulty, which are based on the expansions of the squared normalization constant described in Equations 5.7 and 5.9: one is a Fourier series expansion, and the other is a partial fraction expansion used in the theory of complex variables. The former gives exact equations, with terms that are mathematically tractable, but infinite in number. On the other hand, the latter gives poles of the normalization constants, suggesting use of a contour integral, which is also exact as long as we take all the poles into account. However, since each pole represents a cavity resonant mode, it is sometimes appropriate to treat only one pole, rather than all the poles. This method of taking into account only one pole is thus an approximation.

In this chapter we try the contour integral method based on the expansion in Equation 5.9. We take into account only one cavity mode and see how far we can go by this method, which, as stated above, involves an approximation.

6.1
Contour Integral Method: Semiclassical Linear Gain Analysis

Here we solve Equation 5.33a with the Langevin noise terms discarded and the atomic inversion replaced by a constant σ_m:

$$\hat{E}^{(+)}(z,t) = \sum_m \left[\frac{|p_m|^2 v_m^2 \sigma_m}{2\hbar\omega} \int_0^t \sum_j U_j(z) U_j(z_m)\, e^{-i\omega_j(t-t')} \right. \\ \left. \times \int_0^{t'} e^{-(i\nu_m+\gamma_m)(t'-t'')} \hat{E}^{(+)}(z_m,t'')\, dt''\, dt' \right] \tag{6.1}$$

We concentrate on the self-consistent equation for inside the cavity. Therefore, we have $-d \le z \le 0$ and $-d < z_m < 0$. We first evaluate the sum over the modes of the "universe" j:

Output Coupling in Optical Cavities and Lasers: A Quantum Theoretical Approach
Kikuo Ujihara

ISBN: 978-3-527-40763-7

$$\sum_j U_j(z)\,U_j(z_m)\,e^{-i\omega_j(t-t')} = \frac{2}{\varepsilon_1 L}\int_0^{\infty} \rho(\omega_j)\frac{\sin k_{1j}z \sin k_{1j}z_m}{1-K\sin^2 k_{1j}d} e^{-i\omega_j(t-t')} d\omega_j \quad (6.2)$$

where $\rho(\omega_j) = L/c_0\pi$ as given in Equation 5.4. Using the expansion in Equation 5.9 and taking only one pole at $\omega_j = \Omega_c = \omega_c - i\gamma_c$, where ω_c is one of the ω_{cm}, we have

$$\sum_j U_j(z)\,U_j(z_m)\,e^{-i\omega_j(t-t')} = \frac{1}{\varepsilon_1 d\pi}\int_0^{\infty} \frac{\sin k_{1j}(z+d)\sin k_{1j}(z_m+d)}{\gamma_c - i(\omega_j-\omega_c)} \times e^{-i\omega_j(t-t')} d\omega_j \quad (6.3)$$

If we expand the numerator in exponential functions, we will have exponents with $-i\omega_j[(t-t') \pm \{(z+d) \pm (z_m+d)\}/c_1]$. For simplicity, we assume that we are interested in phenomena that change slowly in a time of order $|(z+d)\pm (z_m+d)|/c_1 \le 2d/c_1$, that is, we concentrate on the changes on a time scale that is greater than the round-trip time in the cavity. Since this assumption requires an optical spectrum that is narrower than $c_1/(2d)$, this is consistent with the choice of only one cavity mode, which in turn requires an optical spectrum that is narrower than the cavity mode spacing $\Delta\omega_c = c_1\pi/d$. Then, we may decide on the contour of integration by the fact that $t \ge t'$. In this case, the contour of integration may be taken in the lower half-plane of the variable ω_j wherein a pole exists at $\omega_j = \Omega_c = \omega_c - i\gamma_c$. See Figure 6.1 for the arrangement of the poles and the contour of integration. Provided $\Delta\omega_c(t-t') \gg 1$ and $\Delta\omega_c \gg \gamma_c$, the contour simulates an infinitely large semicircle in the lower half-plane. Then the result is

$$\sum_j U_j(z)\,U_j(z_m)\,e^{-i\omega_j(t-t')} = \frac{2}{\varepsilon_1 d}\sin(\Omega_c z/c_1)\sin(\Omega_c z_m/c_1)e^{-i\Omega_c(t-t')} \quad (6.4)$$

For later convenience, we define the "normalized" cavity resonant mode, which is proportional to the spatial part of the outgoing mode in Equation 1.21b:

$$\mathcal{U}_c(z) \equiv \sqrt{\frac{2}{\varepsilon_1 d}}\sin\frac{\Omega_c(z+d)}{c_1} \quad (6.5)$$

Therefore, we have

$$\sum_j U_j(z)\,U_j(z_m)\,e^{-i\omega_j(t-t')} = \mathcal{U}_c(z)\mathcal{U}_c(z_m)e^{-i\Omega_c(t-t')} \quad (6.6)$$

Then Equation 6.1 becomes

$$\hat{E}^{(+)}(z,t) = \sum_m \left[\frac{|p_m|^2 v_m^2 \sigma_m}{2\hbar\omega}\int_0^t \mathcal{U}_c(z)\mathcal{U}_c(z_m)e^{-i(\omega_c - i\gamma_c)(t-t')} \times \int_0^{t'} e^{-(i\nu_m+\gamma_m)(t'-t'')}\hat{E}^{(+)}(z_m,t'')dt''dt'\right] \quad (6.7)$$

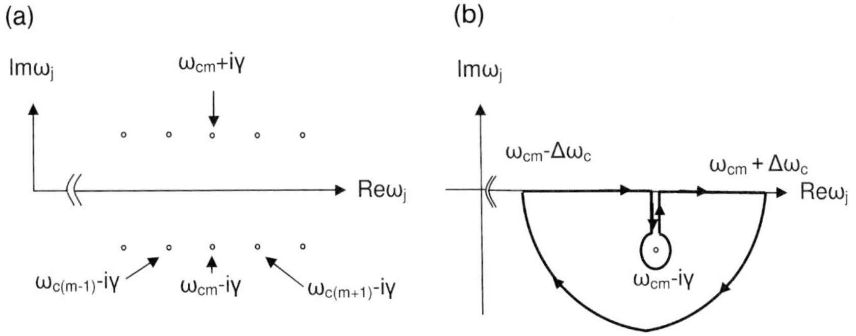

Figure 6.1 (a) Arrangement of poles given by Equation 5.9. (b) Contour of integration for a single pole.

This form suggests that we put

$$\hat{E}^{(+)}(z,t) = i\sqrt{\frac{\hbar\omega_c}{2}}\mathcal{U}_c(z)\hat{a}(t) \tag{6.8}$$

For simplicity, we go to the case of homogeneously broadened atoms and homogeneous pumping: $\nu_m = \nu_0$, $p_m = p_a$, $\gamma_m = \gamma$, $\sigma_m = \sigma$. Then we have

$$\hat{a}(t) = k_{CSL}^2 N\sigma \int_0^t e^{-i(\omega_c - i\gamma_c)(t-t')} \int_0^{t'} e^{-i(\nu_0 - i\gamma)(t'-t'')}\hat{a}(t'')dt''dt' \tag{6.9}$$

where

$$k_{CSL}^2 N\sigma = \frac{|p_a|^2 \nu_0^2 N\sigma}{2\hbar\omega}\int_{-d}^{0}\mathcal{U}_c{}^2(z_m)dz_m \simeq \frac{|p_a|^2\nu_0^2 N\sigma}{2\hbar\omega\varepsilon_1} \tag{6.10}$$

In the last approximate equality, in the integration of $\mathcal{U}_c{}^2(z_m)$, we have discarded $(c_1/4)\sin(2\Omega_c d/c_1)/\Omega_c \sim \lambda_c/(8\pi)$ as compared to $d/2$, assuming that the cavity length d is much larger than λ_c, the intra-cavity resonant wavelength of the cavity mode. Because the parameter k_{CSL}^2 in Equation 6.10 is nearly equal to the parameter k^2 in Equation 4.9, Equation 6.9 is essentially the same as Equation 4.8 for the semiclassical linear gain analysis for the quasimode laser. Thus Equations 4.10 to 4.14 to obtain the threshold oscillation frequency and the threshold atomic inversion for the quasimode laser apply also in this case of semiclassical linear gain analysis of a laser with output coupling. We have

$$\omega_{th} = \frac{\gamma\omega_c + \gamma_c\nu_0}{\gamma + \gamma_c} \tag{6.11}$$

$$\sigma_{th} = \frac{2\hbar\varepsilon_1\gamma\gamma_c}{|p_a|^2\nu_0 N}(1+\delta^2) \tag{6.12}$$

The mode function excited here is a complex function in Equation 6.5 in the form of the spatial part of the cavity resonant mode given in Equation 1.21b, compared to the real function given by Equation 3.1 in the quasimode analysis. This new result may be regarded as an advantage over the quasimode analysis. Thus the contour integral method is successful in this case.

6.2
Contour Integral Method: Semiclassical Nonlinear Gain Analysis

In this case, Equation 6.7 can be used under the condition that the atomic inversion is made a function of the location of the atom, the location determining the local field amplitude:

$$\begin{aligned}\hat{E}^{(+)}(z,t) = \sum_m \Bigg[\frac{|p_m|^2 \nu_m^2 \sigma_m(z_m)}{2\hbar\omega} \int_0^t \mathcal{U}_c(z)\mathcal{U}_c(z_m) e^{-i(\omega_c - i\gamma_c)(t-t')} \\ \times \int_0^{t'} e^{-(i\nu_m+\gamma_m)(t'-t'')}\hat{E}^{(+)}(z_m,t'')dt''dt' \Bigg]\end{aligned} \tag{6.13}$$

If we go to the case of homogeneously broadened atoms $\nu_m = \nu_0$, $p_m = p_a$, $\gamma_m = \gamma$, this equation becomes

$$\begin{aligned}\hat{E}^{(+)}(z,t) = \sum_m \Bigg[\frac{|p_a|^2 \nu_0^2 \sigma_m(z_m)}{2\hbar\omega} \int_0^t \mathcal{U}_c(z)\mathcal{U}_c(z_m) e^{-i(\omega_c - i\gamma_c)(t-t')} \\ \times \int_0^{t'} e^{-i(\nu_0-i\gamma)(t'-t'')}\hat{E}^{(+)}(z_m,t'')dt''dt' \Bigg]\end{aligned} \tag{6.14}$$

If we assume as in Equation 6.8 that

$$\hat{E}^{(+)}(z,t) = i\sqrt{\frac{\hbar\omega_c}{2}}\mathcal{U}_c(z)\hat{a}(t) \tag{6.15}$$

we will formally have

$$\hat{a}(t) = \left[k_{CSN}^2 N\sigma_{CSN} \int_0^t e^{-i(\omega_c-i\gamma_c)(t-t')} \int_0^{t'} e^{-i(\nu_m-i\gamma_m)(t'-t'')}\hat{a}(t'')dt''dt' \right] \tag{6.16}$$

where

$$\begin{aligned} k_{CSN}^2 N\sigma_{CSN} &= \frac{|p_a|^2\nu_0^2 N}{2\hbar\omega}\int_{-d}^0 \sigma(z_m)\mathcal{U}_c^2(z_m)dz_m \\ &= \frac{|p_a|^2\nu_0^2 N}{2\hbar\omega}\int_{-d}^0 \frac{\sigma^0}{1+\left|\tilde{E}^{(+)}(z_m)/E_s\right|^2}\mathcal{U}_c^2(z_m)dz_m \end{aligned} \tag{6.17}$$

Equation 4.29 has been used in the second line. Here we are regarding the field amplitude as a classical value, and the saturation parameter is given by Equation 4.35. We have already assumed in Equation 6.15 that $\hat{E}^{(+)}(z) \propto \mathcal{U}_c(z)$. However, there is no guarantee that the field distribution is still in the form of $\mathcal{U}_c(z)$ when the inversion distribution is in the form of $\sigma(z_m) = \sigma^0 \Big/ \left\{1 + \left|\tilde{E}^{(+)}(z_m)/E_s\right|^2\right\}$. Since $\mathcal{U}_c(z)$ is proportional to the field mode amplitude of an empty cavity, it is highly unlikely that the field distribution remains in the same form when there exists non-uniform gain. Thus the assumption made in Equation 6.15 is invalid. This point will be treated in Chapters 8 and 10, where the field distribution that is consistent with the saturated inversion distribution will be rigorously considered. Thus what we can get here is only the threshold condition, which is obtained by setting $\tilde{E}^{(+)}(z_m) = 0$, and going back to Equations 6.9 and 6.10, which will only give the results obtained in the previous section on semiclassical linear gain analysis. Another thing we can do is to forget about the field distribution by setting $\mathcal{U}_c(z) =$ constant and assuming a uniform field distribution. But this is just what was done in the quasimode analysis. Therefore, we cannot go further for new results. Thus, the contour integral method is a failure in the case of the saturated, nonlinear gain analysis.

6.3 Contour Integral Method: Quantum Linear Gain Analysis

We consider Equations 5.33a–5.33c with the assumption of a constant atomic inversion. Thus we consider

$$\hat{E}^{(+)}(z,t) = \hat{F}_t(z,t) + \hat{F}_q(z,t) + \sum_m \left[\frac{|p_m|^2 v_m^2 \sigma_m}{2\hbar\omega} \right. \\ \left. \times \int_0^t \sum_j U_j(z)\, U_j(z_m)\, e^{-i\omega_j(t-t')} \int_0^{t'} e^{-(i\nu_m+\gamma_m)(t'-t'')}\, \hat{E}^{(+)}(z_m, t'')dt''dt' \right] \tag{6.18}$$

The last term can be modified by use of the contour integral just as in Section 6.1 in the form of Equation 6.7. The quantum noise term $\hat{F}_q(z,t)$ is treated in just the same way, since also in this term we can set (see Equation 6.6)

$$\sum_j U_j(z)\, U_j(z_m)\, e^{-i\omega_j(t-t')} = \mathcal{U}_c(z)\mathcal{U}_c(z_m)e^{-i\Omega_c(t-t')} \tag{6.19}$$

and get

$$\hat{F}_q(z,t) = \sum_m \frac{ip_m v_m}{2} \mathcal{U}_c(z)\mathcal{U}_c(z_m) \int_0^t e^{(-i\omega_c-\gamma_c)(t-t')} \\ \left. \times \int_0^{t'} e^{-(i\nu_m+\gamma_m)(t'-t'')}\hat{\Gamma}(t'')dt''dt' \right] \tag{6.20}$$

For the thermal noise term we remember Equation 2.63, which is applicable to the thermal noise described by Equation 2.35a, which is the same as $\hat{F}_t(z,t)$ in Equation 5.33b:

$$\frac{d}{dt}\hat{E}^{(+)}(z,t) = -(\gamma_c + i\omega_c)\hat{E}^{(+)}(z,t) + \hat{f}(z,t) \tag{6.21}$$

with Equation 2.70b

$$\left\langle \hat{f}^{\dagger}(z',t')\hat{f}(z,t)\right\rangle = \frac{2\gamma_c \hbar\omega_c \langle n_{\omega_c}\rangle}{\varepsilon_1 d} u_{\omega_c}(z')u_{\omega_c}(z)\delta(t-t') \tag{6.23}$$

Thus

$$\left\langle \hat{f}^{\dagger}(z',t')\hat{f}(z,t)\right\rangle = \gamma_c \hbar\omega_c \langle n_{\omega_c}\rangle \mathcal{U}_c(z')\mathcal{U}_c(z)\delta(t-t') \tag{6.24}$$

The $\hat{F}_t(z,t)$ in Equation 5.33b can be expressed as

$$\hat{F}_t(z,t) = \hat{F}_t(z,0)e^{-(\gamma_c+i\omega_c)t} + \int_0^t e^{-(\gamma_c+i\omega_c)(t-t')}\hat{f}(z,t')dt' \tag{6.25}$$

assuming the same z-dependences for both $\hat{F}_t(z,t)$ and $\hat{f}(z,t)$. The first term on the right-hand side can be ignored for the steady-state analysis. Substituting Equation 6.20 and the second term of Equation 6.25 into Equation 6.18 we obtain

$$\begin{aligned}\hat{E}^{(+)}(z,t) =& \sum_m \frac{|p_m|^2 \nu_m^2 \sigma_m}{2\hbar\omega}\mathcal{U}_c(z)\mathcal{U}_c(z_m)\int_0^t e^{-i(\omega_c - i\gamma_c)(t-t')} \\ &\times \int_0^{t'} e^{-(i\nu_m+\gamma_m)(t'-t'')}\hat{E}^{(+)}(z_m,t'')dt''dt' + \sum_m \frac{ip_m\nu_m}{2}\mathcal{U}_c(z)\mathcal{U}_c(z_m) \\ &\times \int_0^t e^{(-i\omega_c-\gamma_c)(t-t')}\int_0^{t'} e^{-(i\nu_m+\gamma_m)(t'-t'')}\hat{\Gamma}_m(t'')dt''dt' \\ &+ \int_0^t e^{-(\gamma_c+i\omega_c)(t-t')}\hat{f}(z,t')dt'\end{aligned} \tag{6.26}$$

We assume the spatial dependence of the field and the thermal noise in the forms

$$\hat{E}^{(+)}(z,t) = i\sqrt{\frac{\hbar\omega_c}{2}}\mathcal{U}_c(z)\hat{a}(t) \tag{6.27}$$

$$\hat{f}(z,t) = i\sqrt{\frac{\hbar\omega_c}{2}}\mathcal{U}_c(z)\hat{g}(t) \tag{6.28}$$

with

$$\langle \hat{g}^{\dagger}(t)g(t')\rangle = 2\gamma_c\langle n_{\omega_c}\rangle\delta(t-t') \tag{6.29}$$

Equation 6.26 is then almost equivalent to Equation 4.41 for the quasimode quantum linear gain analysis provided that the mode function $U(z)$ is replaced by $\mathcal{U}_c(z)$, except for the thermal noise, and provided that the Langevin force $\hat{\Gamma}_f(t)$ is replaced by $\hat{g}(t)$. Note that the correlation function of $\hat{g}(t)$ is the same as that for $\hat{\Gamma}_f(t)$ described in Equation 3.36. Going to homogeneously broadened atoms and homogeneous pumping $\nu_m = \nu_0$, $p_m = p_a$, $\gamma_m = \gamma$, $\sigma_m = \sigma$, we multiply both sides of Equation 6.26 by $\mathcal{U}_c{}^*(z)$, integrate with respect to z, and then divide both sides by

$$\int_{-d}^{0} \mathcal{U}_c^*(z)\mathcal{U}_c(z)\,dz = (1/\varepsilon_1)[(1-r^2)/\{2r\ln(1/r)\}]$$

The factor $\hat{g}(t')$ will then be multiplied by

$$\int_{-d}^{0} \mathcal{U}_c^*(z)U_c(z)\,dz \Big/ \int_{-d}^{0} \mathcal{U}_c^*(z)\mathcal{U}_c(z)\,dz = 2\sqrt{r}/(1+r) \equiv h(r)$$

We obtain

$$\begin{aligned}\hat{a}(t) &= \sum_m \frac{|p_a|^2\nu_0^2\sigma}{2\hbar\omega}\mathcal{U}_c^2(z_m)\int_0^t e^{-i(\omega_c - i\gamma_c)(t-t')}\int_0^{t'} e^{-(i\nu_0+\gamma)(t'-t'')}\hat{a}(t'')dt''dt' \\ &+ \sum_m \sqrt{\frac{2}{\hbar\omega_c}}\frac{p_a\nu_0}{2}\mathcal{U}_c(z_m)\int_0^t e^{(-i\omega_c-\gamma_c)(t-t')} \\ &\times \int_0^{t'} e^{-(i\nu_0+\gamma)(t'-t'')}\hat{\Gamma}_m(t'')dt''dt' \\ &+ h(r)\int_0^t e^{-(\gamma_c+i\omega_c)(t-t')}\hat{g}(t')dt'\end{aligned} \tag{6.30}$$

We go to the slowly varying amplitudes by writing $\hat{a}(t) = \tilde{a}(t)e^{-i\omega t}$, $\hat{g}(t) = \tilde{g}(t)e^{-i\omega t}$, and $\hat{\Gamma}_m(t) = \tilde{\Gamma}_m(t)e^{-i\omega t}$. Then differentiating twice with respect to time we have

$$\begin{aligned}&\ddot{\tilde{a}} + \{i(\omega_c + \nu_0 - 2\omega) + \gamma_c + \gamma\}\dot{\tilde{a}} \\ &\quad - [k_{CQL}^2 N\sigma - \{i(\nu_0 - \omega) + \gamma\}\{i(\omega_c - \omega) + \gamma_c\}]\tilde{a} \\ &= h(r)[\dot{\tilde{g}} + \{i(\nu_0 - \omega) + \gamma)\tilde{g}\}] - i\sum_m \kappa_{mC}\tilde{\Gamma}_m\end{aligned} \tag{6.31}$$

where

$$k_{CQL}^2 N\sigma = k_{CSL}^2 N\sigma = \frac{|p_a|^2\nu_0^2 N\sigma}{2\hbar\omega}\int_{-d}^{0}\mathcal{U}_c^2(z_m)dz_m \simeq \frac{|p_a|^2\nu_0^2 N\sigma}{2\hbar\omega\varepsilon_1} \tag{6.32}$$

and

$$\kappa_{mC} = i\sqrt{\frac{1}{2\hbar\omega_c}}p_a\nu_0\mathcal{U}_c(z_m) \tag{6.33}$$

The square of κ_{mC} has the property that

$$\sum_m |\kappa_{mC}|^2 = \frac{|p_a|^2 \nu_0^2 N}{2\hbar\omega} \int_{-d}^{0} |\mathcal{U}_c(z_m)|^2 dz_m$$

$$= \frac{|p_a|^2 \nu_0^2 N}{2\hbar\omega} \frac{2}{\varepsilon_1 d} \int_{-d}^{0} \left| \frac{e^{(i\omega_c+\gamma_c)(z_m+d)/c_1} - e^{-(i\omega_c+\gamma_c)(z_m+d)/c_1}}{2i} \right|^2 dz_m \quad (6.34)$$

$$= \frac{|p_a|^2 \nu_0^2 N}{2\hbar\omega\varepsilon_1} \frac{(1-r^2)/(2r)}{\ln(1/r)} = k_{CQL}^2 N \frac{\beta_c}{\gamma_c}$$

where we note that $\gamma_c = (c_1/2d)\ln(1/r)$ (see Equation 1.18). Here we have defined

$$\beta_c \equiv \frac{c_1}{2d} \frac{1-r^2}{2r} \quad (6.35)$$

Because, as stated below Equation 6.10, the factor in Equation 6.32 is numerically the same as $k^2 N\sigma$ in the quantum linear gain analysis of the quasimode laser model in Equation 4.9, we are dealing with almost the same equation as in Equation 4.42 despite the appearance of the resonant, outgoing mode function $\mathcal{U}_c(z)$ instead of the perfect cavity mode function $U_c(z)$. Thus, the resulting time dependence of the amplitude $\hat{a}(t)$ is the same as Equation 4.45 with $\hat{\Gamma}_f(t)$ replaced by $h(r)\hat{g}(t)$ and κ_m replaced by κ_{mC}:

$$\hat{a}(t) = \hat{a}(0)e^{(s_0 - i\omega)t} + \frac{1}{i(\omega_c + \nu_0 - 2\omega) + \gamma_c + \gamma}$$
$$\times \int_0^t e^{(s_0 - i\omega)(t-t')} \left\{ \{i(\nu_0 - \omega) + \gamma\} h(r)\hat{g}(t') - i\sum_m \kappa_{mC}\hat{\Gamma}_m \right\} dt' \quad (6.36)$$

Here the decay constant s_0 is the same as that in Equation 4.44 with the k^2 replaced by k_{CQL}^2 and is numerically the same as that in Equation 4.44:

$$s_0 = \frac{k_{CQL}^2 N\sigma - \{i(\nu_0 - \omega) + \gamma\}\{i(\omega_c - \omega) + \gamma_c\}}{i(\omega_c + \nu_0 - 2\omega) + \gamma_c + \gamma} \quad (6.37)$$

We have the threshold conditions

$$\omega_{th} = \frac{\gamma\omega_c + \gamma_c \nu_0}{\gamma + \gamma_c} \quad (6.38)$$

$$\sigma_{th} = \frac{\gamma\gamma_c}{k_{CQL}^2 N} \left\{ 1 + \frac{(\nu_0 - \omega_c)^2}{(\gamma + \gamma_c)^2} \right\} \simeq \frac{2\hbar\omega\varepsilon_1 \gamma\gamma_c}{|p_a|^2 \nu_0^2 N} (1 + \delta^2) \quad (6.39)$$

Using the properties of the Langevin forces described in Equations 4.50 and 6.29, we have the correlation function for the field amplitude for large t as

$$\langle \hat{a}^\dagger(t+\tau)\hat{a}(t)\rangle = \frac{2\{(\nu_0-\omega_o)^2+\gamma^2\}\gamma_c h^2(r)\langle n_c\rangle + k_{CQL}^2 N(\beta_c/\gamma_c)\gamma(1+\sigma)}{(\omega_c+\nu_0-2\omega_o)^2+(\gamma_c+\gamma)^2} \times \begin{cases} \dfrac{e^{(s_0^*+i\omega_o)\tau}}{2|\mathrm{Re}\, s_0|}, & \tau>0 \\ \dfrac{e^{-(s_0-i\omega_o)\tau}}{2|\mathrm{Re}\, s_0|}, & \tau<0 \end{cases} \tag{6.40}$$

where Equation 6.34 has been used. Here ω_o is the center frequency defined in Equation 4.52. We have the laser linewidth (FWHM)

$$\Delta\omega = 2|\mathrm{Re}\, s_0| = \frac{2(\gamma+\gamma_c)[\gamma\gamma_c(1+\delta^2) - k_{CQL}^2 N\sigma]}{(\gamma+\gamma_c)^2 + \delta^2(\gamma-\gamma_c)^2} \tag{6.41}$$

The linewidth is numerically the same as that in Equation 4.53. However, the quantum noise part of the correlation function differs by the factor β_c/γ_c from that in Equation 4.52. Also, the thermal noise part differs by $h^2(r)$.

A strange difference from the quasimode laser model arises when we try to express the linewidth in terms of the power output P. The difference originates from the field distribution $\mathcal{U}_c(z)$ versus $U_c(z)$. In the quasimode model the stored energy was calculated in Equation 4.59 as

$$\begin{aligned} W &= \int_{-d}^{0} \varepsilon_1 |U_c(z)|^2 (\hbar\omega_c/2)\{\langle \hat{a}(t)\hat{a}^\dagger(t)\rangle + \langle \hat{a}^\dagger(t)\hat{a}(t)\rangle\} dz \\ &= \varepsilon_1(\hbar\omega_c/2)\{\langle \hat{a}(t)\hat{a}^\dagger(t)\rangle + \langle \hat{a}^\dagger(t)\hat{a}(t)\rangle\}\frac{2}{\varepsilon_1 d}\frac{d}{2} \\ &= \hbar\omega_c\left\{\langle \hat{a}^\dagger(t)\hat{a}(t)\rangle + \frac{1}{2}\right\} \end{aligned} \tag{6.42}$$

Here the mode function is $\mathcal{U}_c(z)$. Thus we have

$$\begin{aligned} W &= \int_{-d}^{0} \varepsilon_1 |\mathcal{U}_c(z)|^2 (\hbar\omega_c/2)\{\langle \hat{a}(t)\hat{a}^\dagger(t)\rangle + \langle \hat{a}^\dagger(t)\hat{a}(t)\rangle\} dz \\ &= \varepsilon_1(\hbar\omega_c/2)\{\langle \hat{a}(t)\hat{a}^\dagger(t)\rangle + \langle \hat{a}^\dagger(t)\hat{a}(t)\rangle\}\frac{2}{\varepsilon_1 d}\frac{d}{2}\frac{(1-r^2)/(2r)}{\ln(1/r)} \\ &= \hbar\omega_c\left\{\langle \hat{a}^\dagger(t)\hat{a}(t)\rangle + \frac{1}{2}\right\}\frac{\beta_c}{\gamma_c} \end{aligned} \tag{6.43}$$

Assuming that $2\gamma_c$ is the correct damping factor also in this case, and discarding the zero-point energy $\left(\frac{1}{2}\right)$, we have

$$P = 2\gamma_c W = 2\gamma_c \hbar\omega_c \langle \hat{a}^\dagger(t)\hat{a}(t)\rangle \frac{\beta_c}{\gamma_c} \tag{6.44}$$

Using Equation 6.40 with $\tau = 0$ we have

$$\begin{aligned}\Delta\omega &= 2|\mathrm{Re}\, s_0| \\ &= \frac{2\gamma_c\hbar\omega_c}{P}\frac{2\gamma^2\gamma_c(1+\delta^2)h^2(r)\langle n_c\rangle + k_{CQL}^2 N(\beta_c/\gamma_c)\gamma(1+\sigma)}{\delta^2(\gamma-\gamma_c)^2+(\gamma_c+\gamma)^2}\frac{\beta_c}{\gamma_c} \\ &= \frac{4\hbar\omega_c\gamma_c^2}{P}\frac{\gamma^2(1+\delta^2)}{(\gamma_c+\gamma)^2+\delta^2(\gamma-\gamma_c)^2}\left\{h^2(r)\langle n_c\rangle + \frac{N_2}{N\sigma_{th}}\frac{\beta_c}{\gamma_c}\right\}\frac{\beta_c}{\gamma_c}\end{aligned} \tag{6.45}$$

where use has been made of Equation 6.39 in the second line.

Thus, in this form of the linewidth, we have, as compared to the formula Equation 4.62a for the quasimode model, a correction factor $(\beta_c/\gamma_c)^2$ for the quantum noise. The correction factor is $h^2(r)(\beta_c/\gamma_c) = 2\{(1-r)/(1+r)\}/\ln(1/r)$ for the thermal noise. The ratio $\beta_c/\gamma_c = \{(1-r^2)/(2r)\}/\ln(1/r)$ is always larger than unity and is large especially when the reflectivity r of the coupling surface is small. Note that the correction factor originates essentially from the field distribution $\mathcal{U}_c(z)$, which is non-uniform along the z-axis. In Chapters 9 and 10, we will show that the correction factor $(\beta_c/\gamma_c)^2$ appears for both the thermal and the quantum noise. The reason why the thermal noise term here has a different correction factor seems to be the absence of amplification associated with spatial propagation for the thermal noise in this contour integral method. In Chapter 9 it will be shown that both the spatial field distribution and the amplification with propagation contribute to the factor $(\beta_c/\gamma_c)^2$.

6.4 Contour Integral Method: Quantum Nonlinear Gain Analysis

In this case, Equation 6.26 can be used under the condition that the atomic inversion is a function of the location of the atom through the local field amplitude:

$$\begin{aligned}\hat{E}^{(+)}(z,t) &= \sum_m \frac{|p_m|^2 v_m^2 \sigma(z_m)}{2\hbar\omega}\mathcal{U}_c(z)\mathcal{U}_c(z_m)\int_0^t e^{-i(\omega_c - i\gamma_c)(t-t')} \\ &\quad\times\int_0^{t'} e^{-(iv_m+\gamma_m)(t'-t'')}\hat{E}^{(+)}(z_m,t'')dt''dt' \\ &\quad+\sum_m \frac{ip_m v_m}{2}\mathcal{U}_c(z)\mathcal{U}_c(z_m)\int_0^t e^{(-i\omega_c-\gamma_c)(t-t')} \\ &\quad\times\int_0^{t'} e^{-(iv_m+\gamma_m)(t'-t'')}\hat{\Gamma}_m(t'')dt''dt' \\ &\quad+\int_0^t e^{-(\gamma_c+i\omega_c)(t-t')}\hat{f}(z,t')dt'\end{aligned} \tag{6.45}$$

with

$$\sigma(z_m) = \sigma^0\Big/\left\{1+|E(z_m)/E_s|^2\right\} \tag{6.46}$$

where $|E(z_m)|^2$ is the reservoir average of $\hat{E}^{(-)}(z_m)\hat{E}^{(+)}(z_m)$. As discussed in Section 6.2 for the semiclassical analysis, using this equation, it is difficult to obtain the correct field distribution because of the nonlinear dependence of the inversion on the field amplitude. The addition of the noise terms further complicates the problem. So, we refrain from going further with this contour integral method. The correct treatment will be given in Chapter 10.

7
A One-Dimensional Laser with Output Coupling: Semiclassical Linear Gain Analysis

In Chapters 7 and 8 we solve the laser equation of motion (Equation 5.33a) ignoring the Langevin noise forces $\hat{F}_t(z,t)$ and $\hat{F}_q(z,t)$ for the one-sided cavity model described in Section 1.3.1. In Chapter 6, the spatial field distribution was inferred from the results of contour integration with respect to the continuous mode frequency. In contrast, the treatment in this chapter relies on the Fourier series expansion of the normalization constant of the mode function. This allows one to follow the variation of the field amplitude along the laser cavity axis and at the cavity end surfaces. Moreover, an explicit expression for the output field is obtained using the present continuous mode analysis, which takes the output coupling into account exactly. Especially, one can obtain, in principle, the field distribution even in the case of the saturated, nonlinear gain case, which we failed to obtain by use of the contour integral method. As in the case of the quasimode laser, we divide the analysis into two categories: linear gain analysis applicable to operation below threshold, and nonlinear, saturated gain analysis applicable to operation above threshold. In the former case, we take the atomic inversion $\hat{\sigma}_m(t)$ as a constant σ_m that is determined by the pumping process only, ignoring the saturation effects of the field on the inversion. In the nonlinear gain analysis, which is described in the next chapter, we take $\hat{\sigma}_m(t)$ as a scalar $\sigma_m(t)$ that is dependent on the average field intensity at the atomic location. The essence of the content of this chapter was published in [1] (where the negative frequency part of the electric field was considered).

We here concentrate on the linear gain analysis. The equation to be solved reads, from Equation 5.33a, for the entire region $-d < z < L$,

$$\hat{E}^{(+)}(z,t) = \sum_m \left[\frac{|p_m|^2 v_m^2 \sigma_m}{2\hbar\omega} \int_0^t \sum_j U_j(z) U_j(z_m)\, e^{-i\omega_j(t-t')} \right. \\ \left. \times \int_0^{t'} e^{-(i\nu_m+\gamma_m)(t'-t'')}\, \hat{E}^{(+)}(z_{m'} t'')\, dt'' dt' \right] \tag{7.1}$$

Because this equation has no driving force for the electric field, we arbitrarily add an initial field $I(z)\delta(t)$, which is a delta function of time t. For later convenience, we truncate the oscillation in the optical frequency from the electric field

Output Coupling in Optical Cavities and Lasers: A Quantum Theoretical Approach
Kikuo Ujihara

ISBN: 978-3-527-40763-7

$$\hat{E}^{(+)}(z,t) = \tilde{E}^{(+)}(z,t)e^{-i\omega t} \tag{7.2}$$

where ω is the center frequency of oscillation to be determined. Then we have

$$\begin{aligned} &\tilde{E}^{(+)}(z,t) - I(z)\delta(t) \\ &= \sum_m \left[\frac{|p_m|^2 v_m^2 \sigma_m}{2\hbar\omega} \int_0^t \sum_j U_j(z) U_j(z_m)\, e^{i(\omega-\omega_j)(t-t')} \right. \\ &\quad \left. \times \int_0^{t'} e^{\{i(\omega-\nu_m)-\gamma_m\}(t'-t'')} \tilde{E}^{(+)}(z_m,t'')dt''dt' \right] \end{aligned} \tag{7.3}$$

where $I(z)\delta(t)$ appears unchanged because of the delta function. The equation is a self-consistency equation for inside the cavity $-d < z < 0$, where the atoms are located. For the field outside the cavity, $z > 0$, we have only to carry out the summation and the integration once $\tilde{E}^{(+)}(z_m,t)$ is known.

7.1 The Field Equation Inside the Cavity

In order to treat the sum $\sum_j U_j(z)U_j(z_m)e^{-i\omega_j(t-t')}$, we use the first expansion of the squared normalization constant in Equation 1.70a:

$$\frac{1}{1-K\sin^2 k_{1j}d} = \frac{2c_0}{c_1}\left\{ \sum_{n=0}^{\infty} \frac{1}{1+\delta_{0,n}}(-r)^n \cos 2nk_{1j}d \right\} \tag{7.4}$$

Then from the sine functions of z and z_m, from the cosine function in the expansion, and from the exponential function of the time difference, we have eight types of infinite series of sums of exponential functions over the frequency ω_j:

$$\begin{aligned} &\sum_j U_j(z)U_j(z_m)e^{-i(\omega_j-\omega)(t-t')} \\ &= \sum_j \frac{2}{\varepsilon_1 L}\frac{1}{1-K\sin^2 k_{1j}d}\sin k_{1j}(z+d)\sin k_{1j}(z_m+d)e^{-i(\omega_j-\omega)(t-t')} \\ &= \frac{c_0}{2L\varepsilon_1 c_1}\sum_{n=0}^{\infty}\frac{1}{1+\delta_{0,n}}(-r)^n\sum_{\rho=1}^{4}\alpha_\rho\sum_j\left(e^{i\tau_{\rho n}\omega_j}+e^{-i\tau_{\rho n}\omega_j}\right)e^{i(\omega_j-\omega)(t-t')} \end{aligned} \tag{7.5}$$

where the factors $\alpha_1 = \alpha_2 = 1$ and $\alpha_3 = \alpha_4 = -1$. The delay times are

$$\begin{aligned} \tau_{1n} &= \frac{2nd+z-z_m}{c_1}, & \tau_{2n} &= \frac{2nd-z+z_m}{c_1} \\ \tau_{3n} &= \frac{2nd+2d+z+z_m}{c_1}, & \tau_{4n} &= \frac{2nd-2d-z-z_m}{c_1} \end{aligned} \tag{7.6}$$

Note that the delay times $\tau_{\rho n}$ depend on the atomic location z_m, but we have omitted the suffix m for simplicity.

Let us consider the summation

$$S_{\rho n}^{\pm} = \sum_{j} \exp\left[i\{\pm\tau_{\rho n}\omega_j + (\omega_j - \omega)(t - t')\}\right] \tag{7.7}$$

We go to an integration using the density of modes in Equation 1.64 and setting $X = \omega_j - \omega$:

$$S_{\rho n}^{\pm} = \left(\frac{L}{c_0 \pi}\right) \exp(\pm i\tau_{\rho n}\omega) \int_{-\omega}^{\infty} \exp\{i(\pm\tau_{\rho n} + t - t')X\} dX \tag{7.8}$$

Since the frequency is very high in the optical region of the spectrum, the lower limit of the integration may be replaced by $-\infty$. This approximation yields a delta function:

$$S_{\rho n}^{\pm} = \left(\frac{2L}{c_0}\right) \exp(\pm i\tau_{\rho n}\omega)\delta(\pm\tau_{\rho n} + t - t') \tag{7.9}$$

Thus we have

$$\begin{aligned} \sum_{j} \left\{ U_j(z)\, U_j(z_m) e^{-i(\omega_j - \omega)(t - t')} \right\} &= \frac{1}{\varepsilon_1 c_1} \sum_{n=0}^{\infty} \frac{1}{1 + \delta_{0,n}} (-r)^n \\ &\times \sum_{\rho=1}^{4} \alpha_\rho \left\{ e^{i\tau_{\rho n}\omega} \delta(-\tau_{\rho n} + t - t') + e^{-i\tau_{\rho n}\omega} \delta(\tau_{\rho n} + t - t') \right\} \end{aligned} \tag{7.10}$$

Substitution of Equation 7.10 into Equation 7.3 yields

$$\begin{aligned} \tilde{E}^{(+)}(z,t) &= I(z)\delta(t) \\ &+ \sum_{m} G_m \left\{ \int_0^{t - |z - z_m|/c_1} \right. \\ &\times \exp\left[\{-i(\nu_m - \omega) - \gamma_m\}(t - t') + (i\nu_m + \gamma_m)\frac{|z - z_m|}{c_1}\right] \\ &\times \tilde{E}^{(+)}(z_m, t') dt' - \int_0^{t - (2d + z + z_m)/c_1} \\ &\times \exp\left[\{-i(\nu_m - \omega) - \gamma_m\}(t - t') + (i\nu_m + \gamma_m)\frac{2d + z + z_m}{c_1}\right] \\ &\times \left. \tilde{E}^{(+)}(z_m, t') dt' + \sum_{n=1}^{n_M} (-r)^n (I_{1n} + I_{2n} - I_{3n} - I_{4n}) \right\} \end{aligned} \tag{7.11}$$

where

$$I_{\rho n} = \int_0^{t - \tau_{\rho n}} \exp\left[\{-i(\nu_m - \omega) - \gamma_m\}(t - t') + (i\nu_m + \gamma_m)\tau_{\rho n}\right] \tilde{E}^{(+)}(z_m, t')\, dt' \tag{7.12}$$

and

$$G_m = \frac{|p_m|^2 \nu_m^2 \sigma_m}{2\hbar\omega\varepsilon_1 c_1} \tag{7.13}$$

Note that the absolute sign in the first integral in Equation 7.11 appears from the $n=0$ terms in the expansion 7.10 on using the delta functions $\delta(\pm\tau_{10}+t-t')$ and $\delta(\pm\tau_{20}+t-t')$ because of the constraint that $t\geq t'$ in the double integral in Equation 7.3. The second integral in Equation 7.11 comes from the terms of τ_{30} and τ_{40} in Equation 7.10. The integer n_M is the maximum value of n for which $t>\tau_{\rho n}$ and may differ for different τ_ρ. For simplicity, we have written the equations here as if all the τ_ρ have the same n_M. In the steady state, $t\to\infty$, we can make n_M go to infinity.

7.2
Homogeneously Broadened Atoms and Uniform Atomic Inversion

We specialize to the case of homogeneously broadened atoms and uniform atomic inversion by setting

$$\nu_m=\nu_0,\qquad p_m=p_a,\qquad \gamma_m=\gamma,\qquad \sigma_m=\sigma \tag{7.14}$$

Using Equation 7.14 and differentiating Equation 7.11 with respect to time t, we have

$$\begin{aligned}\frac{\partial}{\partial t}&\left\{\tilde{E}^{(+)}(z,\,t)-I(z)\delta(t)\right\}\\&=\{-i(\nu_0-\omega)-\gamma\}\left\{\tilde{E}^{(+)}(z,\,t)-I(z)\delta(t)\right\}\\&+\sum_m G\left[\exp\left(i\omega\frac{|z-z_m|}{c_1}\right)\tilde{E}^{(+)}\left(z_m,\,t-\frac{|z-z_m|}{c_1}\right)\right.\\&-\exp\left(i\omega\frac{2d+z+z_m}{c_1}\right)\tilde{E}^{(+)}\left(z_m,\,t-\frac{2d+z+z_m}{c_1}\right)\\&\left.+\sum_{n=1}^{n_M}(-r)^n\left\{\sum_{\rho=1}^{4}\alpha_\rho\exp\left(i\omega\tau_{\rho n}\right)\tilde{E}^{(+)}\left(z_m,\,t-\tau_{\rho n}\right)\right\}\right]\end{aligned} \tag{7.15}$$

where

$$G=\frac{|p_a|^2\nu_0^2\sigma}{2\hbar\omega\varepsilon_1 c_1} \tag{7.16}$$

The first term on the right-hand side represents damping of the field via the damping of the atomic polarization. The other terms represent the net increase in the field amplitude at location z at time t. Examination of the latter terms reveals the following amplification processes in the cavity. The atoms emit, by the stimulated process, increments of waves to the positive and negative directions that are proportional to the instantaneous field intensity at the location of the respective atoms. The increment of the field that is proportional to $G\tilde{E}(z_m)$ is emitted to both directions by the mth atom and transmitted without changing its amplitude but with proper phase changes. At the boundaries, the increment is reflected with changes in amplitude or phase. As a result, the instantaneous increase of the field amplitude at a particular spatial point z is given by a sum of such increments that have just reached the position z. Figure 7.1 shows the time charts for these contributions by two representative atoms at z_m and $z_m{}'$. The four kinds of retarded times are depicted. For example, the second term in Equation 7.15 gives, except for the phase associated with the propagation, the increment of the field emitted at the mth atom that is proportional to the field strength at z_m at the proper retarded time and reached at z at time t by direct propagation from z_m to z with the distance of propagation $|z - z_m|$. The retarded time is $|z - z_m|/c_1$. The third term, coming from the τ_{30} term, on the other hand, gives an increment emitted at z_m and first propagated to the perfect conductor mirror at $z = -d$ and then to the location z. The net distance for this folded propagation is $\{z - (-d)\} + \{z_m - (-d)\} = z + z_m + 2d$. The minus sign of this term represents the extra phase change of π on reflection at the perfect conductor. The τ_{41} term represents another increment emitted at z_m and first propagated to the coupling surface at $z = 0$ and then to the location z. The distance of propagation is $(0 - z) + (0 - z_m) = -z - z_m$. There is a doubly folded route for an increment to reach to the position z from the initial position z_m. For example, the τ_{11} term with $z_m > z$ gives a route from z_m to the coupling surface, then to the perfect conductor, and finally to the position z. The propagation distance for this route is $2d - (z_m - z)$. For all the above routes, there are associated routes with integer number of added round trips in the cavity. The members in the fourth term represent such routes with round trips. The factor $(-r)^n$ represents the phase change and the reduction in amplitude at the end surfaces associated with n round trips after emission by the mth atom. We see that even a single atom contributes many times to the field increase at a particular location, with decreasing weight for increasing retarded time.

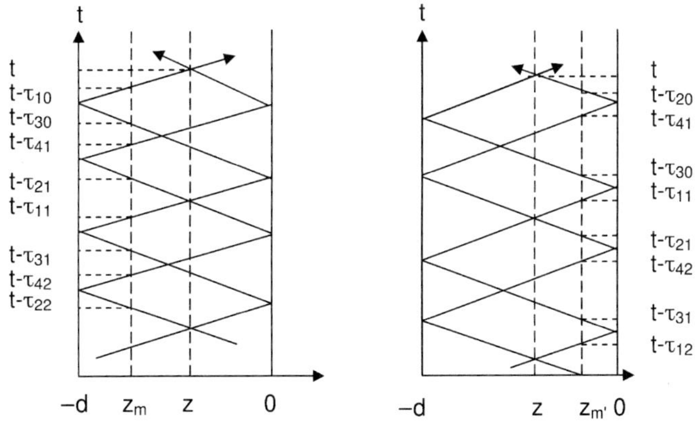

Figure 7.1 The time charts showing the contributions of an atom at z_m or $z_{m'}$ to the time derivative of the field amplitude at z at time t. The contributions are composed of the field values at the retarded times indicated.

We note that the increments, once emitted, propagate with velocity determined by the passive dielectric and are never amplified nor absorbed. They undergo amplitude or phase change at the cavity boundaries. In another words, the increments propagate as if in an empty cavity. However, they stimulate the atoms as they pass them to emit new increments that are in phase with them and proportional to the inducing increments in magnitude. The effects of stimulated absorption by non-inverted atoms are also taken into account in Equation 7.15 through the appearance of the atomic inversion σ in the gain coefficient G for the increments. This coefficient describes the net effect of stimulated emission and stimulated absorption. This picture of laser amplification described by Equation 7.15 gives a clear space-time structure of the laser action in the linear gain regime.

7.3
Solution of the Laser Equation of Motion

Equation 7.15 was derived for the field inside the cavity and for homogeneously broadened atoms with uniform atomic inversion.

7.3.1
The Field Equation for Inside the Cavity

We assume that the field inside the cavity can be divided into two oppositely traveling waves as

$$\tilde{E}^{(+)}(z, t) = e^{+}(z, t)\exp\{+i\omega(z+d)/c_1\} + e^{-}(z, t) \times \exp\{-i\omega(z+d)/c_1\} \tag{7.17}$$

We also assume that the envelope functions $e^{+}(z, t)$ and $e^{-}(z, t)$ are slowly varying in the z-direction. They are also slowly varying with time. Substituting Equation 7.17 into Equation 7.15 and comparing the coefficients of $\exp\{+i\omega(z+d)/c_1\}$ and $\exp\{-i\omega(z+d)/c_1\}$, we have

$$\begin{aligned}\left(\frac{\partial}{\partial t}+\gamma'\right)&\{e^{+}(z, t) - v^{+}(z, t)\} \\ &= \sum_m G\Bigg[H(z-z_m)e^{+}(z_m, t-\tau_{10}) - e^{-}(z_m, t-\tau_{30}) \\ &\quad + \sum_{n=1}^{n_M}(r')^n\{e^{+}(z_m, t-\tau_{1n}) - e^{-}(z_m, t-\tau_{3n})\}\Bigg]\end{aligned} \tag{7.18}$$

and

$$\left(\frac{\partial}{\partial t}+\gamma'\right)\{e^-(z,\ t)-\upsilon^-(z,\ t)\} = \sum_m G\left[H(z_m-z)e^-(z_m,\ t-\tau_{20}) + \sum_{n=1}^{n_M}(r')^n\{e^-(z_m,\ t-\tau_{2n})-e^+(z_m,\ \tau_{4n})\}\right] \quad (7.19)$$

Here H is the Heaviside unit step function, and

$$\upsilon^+(z,t)=\theta^+(z)\delta(t), \qquad \upsilon^-(z,t)=\theta^-(z)\delta(t) \quad (7.20)$$

where $\theta^+(z)$ and $\theta^-(z)$ are the components of the initial field $I(z)$ varying as $\exp\{+i\omega(z+d)/c_1\}$ and $\exp\{-i\omega(z+d)/c_1\}$, respectively. The constants γ' and r' are respectively defined as

$$\gamma'=\gamma+i(\nu_0-\omega) \quad (7.21)$$

and

$$r'=-r\exp(2id\omega/c_1) \quad (7.22)$$

In deriving Equations 7.18 and 7.19 we have neglected those rapidly oscillating terms with a factor $\exp(+2i\omega z/c_1)$ or $\exp(-2i\omega z/c_1)$. The two oppositely traveling waves are coupled to each other.

7.3.2 Laplace-Transformed Equations

In order to solve the coupled equations involving space variable z and time variable t, we Laplace-transform Laplace transform them with respect to time and concentrate on the spatial region:

$$\begin{aligned} e^+(z,\ t) &\rightarrow L^+(z,\ s) \\ e^-(z,\ t) &\rightarrow L^-(z,\ s) \\ \upsilon^+(z,\ t) &\rightarrow V^+(z,\ s)=\theta^+(z) \\ \upsilon^-(z,\ t) &\rightarrow V^-(z,\ s)=\theta^-(z) \end{aligned} \quad (7.23)$$

Since the Laplace transform of $e^+(z_m,\ t-\tau)$ is $\exp(-\tau s)L^+(z_m,s)$, the summations over n in Equations 7.18 and 7.19 reduce to geometrical progressions, which can be easily evaluated. Here we assume that the time t is so large that the upper limit of the summation can go to infinity. Also, we again assume enough density of atoms Ndz_m for the summation over m to go to integration over z_m. Then we have

$$(s+\gamma')\{L^+(z,\ s) - V^+(z,s)\}$$
$$= GN\left[\int_{-d}^{z} \exp\{-(z-z_m)s/c_1\}L^+(z_m,s)dz_m - \frac{1}{1-r''(s)}\int_{-d}^{0} \exp\{-(z+z_m+2d)s/c_1\}L^-(z_m,s)dz_m + \frac{r''(s)}{1-r''(s)}\int_{-d}^{0} \exp\{-(z-z_m)s/c_1\}L^+(z_m,s)dz_m\right] \tag{7.24a}$$

and

$$(s+\gamma')\{L^-(z,\ s) - V^-(z,s)\}$$
$$= GN\left[\int_{z}^{0} \exp\{(z-z_m)s/c_1\}L^-(z_m,s)dz_m - \frac{r''(s)}{1-r''(s)}\int_{-d}^{0} \exp\{(z+z_m+2d)s/c_1\}L^+(z_m,s)dz_m + \frac{r''(s)}{1-r''(s)}\int_{-d}^{0} \exp\{(z-z_m)s/c_1\}L^-(z_m,s)dz_m\right] \tag{7.24b}$$

where

$$r''(s) = r'\exp(-2ds/c_1) = -r\exp\{(i\omega - s)2d/c_1\} \tag{7.25}$$

The initial values $e^{\pm}(z,0) - v^{\pm}(z,0)$ associated with the Laplace transform vanish, as can be shown by setting $t=0$ in Equation 7.3 with the aid of Equation 7.17. Differentiation with respect to z and division by $(s+\gamma')$ yields

$$\frac{d}{dz}\{L^+(z,s) - V^+(z,s)\} = -\frac{s}{c_1}\{L^+(z,s) - V^+(z,s)\} + \frac{GNL^+(z,s)}{(s+\gamma')} \tag{7.26a}$$

$$\frac{d}{dz}\{L^-(z,s) - V^-(z,s)\} = \frac{s}{c_1}\{L^-(z,s) - V^-(z,s)\} - \frac{GNL^-(z,s)}{(s+\gamma')} \tag{7.26b}$$

Rearranging the terms we have

$$\frac{d}{dz}\{L^+(z,s) - V^+(z,s)\} = \left\{-\frac{s}{c_1} + \frac{GN}{(s+\gamma')}\right\}\{L^+(z,s) - V^+(z,s)\} + \frac{GN}{(s+\gamma')}V^+(z,s) \tag{7.27a}$$

$$\frac{d}{dz}\{L^-(z,s) - V^-(z,s)\} = \left\{\frac{s}{c_1} - \frac{GN}{(s+\gamma')}\right\}\{L^-(z,s) - V^-(z,s)\} - \frac{GN}{(s+\gamma')}V^-(z,s) \tag{7.27b}$$

Integrating these for $L^{\pm}(z,s) - V^{\pm}(z,s)$ we have

$$
\begin{aligned}
L^{+}(z,s) - V^{+}(z,s) = {} & \int_{-d}^{z} \exp\left[\left\{-\frac{s}{c_1} + \frac{GN}{(s+\gamma')}\right\}(z-z_m)\right] \\
& \times \frac{GN}{(s+\gamma')} V^{+}(z_m, s) dz_m \\
& + \exp\left[\left\{-\frac{s}{c_1} + \frac{GN}{(s+\gamma')}\right\}(z+d)\right] \\
& \times \{L^{+}(-d,s) - V^{+}(-d,s)\}
\end{aligned}
\tag{7.28a}
$$

$$
\begin{aligned}
L^{-}(z,s) - V^{-}(z,s) = {} & -\int_{-d}^{z} \exp\left[\left\{\frac{s}{c_1} - \frac{GN}{(s+\gamma')}\right\}(z-z_m)\right] \\
& \times \frac{GN}{(s+\gamma')} V^{-}(z_m, s) dz_m \\
& + \exp\left[\left\{\frac{s}{c_1} - \frac{GN}{(s+\gamma')}\right\}(z+d)\right] \\
& \times \{L^{-}(-d,s) - V^{-}(-d,s)\}
\end{aligned}
\tag{7.28b}
$$

where $L^{\pm}(-d,s)$ are undetermined constants. In order to determine these constants, we first set $z=-d$ and $z=0$ in Equations 7.24a and 7.24b to obtain the boundary conditions at the two ends of the cavity. Then we set $z=0$ in Equations 7.28a and 7.28b. Then we have four coupled equations for $L^{\pm}(-d,s)$ and $L^{\pm}(0,s)$, which can be solved easily (see Appendix D for the details). We obtain

$$
\begin{aligned}
L^{\pm}(-d,\ s) = {} & \theta^{\pm}(-d) \pm \frac{GN}{s+\gamma'} \\
& \times \frac{\int_{-d}^{0}\left\{r'\theta^{+}(z_m)\exp\left(\frac{z_m-d}{c_1}s - \frac{GNz_m}{s+\gamma'}\right) - \theta^{-}(z_m)\exp\left(-\frac{z_m+d}{c_1}s + \frac{GNz_m}{s+\gamma'}\right)\right\}dz_m}{\exp\left(-\frac{GNd}{s+\gamma'}\right) - r'\exp\left(\frac{GNd}{s+\gamma'} - \frac{2ds}{c_1}\right)}
\end{aligned}
\tag{7.29}
$$

When substituted into Equations 7.28a and 7.28b, the terms with the denominator in Equation 7.29 give the main pole, yielding slowly decaying terms as compared with the first terms, which have a pole at $s=-\gamma'=-\gamma-i(\nu_0-\omega)$, leading to fast decays. Let us examine the pole mentioned above. Set

$$
\exp\left(-\frac{GNd}{s+\gamma'}\right) - r'\exp\left(\frac{GNd}{s+\gamma'} - \frac{2ds}{c_1}\right) = 0
\tag{7.30}
$$

or with Equation 7.22 for r'

$$
1 + r\exp\left(\frac{2id\omega}{c_1}\right)\exp\left(\frac{2GNd}{s+\gamma'} - \frac{2ds}{c_1}\right) = 0
\tag{7.31}
$$

Thus we have

$$1-\exp\left\{\ln r+\frac{2id\omega}{c_1}+\frac{2GNd}{s+\gamma'}-\frac{2ds}{c_1}-(2m+1)\pi i\right\}=0 \tag{7.32}$$

or

$$\begin{aligned}
&1-\exp\left[\frac{2d}{c_1}\left\{-\gamma_c+i\omega+\frac{GNc_1}{s+\gamma+i(\nu_0-\omega)}-s-i\omega_c\right\}\right]\\
&=1-\exp\left[\frac{2d}{c_1}\left\{\frac{GNc_1-\{s+\gamma+i(\nu_0-\omega)\}\{s+\gamma_c+i(\omega_c-\omega)\}}{s+\gamma+i(\nu_0-\omega)}\right\}\right]=0
\end{aligned} \tag{7.33}$$

where m is an integer and we have used Equation 1.18a, writing $\omega_{cm}=\omega_c$. Also, Equation 7.21 has been used. This equation yields

$$\begin{aligned}
&s^2+\{\gamma+\gamma_c+i(\nu_0+\omega_c-2\omega)\}s+\gamma_c\gamma\\
&\quad+(\nu_0-\omega)(\omega-\omega_c)-GNc_1-i\{\gamma(\omega-\omega_c)+\gamma_c(\omega-\nu_0)\}=0
\end{aligned} \tag{7.34}$$

As in the previous chapters, we discard s^2, anticipating a slow decay. Writing the pole satisfying Equations 7.30–7.34 as s_0, we have the main pole

$$s_0=-\frac{\gamma\gamma_c+(\nu_0-\omega)(\omega-\omega_c)-GNc_1-i\{\gamma(\omega-\omega_c)+\gamma_c(\omega-\nu_0)\}}{\gamma+\gamma_c+i(\nu_0+\omega_c-2\omega)} \tag{7.35}$$

where

$$GNc_1=\frac{|p_a|^2\nu_0^2N\sigma}{2\hbar\omega\varepsilon_1}=k^2N\sigma \tag{7.36}$$

The factor k^2 was defined in Equation 4.9 for the quasimode analysis. Here we have ignored the small difference between ω and ν_0. Thus the denominator in Equation 7.29 for s around the main pole s_0 can be rewritten as

$$\begin{aligned}
&\exp\left(-\frac{GNd}{s+\gamma'}\right)-r'\exp\left(\frac{GNd}{s+\gamma'}-\frac{2ds}{c_1}\right)\\
&=\exp\left(-\frac{GNd}{s_0+\gamma'}\right)\frac{2d/c_1}{s_0+\gamma'}\{\gamma+\gamma_c+i(\nu_0+\omega_c-2\omega)\}(s-s_0)
\end{aligned} \tag{7.37}$$

Here we briefly mention the cavity decay constant γ_c for the one-sided cavity. This was defined in Equation 1.18a and the modified cavity model with a perfect mirror at $z=L$ is being used in this chapter. In Chapter 4 analyzing the quasimode laser, we introduced the decay constant using the same symbol γ_c without any concrete model for the cavity decay. Now comparing the form of Equation 7.35 for the one-sided cavity and the decay equation in Equation 4.11 for the quasimode cavity, we see that the cavity decay constant γ_c in this chapter replaces the role of that in Chapter 4.

Returning to the topic of the pole, for later use we derive from Equation 7.33 an equation that is equivalent to Equation 7.34. Since the quantity in the curly bracket in the first line is zero, we have

$$\frac{s_0}{c_1} - \frac{GN}{s_0 + \gamma'} = -\frac{i(\omega_c - \omega) + \gamma_c}{c_1} \tag{7.38}$$

where we have written s_0 instead of s.

7.3.3 The Field Inside the Cavity

Substituting Equation 7.29 into Equations 7.28a and 7.28b and inverse Laplace-transforming for the main poles, we have

$$e^+(z,t) = \exp\left[\left\{-\frac{s_0}{c_1} + \frac{GN}{(s_0+\gamma')}\right\}(z+d)\right] C\exp(s_0 t) \tag{7.39a}$$

$$e^-(z,t) = -\exp\left[\left\{\frac{s_0}{c_1} - \frac{GN}{(s_0+\gamma')}\right\}(z+d)\right] C\exp(s_0 t) \tag{7.39b}$$

where the constant C is given by

$$\begin{aligned} C = &-\frac{(GNc_1/2d)\exp\left[GNd/(s_0+\gamma')\right]}{\gamma+\gamma_c+i(\nu_0+\omega_c-2\omega)} \\ &\times \int_{-d}^{0}\left\{\theta^+(z_m)r'\exp\left[-\{\gamma_c - i(\omega-\omega_c)\}\frac{z_m}{c_1} - \frac{s_0 d}{c_1}\right]\right. \\ &\left.-\theta^-(z_m)\exp\left[\{\gamma_c - i(\omega-\omega_c)\}\frac{z_m}{c_1} - \frac{s_0 d}{c_1}\right]\right\}dz_m \\ = &-\frac{(GNc_1/2d)}{\gamma+\gamma_c+i(\nu_0+\omega_c-2\omega)} \\ &\times \int_{-d}^{0}\left\{\theta^+(z_m)\exp\left[-\{\gamma_c - i(\omega-\omega_c)\}\frac{z_m+d}{c_1}\right]\right. \\ &\left.-\theta^-(z_m)\exp\left[\{\gamma_c - i(\omega-\omega_c)\}\frac{z_m+d}{c_1}\right]\right\}dz_m \end{aligned} \tag{7.40}$$

and where Equation 7.38 has been used for the coefficients of z_m/c_1 in the exponentials. Also, Equation 7.32 has been used to eliminate r'. Here, the driving forces $\theta^+(z)$ and $\theta^-(z)$ are the components of the initial field $I(z)$ varying as $\exp\{+i\omega(z+d)/c_1\}$ and $\exp\{-i\omega(z+d)/c_1\}$, respectively. Note that this right-going (left-going) part of the initial field is projected onto the decreasing (increasing) function of z_m. The latter functions are the left-going (right-going) parts of the function adjoint to the cavity mode function, as will be discussed in Chapter 14.

Remembering Equation 7.38 above and going back to Equation 7.2 via Equation 7.17, we have the main terms

$$\hat{E}^{(+)}(z,t) = C[\exp\{(\gamma_c + i\omega_c)(z+d)/c_1\} - \exp\{-(\gamma_c + i\omega_c)(z+d)/c_1\}] \times \exp\{(s_0 - i\omega)t\} = 2iC\sin\{\Omega_c(z+d)/c_1\}\exp\{(s_0 - i\omega)t\} \tag{7.41}$$

for inside the cavity, $-d < z < 0$. The cavity resonant mode in Equation 1.21b is excited with the decay constant s_0. The oscillating frequency ω may be determined if we assume a near-threshold behavior as in Equation 4.12 for the quasimode laser or as in Equation 6.11 for the contour integral method for the laser with output coupling. At threshold, setting $s_0 = 0$ in Equation 7.35, we have

$$\omega_{th} = \frac{\gamma\omega_c + \gamma_c\nu_0}{\gamma + \gamma_c} \tag{7.42}$$

and

$$\frac{G_{th}Nc_1}{\gamma\left\{1 + (\omega_c - \nu_0)^2 / (\gamma + \gamma_c)^2\right\}} = \gamma_c \tag{7.43}$$

or by Equation 7.16

$$\left\{\frac{|p_a|^2 \nu_0^2}{2\hbar\omega\varepsilon_1\gamma(1+\delta^2)}\right\} N\sigma_{th} = \gamma_c \tag{7.44a}$$

where δ was defined in Equation 4.13c as

$$\delta^2 = \left(\frac{\omega_c - \nu_0}{\gamma + \gamma_c}\right)^2 \tag{7.44b}$$

This can be written, using the amplitude gain per unit density of inverted atoms per unit time, g, defined in Equation 4.14, as

$$gN\sigma_{th} = \gamma_c \tag{7.44c}$$

7.3.4 The Field Outside the Cavity

The power of the continuous mode expansion in terms of the modes of the "universe" in laser analysis is that it allows for the exact expression for the field outside the cavity. This expression is also required for the calculation of the output power. Now that we know the expression for the field inside the cavity, under the constraint of homogeneous broadening of the atoms and uniform pumping as expressed by Equation 7.14, we can use Equation 7.1 to obtain the main part of the field outside the cavity. Here, for $U_j(z)$, we need to use the expression for the field outside in the second line of Equation 1.41b and, for $U_j(z_m)$, that for inside in the first line of Equation 1.41b. So, through a similar procedure to obtaining Equation 7.10, we have

$$\begin{aligned}
&\sum_j \left\{ U_j(z)\, U_j(z_m) e^{-i\omega_j(t-t')} \right\} \\
&= \sum_j \frac{2}{\varepsilon_1 L} \frac{1}{1-K\sin^2 k_{1j}d} \left(\frac{k_{1j}}{k_{0j}} \cos k_{1j}d \sin k_{0j}z + \sin k_{1j}d \cos k_{0j}z \right) \\
&\quad \times \sin k_{1j}(z_m+d) e^{-i\omega_j(t-t')} \\
&= \frac{1}{\varepsilon_1 c_1} \frac{2c_0}{(c_1+c_0)} \sum_{n=0}^{\infty} (-r)^n \{ \delta(t-t'+\tau_{5n}) \\
&\quad + \delta(t-t'-\tau_{5n}) - \delta(t-t'+\tau_{6n}) - \delta(t-t'-\tau_{6n}) \}
\end{aligned} \tag{7.45}$$

where

$$\tau_{5n} = \frac{z}{c_0} + \frac{2nd - z_m}{c_1}, \qquad \tau_{6n} = \frac{z}{c_0} + \frac{2nd + 2d + z_m}{c_1} \tag{7.46}$$

The procedure to obtain the delta functions in Equation 7.45 is just like that from Equations 7.4 to 7.10. Here the delay time τ_{5n} expresses the time required for a signal emitted at z_m inside the cavity to reach z outside the cavity after traveling from z_m towards the positive z-direction followed by n round trips in the cavity and then to z in the outer space. Similarly, τ_{6n} is the time required for a signal that first goes to the negative z-direction with subsequent n round trips and a single travel to z. We have used the fact that $k_{1j}/k_{0j} = c_0/c_1$ to eliminate the k in favor of the c. Substitution of the last expression in Equation 7.45 into Equation 7.1 yields

$$\begin{aligned}
\hat{E}^{(+)}(z,t) &= \sum_m \frac{|p_a|^2 v_0^2 \sigma}{2\hbar\omega} \frac{1}{\varepsilon_1 c_1} \frac{2c_0}{(c_1+c_0)} \sum_{n=0}^{\infty} (-r)^n \int_0^t \{ \delta(t-t'+\tau_{5n}) + \delta(t-t'-\tau_{5n}) \\
&\qquad - \delta(t-t'+\tau_{6n}) - \delta(t-t'-\tau_{6n}) \} \int_0^{t'} e^{-(i\nu_0+\gamma)(t'-t'')} \hat{E}^{(+)}(z_m,t'') dt'' dt' \\
&= \sum_m \frac{|p_a|^2 v_0^2 \sigma}{2\hbar\omega\varepsilon_1 c_1} \frac{2c_0}{(c_1+c_0)} \sum_{n=0}^{\infty} (-r)^n \left\{ \int_0^{t-\tau_{5n}} e^{-(i\nu_0+\gamma)(t-\tau_{5n}-t'')} \hat{E}^{(+)}(z_m,t'') dt'' \right. \\
&\qquad \left. - \int_0^{t-\tau_{6n}} e^{-(i\nu_0+\gamma)(t-\tau_{6n}-t'')} \hat{E}^{(+)}(z_m,t'') dt'' \right\}
\end{aligned} \tag{7.47}$$

Substituting Equation 7.41 for $\hat{E}^{(+)}(z_m)$, using Equation 7.16, and noting that the transmission coefficient at the coupling surface for the wave incident from inside is

$$T = 1 + r = \frac{2c_0}{c_0 + c_1} \tag{7.48}$$

we have

$$\begin{aligned}\hat{E}^{(+)}(z,t) = CGT \sum_m \sum_{n=0}^{\infty} (-r)^n \Bigg\{ \int_0^{t-\tau_{5n}} & e^{-(i\nu_0+\gamma)(t-\tau_{5n}-t'')} [\exp\{(\gamma_c + i\omega_c)(z_m + d)/c_1\} \\ & - \exp\{-(\gamma_c + i\omega_c)(z_m + d)/c_1\}] \exp\{(s_0 - i\omega)t''\} dt'' \\ & - \int_0^{t-\tau_{6n}} e^{-(i\nu_0+\gamma)(t-\tau_{6n}-t'')} [\exp\{(\gamma_c + i\omega_c)(z_m + d)/c_1\} \\ & - \exp\{-(\gamma_c + i\omega_c)(z_m + d)/c_1\}] \exp\{(s_0 - i\omega)t''\} dt'' \Bigg\}\end{aligned} \tag{7.49}$$

The integrations over t'' can easily be performed:

$$\begin{aligned}\hat{E}^{(+)}(z,t) = & \frac{CGT}{\gamma + s_0 + i(\nu_0 - \omega)} \\ & \times \sum_m [\exp\{(\gamma_c + i\omega_c)(z_m + d)/c_1\} - \exp\{-(\gamma_c + i\omega_c)(z_m + d)/c_1\}] \\ & \times \sum_{n=0}^{\infty} (-r)^n [\exp\{(s_0 - i\omega)(t - \tau_{5n})\} - \exp\{-(i\nu_0 + \gamma)(t - \tau_{5n})\} \\ & - \exp\{(s_0 - i\omega)(t - \tau_{6n})\} + \exp\{-(i\nu_0 + \gamma)(t - \tau_{6n})\}]\end{aligned} \tag{7.50}$$

Also, the summations over n, which are simple geometrical progressions, can be easily evaluated:

$$\begin{aligned}\hat{E}^{(+)}(z,t) = & \frac{CGT}{\gamma + s_0 + i(\nu_0 - \omega)} \\ & \times \sum_m [\exp\{(\gamma_c + i\omega_c)(z_m + d)/c_1\} - \exp\{-(\gamma_c + i\omega_c)(z_m + d)/c_1\}] \\ & \times \left[\frac{\exp\{(s_0 - i\omega)(t - z/c_0 + z_m/c_1)\}}{1 + r\exp\{-(s_0 - i\omega)(2d/c_1)\}} \right. \\ & - \frac{\exp\{-(i\nu_0 + \gamma)(t - z/c_0 + z_m/c_1)\}}{1 + r\exp\{(i\nu_0 + \gamma)(2d/c_1)\}} \\ & - \frac{\exp\{(s_0 - i\omega)[t - z/c_0 - (2d + z_m)/c_1]\}}{1 + r\exp\{-(s_0 - i\omega)(2d/c_1)\}} \\ & \left. + \frac{\exp\{-(i\nu_0 + \gamma)[t - z/c_0 - (2d + z_m)/c_1]\}}{1 + r\exp\{(i\nu_0 + \gamma)(2d/c_1)\}} \right]\end{aligned} \tag{7.51}$$

The summations over m are evaluated by going to integrations with the density of atoms N. There appear eight terms to be integrated. The result is

$$\begin{aligned}
&\hat{E}^{(+)}(z,t)\\
&= \frac{CGNc_1 T}{\gamma+s_0+i(\nu_0-\omega)}\left[\frac{\exp\{(s_0-i\omega)(t-z/c_0)\}\exp\{-(s_0-i\omega)(d/c_1)\}}{1+r\exp\{-(s_0-i\omega)(2d/c_1)\}}\right.\\
&\quad\times\left\{\frac{2\sin h\{(\gamma_c+i\omega_c+s_0-i\omega)(d/c_1)\}}{\gamma_c+i\omega_c+s_0-i\omega}-\frac{2\sin h\{(\gamma_c+i\omega_c-s_0+i\omega)(d/c_1)\}}{\gamma_c+i\omega_c-s_0+i\omega}\right\}\\
&\quad-\frac{\exp\{-(i\nu_0+\gamma)(t-z/c_0)\}\exp\{(i\nu_0+\gamma)(d/c_1)\}}{1+r\exp\{(i\nu_0+\gamma)(2d/c_1)\}}\\
&\quad\left.\times\left\{\frac{2\sin h\{(\gamma_c+i\omega_c-i\nu_0-\gamma)(d/c_1)\}}{\gamma_c+i\omega_c-i\nu_0-\gamma}-\frac{2\sin h\{(\gamma_c+i\omega_c+i\nu_0+\gamma)(d/c_1)\}}{\gamma_c+i\omega_c+i\nu_0+\gamma}\right\}\right]
\end{aligned} \tag{7.52}$$

Now the second term in the square bracket decays fast as exp $(-\gamma t)$ and may be ignored. The second term in the first large curly bracket is small compared to the first term because of the sum, as compared to the difference, of two high frequencies in the denominator and may also be ignored. Thus we have

$$\begin{aligned}
\hat{E}^{(+)}(z,t) &= \frac{2CGNc_1 T}{\gamma+s_0+i(\nu_0-\omega)}\,\frac{\exp\{-(s_0-i\omega)(d/c_1)\}}{1+r\exp\{-(s_0-i\omega)(2d/c_1)\}}\\
&\times\frac{\sin h\{(\gamma_c+i\omega_c+s_0-i\omega)(d/c_1)\}}{s_0+\gamma_c+i(\omega_c-\omega)}\exp\{(s_0-i\omega)(t-z/c_0)\}
\end{aligned} \tag{7.53}$$

This can be simplified as follows. First, let us remember the transformation of r to $-\exp\{\ln r-(2m+1)\pi i\}=-\exp\{(-\gamma_c-i\omega_c)(2d/c_1)\}$ in Equation 7.32. By a similar modification of r in the denominator of the second factor, we see that

$$\begin{aligned}
&1+r\exp\{-(s_0-i\omega)(2d/c_1)\}=1-\exp\{-(\gamma_c+i\omega_c+s_0-i\omega)(2d/c_1)\}\\
&=\exp\{-(\gamma_c+i\omega_c)(d/c_1)-(s_0-i\omega)(d/c_1)\}2\sin h\{(\gamma_c+i\omega_c+s_0-i\omega)(d/c_1)\}
\end{aligned} \tag{7.54}$$

Therefore we have

$$\begin{aligned}
&\frac{2\exp\{-(s_0-i\omega)(d/c_1)\}\sin h\{(\gamma_c+i\omega_c+s_0-i\omega)(d/c_1)\}}{1+r\exp\{-(s_0-i\omega)(2d/c_1)\}}\\
&\quad=\exp\{(\gamma_c+i\omega_c)(d/c_1)\}\\
&\quad=\exp\{i\Omega_c(d/c_1)\}
\end{aligned} \tag{7.55}$$

Next, the product of the first and the third factors in the denominator is easily seen to be equal to GNc_1 from Equation 7.34. Therefore we have

$$\hat{E}^{(+)}(z,\ t) = CT\exp\left(i\Omega_c\frac{d}{c_1}\right)\exp\left\{(s_0-i\omega)\left(t-\frac{z}{c_0}\right)\right\} \tag{7.56}$$

This is the desired result for the field coupled out of the cavity to the region $0<z$. The field outside has only an outgoing wave with proper wave velocity. The above derivation procedure of this result is logical, because we have used solely the basic equation (Equation 7.1) for the case of homogeneous broadening and uniform pumping for the derivation of both the field inside the cavity and the field outside.

There is, however, an *ad hoc* means to derive Equation 7.56. This is to use the transmission coefficient T for the amplitude of the right-traveling wave inside the cavity at the coupling surface $z = 0$. The last quantity is, from Equation 7.41,

$$\hat{E}^{(+)}(0,t)\Big|_{\text{right-going}} = C\exp\{(\gamma_c + i\omega_c)(d/c_1)\}\exp\{(s_0 - i\omega)t\} \tag{7.57}$$

So, if we multiply this amplitude by the transmission coefficient T and add the correct translational shift with the velocity of light c_0 in the outside region, we recover Equation 7.56. This second derivation is arbitrary, however, and not necessarily logical. Although we have used the boundary condition at the coupling surface, the transmission coefficient appeared as a consequence of the appropriate use of the mode functions of the "universe," but not from the boundary conditions directly. Also, the fast-decaying component in Equation 7.52 may not be inferred from the *ad hoc* method. In spite of this caution on the *ad hoc* method, it is well known that many rules of optical wave phenomena prevail also in quantum mechanics, and an *ad hoc* compromise of quantum mechanics and classical optics is sometimes used in problems where this compromise gives a convenient method of analysis. We will see examples of such a method in Chapter 11 concerning the derivation of quantum excess noise.

Finally we rewrite Equation 7.56, using Equations 1.18a and 7.48, in the form

$$\hat{E}^{(+)}(z,\, t) = 2iC\sin\left(\Omega_c\frac{d}{c_1}\right)\exp\left\{(s_0 - i\omega)\left(t - \frac{z}{c_0}\right)\right\} \tag{7.58}$$

which shows the form of the outer field in Equation 1.21b when the inner field is Equation 7.41.

Reference

1 Ujihara, K. (1976) *Jpn. J. Appl. Phys.*, *15*, 1529–1541.

8
A One-Dimensional Laser with Output Coupling: Semiclassical Nonlinear Gain Analysis

Here we take into account the gain saturation or saturation in the atomic inversion but still ignore noise forces. The steady-state operation above threshold is examined. The essence of the contents of this chapter was published in [1].

We use Equation 5.33a, discarding the noise terms:

$$\hat{E}^{(+)}(z,t) = \sum_m \left[\frac{|p_m|^2 v_m^2}{2\hbar\omega} \int_0^t \sum_j U_j(z) U_j(z_m) e^{-i\omega_j(t-t')} \times \int_0^{t'} e^{-(i\nu_m+\gamma_m)(t'-t'')} \hat{E}^{(+)}(z_m,t'')\hat{\sigma}_m(t'')dt''dt' \right] \tag{8.1}$$

We go to slowly varying amplitude by setting

$$\hat{E}^{(+)}(z,t) = \tilde{E}^{(+)}(z,t)e^{-i\omega t} \tag{8.2}$$

We obtain

$$\tilde{E}^{(+)}(z,t) = \sum_m \frac{|p_m|^2 v_m^2}{2\hbar\omega} \int_0^t \sum_j U_j(z) U_j(z_m) \times e^{i(\omega-\omega_j)(t-t')} e^{-\{i(\nu_m-\omega)+\gamma_m\}t'} \times \int_0^{t'} e^{\{i(\nu_m-\omega)+\gamma_m\}t''} \hat{\sigma}_m(t'')\tilde{E}^{(+)}(z_m,t'')dt'dt'' \tag{8.3}$$

8.1 The Field Equation Inside the Cavity

For inside the cavity, $-d < z < 0$, the summation over j was calculated in Equation 7.10:

Output Coupling in Optical Cavities and Lasers: A Quantum Theoretical Approach
Kikuo Ujihara

ISBN: 978-3-527-40763-7

$$\begin{aligned}
&\sum_j U_j(z)U_j(z_m)\exp\{-i(\omega_j-\omega)(t-t')\}\\
&=\frac{1}{\varepsilon_1 c_1}\sum_{n=0}^{\infty}\frac{1}{1+\delta_{0,n}}(-r)^n\{\exp(i\omega\tau_{1n}^m)\delta(t-t'-\tau_{1n}^m)\\
&+\exp(-i\omega\tau_{1n}^m)\delta(t-t'+\tau_{1n}^m)\\
&+\exp(i\omega\tau_{2n}^m)\delta(t-t'-\tau_{2n}^m)+\exp(-i\omega\tau_{2n}^m)\delta(t-t'+\tau_{2n}^m)\\
&-\exp(i\omega\tau_{3n}^m)\delta(t-t'-\tau_{3n}^m)-\exp(-i\omega\tau_{3n}^m)\delta(t-t'+\tau_{3n}^m)\\
&-\exp(i\omega\tau_{4n}^m)\delta(t-t'-\tau_{4n}^m)-\exp(-i\omega\tau_{4n}^m)\delta(t-t'+\tau_{4n}^m)\}
\end{aligned}\tag{8.4}$$

where

$$\begin{aligned}
\tau_{1n}^m&=\frac{z-z_m+2nd}{c_1}, &\quad \tau_{2n}^m&=\frac{z_m-z+2nd}{c_1},\\
\tau_{3n}^m&=\frac{2d+z+z_m+2nd}{c_1}, &\quad \tau_{4n}^m&=\frac{-(2d+z+z_m)+2nd}{c_1}
\end{aligned}\tag{8.5}$$

(In equation 18.1 of Ref. [1], $\delta(t'-t-\tau)$ and $\delta(t'-t+\tau)$ should be interchanged.) Here the superscript m for the delay times τ_{pn}^m indicates the dependence of the delay times on z_m, which we omitted in Equations 7.5 and 7.6. Substituting Equation 8.4 into Equation 8.3 we have

$$\begin{aligned}
\tilde{E}^{(+)}(z,t)=\sum_m g_m e^{-\{i(\nu_m-\omega)+\gamma_m\}t}&\left[e^{i(\omega_m+\gamma_m)|z-z_m|/c_1}\int_0^{t-(|z-z_m|/c_1)}f_m(t'')dt''\right.\\
&-e^{(i\nu_m+\gamma_m)(2d+z+z_m)/c_1}\int_0^{t-\{(2d+z+z_m)/c_1\}}f_m(t'')dt''\\
&\left.+\sum_{n\geq 1}(-r)^n\{I_1+I_2-I_3-I_4\}\right]
\end{aligned}\tag{8.6}$$

where

$$f_m(t)=e^{\{i(\nu_m-\omega)+\gamma_m\}t}\hat{\sigma}_m(t)\tilde{E}^{(+)}(z_m,t)\tag{8.7}$$

and

$$I_p=e^{(i\nu_m+\gamma_m)\tau_{pn}^m}\int_0^{t-\tau_{pn}^m}f_m(t'')dt''\tag{8.8}$$

$$g_m=\frac{\nu_m^2|p_m|^2}{2\hbar\omega\varepsilon_1 c_1}\tag{8.9}$$

Equation 8.6 has a similar structure to that of Equation 7.11. The difference is that the atomic inversion $\hat{\sigma}_m(t)$ appears next to the electric field $\tilde{E}^{(+)}(z_m,t)$ instead of within the constant G_m as in Equation 7.13. Here a new constant g_m appears instead of G_m. Note that the absolute sign in the first term in Equation 8.6 appears from the $n = 0$ terms in the expansion in Equation 8.4 on using the delta functions $\delta(\pm\tau_{10}^m + t - t')$ and $\delta(\pm\tau_{20}^m + t - t')$ because of the constraint that $t \geq t'$ in the double integral in Equation 8.3. In Equation 8.6 we have not explicitly shown the upper limit of the summation over n, which is limited by the constraint that the delay times $\tau_{\rho n}^m$ should not exceed time t because of the delta functions. However, for a sufficiently long time t, the contribution from the final term in the sum becomes negligibly small, so that the upper limit can safely be taken to be infinity.

8.2 Homogeneously Broadened Atoms and Uniform Pumping

We go to the case of homogeneously broadened atoms, that is,

$$\nu_m = \nu_0, \qquad p_m = p_a, \qquad \gamma_m = \gamma, \qquad g_m = g = \frac{\nu_0^2 |p_a|^2}{2\hbar\omega\varepsilon_1 c_1} \tag{8.10a}$$

and to uniform pumping and uniform unsaturated atomic inversion, that is,

$$\Gamma_{mp} = \Gamma_p, \qquad \hat{\sigma}_m^0 = \sigma^0 \tag{8.10b}$$

The inversion of the mth atom is now dependent on the field strength at the atom, which, in turn, is dependent on z. Differentiation of Equation 8.6 with respect to time t yields

$$\begin{aligned}
&(\partial/\partial t)\tilde{E}^{(+)}(z,t) \\
&= -\{i(\nu_0 - \omega) + \gamma\}\tilde{E}^{(+)}(z,t) \\
&\quad + g\sum_m \left[\exp\left(\frac{i\omega|z - z_m|}{c_1}\right)\hat{\sigma}_m\left(t - \frac{|z - z_m|}{c_1}\right)\tilde{E}^{(+)}\left(z_m, t - \frac{|z - z_m|}{c_1}\right)\right. \\
&\quad - \exp\left(\frac{i\omega(2d + z + z_m)}{c_1}\right)\hat{\sigma}_m\left(t - \frac{2d + z + z_m}{c_1}\right)\tilde{E}^{(+)}\left(z_m, t - \frac{2d + z + z_m}{c_1}\right) \\
&\quad + \sum_{n=1}^{\infty}(-r)^n\{\exp(i\omega\tau_{1n}^m)\hat{\sigma}_m(t - \tau_{1n}^m)\tilde{E}^{(+)}(z_m, t - \tau_{1n}^m) \\
&\quad + \exp(i\omega\tau_{2n}^m)\hat{\sigma}_m(t - \tau_{2n}^m)\tilde{E}^{(+)}(z_m, t - \tau_{2n}^m) \\
&\quad - \exp(i\omega\tau_{3n}^m)\hat{\sigma}_m(t - \tau_{3n}^m)\tilde{E}^{(+)}(z_m, t - \tau_{3n}^m) \\
&\quad \left. - \exp(i\omega\tau_{4n}^m)\hat{\sigma}_m(t - \tau_{4n}^m)\tilde{E}^{(+)}(z_m, t - \tau_{4n}^m)\}\right]
\end{aligned} \tag{8.11}$$

We have explicitly written the upper limit ∞ for the summation over n according to the discussion given below Equation 8.9.

8.3 The Steady State

We go to the steady state, where we can forget the time dependences of $\hat{\sigma}_m(t)$ and $\tilde{E}^{(+)}(z_m, t)$. Then the equation simplifies to

$$\begin{aligned}\tilde{E}^{(+)}(z) = g' \sum_m \hat{\sigma}_m \tilde{E}^{(+)}(z_m) \Big[& \exp\{i\omega|z - z_m|/c_1\} \\ & - \exp\{i\omega(2d + z + z_m)/c_1\} \\ & + \sum_{n=1}^{\infty} (-r)^n \{\exp(i\omega\tau_{1n}^m) + \exp(i\omega\tau_{2n}^m) \\ & - \exp(i\omega\tau_{3n}^m) - \exp(i\omega\tau_{4n}^m)\}\Big]\end{aligned} \tag{8.12}$$

where

$$g' = \frac{g}{i(\nu_0 - \omega) + \gamma} = \frac{g}{\gamma'} \tag{8.13a}$$

Note that

$$\mathrm{Re}\, g' = \frac{\mathcal{g}}{c_1} \tag{8.13b}$$

that is, Re g' is equal to the amplitude gain per unit density of inverted atoms per unit length, where $\mathcal{g}$ was defined in Equation 4.14.

Now we consider the steady-state atomic inversion. Utilizing Equation 5.32 in Equation 5.26, discarding the noise term, for a steady state we have

$$(\hat{b}_{m1}^{\dagger}\hat{b}_{m2})_{sve} = \frac{-i\nu_0 p_a^* \hat{\sigma}_m/\hbar\omega}{i(\nu_0 - \omega) + \gamma} \tilde{E}^{(+)}(z_m) \tag{8.14}$$

where the suffix *sve* signifies the slowly varying envelope: $(\hat{b}_{m1}^{\dagger}\hat{b}_{m2})_{sve} = (\hat{b}_{m1}^{\dagger}\hat{b}_{m2})e^{i\omega t}$. Then using Equations 5.27 and 5.32 and again discarding the noise term, we obtain

$$\Gamma_p\{\hat{\sigma}_m - \sigma^0\} = -4\frac{\gamma|\nu_0 p_a/\hbar\omega|^2 \left|\tilde{E}^{(+)}(z_m)\right|^2 \hat{\sigma}_m}{\gamma^2 + (\nu_0 - \omega)^2} \tag{8.15}$$

Thus the steady-state atomic inversion is

$$\hat{\sigma}_m = \frac{\sigma^0}{1 + \left|\tilde{E}^{(+)}(z_m)\right|^2 / |E_s|^2} \tag{8.16}$$

where the saturation parameter

$$|E_s|^2 = \left\{ \frac{4\gamma \nu_0^2 |p_a|^2}{\Gamma_p \hbar^2 \omega^2} \frac{1}{(\nu_0 - \omega)^2 + \gamma^2} \right\}^{-1} \tag{8.17}$$

On substitution of Equation 8.16 into Equation 8.12, regarding the summation over m as an integration with respect to z_m, we have a nonlinear integral equation. We solve this equation, again assuming the decomposition of the electric field into two oppositely traveling waves. Beforehand we perform the summation over n in Equation 8.12:

$$\begin{aligned}\tilde{E}^{(+)}(z) = g' \sum_m \frac{\sigma^0}{1+\{|E(z_m)|/|E_s|\}^2} \tilde{E}^{(+)}(z_m) \Bigg\{ \exp(ik|z-z_m|) \\ - \exp\{ik(2d+z+z_m)\} + \frac{-r\exp(2ikd)}{1+r\exp(2ikd)} [\exp\{ik(z-z_m)\} \\ + \exp\{-ik(z-z_m)\} - \exp\{ik(2d+z+z_m)\} - \exp\{-ik(2d+z+z_m)\}] \Bigg\}\end{aligned} \tag{8.18}$$

where

$$k = \omega / c_1 \tag{8.19}$$

and we have written $E(z_m)$ for $\tilde{E}^{(+)}(z_m)$ in the denominator for simplicity. Then we set

$$\tilde{E}^{(+)}(z,t) = e^+(z,t)\exp\{+ik(z+d)\} + e^-(z,t)\exp\{-ik(z+d)\} \tag{8.20}$$

Comparing the terms of $\exp\{+i\omega(z+d)/c_1\}$ and $\exp\{-i\omega(z+d)/c_1\}$, and discarding rapidly oscillating (spatially) terms in Equation 8.18, we have

$$\begin{aligned}e^+(z) = \sum_{z>z_m} \frac{g'\sigma^0}{1+\{|E(z_m)|/|E_s|\}^2} e^+(z_m) \\ - \sum_m \frac{g'\sigma^0}{1+\{|E(z_m)|/|E_s|\}^2} \frac{\{r\exp(2ikd)e^+(z_m) + e^-(z_m)\}}{1+r\exp(2ikd)}\end{aligned} \tag{8.21a}$$

$$\begin{aligned}e^-(z) = \sum_{z<z_m} \frac{g'\sigma^0}{1+\{|E(z_m)|/|E_s|\}^2} e^-(z_m) \\ + \sum_m \frac{g'\sigma^0 r\exp(2ikd)}{1+\{|E(z_m)|/|E_s|\}^2} \frac{\{e^+(z_m) - e^-(z_m)\}}{1+r\exp(2ikd)}\end{aligned} \tag{8.21b}$$

The squared amplitude in the denominator is

$$|E(z_m)|^2 = \tilde{E}^{(+)}(z_m)\tilde{E}^{(+)*}(z_m) \simeq |e^+(z_m)|^2 + |e^-(z_m)|^2 \tag{8.22}$$

Here rapidly oscillating terms in z_m have been ignored. Equations 8.21a and 8.21b are rewritten as

$$e^+(z) = \int_{-d}^{z} \frac{\alpha^0 e^+(z_m)dz_m}{1+|E_{m/s}|^2} + \int_{-d}^{0} \frac{-\alpha^0}{1+|E_{m/s}|^2} \frac{r\exp(2ikd)e^+(z_m) + e^-(z_m)}{1+r\exp(2ikd)} dz_m \tag{8.23a}$$

$$e^-(z) = \int_{z}^{0} \frac{\alpha^0 e^-(z_m)dz_m}{1+|E_{m/s}|^2} + \int_{-d}^{0} \frac{\alpha^0}{1+|E_{m/s}|^2} \frac{r\exp(2ikd)}{1+r\exp(2ikd)} \{e^+(z_m) - e^-(z_m)\} dz_m \tag{8.23b}$$

$$\alpha^0 = g'\sigma^0 N = \frac{GN}{\gamma'} = \frac{\nu_0^2|p_a|^2}{2\hbar\omega\varepsilon_1 c_1} \frac{N\sigma^0}{i(\nu_0-\omega)+\gamma} \tag{8.23c}$$

where G is given by Equation 7.16 with the understanding that the atomic inversion σ in the linear gain analysis is the same as the unsaturated inversion σ^0 in the nonlinear gain analysis. The parameter α^0 is the amplitude gain per unit length, and G / γ' is the gain per atom. Note that α^0 is related to $\mathcal{g}$, the amplitude gain per unit density of inverted atoms per unit time, as

$$\text{Re } \alpha^0 = \text{Re } g'\sigma^0 N = \frac{\mathcal{g} N\sigma^0}{c_1} \tag{8.23d}$$

In Equations 8.23a and 8.23b $|E_{m/s}|^2$ abbreviates $\{|\tilde{E}^{(+)}(z_m)|/|E_s|\}^2$. It can easily be shown that Equations 8.23a and 8.23b are equivalent to the following four equations:

$$(d/dz)e^+(z) = \frac{\alpha^0}{1+|E_{z/s}|^2} e^+(z) \tag{8.24a}$$

$$(d/dz)e^-(z) = \frac{-\alpha^0}{1+|E_{z/s}|^2} e^-(z) \tag{8.24b}$$

$$e^-(-d) = -e^+(-d) \tag{8.24c}$$

$$e^-(0) = r\exp(2ikd)e^+(0) \tag{8.24d}$$

where $|E_{z/s}|^2$ abbreviates $\{|\tilde{E}^{(+)}(z)|/|E_s|\}^2$. Note that Equations 8.24a and 8.24b are coupled equations because both $e^+(z)$ and $e^-(z)$ exist in the denominators as described by Equation 8.22.

8.4
Solution of the Coupled Nonlinear Equations

We can solve for $e^+(z)$ and $e^-(z)$ by the unique structure of the coupled equations as follows. Integrating Equations 8.24a and 8.24b we have

$$e^+(z) = e^+(-d)\exp\{I(z)\} \tag{8.25a}$$

$$e^-(z) = e^-(-d)\exp\{-I(z)\} \tag{8.25b}$$

$$I(z) = \int_{-d}^{z} \frac{\alpha^0 dz'}{1 + \left|E_{z'/s}\right|^2} \tag{8.25c}$$

so that

$$e^+(z)e^-(z) = \text{const} = e^+(-d)e^-(-d) = -\{e^+(-d)\}^2 \tag{8.26}$$

We have used Equation 8.24c in the last equality. This equation is the key to solving the nonlinear equations.

We first look for the steady-state oscillation frequency. For this purpose we set $z = 0$ in Equations 8.25a and 8.25a to obtain

$$e^+(0) = e^+(-d)\exp\{I(0)\} \tag{8.27a}$$

$$e^-(0) = e^-(-d)\exp\{-I(0)\} \tag{8.27b}$$

where $I(0)$ may be written as

$$I(0) = \int_{-d}^{0} \frac{\alpha^0 dz'}{1 + \left|E_{z'/s}\right|^2} = \alpha^0 I \tag{8.28a}$$

where

$$I = \int_{-d}^{0} \frac{dz'}{1 + \left|E_{z'/s}\right|^2} \tag{8.28b}$$

Because we know the ratio of $e^+(0)$ and $e^-(0)$ from Equation 8.24d, and that for $e^+(-d)$ and $e^-(-d)$ from Equation 8.24c, we have from Equations 8.27a and 8.27b

$$\frac{1}{r\exp(2ikd)} = -\exp\{2\alpha^0 I\} \tag{8.29}$$

As we saw in Equation 7.33, $-r^{-1}\exp(-2ikd)$ is equal to $\exp\{(2d/c_1)(\gamma_c - i\omega + i\omega_c)\}$. Therefore, noting that $\alpha^0 = GN/\gamma'$ from Equation 8.23c and that $\gamma' = \gamma + i(\nu_0 - \omega)$ from Equation 8.13a, and comparing the phase and the magnitude of both sides, we have

$$\frac{2d}{c_1}(\omega_c - \omega) = -\frac{2GNI(\nu_0 - \omega)}{\gamma^2 + (\nu_0 - \omega)^2} \tag{8.30}$$

$$\gamma_c = \frac{c_1}{2d}\frac{2GNI\gamma}{\gamma^2 + (\nu_0 - \omega)^2} \tag{8.31}$$

Thus eliminating GNI, we have

$$\omega = \frac{\gamma\omega_c + \gamma_c\nu_0}{\gamma + \gamma_c} \tag{8.32}$$

Equation 8.31 gives the necessary gain

$$GN = \frac{\gamma^2 + (\nu_0 - \omega)^2}{2I\gamma}\frac{2d}{c_1}\gamma_c \tag{8.33}$$

We do not know the value of I as yet. But, at threshold, $\tilde{E}^{(+)}(z) = 0$ and $I = d$. Therefore, we have the threshold population inversion

$$N\sigma_{th}^0 = \frac{2\varepsilon_1\hbar\omega\gamma\gamma_c}{\nu_0^2|p_a|^2}\left\{1 + \frac{(\nu_0 - \omega_c)^2}{(\gamma + \gamma_c)^2}\right\} \tag{8.34}$$

where we have used Equations 7.16 and 8.32. This is the same as that in Equation 7.44a for the linear gain analysis

In order to solve for the field amplitude, we need to consider the absolute squares of the amplitudes because of their appearance in the denominator in Equation 8.25c. Thus multiplying Equation 8.24a by $\{e^+(z)\}^*$ and its complex conjugate by $e^+(z)$ and adding, we have

$$(d/dz)|e^+(z)|^2 = \frac{\alpha^0 + \alpha^{0^*}}{1 + |E_{z/s}|^2}|e^+(z)|^2 \tag{8.35a}$$

Similarly, from Equation 8.24b we have

$$(d/dz)|e^-(z)|^2 = \frac{-(\alpha^0 + \alpha^{0^*})}{1 + |E_{z/s}|^2}|e^-(z)|^2 \tag{8.35b}$$

These equations were derived by Rigrod [2]. Using Equation 8.22 in Equation 8.35a with z_m replaced by z, and eliminating $e^-(z)$ by Equation 8.26, we have

$$\frac{1 + \left(|e^+(z)|^2 + |\text{const}|^2|e^+(z)|^{-2}\right)/|E_s|^2}{|e^+(z)|^2}d|e^+(z)|^2 = (\alpha^0 + \alpha^{0^*})dz \tag{8.36}$$

Integrating both sides we have

$$\ln|e^+(z)|^2 + \frac{|e^+(z)|^2 - |\text{const}|^2|e^+(z)|^{-2}}{|E_s|^2} = (\alpha^0 + \alpha^{0^*})z + C \tag{8.37}$$

We determine the constant C by setting $z = -d$:

$$C = \ln|e^+(-d)|^2 + \frac{|e^+(-d)|^2 - |\text{const}|^2 |e^+(-d)|^{-2}}{|E_s|^2} + (\alpha^0 + \alpha^{0^*})d \tag{8.38}$$

Since in the second term on the right-hand side $|\text{const}|^2|e^+(-d)|^{-2} = |e^-(-d)|^2$ by Equation 8.26 and it is equal to $|e^+(-d)|^2$ by Equation 8.24c, the second term vanishes. Thus we have

$$\begin{aligned} &\ln\{|e^+(z)|/|e^+(-d)|\}^2 + \left\{|e^+(z)|^2 - |\text{const}|^2|e^+(z)|^{-2}\right\}/|E_s|^2 \\ &= \left(\alpha^0 + \alpha^{0^*}\right)(z+d) \end{aligned} \tag{8.39}$$

Note that, if the constant or $\{e^+(-d)\}^2$ is known, the z dependence of $|e^+(z)|^2$ is known from this equation. In particular, the left-hand side is a monotonically increasing function of $|e^+(z)|^2$. On the other hand, taking the logarithm of the absolute square of Equation 8.25a, we have

$$\ln\{|e^+(z)|/e^+(-d)\}^2 = I(z) + I^*(z) = \int_{-d}^{z} \frac{\left(\alpha^0 + \alpha^{0^*}\right)dz'}{1 + \left|E_{z'/s}\right|^2} \tag{8.40}$$

where we have used Equation 8.25c in the last equality. Comparing Equations 8.39 and 8.40 we have

$$I(z) = \frac{-\alpha^0}{\alpha^0 + \alpha^{0^*}} \frac{|e^+(z)|^2 - |e^-(z)|^2}{|E_s|^2} + \alpha^0(z+d) \tag{8.41}$$

where we have replaced $|\text{const}|^2|e^+(z)|^{-2}$ by $|e^-(z)|^2$ using Equation 8.26. Substituting this into Equation 8.25a and setting $z=0$, we have

$$e^+(0) = e^+(-d)\exp\left\{\frac{-\alpha^0}{\alpha^0 + \alpha^{0^*}} \frac{|e^+(0)|^2 - |e^-(0)|^2}{|E_s|^2} + \alpha^0 d\right\} \tag{8.42a}$$

Similarly, from Equation 8.25a we have

$$e^-(0) = e^-(-d)\exp\left\{\frac{\alpha^0}{\alpha^0 + \alpha^{0^*}} \frac{|e^+(0)|^2 - |e^-(0)|^2}{|E_s|^2} - \alpha^0 d\right\} \tag{8.42b}$$

Taking the ratios of both sides of Equations 8.42a and 8.42b, because we know the ratio of $e^+(0)$ and $e^-(0)$ from Equation 8.24d, and that for $e^+(-d)$ and $e^-(-d)$ from Equation 8.24c, we obtain

$$\frac{-1}{r\exp(2ikd)} = \exp\left\{\frac{-2\alpha^0}{\alpha^0 + \alpha^{0^*}} \frac{|e^+(0)|^2(1-r^2)}{|E_s|^2} + 2\alpha^0 d\right\} \tag{8.43}$$

As in Equation 8.29, comparing the phase and magnitude of both sides we have

$$\frac{2d}{c_1}\gamma_c = -(1-r^2)\left\{\frac{|e^+(0)|}{|E_s|}\right\}^2 + (\alpha^0 + \alpha^{0^*})d \tag{8.44}$$

$$\frac{2d}{c_1}(\omega-\omega_c) = -i\frac{(\alpha^0-\alpha^{0^*})}{(\alpha^0+\alpha^{0^*})}\left\{\frac{(1-r^2)|e^+(0)|^2}{|E_s|^2} - d(\alpha^0+\alpha^{0^*})\right\} \tag{8.45}$$

Noting that $i(\alpha^0-\alpha^{0^*})/(\alpha^0+\alpha^{0^*}) = (\nu_0-\omega)/\gamma$ from Equation 8.23c, we have again

$$\omega = \frac{\gamma\omega_c+\gamma_c\nu_0}{\gamma+\gamma_c} \tag{8.46}$$

and from Equation 8.44 we have the absolute square of the amplitude of the right-traveling wave at the output surface:

$$\begin{aligned}|e^+(0)|^2 &= \frac{|E_s|^2}{1-r^2}\left\{\frac{\nu_0^2|\rho_a|^2 N\sigma^0 d}{\varepsilon_1 c_1\hbar\omega}\frac{\gamma}{(\nu_0-\omega)^2+\gamma^2} - \ln(1/r)\right\}\\ &= \frac{|E_s|^2}{1-r^2}\left\{\frac{\nu_0^2|\rho_a|^2 N\sigma^0 d}{\varepsilon_1 c_1\hbar\omega}\frac{1}{\gamma(1+\delta^2)} - \ln(1/r)\right\}\end{aligned} \tag{8.47}$$

Setting $e^+(0)=0$ and using Equation 8.46 we have Equation 8.34 again for the threshold atomic inversion:

$$\sigma_{th}^0 = \frac{2\varepsilon_1\hbar\omega\gamma\gamma_c}{\nu_0^2|\rho_a|^2 N}(1+\delta^2) = \frac{\gamma_c}{gN} \tag{8.48}$$

Now integrating Equation 8.16 and dividing by d, we examine the average atomic inversion for steady state:

$$\bar{\sigma}_{ss} \equiv \frac{1}{d}\int_{-d}^{0}\sigma_m dz_m = \frac{\sigma^0}{d}\int_{-d}^{0}\frac{dz'}{1+|E_{z'/s}|^2} = \frac{\sigma^0}{d}\frac{\ln\{|e^+(0)|/|e^+(-d)|\}^2}{(\alpha^0+\alpha^{0^*})} \tag{8.49}$$

where we have used Equations 8.40 in the last expression. Now comparing the square of Equations 8.42a and 8.43 we have

$$\left\{\frac{e^+(0)}{e^+(-d)}\right\}^2 = \frac{-1}{r\exp(2ikd)} \tag{8.50}$$

Thus

$$\left|\frac{e^+(0)}{e^+(-d)}\right|^2 = \frac{1}{r} \tag{8.51}$$

Using this relation and the expression in Equation 8.23c for α^0 we have

$$\bar{\sigma}_{ss} = \frac{2\hbar\omega\varepsilon_1}{\nu_0^2|\rho_a|^2 N}\gamma\gamma_c(1+\delta^2) = \frac{\gamma_c}{gN} \tag{8.52a}$$

where

$$\delta^2 = \left(\frac{\omega_c - \nu_0}{\gamma + \gamma_c}\right)^2 \tag{8.52b}$$

Thus, comparing it with Equation 8.48, we confirm that

$$\bar{\sigma}_{ss} = \sigma^0_{th} \tag{8.53}$$

as is usual for an oscillator operating above threshold. But here what is equal to the threshold inversion is the space average of the location-dependent inversion.

Now that $|e^+(0)|^2$ is known, we can determine the distribution of the field amplitude as follows. First, we have Equations 8.25a and 8.25a for $e^+(z)$ and $e^-(z)$, which are determined by the integral in Equation 8.25c. The integral is known, in principle, if we know $|e^+(z)|^2$ and $|e^-(z)|^2$. The former $|e^+(z)|^2$ is determined by Equation 8.39 completely if we know $|e^+(-d)|^2$. But it is given by Equation 8.51 in terms of $|e^+(0)|^2$, which is given by Equation 8.47. The latter $|e^-(z)|^2$ is determined by the relation in Equation 8.26. Thus we can, in principle, determine $e^+(z)$ and $e^-(z)$ completely, except for undetermined phases.

Here we briefly discuss how the field distribution in this nonlinear gain analysis is related to that in the linear gain analysis. The present analysis goes to the linear gain analysis in the limit of infinitely large saturation parameter $E_s \to \infty$. That is to say, in the limit $E_{z/s} \to 0$. In this limit, Equations 8.25a–8.25c show that $e^\pm(z) = e^\pm(-d)\exp\{\pm\alpha^0(z+d)\}$. But Equation 8.44 shows that, in this limit, $\alpha^0 \simeq \ln(1/r)/(2d)$. Therefore, we have $e^\pm(z) \simeq e^\pm(-d)\exp\{\pm\gamma_c(z+d)/c_1\}$, which is just the field distribution consistent with the cavity resonant field that appeared in Equation 7.41 of the linear gain analysis.

8.5
The Field Outside the Cavity

Similarly to Equation 7.45, we calculate the summation over j in Equation 8.3 for the field outside the cavity:

$$\begin{aligned}
&\sum_j \left\{U_j(z)U_j(z_m)e^{-i(\omega_j-\omega)(t-t')}\right\} \\
&= \sum_j \frac{2}{\varepsilon_1 L}\frac{1}{1-K\sin^2 k_{1j}d}\left(\frac{k_{1j}}{k_{0j}}\cos k_{1j}d\sin k_{0j}z + \sin k_{1j}d\cos k_{0j}z\right) \\
&\times \sin k_{1j}(z_m+d)e^{-i\omega_j(t-t')} = \frac{1}{\varepsilon_1 c_1}\frac{2c_0}{c_0+c_1}\sum_{n=0}^{\infty}(-r)^n \\
&\times \left\{\exp(i\omega\tau^m_{5n})\delta(t'-t+\tau^m_{5n}) + \exp(-i\omega\tau^m_{5n})\delta(t'-t-\tau^m_{5n})\right. \\
&\left. - \exp(i\omega\tau^m_{6n})\delta(t'-t+\tau^m_{6n}) - \exp(-i\omega\tau^m_{6n})\delta(t'-t-\tau^m_{6n})\right\}
\end{aligned} \tag{8.54}$$

where

9
A One-Dimensional Laser with Output Coupling: Quantum Linear Gain Analysis

In Chapters 9 and 10 we solve the laser equation of motion (Equation 5.33a) quantum mechanically for the one-sided cavity model described in Sections 1.3.1 and 1.4. As in the case of the quasimode laser, we divide the analysis into two categories: linear gain analysis, applicable to operation below threshold; and nonlinear gain analysis, applicable to above-threshold operation. In this chapter, the atomic inversion is assumed to be constant, but it will be allowed to be field dependent in the next chapter. In Chapter 7, the linear gain analysis was performed semiclassically, with the atomic inversion assumed to have a fixed value and the noise terms being ignored. There, instead of the noise terms, an initial field distribution with temporal delta-function property was assumed. The temporal decay of the field from the initial value was derived using the Laplace transform method, which yielded a decay constant that represents the net effect of the field gain associated with propagation along the gain medium and reflections at the end surfaces. The explicit expression for the output field was also obtained by virtue of the continuous mode expansion of the field. Here, we take into account the noise terms for the atoms. However, the thermal noise is derived automatically by the field expansion using the continuous modes of the "universe." These noise terms act as incessant driving forces for the field. Thus, because of the linear nature of the assumed equation, the field becomes a superposition of the decaying components, each excited at the atoms or in the cavity at random instants, all the decay constants being the same as that obtained in Chapter 7. The spatial dependence of the excited field will be that of the cavity resonant mode as in Chapter 7. An important quantum result is that the expression for the linewidth of the output field has a correction factor compared with the conventional formula obtained for the quasimode laser. This factor is $(\beta_c/\gamma_c)^2 = \{1 - r^2/(2r)\}^2/\{\ln(1/r)\}^2$, which is determined solely by the reflection coefficient at the cavity end surface. Similar factors appeared in Chapter 6, where the contour integral method was used. The nonlinear, saturated gain analysis applicable to operation above threshold will be covered in the next chapter, where correction factors for the linewidth will also be derived. The essence of the contents of this chapter was published in [1].

Output Coupling in Optical Cavities and Lasers: A Quantum Theoretical Approach
Kikuo Ujihara

ISBN: 978-3-527-40763-7

9.1 The Equation for the Quantum Linear Gain Analysis

Here we concentrate on the linear gain analysis. The equation to be solved reads, from Equation 5.33, for the entire region $-d < z < L$,

$$\hat{E}^{(+)}(z,t) = \hat{F}_t(z,t) + \hat{F}_q(z,t) + \sum_m \left[\frac{|p_m|^2 v_m^2 \sigma_m}{2\hbar\omega} \int_0^t \sum_j U_j(z)\, U_j(z_m)\, e^{-i\omega_j(t-t')} \times \int_0^{t'} e^{-(i v_m + \gamma_m)(t'-t'')} \hat{E}^{(+)}(z_m, t'')\, dt''\, dt' \right] \tag{9.1}$$

where

$$\hat{F}_t(z,t) = i \sum_j \sqrt{\frac{\hbar\omega_j}{2}}\, U_j(z)\hat{a}_j(0) e^{-i\omega_j t} \tag{9.2}$$

$$\hat{F}_q(z,t) = \sum_m \left[\frac{i p_m v_m}{2} \int_0^t \sum_j U_j(z)\, U_j(z_m)\, e^{-i\omega_j(t-t')} \int_0^{t'} e^{-(i v_m + \gamma_m)(t'-t'')} \hat{\Gamma}_m(t'')\, dt''\, dt' \right] \tag{9.3}$$

Here we have taken the atomic inversion $\hat{\sigma}_m(t)$ as a constant σ_m that is determined by the pumping process only. The strategy for solving the equation is to seek the expression for the field in terms of the noise forces and to construct the correlation function. The correlation functions of the noise forces will determine the correlation function of the field. Then we will obtain the field power spectrum as the Fourier transform of the field correlation function. From Equations 5.36, 5.38, and 5.39, the correlation functions to be used are

$$\left\langle a_i^\dagger(0) a_j(0) \right\rangle = \langle n_j \rangle \delta_{ij} \tag{9.4a}$$

$$\left\langle a_i(0) a_j^\dagger(0) \right\rangle = (\langle n_j \rangle + 1) \delta_{ij} \tag{9.4b}$$

and

$$\langle \hat{\Gamma}_m^\dagger(t) \hat{\Gamma}_{m'}(t') \rangle = G_{21,12}^m \delta(t-t') \delta_{mm'}$$

$$G_{21,12}^m = \frac{1}{2}\Gamma_{mp}(1+\sigma_m^0)\left\langle \frac{1}{2}(1-\sigma_m) \right\rangle - \frac{1}{2}\Gamma_{mp}(1-\sigma_m^0)\left\langle \frac{1}{2}(1+\sigma_m) \right\rangle + 2\gamma_m \left\langle \frac{1}{2}(1+\sigma_m) \right\rangle \tag{9.5a}$$

$$\left\langle \hat{\Gamma}_m(t)\hat{\Gamma}^\dagger_{m'}(t')\right\rangle = G^m_{12,21}\delta(t-t')\delta_{mm'}$$

$$G^m_{12,21} = \frac{1}{2}\Gamma_{mp}(1-\sigma^0_m)\left\langle\frac{1}{2}(1+\sigma_m)\right\rangle - \frac{1}{2}\Gamma_{mp}(1+\sigma^0_m)\left\langle\frac{1}{2}(1-\sigma_m)\right\rangle + 2\gamma_m\left\langle\frac{1}{2}(1-\sigma_m)\right\rangle \tag{9.5b}$$

Here we have assumed that the different modes of the universe are not correlated initially. We have also assumed that the reservoirs for the dipoles of different atoms are also not correlated. In Equations 9.4a and 9.4b $\langle n_j\rangle$ is the expectation value of the number of thermal photons, that is, the Planck distribution, in the jth "universal" mode. If the pumping and damping of the atoms are non-uniform, the parameters in Equations 9.5a and 9.5b are different for different atoms. Thus we have added the suffix m to indicate the individual atoms. In this chapter, the saturated and unsaturated atomic inversion are the same constants, that is $\sigma_m = \sigma^0_m$. Thus we have

$$\langle \hat{\Gamma}^\dagger_m(t)\hat{\Gamma}_{m'}(t')\rangle = \gamma_m(1+\sigma_m)\delta_{mm'}\delta(t-t') \tag{9.5c}$$

$$\left\langle \hat{\Gamma}_m(t)\hat{\Gamma}^\dagger_{m'}(t')\right\rangle = \gamma_m(1-\sigma_m)\delta_{mm'}\delta(t-t') \tag{9.5d}$$

We have removed the sign of the ensemble average because we are assuming here that the atomic inversion is a constant. In order to utilize the calculations in Chapter 7 as far as possible, we truncate the oscillation in the optical frequency from the electric field and the noise, as in Chapter 7,

$$\hat{E}^{(+)}(z,t) = \tilde{E}^{(+)}(z,t)e^{-i\omega t}, \quad \hat{F}_t(z,t) = \tilde{F}_t(z,t)e^{-i\omega t}, \quad \hat{F}_q(z,t) = \tilde{F}_q(z,t)e^{-i\omega t} \tag{9.6}$$

and obtain

$$\tilde{E}^{(+)}(z,t) = \sum_m\left[\frac{|p_m|^2 v_m^2\sigma_m}{2\hbar\omega}\int_0^t\sum_j U_j(z)\,U_j(z_m)\,e^{i(\omega-\omega_j)(t-t')} \times\int_0^{t'} e^{\{i(\omega-\nu_m)-\gamma_m\}(t'-t'')}\,\tilde{E}^{(+)}(z_m,t'')dt''dt'\right] + \tilde{F}_t(z,t) + \tilde{F}_q(z,t) \tag{9.7}$$

This form of the equation is the same as Equation 7.3 except that the initial driving force term in Equation 7.3 is replaced by the noise forces. Thus the analysis in Sections 7.1–7.3 can be applied here with the proper replacement of the driving force by the noise forces. Let us follow the procedure of the analysis in Sections 7.1–7.3. We first evaluate the summation over j and find delta functions of time involving delay times that correspond to the possible routes for a light signal to go from position z_m to position z:

$$\begin{aligned}\sum_j \left\{ U_j(z) U_j(z_m) e^{-i(\omega_j-\omega)(t-t')} \right\} \\ = \frac{1}{\varepsilon_1 c_1} \sum_{n=0}^{\infty} \frac{1}{1+\delta_{0,n}} (-r)^n \\ \times \sum_{\rho=1}^{4} \alpha_\rho \left\{ e^{i\tau_{\rho n}\omega} \delta(-\tau_{\rho n} + t - t') + e^{-i\tau_{\rho n}\omega} \delta(\tau_{\rho n} + t - t') \right\}\end{aligned} \tag{9.8}$$

where the factors $\alpha_1 = \alpha_2 = 1$ and $\alpha_3 = \alpha_4 = -1$. The delay times are

$$\begin{aligned}\tau_{1n} &= \frac{2nd + z - z_m}{c_1}, \qquad \tau_{2n} = \frac{2nd - z + z_m}{c_1} \\ \tau_{3n} &= \frac{2nd + 2d + z + z_m}{c_1}, \qquad \tau_{4n} = \frac{2nd - 2d - z - z_m}{c_1}\end{aligned} \tag{9.9}$$

After performing the integration using the delta functions, we obtain a temporal integral equation containing a sum over the field values at each atom at various retarded times:

$$\begin{aligned}\tilde{E}^{(+)}(z,t) &= \tilde{F}_t(z,t) + \tilde{F}_q(z,t) + \sum_m G_m \left\{ \int_0^{t-|z-z_m|/c_1} \right. \\ &\times \exp\left[\{-i(\nu_m - \omega) - \gamma_m\}(t-t') + (i\nu_m + \gamma_m)\frac{|z-z_m|}{c_1} \right] \tilde{E}(z_m, t')dt' \\ &- \int_0^{t-(2d+z+z_m)/c_1} \exp\left[\{-i(\nu_m - \omega) - \gamma_m\}(t-t') + (i\nu_m + \gamma_m)\frac{2d+z+z_m}{c_1} \right] \\ &\times \left. \tilde{E}(z_m, t')dt' + \sum_{n=1}^{n_M} (-r)^n (I_{1n} + I_{2n} - I_{3n} - I_{4n}) \right\}\end{aligned} \tag{9.10}$$

where

$$I_{\rho n} = \int_0^{t-\tau_{\rho n}} \exp\left[\{-i(\nu_m - \omega) - \gamma_m\}(t-t') + (i\nu_m + \gamma_m)\tau_{\rho n} \right] \tilde{E}(z_m, t')\, dt' \tag{9.11a}$$

and

$$G_m = \frac{|p_m|^2 \nu_m^2 \sigma_m}{2\hbar\omega\varepsilon_1 c_1} \tag{9.11b}$$

9.2 Homogeneously Broadened Atoms and Uniform Atomic Inversion

To go further, we have to assume homogeneously broadened atoms and homogeneous pumping:

$$\nu_m = \nu_0, \qquad p_m = p_a, \qquad \gamma_m = \gamma, \qquad \sigma_m = \sigma \tag{9.12}$$

Then, differentiating with respect to time t, the integral equation is converted to a simplified differential equation:

$$\begin{aligned}
&\frac{\partial}{\partial t}\left\{\tilde{E}^{(+)}(z,t) - \tilde{F}_t(z,t) - \tilde{F}_q(z,t)\right\} \\
&\quad = \{-i(\nu_0 - \omega) - \gamma\}\left\{\tilde{E}^{(+)}(z,t) - \tilde{F}_t(z,t) - \tilde{F}_q(z,t)\right\} \\
&\qquad + \sum_m G\left[\exp\left(i\omega\frac{|z-z_m|}{c_1}\right)\tilde{E}^{(+)}\left(z_m, t - \frac{|z-z_m|}{c_1}\right)\right. \\
&\qquad - \exp\left(i\omega\frac{2d+z+z_m}{c_1}\right)\tilde{E}^{(+)}\left(z_m, t - \frac{2d+z+z_m}{c_1}\right) \\
&\qquad \left. + \sum_{n=1}^{n_M}(-r)^n\left\{\sum_{\rho=1}^{4}\alpha_\rho \exp(i\omega\tau_{\rho n})\tilde{E}^{(+)}(z_m, t-\tau_{\rho n})\right\}\right]
\end{aligned} \tag{9.13}$$

with

$$G = \frac{|p_a|^2 \nu_0^2 \sigma}{2\hbar\omega\varepsilon_1 c_1} \tag{9.14}$$

Here we introduce the assumption that the field waves and the driving forces are both divided into right- and left-going waves, respectively:

$$\begin{aligned}
\tilde{E}^{(+)}(z,t) &= \hat{e}^+(z,t)\exp\{+i\omega(z+d)/c_1\} \\
&\quad + \hat{e}^-(z,t)\exp\{-i\omega(z+d)/c_1\}
\end{aligned} \tag{9.15}$$

and similarly for the noise force terms. Here we have stressed that the field amplitudes are operators, in contrast to those in Chapter 7, where they were classical variables. Then we get two coupled temporal differential equations for the two traveling waves, which still contain various retarded times corresponding to the number of round trips that the waves make until they arrive at location z after having started from location z_m of the mth atom. The equations also contain summations over the atoms:

$$\begin{aligned}
\left(\frac{\partial}{\partial t} + \gamma'\right)\{\hat{e}^+(z,t) - \hat{\upsilon}^+(z,t)\} &= \sum_m G\left[H(z-z_m)\hat{e}^+(z_m, t-\tau_{10}) - \hat{e}^-(z_m, t-\tau_{30})\right. \\
&\quad \left. + \sum_{n=1}^{n_M}(r')^n\{\hat{e}^+(z_m, t-\tau_{1n}) - \hat{e}^-(z_m, t-\tau_{3n})\}\right]
\end{aligned} \tag{9.16a}$$

$$\left(\frac{\partial}{\partial t}+\gamma'\right)\{\hat{e}^-(z,t)-\hat{v}^-(z,t)\}=\sum_m G\left[H(z_m-z)\hat{e}^-(z_m,t-\tau_{20})\right.$$
$$\left.+\sum_{n=1}^{n_M}(r')^n\{\hat{e}^-(z_m,t-\tau_{2n})-\hat{e}^+(z_m,\tau_{4n})\}\right] \quad (9.16b)$$

where

$$\hat{v}^+(z,t)=\hat{f}_t^+(z,t)+\hat{f}_q^+(z,t), \qquad \hat{v}^-(z,t)=\hat{f}_t^-(z,t)+\hat{f}_q^-(z,t) \quad (9.17)$$

Here $\hat{f}_t^{\pm}(z,t)$ are the right- and left-traveling parts, respectively, of the thermal noise operator $\tilde{F}_t(z,t)$, varying as $\exp\{\pm i\omega(z+d)/c_1\}$, respectively, and $\hat{f}_q^{\pm}(z,t)$ are the right- and left-traveling parts of the quantum noise operator $\tilde{F}_q(z,t)$. The constants γ' and r', respectively, are defined as

$$\gamma'=\gamma+i(\nu_0-\omega) \quad (9.18)$$

and

$$r'=-r\exp(2id\omega/c_1) \quad (9.19)$$

9.3 Laplace-Transformed Equations

In order to solve the coupled equations 9.16a and 9.16b involving space variable z and time variable t, we Laplace-transform the field and the noise operators as follows and concentrate on the spatial region:

$$\begin{aligned}\hat{e}^+(z,t)&\rightarrow\hat{L}^+(z,s)\\ \hat{e}^-(z,t)&\rightarrow\hat{L}^-(z,s)\\ \hat{v}^+(z,t)&\rightarrow\hat{V}^+(z,s)\\ \hat{v}^-(z,t)&\rightarrow\hat{V}^-(z,s)\end{aligned} \quad (9.20)$$

Proceeding as in Chapter 7, after performing the summation over n, with $n_M\rightarrow\infty$, the transformed equations become

$$\begin{aligned}&(s+\gamma')\{\hat{L}^+(z,s)-\hat{V}^+(z,s)\}\\ &=GN\left[\int_{-d}^{z}\exp\{-(z-z_m)s/c_1\}\hat{L}^+(z_m,s)dz_m\right.\\ &\quad-\frac{1}{1-r''(s)}\int_{-d}^{0}\exp\{-(z+z_m+2d)s/c_1\}\hat{L}^-(z_m,s)dz_m\\ &\quad\left.+\frac{r''(s)}{1-r''(s)}\int_{-d}^{0}\exp\{-(z-z_m)s/c_1\}\hat{L}^+(z_m,s)dz_m\right]\end{aligned} \quad (9.21a)$$

and

$$(s+\gamma')\{\hat{L}^-(z,s)-\hat{V}^-(z,s)\}$$
$$= GN\left[\int_z^0 \exp\{(z-z_m)s/c_1\}\hat{L}^-(z_m,s)dz_m - \frac{r''(s)}{1-r''(s)}\int_{-d}^0 \exp\{(z+z_m+2d)s/c_1\}\hat{L}^+(z_m,s)dz_m + \frac{r''(s)}{1-r''(s)}\int_{-d}^0 \exp\{(z-z_m)s/c_1\}\hat{L}^-(z_m,s)dz_m\right] \quad (9.21b)$$

where

$$r''(s) = r'\exp(-2ds/c_1) = -r\exp\{(i\omega - s)2d/c_1\} \quad (9.22)$$

The initial values accompanying the Laplace transforms of the time derivatives, $\hat{e}^{\pm}(z,0)-\hat{v}^{\pm}(z,0)$, vanish, as can be seen by inspection of Equations 9.7 and 9.15 and the definition of $\hat{v}^{\pm}(z,t)$. Differentiation with respect to z yields the following coupled differential equations:

$$\frac{d}{dz}\{\hat{L}^+(z,s)-\hat{V}^+(z,s)\} = \left\{-\frac{s}{c_1}+\frac{GN}{(s+\gamma')}\right\}\{\hat{L}^+(z,s)-\hat{V}^+(z,s)\}+\frac{GN}{(s+\gamma')}\hat{V}^+(z,s) \quad (9.23)$$

$$\frac{d}{dz}\{\hat{L}^-(z,s)-\hat{V}^-(z,s)\} = \left\{\frac{s}{c_1}-\frac{GN}{(s+\gamma')}\right\}\{\hat{L}^-(z,s)-\hat{V}^-(z,s)\}-\frac{GN}{(s+\gamma')}\hat{V}^-(z,s) \quad (9.24)$$

Integrating, we have the following formal solutions:

$$\hat{L}^+(z,s)-\hat{V}^+(z,s)=\int_{-d}^{z}\exp\left[\left\{-\frac{s}{c_1}+\frac{GN}{(s+\gamma')}\right\}(z-z')\right]\frac{GN}{(s+\gamma')}\hat{V}^+(z',s)dz' + \exp\left[\left\{-\frac{s}{c_1}+\frac{GN}{(s+\gamma')}\right\}(z+d)\right]\{\hat{L}^+(-d,s)-\hat{V}^+(-d,s)\} \quad (9.25)$$

$$\hat{L}^-(z,s)-\hat{V}^-(z,s) = -\int_{-d}^{z}\exp\left[\left\{\frac{s}{c_1}-\frac{GN}{(s+\gamma')}\right\}(z-z')\right]\frac{GN}{(s+\gamma')}\hat{V}^-(z',s)dz' + \exp\left[\left\{\frac{s}{c_1}-\frac{GN}{(s+\gamma')}\right\}(z+d)\right]\{\hat{L}^-(-d,s)-\hat{V}^-(-d,s)\} \quad (9.26)$$

These are formally equivalent to Equations 7.28a and 7.28b. The undetermined factors $\hat{L}^{\pm}(-d,s)-\hat{V}^{\pm}(-d,s)$ are obtained by using the results of Appendix D and expressed in terms of $\hat{V}^{\pm}(z,s)$ as

$$\begin{aligned}&\hat{L}^{\pm}(-d,s)-\hat{V}^{\pm}(-d)\\&\quad=\pm\frac{GN}{s+\gamma'}\\&\quad\times\frac{\int_{-d}^{0}\left\{r'\hat{V}^{+}(z',s)\exp\left(\frac{z'-d}{c_1}s-\frac{GNz'}{s+\gamma'}\right)-\hat{V}^{-}(z',s)\exp\left(-\frac{z'+d}{c_1}s+\frac{GNz'}{s+\gamma'}\right)\right\}dz'}{\exp\left(-\frac{GNd}{s+\gamma'}\right)-r'\exp\left(\frac{GNd}{s+\gamma'}-\frac{2ds}{c_1}\right)}\end{aligned}\tag{9.27}$$

Equations 9.25 and 9.26 together with Equation 9.27 are the formal solutions in the Laplace-transformed domain.

9.4
Laplace-Transformed Noise Forces

In order to evaluate Equations 9.25–9.27, we need the Laplace transforms of the noise forces. From Equations 9.17 and 9.20 we have

$$\begin{aligned}\hat{V}^{\pm}(z,s)&=\hat{V}_t^{\pm}(z,s)+\hat{V}_q^{\pm}(z,s)\\\hat{V}_t^{\pm}(z,s)&=\mathscr{L}\left\{\hat{f}_t^{\pm}(z,t)\right\}\\\hat{V}_q^{\pm}(z,s)&=\mathscr{L}\left\{\hat{f}_q^{\pm}(z,t)\right\}\end{aligned}\tag{9.28}$$

where the letter $\mathscr{L}$ signifies the Laplace transform. From Equations 9.2 and 1.62b, the thermal noise term reads

$$\begin{aligned}\tilde{F}_t(z,t)&=i\sum_j\sqrt{\frac{\hbar\omega_j}{2}}U_j(z)\hat{a}_j(0)e^{-i(\omega_j-\omega)t}\\&=i\sum_j\sqrt{\frac{\hbar\omega_j}{2}}\sqrt{\frac{2}{\varepsilon_1L(1-K\sin^2k_{1j}^2d)}}\sin k_{1j}(z+d)\hat{a}_j(0)e^{-i(\omega_j-\omega)t}\end{aligned}\tag{9.29}$$

Thus

$$\hat{f}_t^{\pm}(z,t)=\pm\sum_j\frac{1}{2}\sqrt{\frac{\hbar\omega_j}{\varepsilon_1L(1-K\sin^2k_{1j}^2d)}}\,e^{\pm i(k_{1j}-k)(z+d)}\hat{a}_j(0)e^{-i(\omega_j-\omega)t}\tag{9.30}$$

and

$$\begin{aligned}\hat{V}_t^{\pm}(z,s)&=\pm\sum_j\frac{1}{2}\sqrt{\frac{\hbar\omega_j}{\varepsilon_1L(1-K\sin^2k_{1j}^2d)}}\\&\quad\times\ e^{\pm i(k_{1j}-k)(z+d)}\hat{a}_j(0)\frac{1}{s+i(\omega_j-\omega)}\end{aligned}\tag{9.31}$$

For the quantum noise part, the noise force $\tilde{F}_q(z,t)$ defined by Equations 9.3 and 9.6 reads

$$\tilde{F}_q(z,t) = \sum_m \left[\frac{ip_m \nu_m}{2} \int_0^t \sum_j U_j(z)\, U_j(z_m)\, e^{i(\omega-\omega_j)(t-t')} \times \int_0^{t'} e^{\{i(\omega-\nu_m)-\gamma_m\}(t'-t'')} \tilde{\Gamma}_m(t'') dt'' dt' \right] \tag{9.32}$$

It is easy to see that the right-hand side has the same structure as the second line of Equation 9.7 except that the constant factor $|p_m|^2 \nu_m^2 \sigma_m/(2\hbar\omega)$ is replaced by $ip_m\nu_m/2$ and that the field amplitude $\tilde{E}^{(+)}(z_m, t'')$ is replaced by the Langevin force $\tilde{\Gamma}_m(t'')$. Therefore, the evaluation of the $\tilde{F}_q(z,t)$ goes just the same as for the second line of Equation 9.7. For homogeneous broadening and uniform pumping, referring to Equations 9.8–9.10, we have

$$\begin{aligned} \tilde{F}_q(z,t) = \sum_m h \Bigg\{ &\int_0^{t-|z-z_m|/c_1} \exp\left[\{-i(\nu_0-\omega)-\gamma\}(t-t') + (i\nu_0+\gamma)\frac{|z-z_m|}{c_1} \right] \tilde{\Gamma}_m(t')dt' \\ &- \int_0^{t-(2d+z+z_m)/c_1} \exp\left[\{-i(\nu_0-\omega)-\gamma\}(t-t') + (i\nu_0+\gamma)\frac{2d+z+z_m}{c_1} \right] \tilde{\Gamma}_m(t')dt' \\ &+ \sum_{n=1}^{n_M} (-r)^n (I_{1n}+I_{2n}-I_{3n}-I_{4n}) \Bigg\} \end{aligned} \tag{9.33a}$$

where

$$I_{\rho n} = \int_0^{t-\tau_{\rho n}} \exp\left[\{-i(\nu_0-\omega)-\gamma\}(t-t') + (i\nu_0+\gamma)\tau_{\rho n} \right] \tilde{\Gamma}_m(t')\, dt' \tag{9.33b}$$

and

$$h = \frac{ip_a \nu_0}{2\varepsilon_1 c_1} \tag{9.33c}$$

As right-going waves, we choose those terms containing the factor $\exp(i\nu_0 z/c_1)$, and as left-going wave those containing $\exp(-i\nu_0 z/c_1)$. Thus, using Equation 9.9, we obtain

$$\begin{aligned} &\hat{f}_q^+(z,t) \exp\{i\omega(z+d)/c_1\} \\ &= h \sum_m e^{-\gamma' t} \Bigg[H(z-z_m) \exp\{(i\nu_0+\gamma)(z-z_m)/c_1\} \\ &\times \int_0^{t-(z-z_m)/c_1} \exp(\gamma' t')\tilde{\Gamma}_m(t')dt' - \exp\left\{ (i\nu_0+\gamma)\frac{2d+z+z_m}{c_1} \right\} \\ &\times \int_0^{t-(2d+z+z_m)/c_1} \exp(\gamma' t')\tilde{\Gamma}_m(t')dt' + \sum_{n=1}^{\infty} (-r)^n \Big\{ \exp\{(i\nu_0+\gamma)\tau_{1n}\} \end{aligned}$$

$$\times \int_0^{t-\tau_{1n}} \exp(\gamma' t')\tilde{\Gamma}_m(t')dt' - \exp\{(i\nu_0+\gamma)\tau_{3n}\}$$
$$\times \int_0^{t-\tau_{3n}} \exp(\gamma' t')\tilde{\Gamma}_m(t')dt' \Bigg\}\Bigg] \tag{9.34a}$$

$$\begin{aligned}
&\hat{f}_q^-(z,t)\exp\{-i\omega(z+d)/c_1\} \\
&= h\sum_m e^{-\gamma' t}\Bigg[H(z_m-z)\exp\{(i\nu_0+\gamma)(z_m-z)/c_1\} \\
&\quad\times \int_0^{t-(z_m-z)/c_1} \exp(\gamma' t')\tilde{\Gamma}_m(t')dt' \\
&\quad+\sum_{n=1}^{\infty}(-r)^n\Bigg\{\exp\{(i\nu_0+\gamma)\tau_{2n}\}\int_0^{t-\tau_{2n}} \exp(\gamma' t')\tilde{\Gamma}_m(t')dt' \\
&\quad-\exp\{(i\nu_0+\gamma)\tau_{4n}\}\int_0^{t-\tau_{4n}} \exp(\gamma' t')\tilde{\Gamma}_m(t')dt'\Bigg\}\Bigg]
\end{aligned} \tag{9.34b}$$

where H is the Heaviside unit step function and γ' was defined in Equation 9.18.

Now, combining the rules of the Laplace transform, we have

$$\begin{aligned}
&\mathscr{L}\left\{e^{-\gamma' t}\int_0^{t-(z-z_m)/c_1} \exp(\gamma' t')\tilde{\Gamma}_m(t')dt'\right\} \\
&= \frac{1}{s+\gamma'}\exp\left\{\frac{-(z-z_m)(s+\gamma')}{c_1}\right\}\tilde{\Gamma}_m(s)
\end{aligned} \tag{9.35a}$$

where

$$\tilde{\Gamma}_m(s) = \mathscr{L}\{\tilde{\Gamma}_m(t)\} \tag{9.35b}$$

Thus we have

$$\begin{aligned}
&\hat{V}_q^+(z,s)\exp\{i\omega(z+d)/c_1\} \\
&= h\sum_m \frac{\tilde{\Gamma}_m(s)}{s+\gamma'}\Bigg[H(z-z_m)\exp\left\{\frac{(z-z_m)(i\omega-s)}{c_1}\right\} \\
&\quad-\exp\left\{(i\omega-s)\frac{2d+z+z_m}{c_1}\right\} \\
&\quad+\sum_{n=1}^{\infty}(-r)^n\Bigg\{\exp\left\{\frac{2nd+z-z_m}{c_1}(i\omega-s)\right\} \\
&\quad-\exp\left\{\frac{2nd+2d+z+z_m}{c_1}(i\omega-s)\right\}\Bigg\}\Bigg]
\end{aligned}$$

$$= h\sum_m \frac{\tilde{\Gamma}_m(s)}{s+\gamma'}\left[H(z-z_m)\exp\left\{\frac{(z-z_m)(i\omega-s)}{c_1}\right\}\right.$$
$$-\frac{1}{1-r''(s)}\exp\left\{(i\omega-s)\frac{2d+z+z_m}{c_1}\right\}$$
$$\left.+\frac{r''(s)}{1-r''(s)}\exp\left\{\frac{(z-z_m)(i\omega-s)}{c_1}\right\}\right] \tag{9.36a}$$

and

$$\hat{V}_q^-(z,s)\exp\{-i\omega(z+d)/c_1\}$$
$$= h\sum_m \frac{\tilde{\Gamma}_m(s)}{s+\gamma'}\left[H(z_m-z)\exp\left\{\frac{(z_m-z)(i\omega-s)}{c_1}\right\}\right.$$
$$+\sum_{n=1}^{\infty}(-r)^n\left\{\exp\left\{\frac{2nd-z+z_m}{c_1}(i\omega-s)\right\}\right.$$
$$\left.\left.-\exp\left\{\frac{2nd-2d-z-z_m}{c_1}(i\omega-s)\right\}\right\}\right] \tag{9.36b}$$
$$= h\sum_m \frac{\tilde{\Gamma}_m(s)}{s+\gamma'}\left[H(z_m-z)\exp\left\{\frac{(z_m-z)(i\omega-s)}{c_1}\right\}\right.$$
$$+\frac{r''(s)}{1-r''(s)}\exp\left\{(i\omega-s)\frac{z_m-z}{c_1}\right\}$$
$$\left.-\frac{r''(s)}{1-r''(s)}\exp\left\{\frac{-(2d+z+z_m)(i\omega-s)}{c_1}\right\}\right]$$

where we have used Equation 9.18 and set $n_M \to \infty$ as we did in Equation 9.21a and 9.21b.

Finally we have

$$\hat{V}_q^+(z,s) = h\sum_m \frac{\tilde{\Gamma}_m(s)}{s+\gamma'}\left[H(z-z_m)\exp\left\{-s\frac{z-z_m}{c_1}-i\omega\frac{z_m+d}{c_1}\right\}\right.$$
$$-\frac{1}{1-r''(s)}\exp\left\{-s\frac{2d+z+z_m}{c_1}+i\omega\frac{d+z_m}{c_1}\right\} \tag{9.37a}$$
$$\left.+\frac{r''(s)}{1-r''(s)}\exp\left\{-s\frac{z-z_m}{c_1}-i\omega\frac{z_m+d}{c_1}\right\}\right]$$

$$\hat{V}_q^-(z,s) = h\sum_m \frac{\tilde{\Gamma}_m(s)}{s+\gamma'}\left[H(z_m-z)\exp\left\{-s\frac{z_m-z}{c_1}+i\omega\frac{z_m+d}{c_1}\right\}\right.$$
$$+\frac{r''(s)}{1-r''(s)}\exp\left\{-s\frac{z_m-z}{c_1}+i\omega\frac{z_m+d}{c_1}\right\} \tag{9.37b}$$
$$\left.-\frac{r''(s)}{1-r''(s)}\exp\left\{s\frac{2d+z+z_m}{c_1}-i\omega\frac{z_m+d}{c_1}\right\}\right]$$

where $r''(s)$ was defined in Equation 9.22.

9.5
The Field Inside the Cavity

Up to now, in this chapter, we have been considering the field inside the cavity. Here, we inverse Laplace-transform Equations 9.25 and 9.26 using Equation 9.27. The main pole, yielding a slowly decaying time function, appears in the denominator in Equation 9.27, which was derived in Equations 7.35 and 7.37. Equation 9.27 contains the main pole in the form of Equation 7.37:

$$\exp\left(-\frac{GNd}{s+\gamma'}\right) - r'\exp\left(\frac{GNd}{s+\gamma'} - \frac{2ds}{c_1}\right) = \exp\left(-\frac{GNd}{s_0+\gamma'}\right)\frac{2d/c_1}{s_0+\gamma'}\{\gamma+\gamma_c+i(\nu_0+\omega_c-2\omega)\}(s-s_0) \tag{9.38}$$

where s_0 is the solution of Equation 7.34 and to a good approximation is given by Equation 7.35, which was

$$s_0 = -\frac{\gamma\gamma_c + (\nu_0-\omega)(\omega-\omega_c) - GNc_1 - i\{\gamma(\omega-\omega_c)+\gamma_c(\omega-\nu_0)\}}{\gamma+\gamma_c+i(\nu_0+\omega_c-2\omega)}$$

By setting $s_0 = 0$, we have the threshold oscillation frequency and the threshold atomic inversion in the forms of Equations 7.42 and 7.44a, which were, respectively,

$$\omega_{th} = \frac{\gamma\omega_c + \gamma_c\nu_0}{\gamma+\gamma_c}$$

and

$$\left\{\frac{|p_a|^2\nu_0^2}{2\hbar\omega\varepsilon_1\gamma(1+\delta^2)}\right\}N\sigma_{th} = \gamma_c$$

In order to obtain the Laplace transforms $L^{\pm}(z,s)$ of $\hat{e}^{\pm}(z,t)$ in concrete form, the Laplace-transformed noise forces in Equations 9.31, 9.37a, and 9.37b should be substituted into Equation 9.27 and further into Equations 9.25 and 9.26. For simplicity, we rewrite Equation 9.38 as

$$\exp\left(-\frac{GNd}{s+\gamma'}\right) - r'\exp\left(\frac{GNd}{s+\gamma'} - \frac{2ds}{c_1}\right) = M(s_0)(s-s_0)$$
$$M(s_0) = \exp\left(-\frac{GNd}{s_0+\gamma'}\right)\frac{2d/c_1}{s_0+\gamma'}\{\gamma+\gamma_c-i(\nu_0+\omega_c-2\omega)\} \tag{9.39}$$

Then Equation 9.27 reads

$$\hat{L}^{\pm}(-d,s) - \hat{V}^{\pm}(-d) = \pm\frac{GN}{s+\gamma'}\frac{1}{M(s_0)(s-s_0)} \times \int_{-d}^{0}\left\{r'\hat{V}^{+}(z',s)\exp\left(\frac{z'-d}{c_1}s - \frac{GNz'}{s+\gamma'}\right) - \hat{V}^{-}(z',s)\exp\left(-\frac{z'+d}{c_1}s + \frac{GNz'}{s+\gamma'}\right)\right\}dz' \tag{9.40}$$

Now, in Equations 9.25 and 9.26, the first integral terms, with a pole at $s=-\gamma'$, will give a rapidly decaying field as compared to the second terms, with the pole at $s=s_0$, which will yield slowly decaying fields. Thus these first terms will be ignored. Also, the noise terms $\hat{V}^{\pm}(z,s)$ on the left-hand sides of these equations will simply give lasting noise fields, which are small compared to the amplified terms of the main pole. These noise terms will also be ignored. Thus the main contributions to the Laplace transforms $\hat{L}^{\pm}(z,s)$ read

$$\hat{L}^{+}(z,s) = \exp\left[\left\{-\frac{s}{c_1} + \frac{GN}{(s+\gamma')}\right\}(z+d)\right]\frac{GN}{s+\gamma'}\frac{1}{M(s_0)(s-s_0)} \times \int_{-d}^{0}\left\{r'\hat{V}^{+}(z',s)\exp\left(\frac{z'-d}{c_1}s - \frac{GNz'}{s+\gamma'}\right) - \hat{V}^{-}(z',s)\exp\left(-\frac{z'+d}{c_1}s + \frac{GNz'}{s+\gamma'}\right)\right\}dz' \tag{9.41}$$

and

$$\hat{L}^{-}(z,s) = -\exp\left[\left\{\frac{s}{c_1} - \frac{GN}{(s+\gamma')}\right\}(z+d)\right]\frac{GN}{s+\gamma'}\frac{1}{M(s_0)(s-s_0)} \times \int_{-d}^{0}\left\{r'\hat{V}^{+}(z',s)\exp\left(\frac{z'-d}{c_1}s - \frac{GNz'}{s+\gamma'}\right) - \hat{V}^{-}(z',s)\exp\left(-\frac{z'+d}{c_1}s + \frac{GNz'}{s+\gamma'}\right)\right\}dz' \tag{9.42}$$

As can be seen from Equation 7.38, we have the following relation at the main pole for a cavity resonant mode of angular frequency ω_c:

$$\frac{s_0}{c_1} - \frac{GN}{s_0+\gamma'} = -\frac{i(\omega_c-\omega)+\gamma_c}{c_1} \tag{9.43}$$

Thus, rewriting the first exponential functions, we obtain

$$\hat{L}^{\pm}(z,s) = \pm\exp\left[\pm\left\{\frac{i(\omega_c-\omega)+\gamma_c}{c_1}\right\}(z+d)\right]\frac{GN}{s+\gamma'}\frac{1}{M(s_0)(s-s_0)} \times \int_{-d}^{0}\left\{r'\hat{V}^{+}(z',s)\exp\left(\frac{-d}{c_1}s - \frac{i(\omega_c-\omega)+\gamma_c}{c_1}z'\right) - \hat{V}^{-}(z',s)\exp\left(-\frac{d}{c_1}s + \frac{i(\omega_c-\omega)+\gamma_c}{c_1}z'\right)\right\}dz' \tag{9.44}$$

Here we note that the spatial functions excited by the noise are the right- and left-going waves of the cavity mode in Equation 1.21b. We also note that, as stated in Chapter 7, the right-traveling (left-traveling) noise is multiplied by a decreasing (increasing) function of z. The latter functions are the left-going (right-going) part of the adjoint mode function, which will be discussed in Chapter 14. This can be seen more clearly if we use Equation 9.48 below and rewrite the integrand as

$$\exp\left(-\frac{d}{c_1}\{i(\omega_c-\omega)+\gamma_c+s\}\right)$$
$$\times\left[\hat{V}^{+}(z',s)\exp\left\{\frac{i\omega}{c_1}(z'+d)\right\}\exp\left\{-\frac{i\omega_c+\gamma_c}{c_1}(z'+d)\right\}\right.$$
$$\left.-\hat{V}^{-}(z',s)\exp\left\{-\frac{i\omega}{c_1}(z'+d)\right\}\exp\left\{\frac{i\omega_c+\gamma_c}{c_1}(z'+d)\right\}\right]$$

The remaining task before inverse Laplace-transforming is to evaluate the integral in Equation 9.44. We need to substitute the Laplace transforms of the noise terms defined in Equation 9.28. We consider $\hat{V}_t^{\pm}(z,s)$ and $\hat{V}_q^{\pm}(z,s)$ separately to obtain four integrals.

9.5.1 Thermal Noise

First we use Equation 9.31 for the thermal noise:

$$I_1(s)=\int_{-d}^{0}\left\{r'\hat{V}_t^{+}(z',s)\exp\left(\frac{-d}{c_1}s-\frac{i(\omega_c-\omega)+\gamma_c}{c_1}z'\right)\right\}dz'$$
$$=\sum_j\frac{1}{2}\sqrt{\frac{\hbar\omega_j}{\varepsilon_1L(1-K\sin^2k_1^2d)}}\hat{a}_j(0)\frac{1}{s+i(\omega_j-\omega)}$$
$$\times\int_{-d}^{0}\left\{r'e^{i(\omega_j-\omega)(z'+d)/c_1}\exp\left(\frac{-d}{c_1}s-\frac{i(\omega_c-\omega)+\gamma_c}{c_1}z'\right)\right\}dz' \qquad (9.45)$$
$$=\sum_j\frac{1}{2}\sqrt{\frac{\hbar\omega_j}{\varepsilon_1L(1-K\sin^2k_1^2d)}}\hat{a}_j(0)\frac{r'e^{-ds/c_1+i(\omega_j-\omega)d/c_1}}{s+i(\omega_j-\omega)}$$
$$\times\frac{1-e^{-\{i(\omega_j-\omega_c)-\gamma_c\}d/c_1}}{\{i(\omega_j-\omega_c)-\gamma_c\}/c_1}$$

and

$$I_2(s)=\int_{-d}^{0}\left\{-\hat{V}_t^{-}(z',s)\exp\left(-\frac{d}{c_1}s+\frac{i(\omega_c-\omega)+\gamma_c}{c_1}z'\right)\right\}dz'$$
$$=\sum_j\frac{1}{2}\sqrt{\frac{\hbar\omega_j}{\varepsilon_1L(1-K\sin^2k_1^2d)}}\hat{a}_j(0)\frac{1}{s+i(\omega_j-\omega)}$$

$$\times \int_{-d}^{0} \left\{ e^{-i(\omega_j-\omega)(z'+d)/c_1} \exp\left(-\frac{d}{c_1}s + \frac{i(\omega_c-\omega)+\gamma_c}{c_1}z' \right) \right\} dz'$$
$$= \sum_j \frac{1}{2}\sqrt{\frac{\hbar\omega_j}{\varepsilon_1 L(1-K\sin^2 k_1^2 d)}}\hat{a}_j(0)\frac{e^{-ds/c_1-i(\omega_j-\omega)d/c_1}}{s+i(\omega_j-\omega)} \tag{9.46}$$
$$\times \frac{1-e^{\{i(\omega_j-\omega_c)-\gamma_c\}d/c_1}}{\{-i(\omega_j-\omega_c)+\gamma_c\}/c_1}$$

Adding these two we have

$$I_1(s)+I_2(s) = \sum_j \frac{1}{2}\sqrt{\frac{\hbar\omega_j}{\varepsilon_1 L(1-K\sin^2 k_1^2 d)}}\hat{a}_j(0)\frac{e^{-ds/c_1}}{s+i(\omega_j-\omega)}\frac{1}{\{i(\omega_j-\omega_c)-\gamma_c\}/c_1}$$
$$\times \left[r' e^{i(\omega_j-\omega)d/c_1} \times \left\{ 1-e^{-\{i(\omega_j-\omega_c)-\gamma_c\}d/c_1} \right\} \right.$$
$$\left. - e^{-i(\omega_j-\omega)d/c_1}\left\{ 1-e^{\{i(\omega_j-\omega_c)-\gamma_c\}d/c_1} \right\} \right] \tag{9.47}$$
$$= \sum_j \frac{1}{2}\sqrt{\frac{\hbar\omega_j}{\varepsilon_1 L(1-K\sin^2 k_1^2 d)}}\hat{a}_j(0)\frac{e^{-ds/c_1}}{s+i(\omega_j-\omega)}\frac{1}{\{i(\omega_j-\omega_c)-\gamma_c\}/c_1}$$
$$\times \left[r'\left\{ e^{i(\omega_j-\omega)d/c_1} - e^{\{i(\omega_c-\omega)+\gamma_c\}d/c_1} \right\} - \left\{ e^{-i(\omega_j-\omega)d/c_1} - e^{-\{i(\omega_c-\omega)+\gamma_c\}d/c_1} \right\} \right]$$

The second and fourth terms cancel because, as seen from the comparison of Equations 7.31 and 7.33,

$$r' = -r\exp\left(\frac{2id}{c_1}\omega\right) = \exp\left[-\frac{2d}{c_1}\{i(\omega_c-\omega)+\gamma_c\} \right] \tag{9.48}$$

Thus we have

$$I_1(s)+I_2(s) = \sum_j \frac{1}{2}\sqrt{\frac{\hbar\omega_j}{\varepsilon_1 L(1-K\sin^2 k_1^2 d)}}\hat{a}_j(0)\frac{e^{-ds/c_1}}{s+i(\omega_j-\omega)}\,\frac{1}{\{i(\omega_j-\omega_c)-\gamma_c\}/c_1}$$
$$\times \left[r' e^{i(\omega_j-\omega)d/c_1} - e^{-i(\omega_j-\omega)d/c_1} \right]$$
$$= -\sum_j \frac{1}{2}\sqrt{\frac{\hbar\omega_j}{\varepsilon_1 L(1-K\sin^2 k_1^2 d)}}\frac{\hat{a}_j(0)}{s+i(\omega_j-\omega)}\,\frac{e^{(i\omega-s)d/c_1}}{\{i(\omega_j-\omega_c)-\gamma_c\}/c_1} \tag{9.49}$$
$$\times \left\{ re^{i\omega_j d/c_1} + e^{-i\omega_j d/c_1} \right\}$$
$$= -\sum_j \frac{1}{2}\sqrt{\frac{\hbar\omega_j}{\varepsilon_1 L(1-K\sin^2 k_1^2 d)}}\frac{c_1\hat{a}_j(0)}{s+i(\omega_j-\omega)}\,\frac{e^{(i\omega-s-\gamma_c)d/c_1}}{\{i(\omega_j-\omega_c)-\gamma_c\}}$$
$$\times 2\cos h\{(\gamma_c - i\omega_j)d/c_1\}$$

where we have used the relation $r = \exp\{-(2d/c_1)\gamma_c\}$ in the last line.

9.5.2 Quantum Noise

Next we turn to the quantum noise part. The final result for the integral in Equation 9.44 concerning the quantum noise will be found in Equation 9.57. For $\hat{V}_q^{\pm}(z,s)$, it is more convenient to return the factor $-\{i(\omega_c-\omega)+\gamma_c\}z'/c_1$ in Equation 9.44 to the original form $(s_0/c_1)z'-\{GN/(s_0+\gamma')\}z'$ in Equation 9.43. Now the Laplace transforms of the right- and left-traveling quantum noise forces are found in Equations 9.37a and 9.37b, respectively. For the right-traveling part we have

$$\begin{aligned}
I_3(s) &= \int_{-d}^{0} r'\hat{V}_q^{+}(z',s)\exp\left(\frac{-d}{c_1}s+\frac{s}{c_1}z'-\frac{GN}{s+\gamma'}z'\right)dz' \\
&= \int_{-d}^{0} r'h\sum_m \frac{\tilde{\Gamma}_m(s)}{s+\gamma'}\left[H(z'-z_m)\exp\left\{-s\frac{(z'-z_m)}{c_1}-i\omega\frac{(z_m+d)}{c_1}\right\}\right. \\
&\quad -\frac{1}{1-r''(s)}\exp\left\{-s\frac{2d+z'+z_m}{c_1}+i\omega\frac{d+z_m}{c_1}\right\} \\
&\quad \left.+\frac{r''(s)}{1-r''(s)}\exp\left\{-s\frac{(z'-z_m)}{c_1}-i\omega\frac{(z_m+d)}{c_1}\right\}\right]\exp\left(\frac{-d}{c_1}s+\frac{s}{c_1}z'-\frac{GN}{s+\gamma'}z'\right)dz' \\
&= r'h\sum_m \frac{\tilde{\Gamma}_m(s)}{s+\gamma'}\int_{z_m}^{0}\exp\left\{\frac{s(z_m-d)}{c_1}-i\omega\frac{(z_m+d)}{c_1}\right\}.\exp\left(-\frac{GN}{s+\gamma'}z'\right)dz' \\
&\quad +\int_{-d}^{0} r'h\sum_m\frac{\tilde{\Gamma}_m(s)}{s+\gamma'}\left[-\frac{1}{1-r''(s)}\exp\left\{-s\frac{2d+z_m+d}{c_1}+i\omega\frac{d+z_m}{c_1}\right\}\right. \\
&\quad \left.+\frac{r''(s)}{1-r''(s)}\exp\left\{\frac{s(z_m-d)}{c_1}-i\omega\frac{(z_m+d)}{c_1}\right\}\right]\exp\left(-\frac{GN}{s+\gamma'}z'\right)dz' \\
&= -r'h\sum_m\frac{\tilde{\Gamma}_m(s)}{GN}\exp\left\{\frac{-2ds}{c_1}-(i\omega-s)\frac{z_m+d}{c_1}\right\}\left\{1-e^{-GNz_m/(s+\gamma')}\right\} \\
&\quad +r'h\sum_m\frac{\tilde{\Gamma}_m(s)}{GN}\left[\frac{1}{1-r''(s)}\exp\left\{-s\frac{2d}{c_1}+(i\omega-s)\frac{d+z_m}{c_1}\right\}\left\{1-e^{GNd/(s+\gamma')}\right\}\right. \\
&\quad \left.-\frac{r''(s)}{1-r''(s)}\exp\left\{\frac{-2ds}{c_1}-(i\omega-s)\frac{z_m+d}{c_1}\right\}\left\{1-e^{GNd/(s+\gamma')}\right\}\right]
\end{aligned} \tag{9.50}$$

For the left-traveling part of the quantum noise we have

$$\begin{aligned}
I_4(s) &= -\int_{-d}^{0}\left\{\hat{V}_q^{-}(z',s)\exp\left(-\frac{d}{c_1}s-\frac{z'}{c_1}s+\frac{GNz'}{s+\gamma'}\right)\right\}dz' \\
&= -\int_{-d}^{0} h\sum_m\frac{\tilde{\Gamma}_m(s)}{s+\gamma'}H(z_m-z')\exp\left\{-s\frac{z_m-z'}{c_1}+i\omega\frac{z_m+d}{c_1}\right\} \\
&\quad \times\exp\left(-\frac{d}{c_1}s-\frac{z'}{c_1}s+\frac{GNz'}{s+\gamma'}\right)dz'
\end{aligned}$$

$$
\begin{aligned}
&-\int_{-d}^{0} h\sum_m \frac{\tilde{\Gamma}_m(s)}{s+\gamma'}\left[+\frac{r''(s)}{1-r''(s)}\exp\left\{-s\frac{z_m-z'}{c_1}+i\omega\frac{z_m+d}{c_1}\right\}\right.\\
&\left.-\frac{r''(s)}{1-r''(s)}\exp\left\{s\frac{2d+z'+z_m}{c_1}-i\omega\frac{z_m+d}{c_1}\right\}\right]\exp\left(-\frac{d}{c_1}s-\frac{z'}{c_1}s+\frac{GNz'}{s+\gamma'}\right)dz'\\
&=-\int_{-d}^{z_m} h\sum_m \frac{\tilde{\Gamma}_m(s)}{s+\gamma'}\exp\left\{(i\omega-s)\frac{z_m+d}{c_1}\right\}.\exp\left(\frac{GNz'}{s+\gamma'}\right)dz'\\
&-\int_{-d}^{0} h\sum_m \frac{\tilde{\Gamma}_m(s)}{s+\gamma'}\left[+\frac{r''(s)}{1-r''(s)}\exp\left\{(i\omega-s)\frac{z_m+d}{c_1}\right\}\right.\\
&\left.-\frac{r''(s)}{1-r''(s)}\exp\left\{-(i\omega-s)\frac{z_m+d}{c_1}\right\}\right]\exp\left(\frac{GNz'}{s+\gamma'}\right)dz'\\
&=h\sum_m \frac{\tilde{\Gamma}_m(s)}{GN}\left[-\exp\left\{(i\omega-s)\frac{z_m+d}{c_1}\right\}\left\{e^{GNz_m/(s+\gamma')}-e^{-GNd/(s+\gamma')}\right\}\right.\\
&-\frac{r''(s)}{1-r''(s)}\exp\left\{(i\omega-s)\frac{z_m+d}{c_1}\right\}\left\{1-e^{-GNd/(s+\gamma')}\right\}\\
&\left.+\frac{r''(s)}{1-r''(s)}\exp\left\{-(i\omega-s)\frac{z_m+d}{c_1}\right\}\left\{1-e^{-GNd/(s+\gamma')}\right\}\right]
\end{aligned}
\tag{9.51}
$$

Adding the right- and left-going parts we obtain

$$
\begin{aligned}
I_3(s)+I_4(s)&=-r'h\sum_m \frac{\tilde{\Gamma}_m(s)}{GN}\exp\left\{\frac{-2ds}{c_1}-(i\omega-s)\frac{z_m+d}{c_1}\right\}\left\{1-e^{-GNz_m/(s+\gamma')}\right\}\\
&+r'h\sum_m \frac{\tilde{\Gamma}_m(s)}{GN}\left[\frac{1}{1-r''(s)}\exp\left\{-s\frac{2d}{c_1}+(i\omega-s)\frac{d+z_m}{c_1}\right\}\left\{1-e^{GNd/(s+\gamma')}\right\}\right.\\
&\left.-\frac{r''(s)}{1-r''(s)}\exp\left\{\frac{-2ds}{c_1}-(i\omega-s)\frac{z_m+d}{c_1}\right\}\left\{1-e^{GNd/(s+\gamma')}\right\}\right]\\
&+h\sum_m \frac{\tilde{\Gamma}_m(s)}{GN}\left[-\exp\left\{(i\omega-s)\frac{z_m+d}{c_1}\right\}\left\{e^{GNz_m/(s+\gamma')}-e^{-GNd/(s+\gamma')}\right\}\right.\\
&-\frac{r''(s)}{1-r''(s)}\exp\left\{(i\omega-s)\frac{z_m+d}{c_1}\right\}\left\{1-e^{-GNd/(s+\gamma')}\right\}\\
&\left.+\frac{r''(s)}{1-r''(s)}\exp\left\{-(i\omega-s)\frac{z_m+d}{c_1}\right\}\left\{1-e^{-GNd/(s+\gamma')}\right\}\right]
\end{aligned}
\tag{9.52}
$$

Here we note from Equation 7.37 that, for the main pole $s=s_0$,

$$
r''(s)=r'\exp(-2ds/c_1)=\exp\left(-\frac{2GNd}{s+\gamma'}\right)
\tag{9.53}
$$

or

$$r' = \exp\left(\frac{2ds_0}{c_1} - \frac{2GNd}{s_0 + \gamma'}\right) \tag{9.54}$$

Then we find that the terms without the factors $\exp\{\pm GNz_m/(s+\gamma')\}$ cancel each other: the coefficient of $\exp\{(i\omega - s)(z_m + d)/c_1\}$ in the summation $h\sum\tilde{\Gamma}_m(s)/GN$ is

$$\begin{aligned}
&+\exp\left(\frac{2ds_0}{c_1} - \frac{2GNd}{s_0+\gamma'}\right)\frac{1}{1-r''(s)}\exp\left\{-s\frac{2d}{c_1}\right\}\left\{1 - e^{GNd/(s+\gamma')}\right\}\\
&\quad -\left\{e^{GNz_m/(s+\gamma')} - e^{-GNd/(s+\gamma')}\right\}\\
&\quad -\frac{1}{1-r(s)}\exp\left(-\frac{2GNd}{s+\gamma'}\right)\left\{1 - e^{-GNd/(s+\gamma')}\right\}\\
&= \frac{1}{1-r''(s)}\left\{e^{-2GNd/(s+\gamma')} - e^{-GNd/(s+\gamma')} - e^{-2GNd/(s+\gamma')} + r''(s)e^{-GNd/(s+\gamma')}\right\}\\
&\quad -\left\{e^{GNz_m/(s+\gamma')} - e^{-GNd/(s+\gamma')}\right\}\\
&= -e^{GNz_m/(s+\gamma')}
\end{aligned} \tag{9.55a}$$

where we have set $s = s_0$ and used Equation 9.54 in the first line and Equation 9.53 in the second and the third lines. The coefficient of $\exp\{-(i\omega - s)(z_m + d)/c_1\}$ is, similarly,

$$\begin{aligned}
&-\exp\left(\frac{2ds_0}{c_1} - \frac{2GNd}{s_0+\gamma'}\right)\exp\left\{\frac{-2ds}{c_1}\right\}\left\{1 - e^{-GNz_m/(s+\gamma')}\right\}\\
&\quad +\exp\left(\frac{2ds_0}{c_1} - \frac{2GNd}{s_0+\gamma'}\right)\left[-\frac{r''(s)}{1-r''(s)}\exp\left\{\frac{-2ds}{c_1}\right\}\left\{1 - e^{GNd/(s+\gamma')}\right\}\right]\\
&\quad +\frac{r''(s)}{1-r''(s)}\left\{1 - e^{-GNd/(s+\gamma')}\right\}\\
&= -\exp\left(-\frac{2GNd}{s_0+\gamma'}\right)\left\{1 - e^{-GNz_m/(s+\gamma')}\right\}\\
&\quad -\frac{r''(s)}{1-r''(s)}\left\{e^{-2GNd/(s+\gamma')} - e^{-GNd/(s+\gamma')} - 1 + e^{-GNd/(s+\gamma')}\right\}\\
&= e^{-GN(z_m+2d)/(s+\gamma')}
\end{aligned} \tag{9.55b}$$

Therefore, we have a rather simple result:

$$\begin{aligned}
I_3(s) + I_4(s) = h\sum_m \frac{\tilde{\Gamma}_m(s)}{GN}&\left[-\exp\left\{(i\omega - s)\frac{z_m+d}{c_1}\right\}\exp\left\{\frac{GNz_m}{s+\gamma'}\right\}\right.\\
&\left.+\exp\left\{-(i\omega - s)\frac{z_m+d}{c_1}\right\}\exp\left\{\frac{-GN(z_m+2d)}{s+\gamma'}\right\}\right]
\end{aligned} \tag{9.56}$$

The modification and use of Equation 9.43 reveals that the coupling strength of the quantum noise at the mth atom to the field is proportional to the amplitude of the cavity resonant mode at the location of the atom, that is,

$$
\begin{aligned}
I_3(s)+I_4(s) &= h\sum_m \frac{\tilde{\Gamma}_m(s)}{GN}\exp\left\{-\frac{GNd}{s+\gamma'}\right\}\left[-\exp\left\{(i\omega-s)\frac{z_m+d}{c_1}\right\}\exp\left\{\frac{GN(z_m+d)}{s+\gamma'}\right\}\right.\\
&\quad\left.+\exp\left\{-(i\omega-s)\frac{z_m+d}{c_1}\right\}\exp\left\{\frac{-GN(z_m+d)}{s+\gamma'}\right\}\right]\\
&= -2h\sum_m \frac{\tilde{\Gamma}_m(s)}{GN}\exp\left\{-\frac{GNd}{s+\gamma'}\right\}\sin h\left\{\left(\frac{i\omega-s}{c_1}+\frac{GN}{s+\gamma'}\right)(z_m+d)\right\}\\
&= -2h\sum_m \frac{\tilde{\Gamma}_m(s)}{GN}\exp\left\{-\frac{GNd}{s+\gamma'}\right\}\sin h\left\{\frac{i\omega_c+\gamma_c}{c_1}(z_m+d)\right\}\\
&= -2ih\sum_m \frac{\tilde{\Gamma}_m(s)}{GN}\exp\left\{-\frac{GNd}{s+\gamma'}\right\}\sin\left\{\frac{\omega_c-i\gamma_c}{c_1}(z_m+d)\right\}
\end{aligned}
\tag{9.57}
$$

9.5.3
The Total Field

Now we substitute the above results obtained in Equations 9.49 and 9.57 into the integral in Equation 9.44. We also substitute Equation 9.39 for $M(s_0)$ into Equation 9.44 and use Equation 9.33c for h:

$$
\begin{aligned}
\hat{L}^{\pm}(z,s) &= \pm\exp\left[\pm\left\{\frac{i(\omega_c-\omega)+\gamma_c}{c_1}\right\}(z+d)\right]\frac{GN}{s+\gamma'}\frac{1}{M(s_0)(s-s_0)}\\
&\quad\times\left[-\sum_j\frac{1}{2}\sqrt{\frac{\hbar\omega_j}{\varepsilon_1 L(1-K\sin^2 k_1^2 d)}}\frac{c_1\hat{a}_j(0)}{s+i(\omega_j-\omega)}\right.\\
&\quad\times\frac{e^{(i\omega-s-\gamma_c)d/c_1}}{\{i(\omega_j-\omega_c)-\gamma_c\}}2\cos h\left(\frac{(\gamma_c-i\omega_j)d}{c_1}\right)\\
&\quad\left.-2ih\sum_m\frac{\tilde{\Gamma}_m(s)}{GN}\exp\left(-\frac{GNd}{s_0+\gamma'}\right)\sin\left\{\frac{\omega_c-i\gamma_c}{c_1}(z_m+d)\right\}\right]\\
&= \pm\exp\left[\pm\left\{\frac{i(\omega_c-\omega)+\gamma_c}{c_1}\right\}(z+d)\right]\frac{1}{\{\gamma+\gamma_c-i(\nu_0+\omega_c-2\omega)\}}\frac{1}{(s-s_0)}\\
&\quad\times\left[-\sum_j\frac{1}{2}\sqrt{\frac{\hbar\omega_j}{\varepsilon_1 L(1-K\sin^2 k_1^2 d)}}\frac{\{GNc_1^2/(2d)\}\hat{a}_j(0)}{s+i(\omega_j-\omega)}\frac{1}{\{i(\omega_j-\omega_c)-\gamma_c\}}\right.\\
&\quad\times\exp\left(\frac{GNd}{s_0+\gamma'}\right)e^{(i\omega-s-\gamma_c)d/c_1}2\cos h\{(\gamma_c-i\omega_j)d/c_1\}\\
&\quad\left.+\frac{p_a\nu_0}{2\varepsilon_1 d}\sum_m\tilde{\Gamma}_m(s)\sin\left\{\frac{\omega_c-i\gamma_c}{c_1}(z_m+d)\right\}\right]
\end{aligned}
\tag{9.58}
$$

By use of Equation 9.43 again, we can rewrite the product in the second line from the bottom as

$$\exp\left(\frac{GNd}{s_0+\gamma'}\right)e^{(i\omega-s-\gamma_c)d/c_1} = \exp\left\{\frac{s_0 d}{c_1}+\frac{i(\omega_c-\omega)+\gamma_c}{c_1}d\right\}e^{(i\omega-s-\gamma_c)d/c_1}$$
$$= \exp\left\{\frac{i\omega_c}{c_1}d\right\} \tag{9.59}$$

where we have set $s=s_0$. We finally obtain

$$\begin{aligned}\hat{L}^{\pm}(z,s) &= \pm\exp\left[\pm\left\{\frac{i(\omega_c-\omega)+\gamma_c}{c_1}\right\}(z+d)\right]\frac{1}{\{\gamma+\gamma_c-i(\nu_0+\omega_c-2\omega)\}}\frac{1}{(s-s_0)}\\ &\times\left[-\sum_j\frac{1}{2}\sqrt{\frac{\hbar\omega_j}{\varepsilon_1 L(1-K\sin^2 k_1^2 d)}}\frac{(GNc_1^2/d)\hat{a}_j(0)}{s+i(\omega_j-\omega)}\frac{1}{\{i(\omega_j-\omega_c)-\gamma_c\}}\right.\\ &\times\exp\left\{\frac{i\omega_c}{c_1}d\right\}\cosh\left(\frac{(\gamma_c-i\omega_j)d}{c_1}\right)\\ &\left.+\frac{p_a\nu_0}{2\varepsilon_1 d}\sum_m\tilde{\Gamma}_m(s)\sin\left\{\left(\frac{\omega_c-i\gamma_c}{c_1}\right)(z_m+d)\right\}\right]\end{aligned} \tag{9.60}$$

The inverse Laplace transform of a product of two functions of s is a convolution of inverse transformed functions. Thus

$$\begin{aligned}\mathcal{L}^{-1}\left\{\frac{1}{s-s_0}\frac{1}{s+i(\omega_j-\omega)}\right\} &= \int_0^t e^{-i(\omega_j-\omega)\tau}e^{s_0(t-\tau)}d\tau\\ &= e^{i\omega t}\int_0^t e^{-i\omega_j\tau}e^{(s_0-i\omega)(t-\tau)}d\tau\\ \mathcal{L}^{-1}\left\{\frac{1}{s-s_0}\tilde{\Gamma}_m(s)\right\} &= \int_0^t \tilde{\Gamma}_m(\tau)\,e^{s_0(t-\tau)}d\tau\\ &= e^{i\omega t}\int_0^t \tilde{\Gamma}_m(\tau)e^{-i\omega\tau}e^{(s_0-i\omega)(t-\tau)}d\tau\\ &= e^{i\omega t}\int_0^t \hat{\Gamma}_m(\tau)\,e^{(s_0-i\omega)(t-\tau)}d\tau\end{aligned} \tag{9.61}$$

So, going back to Equation 9.15, we have for the field excited by the thermal noise and quantum noise, respectively,

$$\begin{aligned}&\hat{e}^{\pm}_{thermal}(z,t)\exp\left\{\pm\frac{i\omega}{c_1}(z+d)\right\}\\ &\quad=\pm\exp\left[\pm\left\{\frac{i\omega_c+\gamma_c}{c_1}\right\}(z+d)\right]\frac{1}{\{\gamma+\gamma_c-i(\nu_0+\omega_c-2\omega)\}}\end{aligned}$$

$$\times\left[-\sum_j \frac{1}{2}\sqrt{\frac{\hbar\omega_j}{\varepsilon_1 L(1-K\sin^2 k_1^2 d)}}\frac{(GNc_1^2/d)\hat{a}_j(0)}{i(\omega_j-\omega_c)-\gamma_c}\right. \tag{9.62}$$
$$\left.\times\exp\left\{\frac{i\omega_c}{c_1}d\right\}\cosh\{(\gamma_c-i\omega_j)d/c_1\}e^{i\omega t}\int_0^t e^{-i\omega_j\tau}e^{(s_0-i\omega)(t-\tau)}d\tau\right]$$

and

$$\hat{e}^{\pm}_{quantum}(z,t)\exp\left\{\pm\frac{i\omega}{c_1}(z+d)\right\}$$
$$=\pm\exp\left[\pm\left\{\frac{i\omega_c+\gamma_c}{c_1}\right\}(z+d)\right]\frac{1}{\{\gamma+\gamma_c-i(\nu_0+\omega_c-2\omega)\}} \tag{9.63}$$
$$\times\left[\frac{p_a\nu_0}{2\varepsilon_1 d}\sum_m \sin\left\{\left(\frac{\omega_c-i\gamma_c}{c_1}\right)(z_m+d)\right\}e^{i\omega t}\int_0^t \hat{\Gamma}_m(\tau)\,e^{(s_0-i\omega)(t-\tau)}d\tau\right]$$

Thus going further back to Equation 9.6, we have the expression for the field inside the cavity:

$$\hat{E}^{(+)}(z,t)=\frac{\sin\Omega_c(z+d)/c_1}{\gamma+\gamma_c+i(\nu_0+\omega_c-2\omega)}$$
$$\times\left[\sum_j C_j\hat{a}_j(0)\int_0^t e^{-i\omega_j\tau}\exp[(s_0-i\omega)(t-\tau)]d\tau+\frac{i\nu_0 p_a}{\varepsilon_1 d}\right. \tag{9.64}$$
$$\left.\times\sum_m \sin\{\Omega_c(z_m+d)/c_1\}\int_0^t \hat{\Gamma}_m(\tau)\exp[(s_0-i\omega)(t-\tau)]d\tau\right],$$
$$-d\le z\le 0$$

with

$$C_j=-i\left(\frac{\hbar\omega}{\varepsilon_1 L}\frac{1}{1-K\sin^2 k_{1j}d}\right)^{1/2} \tag{9.65}$$
$$\times\frac{GN(c_1)^2}{d}\frac{\exp(i\omega_c d/c_1)\cosh\left[(\gamma_c-i\omega_j)d/c_1\right]}{i(\omega_j-\omega_c)-\gamma_c}$$

We see that the cavity resonant mode is excited by thermal noise coming from the initial fluctuation of every "universal" mode and by the quantum noise coming from damping of every atomic polarization. The strength of the quantum noise at the mth atom is proportional to the amplitude of the pertinent cavity resonant mode at the location of the atom, as noted earlier. Before examining the correlation function of the field inside the cavity, we look for the expression for the field outside the cavity.

9.6
The Field Outside the Cavity

Now that we know the expression for the field inside the cavity, we can use Equation 9.1 to derive the expression for the field outside the cavity, just as we did in Chapter 7 for the semiclassical analysis. In this case, the "universal" mode function $U_j(z)$ is that for outside the cavity, as given by the last line of Equation 1.62b. The function $U_j(z_m)$ is, of course, that for inside the cavity. If we use Equation 7.45, the summation over j in Equation 9.1 reads

$$\begin{aligned}
&\sum_j \left\{ U_j(z)\, U_j(z_m) e^{-i\omega_j(t-t')} \right\} \\
&= \sum_j \frac{2}{\varepsilon_1 L} \frac{1}{1 - K \sin^2 k_{1j}d} \left(\frac{k_{1j}}{k_{0j}} \cos k_{1j}d \sin k_{0j}z + \sin k_{1j}d \cos k_{0j}z \right) \\
&\quad \times \sin k_{1j}(z_m + d) e^{-i\omega_j(t-t')} \\
&= \frac{1}{\varepsilon_1 c_1} \frac{2c_0}{(c_1 + c_0)} \sum_{n=0}^{\infty} (-r)^n \{ \delta(t - t' + \tau_{5n}) \\
&\quad + \delta(t - t' - \tau_{5n}) - \delta(t - t' + \tau_{6n}) - \delta(t - t' - \tau_{6n}) \}
\end{aligned} \tag{9.66}$$

where

$$\tau_{5n} = \frac{z}{c_0} + \frac{2nd - z_m}{c_1}, \qquad \tau_{6n} = \frac{z}{c_0} + \frac{2nd + 2d + z_m}{c_1} \tag{9.67}$$

Substituting Equation 9.66 into Equation 9.1 we have

$$\begin{aligned}
\hat{E}^{(+)}(z,t) &= \hat{F}_t(z,t) + \hat{F}_q(z,t) + \sum_m \frac{|p_a|^2 \nu_0^2 \sigma}{2\hbar\omega} \frac{1}{\varepsilon_1 c_1} \frac{2c_0}{(c_1 + c_0)} \\
&\quad \times \sum_{n=0}^{\infty} (-r)^n \int_0^t \{ \delta(t - t' + \tau_{5n}) + \delta(t - t' - \tau_{5n}) \\
&\quad - \delta(t - t' + \tau_{6n}) - \delta(t - t' - \tau_{6n}) \} \int_0^{t'} e^{-(i\nu_0 + \gamma)(t' - t'')} \hat{E}^{(+)}(z_m, t'') dt'' dt' \\
&= \hat{F}_t(z,t) + \hat{F}_q(z,t) + \sum_m \frac{|p_a|^2 \nu_0^2 \sigma}{2\hbar\omega \varepsilon_1 c_1} \frac{2c_0}{(c_1 + c_0)} \\
&\quad \times \sum_{n=0}^{\infty} (-r)^n \left\{ \int_0^{t - \tau_{5n}} e^{-(i\nu_0 + \gamma)(t - \tau_{5n} - t'')} \hat{E}^{(+)}(z_m, t'') dt'' \right. \\
&\quad \left. - \int_0^{t - \tau_{6n}} e^{-(i\nu_0 + \gamma)(t - \tau_{6n} - t'')} \hat{E}^{(+)}(z_m, t'') dt'' \right\}
\end{aligned} \tag{9.68}$$

We substitute Equation 9.64 for $\hat{E}^{(+)}(z_m, t'')$ and obtain the field outside the cavity. As was the case for the field inside the cavity, the first and second terms in Equation 9.68 represent the lasting small noise terms and will be ignored. (In the next chapter we will show that the net result of including these terms is the appearance of an extra thermal noise term in the expression for the output field.) Then, from the calculations in Chapter 7, we find that the substitution of the field inside the cavity in Equation 7.41 results in the outside field in the form of Equation 7.56. The net effect of the conversion was, except for the neglect of a rapidly decaying term and of a small term (in Equation 7.52), to change from

$$\begin{aligned}\hat{E}^{(+)}(z,t) &= C[\exp\{(\gamma_c + i\omega_c)(z+d)/c_1\} \\ &\quad - \exp\{-(\gamma_c + i\omega_c)(z+d)/c_1\}]\exp\{(s_0 - i\omega)t\} \\ &= 2iC\sin\Omega_c(z+d)\exp\{(s_0 - i\omega)t\}, \qquad -d \le z \le 0\end{aligned} \tag{9.69}$$

to

$$\begin{aligned}\hat{E}^{(+)}(z,t) &= CT\exp\left\{(\gamma_c + i\omega_c)\frac{d}{c_1}\right\}\exp\left\{(s_0 - i\omega)\left(t - \frac{z}{c_0}\right)\right\} \\ &= CT\exp\left\{i\Omega_c\frac{d}{c_1}\right\}\exp\left\{(s_0 - i\omega)\left(t - \frac{z}{c_0}\right)\right\}, \qquad 0 \le z\end{aligned} \tag{9.70}$$

The effect is (i) to get the field value of the right-traveling wave at $z=0$, (ii) to multiply by the transmission coefficient T, and (iii) to add the retarded time z/c_0. Now, if we look back at Equation 9.64 for the field inside the cavity, in spite of the seeming complexity of the expression, the equation is a linear superposition of the form in Equation 9.69: the z dependence $\sin\Omega_c(z+d)/c_1$ and the time dependence $\exp\{(s_0 - i\omega)t\}$ are common to all the terms in the summation over j and over m for a fixed value of the parameter τ. In other words, Equation 9.64 is a superposition of terms of the form in Equation 9.69 summed over j and m and integrated over τ. Thus, applying the three conversion rules stated above to every member of the summation and ingredients of the integration, we find

$$\begin{aligned}\hat{E}^{(+)}(z,t) &= \frac{(1/2i)\exp(i\Omega_c d/c_1)T}{\gamma + \gamma_c + i(\nu_0 + \omega_c - 2\omega)} \\ &\times\left[\sum_j C_j a_j(0)\int_0^t e^{-i\omega_j\tau}\exp[(s_0 - i\omega)(t - z/c_0 - \tau)]d\tau + \frac{i\nu_0 p_a}{\varepsilon_1 d}\right. \\ &\left.\times\sum_m \sin\{\Omega_c(z_m + d)/c_1\}\int_0^t \hat{\Gamma}_m(\tau)\exp[(s_0 - i\omega)(t - z/c_0 - \tau)]d\tau\right], \\ &0 \le z\end{aligned} \tag{9.71}$$

This expression can, of course, be obtained by substituting Equation 9.64 into Equation 9.68 and faithfully performing the integration and related evaluations. But this is simply to repeat the calculations of Equations 7.47–7.56 on every member of the above-mentioned summations and the ingredients of the integration, including the approximations stated below Equation 7.52. A more precise treatment of the thermal noise outside the cavity will be given in the next chapter.

9.7
The Field Correlation Function

Now we are in a position to derive the correlation functions of the field inside and outside the cavity. We are assuming that the "universal" modes are initially mutually independent and the Langevin forces for different atoms are also mutually independent. Thus we will use the correlations (see Equations 9.4a and 9.5c)

$$\left\langle a_i^\dagger(0)a_j(0)\right\rangle = \langle n_j\rangle\delta_{ij} = (e^{\beta\hbar\omega_j}-1)^{-1}\delta_{ij} \tag{9.72}$$

$$\langle \hat{\Gamma}_m^\dagger(t)\hat{\Gamma}_{m'}(t')\rangle = \langle \hat{\Gamma}_{21m}(t)\hat{\Gamma}_{12m'}(t')\rangle = \gamma(1+\sigma)\delta_{mm'}\delta(t-t') \tag{9.73}$$

The reader is referred to Equations 2.42 and 4.50 for these equations. Also, the thermal and quantum noise forces are assumed to be mutually independent. Therefore, we evaluate the correlation function separately for the thermal part and for the quantum part.

First take the thermal part of the field $\hat{E}_t^{(+)}(z,t)$ inside the cavity described by the first term in Equation 9.64. Using Equation 9.72 we calculate

$$\begin{aligned}
&\left\langle \hat{E}_t^{(-)}(z',t')\hat{E}_t^{(+)}(z,t)\right\rangle \\
&= \left|\frac{1}{\gamma+\gamma_c+i(\nu_0+\omega_c-2\omega)}\right|^2 \sin\{\Omega_c^*(z'+d)/c_1\}\sin\{\Omega_c(z+d)/c_1\} \\
&\quad\times\left\langle\sum_j C_j^*\hat{a}_j^\dagger(0)\int_0^{t'} e^{i\omega_j\tau'}\exp\left[(s_0^*+i\omega)(t'-\tau')\right]d\tau'\right. \\
&\quad\left.\times\sum_k C_k\hat{a}_k(0)\int_0^t e^{-i\omega_k\tau}\exp[(s_0-i\omega)(t-\tau)]d\tau\right\rangle \\
&= \left|\frac{1}{\gamma+\gamma_c+i(\nu_0+\omega_c-2\omega)}\right|^2 \sin\{\Omega_c^*(z'+d)/c_1\}\sin\{\Omega_c(z+d)/c_1\} \\
&\quad\times\sum_j\left|C_j\right|^2\langle n_j\rangle\int_0^{t'} e^{i\omega_j\tau'}\exp\left[(s_0^*+i\omega)(t'-\tau')\right]d\tau' \\
&\quad\times\int_0^t e^{-i\omega_j\tau}\exp[(s_0-i\omega)(t-\tau)]d\tau
\end{aligned} \tag{9.74}$$

To go further we need the evaluation of the summation over the "universal" modes j:

$$\begin{aligned}
\sum_j|C_j|^2\langle n_j\rangle e^{i\omega_j(\tau'-\tau)} &= \frac{\hbar\omega}{\varepsilon_1 L}\left(\frac{GN(c_1)^2}{d}\right)^2 \\
&\quad\times\sum_j\langle n_j\rangle e^{i\omega_j(\tau'-\tau)}\frac{1}{(\omega_j-\omega_c)^2+\gamma_c^2}\left[\frac{\left|\cosh\{(\gamma_c-i\omega_j)d/c_1\}\right|^2}{1-K\sin^2 k_{1j}d}\right]
\end{aligned} \tag{9.75}$$

We will shortly show that the quantity in the square bracket yields a constant factor independent of the index j. Also, since $\langle n_j \rangle$ is a very slowly varying function of the frequency ω_j, while important contributions come from the region around $-\gamma_c < \omega_j - \omega_c < \gamma_c$, it can be taken outside the summation sign. So, noting that the density of modes is given by Equation 1.64, we calculate

$$\begin{aligned}
&\sum_j \exp\{i\omega_j(\tau' - \tau)\} \frac{1}{(\omega_j - \omega_c)^2 + \gamma_c^2} \\
&= \int_0^\infty \frac{L}{c_0 \pi} \exp\{i\omega_j(\tau' - \tau)\} \frac{d\omega_j}{(\omega_j - \omega_c)^2 + \gamma_c^2} \\
&= \int_{-\omega_c}^\infty \frac{L}{c_0 \pi} \exp\{i(x + \omega_c)(\tau' - \tau)\} \frac{dx}{x^2 + \gamma_c^2}
\end{aligned} \tag{9.76}$$

where we have set $x = \omega_j - \omega_c$. Here we make the following approximation. That is, since the important contribution to the integral comes from the region around $-\gamma_c < x < \gamma_c$, the lower limit of integration can safely be replaced by $-\infty$ as long as the cavity half-width γ_c is much smaller than the resonance frequency ω_c. Thus we have

$$\begin{aligned}
&\sum_j \exp\{i\omega_j(\tau' - \tau)\} \frac{1}{(\omega_j - \omega_c)^2 + \gamma_c^2} \\
&= \int_{-\infty}^\infty \frac{L}{c_0 \pi} \exp\{i(x + \omega_c)(\tau' - \tau)\} \frac{dx}{x^2 + \gamma_c^2} \\
&= \frac{L}{c_0 \pi} \exp\{i\omega_c(\tau' - \tau)\} \frac{\pi}{\gamma_c} \exp\{-\gamma_c|\tau' - \tau|\}
\end{aligned} \tag{9.77}$$

The last line is obtained by contour integrations in the upper (lower) half region of the complex x-plane for $\tau' - \tau > 0$ $(\tau' - \tau < 0)$ with the pole at $x = i\gamma_c(-i\gamma_c)$. Since we are concentrating on the slowly varying field amplitude corresponding to a narrow laser linewidth, we make the further assumption that the difference $\tau' - \tau$ of interest is much larger than the cavity decay time γ_c^{-1}. This assumption is valid if the laser linewidth is much smaller than the cavity half-width γ_c. Then the exponential function in Equation 9.77 can be taken to be like a delta function. The area below the exponential function is $2/\gamma_c$. Therefore the exponential function is regarded to be equal to $(2/\gamma_c)\delta(\tau' - \tau)$. Thus we have

$$\sum_j \exp\{i\omega_j(\tau' - \tau)\} \frac{1}{(\omega_j - \omega_c)^2 + \gamma_c^2} = \frac{2L}{c_0 \gamma_c^2} \delta(\tau' - \tau) \tag{9.78}$$

Next, using Equation 1.18 for γ_c and Equations 1.17 and 1.43 for K, we evaluate

$$\begin{aligned}
&\left|\cos h\{(\gamma_c - i\omega_j)d/c_1\}\right|^2 \\
&= \frac{1}{4}\left|e^{(\gamma_c - i\omega_j)d/c_1} + e^{-(\gamma_c - i\omega_j)d/c_1}\right|^2 \\
&= \frac{1}{4}\left\{\left(\sqrt{r} + \frac{1}{\sqrt{r}}\right)^2 \cos^2 k_{1j}d + \left(\sqrt{r} - \frac{1}{\sqrt{r}}\right)^2 \sin^2 k_{1j}d\right\}
\end{aligned}$$

$$
\begin{aligned}
&= \frac{(1+r)^2}{4r}\left[1-\left\{1-\left(\frac{1-r}{1+r}\right)^2\right\}\sin^2 k_{1j}d\right] \\
&= \frac{(1+r)^2}{4r}\left(1-K\sin^2 k_{1j}d\right)
\end{aligned}
\tag{9.79}
$$

Thus

$$
\frac{\left|\cos h\{(\gamma_c - i\omega_j)d/c_1\}\right|^2}{1-K\sin^2 k_{1j}d} = \frac{(1+r)^2}{4r}
\tag{9.80}
$$

Summarizing Equations 9.75–9.80 we have

$$
\begin{aligned}
\sum_j |C_j|^2 \langle n_j\rangle e^{i\omega_j(\tau'-\tau)} &= \frac{\hbar\omega}{\varepsilon_1 L}\left(\frac{GN(c_1)^2}{d}\right)^2 \\
&\times \sum_j \langle n_j\rangle e^{i\omega_j(\tau'-\tau)} \frac{1}{(\omega_j-\omega_c)^2+\gamma_c^2}\left[\frac{\left|\cos h\{(\gamma_c - i\omega_j)d/c_1\}\right|^2}{1-K\sin^2 k_{1j}d}\right] \\
&= D\delta(\tau'-\tau)
\end{aligned}
\tag{9.81a}
$$

with

$$
D = \frac{2\hbar\omega}{\varepsilon_1 c_0 \gamma_c^2}\left(\frac{GN(c_1)^2}{d}\right)^2 \frac{(1+r)^2}{4r}\langle n_\omega\rangle
\tag{9.81b}
$$

where $\langle n_\omega\rangle$ is the thermal photon number per "universal" mode at the central oscillation frequency ω, which may be close to the threshold oscillation frequency ω_{ih} in Equation 7.42. The correlation function in Equation 9.74 then becomes

$$
\begin{aligned}
\left\langle \hat{E}_t^{(-)}(z',t')\hat{E}_t^{(+)}(z,t)\right\rangle &= \left|\frac{1}{\gamma+\gamma_c+i(\nu_0+\omega_c-2\omega)}\right|^2 \\
&\times \sin\{\Omega_c^*(z'+d)/c_1\}\sin\{\Omega_c(z+d)/c_1\} D\int_0^{t'} d\tau' \int_0^t d\tau\, \delta(\tau'-\tau) \\
&\times \exp\left[(s_0^*+i\omega)(t'-\tau')\right]\exp[(s_0-i\omega)(t-\tau)]
\end{aligned}
\tag{9.82}
$$

The double integral is, as in Equation 4.51,

$$
\begin{aligned}
&\int_0^{t'} d\tau' \int_0^t d\tau\, \delta(\tau'-\tau)\exp\left[(s_0^*+i\omega)(t'-\tau')\right]\exp[(s_0-i\omega)(t-\tau)] \\
&= \begin{cases} \int_0^t \exp\left[(s_0^*+i\omega)(t'-\tau)+(s_0-i\omega)(t-\tau)\right]d\tau, & t<t' \\ \int_0^{t'} \exp\left[(s_0^*+i\omega)(t'-\tau)+(s_0-i\omega)(t-\tau)\right]d\tau, & t>t' \end{cases} \\
&= \begin{cases} \dfrac{e^{(s_0^*+i\omega)t'+(s_0-i\omega)t} - e^{(s_0^*+i\omega)(t'-t)}}{s_0+s_0^*}, & t<t' \\ \dfrac{e^{(s_0^*+i\omega)t'+(s_0-i\omega)t} - e^{(s_0-i\omega)(t-t')}}{s_0+s_0^*}, & t>t' \end{cases}
\end{aligned}
\tag{9.83}
$$

The first terms decay relatively fast and are unimportant after a long time. The second terms decay according to the time difference. Thus the first terms will be ignored. Then we have, noting that Re $s_0<0$,

$$\left\langle \hat{E}_t^{(-)}(z',\ t+\tau)\hat{E}_t^{(+)}(z,t)\right\rangle = \frac{D}{(\gamma+\gamma_c)^2+(\nu_0+\omega_c-2\omega)^2}$$
$$\times \sin\{\Omega_c^*(z'+d)/c_1\}\sin\{\Omega_c(z+d)/c_1\}$$
$$\times\begin{cases}\dfrac{e^{(s_0^*+i\omega)\tau}}{2|\text{Re}\ s_0|}, & \tau>0\\ \dfrac{e^{-(s_0-i\omega)\tau}}{2|\text{Re}\ s_0|}, & \tau<0\end{cases} \tag{9.84}$$

Using Equation 7.35 for s_0, we have

$$2|\text{Re}\,s_0| = \frac{2}{(\gamma+\gamma_c)^2+\delta^2(\gamma-\gamma_c)^2}\left[(\gamma+\gamma_c)\{\gamma\gamma_c(1+\delta^2)-GNc_1\}\right.$$
$$\left.+(\nu_0+\omega_c-2\omega)\{\gamma(\omega-\omega_c)+\gamma_c(\omega-\nu_0)\}\right] \tag{9.85}$$
$$\simeq \frac{2(\gamma+\gamma_c)\{\gamma\gamma_c(1+\delta^2)-GNc_1\}}{(\gamma+\gamma_c)^2+\delta^2(\gamma-\gamma_c)^2}$$

In the third line, we have, for simplicity, neglected the quantity in the second line, assuming that the oscillation frequency is close to the threshold frequency given by Equation 7.42. Further, using Equation 9.81b for D and Equation 9.14 for G, and referring to Equations 7.44 and 6.35 for σ_{th} and β_c, respectively, we have the constant part in Equation 9.84 as

$$\frac{D}{(\gamma+\gamma_c)^2+(\nu_0+\omega_c-2\omega)^2}\frac{1}{2|\text{Re}\,s_0|}$$
$$=\frac{2\hbar\omega}{\varepsilon_1 c_0\gamma_c^2}\left(\frac{GN(c_1)^2}{d}\right)^2\frac{(1+r)^2}{4r}\langle n_\omega\rangle\frac{1}{2(\gamma+\gamma_c)[\gamma\gamma_c(1+\delta^2)-GNc_1]} \tag{9.86}$$
$$=\frac{\hbar\omega\gamma\beta_c/\gamma_c}{\varepsilon_1 d(\gamma+\gamma_c)(1-\sigma/\sigma_{th})}\left(\frac{\sigma^2}{\sigma_{th}\sigma_{th\,0}}\right)\langle n_\omega\rangle$$

where $\sigma_{th\,0}$ is the threshold atomic inversion at zero detuning. In the last line, we have used the relation $G_{th}Nc_1=\gamma\gamma_c\{1+\delta^2\}$ from Equation 7.43 and the fact that $G/G_{th}=\sigma/\sigma_{th}$. The quantity β_c was defined in Equation 6.35 and $c_1/c_0=(1-r)/(1+r)$ by Equation 1.17. Note that the quantity in Equation 9.86 diverges as the atomic inversion σ approaches the threshold value. Thus we have

$$\left\langle \hat{E}_t^{(-)}(z',\ t+\tau)\hat{E}_t^{(+)}(z,t)\right\rangle$$
$$=\sin\{\Omega_c^*(z'+d)/c_1\}\sin\{\Omega_c(z+d)/c_1\}$$
$$\times\frac{\hbar\omega\gamma\beta_c/\gamma_c}{\varepsilon_1 d(\gamma+\gamma_c)(1-\sigma/\sigma_{th})}\left(\frac{\sigma^2}{\sigma_{th}\sigma_{th\,0}}\right)\langle n_\omega\rangle\times\begin{cases}e^{(s_0^*+i\omega)\tau}, & \tau>0\\ e^{-(s_0-i\omega)\tau}, & \tau<0\end{cases} \tag{9.87}$$

Next we consider the correlation function of the quantum noise part $\hat{E}_q^{(+)}(z,t)$ for the field inside the cavity described by the second term in Equation 9.64. Using Equation 9.73 we have

$$
\begin{aligned}
&\left\langle \hat{E}_q^{(-)}(z',t')\hat{E}_q^{(+)}(z,t)\right\rangle \\
&= \frac{\sin\{\Omega_c^*(z'+d)/c_1\}\sin\{\Omega_c(z+d)/c_1\}}{|\gamma+\gamma_c+i(\nu_0+\omega_c-2\omega)|^2}\left(\frac{\nu_0|p_a|}{\varepsilon_1 d}\right)^2 \\
&\quad\times\left\langle\sum_{m'}\sin\{\Omega_c^*(z_{m'}+d)/c_1\}\int_0^{t'}\hat{\Gamma}_{m'}^{\dagger}(\tau')\exp\left[(s_0^*+i\omega)(t'-\tau')\right]d\tau'\right. \\
&\quad\left.\times\sum_m\sin\{\Omega_c(z_m+d)/c_1\}\int_0^t\hat{\Gamma}_m(\tau)\exp[(s_0-i\omega)(t-\tau)]d\tau\right\rangle \qquad (9.88)\\
&= \sin\{\Omega_c^*(z'+d)/c_1\}\sin\{\Omega_c(z+d)/c_1\}\frac{(\nu_0|p_a|/\varepsilon_1 d)^2}{|\gamma+\gamma_c+i(\nu_0+\omega_c-2\omega)|^2} \\
&\quad\times\sum_m|\sin\{\Omega_c(z_m+d)/c_1\}|^2\gamma(1+\sigma)\int_0^{t'}d\tau'\int_0^t d\tau\,\delta(\tau'-\tau) \\
&\quad\times\exp\left[(s_0^*+i\omega)(t'-\tau')\right]\exp[(s_0-i\omega)(t-\tau)]
\end{aligned}
$$

The double integral is the same as the one evaluated in Equation 9.83. Thus we have

$$
\begin{aligned}
&\left\langle \hat{E}_q^{(-)}(z',t')\hat{E}_q^{(+)}(z,t)\right\rangle \\
&= \sin\{\Omega_c^*(z'+d)/c_1\}\sin\{\Omega_c(z+d)/c_1\}\frac{(\nu_0|p_a|/\varepsilon_1 d)^2}{|\gamma+\gamma_c+i(\nu_0+\omega_c-2\omega)|^2} \\
&\quad\times\sum_m|\sin\{\Omega_c(z_m+d)/c_1\}|^2\gamma(1+\sigma) \qquad (9.89)\\
&\quad\times\begin{cases}\dfrac{e^{(s_0^*+i\omega)t'+(s_0-i\omega)t}-e^{(s_0^*+i\omega)(t'-t)}}{s_0+s_0^*}, & t<t' \\[2ex] \dfrac{e^{(s_0^*+i\omega)t'+(s_0-i\omega)t}-e^{(s_0-i\omega)(t-t')}}{s_0+s_0^*}, & t>t'\end{cases}
\end{aligned}
$$

By the same reasoning as that used below Equation 9.83, we go to

$$
\begin{aligned}
&\left\langle \hat{E}_q^{(-)}(z',t+\tau)\hat{E}_q^{(+)}(z,t)\right\rangle \\
&= \sin\{\Omega_c^*(z'+d)/c_1\}\sin\{\Omega_c(z+d)/c_1\} \\
&\quad\times\frac{(\nu_0|p_a|/\varepsilon_1 d)^2}{|\gamma+\gamma_c+i(\nu_0+\omega_c-2\omega)|^2}\frac{\gamma(1+\sigma)}{2|\mathrm{Re}\,s_0|} \qquad (9.90)\\
&\quad\times\sum_m|\sin\{\Omega_c(z_m+d)/c_1\}|^2\begin{cases}e^{(s_0^*+i\omega)\tau}, & \tau>0 \\ e^{-(s_0-i\omega)\tau}, & \tau<0\end{cases}
\end{aligned}
$$

Using Equation 9.85 again, and also using Equation 7.44, we have the constant part in Equation 9.90 as

$$\begin{aligned}\frac{(\nu_0|p_a|/\varepsilon_1 d)^2}{|\gamma+\gamma_c+i(\nu_0+\omega_c-2\omega)|^2}\frac{\gamma(1+\sigma)}{2|\mathrm{Re}\, s_0|} &= \frac{(\nu_0|p_a|/\varepsilon_1 d)^2\gamma(1+\sigma)}{2(\gamma+\gamma_c)[\gamma\gamma_c(1+\delta^2)-GNc_1]} \\ &= \frac{\hbar\omega\gamma(1+\sigma)}{\varepsilon_1 d^2(\gamma+\gamma_c)\{1-(\sigma/\sigma_{th})\}N\sigma_{th}}\end{aligned} \quad (9.91)$$

The summation over m in Equation 9.90 is evaluated by using Equation 1.18 and going to the integration over z_m:

$$\begin{aligned}\sum_m |\sin\{\Omega_c(z_m+d)/c_1\}|^2 &= \int_{-d}^{0} N\,dz_m|\sin\{\Omega_c(z_m+d)/c_1\}|^2 \\ &= \frac{Nd}{2}\frac{(1-r^2)/(2r)}{\ln(1/r)} \\ &= \frac{Nd}{2}\frac{\beta_c}{\gamma_c}\end{aligned} \quad (9.92)$$

Thus writing the population in the upper level as

$$N(1+\sigma)/2 = N_2 \quad (9.93)$$

we have

$$\begin{aligned}\left\langle \hat{E}_q^{(-)}(z',t+\tau)\hat{E}_q^{(+)}(z,t)\right\rangle &= \sin\{\Omega_c^*(z'+d)/c_1\} \\ &\times \sin\{\Omega_c(z+d)/c_1\}\frac{\beta_c}{\gamma_c}\frac{\hbar\omega\gamma N_2}{\varepsilon_1 d(\gamma+\gamma_c)\{1-(\sigma/\sigma_{th})\}N\sigma_{th}} \\ &\times \begin{cases} e^{(s_0^*+i\omega)\tau}, & \tau>0 \\ e^{-(s_0-i\omega)\tau}, & \tau<0 \end{cases}\end{aligned} \quad (9.94)$$

Adding Equations 9.87 and 9.94 we have the total correlation function for inside the cavity:

$$\begin{aligned}\left\langle \hat{E}^{(-)}(z',\, t+\tau)\hat{E}^{(+)}(z,t)\right\rangle &= \sin\{\Omega_c^*(z'+d)/c_1\}\sin\{\Omega_c(z+d)/c_1\} \\ &\times \frac{\hbar\omega\gamma\beta_c/\gamma_c}{\varepsilon_1 d(\gamma+\gamma_c)(1-\sigma/\sigma_{th})}\left\{\left(\frac{\sigma^2}{\sigma_{th}\sigma_{th\,0}}\right)\langle n_\omega\rangle + \frac{N_2}{N\sigma_{th}}\right\} \\ &\times \begin{cases} e^{(s_0^*+i\omega)\tau}, & \tau>0 \\ e^{-(s_0-i\omega)\tau}, & \tau<0 \end{cases} \\ &-d \le z \le 0\end{aligned} \quad (9.95)$$

Next we turn to the field correlation function for outside the cavity. Using Equation 9.71 we have

$$
\begin{aligned}
&\left\langle \hat{E}^{(-)}(z',t')\hat{E}^{(+)}(z,t)\right\rangle \\
&= \left|\frac{(1/2i)\exp(i\Omega_c d/c_1)T}{\gamma+\gamma_c+i(\nu_0+\omega_c-2\omega)}\right|^2 \left[\sum_{i,j} C_i^* C_j \left\langle \hat{a}_i^\dagger(0)\hat{a}_j(0)\right\rangle\right. \\
&\times \int_0^{t'} d\tau' \int_0^t d\tau\, e^{i(\omega_i\tau'-\omega_j\tau)} \\
&\times \exp\left[\left(s_0^*+i\omega\right)(t'-z'/c_0-\tau')+(s_0-i\omega)(t-z/c_0-\tau)\right] \\
&+ \left|\frac{\nu_0 p_a}{\varepsilon_1 d}\right|^2 \sum_{m',m} \sin\{\Omega_c^*(z_{m'}+d)/c_1\}\sin\{\Omega_c(z_m+d)/c_1\} \\
&\times \int_0^{t'} d\tau' \int_0^t d\tau \left\langle \hat{\Gamma}_{m'}^\dagger(\tau')\hat{\Gamma}_m(\tau)\right\rangle \\
&\times \exp\left[\left(s_0^*+i\omega\right)(t'-z'/c_0-\tau')+(s_0-i\omega)(t-z/c_0-\tau)\right], \\
&\qquad 0\le z',\quad 0\le z
\end{aligned}
\tag{9.96}
$$

If we look back at Equation 9.74, the thermal noise part is obtained by replacing the space functions $\sin\{\Omega_c^*(z'+d)/c_1\}\sin\{\Omega_c(z+d)/c_1\}$ by $|(1/2i)\exp(i\Omega_c d/c_1) T|^2 = T^2/(4r)$ and the time variables t' and t by $t'-(z'/c_0)$ and $t-(z/c_0)$, respectively. Similarly, the quantum noise part is obtained by the same replacements. Thus if we write

$$
z'-z=\Delta z \tag{9.97}
$$

we have (see Equation 9.95)

$$
\begin{aligned}
&\left\langle \hat{E}^{(-)}(z+\Delta z,\ t+\tau)\hat{E}^{(+)}(z,t)\right\rangle \\
&= \frac{T^2\hbar\omega\gamma\beta_c/\gamma_c}{4r\varepsilon_1 d(\gamma+\gamma_c)(1-\sigma/\sigma_{th})}\left\{\left(\frac{\sigma^2}{\sigma_{th}\sigma_{th\,0}}\right)\langle n_\omega\rangle+\frac{N_2}{N\sigma_{th}}\right\} \\
&\times \begin{cases} e^{(s_0^*+i\omega)\{\tau-(\Delta z/c_0)\}}, & \tau-(\Delta z/c_0)>0 \\ e^{-(s_0-i\omega)\{\tau-(\Delta z/c_0)\}}, & \tau-(\Delta z/c_0)<0 \end{cases}
\end{aligned}
\tag{9.98}
$$

9.8
The Laser Linewidth and the Correction Factor

Next, we turn to the laser linewidth below threshold. The correlation function in Equation 9.95 for inside the cavity is in the same form as that for outside the cavity in Equation 9.98 for equal locations $\Delta z = 0$. Thus the power spectra, the Fourier transform of the temporal part of these correlation functions, are both

$$\begin{aligned} I(\omega) &= \int_{-\infty}^{+\infty} \left\langle \hat{E}^{(-)}(z,\ t+\tau)\hat{E}^{(+)}(z,t)\right\rangle e^{-i\omega\tau}\,d\tau \\ &\propto \int_{-\infty}^{0} e^{-(s_0 - i\omega_o)\tau - i\omega\tau}\,d\tau + \int_{0}^{+\infty} e^{(s_0^* + i\omega_o)\tau - i\omega\tau}\,d\tau \\ &= \frac{-2\mathrm{Re}\, s_0}{(\omega_o - \omega - \mathrm{Im}\, s_0)^2 + (\mathrm{Re}\, s_0)^2} \end{aligned} \tag{9.99}$$

where we have rewritten the central frequency ω as ω_o. Thus the power spectrum is a Lorentzian with the full width at half-maximum (FWHM) $\Delta\omega$ given by

$$\begin{aligned} \Delta\omega &= 2|\mathrm{Re}\, s_0| \\ &= \frac{2(\gamma+\gamma_c)[\gamma\gamma_c(1+\delta^2) - GNc_1]}{(\gamma+\gamma_c)^2 + \delta^2(\gamma-\gamma_c)^2} \\ &= \frac{2(\gamma+\gamma_c)\gamma\gamma_c(1+\delta^2)[1-\sigma/\sigma_{th}]}{(\gamma+\gamma_c)^2 + \delta^2(\gamma-\gamma_c)^2} \end{aligned} \tag{9.100}$$

In order to express the laser linewidth in terms of laser output power, we calculate the power output utilizing the correlation function in Equation 9.98. Note that the power output P per unit cross-sectional area is (see discussion on Equation 4.59)

$$P = c_0\varepsilon_0\left\langle \hat{E}^2\right\rangle = 2c_0\varepsilon_0\left\langle \hat{E}^{(-)}\hat{E}^{(+)}\right\rangle \tag{9.101}$$

Thus

$$\begin{aligned} P &= 2c_0\varepsilon_0\left\langle \hat{E}^{(-)}(z,t)\hat{E}^{(+)}(z,t)\right\rangle \\ &= \frac{c_0\varepsilon_0 T^2\hbar\omega\gamma\beta_c/\gamma_c}{2r\varepsilon_1 d(\gamma+\gamma_c)(1-\sigma/\sigma_{th})}\left\{\left(\frac{\sigma^2}{\sigma_{th}\sigma_{th\,0}}\right)\langle n_\omega\rangle + \frac{N_2}{N\sigma_{th}}\right\} \end{aligned} \tag{9.102}$$

Note that this is independent of time t and of location z. Since $T = 1 + r$ and $c_0\varepsilon_0(1+r)^2/(r\varepsilon_1 d) = 4\beta_c$, we have

$$P = \frac{2\hbar\omega\gamma\beta_c^2/\gamma_c}{(\gamma+\gamma_c)(1-\sigma/\sigma_{th})}\left\{\left(\frac{\sigma^2}{\sigma_{th}\sigma_{th\,0}}\right)\langle n_\omega\rangle + \frac{N_2}{N\sigma_{th}}\right\} \tag{9.103}$$

Thus by Equation 9.100 the product $P\Delta\omega$ is

$$P\Delta\omega = \frac{4\hbar\omega\gamma^2\beta_c^2(1+\delta^2)}{(\gamma+\gamma_c)^2 + \delta^2(\gamma-\gamma_c)^2}\left\{\left(\frac{\sigma^2}{\sigma_{th}\sigma_{th\,0}}\right)\langle n_\omega\rangle + \frac{N_2}{N\sigma_{th}}\right\} \tag{9.104}$$

We have the following laser linewidth (FWHM) in angular frequency:

$$\Delta\omega = \frac{4\hbar\omega\beta_c^2}{P}\frac{\gamma^2(1+\delta^2)}{(\gamma+\gamma_c)^2+\delta^2(\gamma-\gamma_c)^2}\left\{\left(\frac{\sigma^2}{\sigma_{th}\sigma_{th\,0}}\right)\langle n_\omega\rangle+\frac{N_2}{N\sigma_{th}}\right\} \tag{9.105}$$

where σ_{th0} is the threshold atomic inversion at zero detuning. Except for the factor before $\langle n_\omega\rangle$, this formula is $(\beta_c/\gamma_c)^2$ times the conventional formula obtained for example by Haken [2] and reproduced in Equation 4.62a for two-level atoms.

Note that, below Equation 7.37, we discussed the validity of replacing the cavity decay constant of the quasimode analysis by that of the present cavity model based on the equivalence of the two in the decay equations for the field amplitude. So, the cavity decay constant γ_c in the above factor $(\beta_c/\gamma_c)^2$ can be replaced by that of the present chapter. This correction factor was reported by Ujihara [1] for the first time. This factor has since been called the noise enhancement factor, excess noise factor, longitudinal Petermann factor, and so on. We shall call this factor the longitudinal excess noise factor K_L. Then we have

$$K_L = \left(\frac{\beta_c}{\gamma_c}\right)^2 = \left\{\frac{(c_1/2d)[(1-r^2)/2r]}{(c_1/2d)\ln(1/r)}\right\}^2 = \left\{\frac{(1-r^2)/2r}{\ln(1/r)}\right\}^2 \tag{9.106}$$

which depends on only the reflection coefficient r and is a decreasing function of r. It approaches unity as the reflection coefficient r goes to unity. The correction becomes important when r is small. We will see in the next chapter that a similar correction factor appears also in the nonlinear gain regime or in operation above threshold.

We will now examine if the same output power as in Equation 9.103 can be derived from the internal field correlation function in Equation 9.95. We have the stored energy W per unit cross-sectional area as

$$\begin{aligned} W &= \int_{-d}^{0} dz\, 2\varepsilon_1\left\langle \hat{E}^{(-)}(z,t)\hat{E}^{(+)}(z,t)\right\rangle \\ &= \left(\int_{-d}^{0} dz\, |\sin\{\Omega_c(z+d)/c_1\}|^2\right)\frac{2\varepsilon_1\hbar\omega\gamma\beta_c/\gamma_c}{\varepsilon_1 d(\gamma+\gamma_c)(1-\sigma/\sigma_{th})} \\ &\quad\times\left\{\left(\frac{\sigma^2}{\sigma_{th}\sigma_{th\,0}}\right)\langle n_\omega\rangle+\frac{N_2}{N\sigma_{th}}\right\} \end{aligned} \tag{9.107}$$

The integral was evaluated in Equation 9.92. Thus we have

$$\begin{aligned} W &= \frac{\beta_c^2}{\gamma_c^2}\frac{\hbar\omega\gamma}{(\gamma+\gamma_c)(1-\sigma/\sigma_{th})}\left\{\left(\frac{\sigma^2}{\sigma_{th}\sigma_{th\,0}}\right)\langle n_\omega\rangle+\frac{N_2}{N\sigma_{th}}\right\} \\ &= \frac{P}{2\gamma_c} \end{aligned} \tag{9.108}$$

where P is given by Equation 9.103. Therefore, we see that the power output is $2\gamma_c$ times the stored energy. This shows that $2\gamma_c$ is the correct power damping factor in this linear gain regime. We will see in the next chapter that this is not the case in the nonlinear gain regime.

Finally, let us consider the mathematical origin of the correction factor $(\beta_c/\gamma_c)^2$ for the laser linewidth. The appearance of this factor is rather direct in the case of the quantum noise and in the case of the evaluation of the laser output through the stored energy in the cavity. Exactly this factor appears in Equation 9.108. The factor (β_c/γ_c) came from Equation 9.95, which in turn came from the integral of the absolute square of the cavity mode function in Equation 9.92. This contribution stems from the quantum noise arising at the location of each atom. Another factor (β_c/γ_c) originates from the integration of the stored energy in Equation 9.107. The correction factor thus seems to come from the square of the integral of the absolute square of the cavity mode function:

$$\left(\int_{-d}^{0} dz\, |\sin\{\Omega_c(z+d)/c_1\}|^2\right)^2 = \left(\frac{d}{2}\frac{\beta_c}{\gamma_c}\right)^2 \tag{9.109}$$

Let us recall from Equation 4.45 that the field in the quasimode cavity in the linear regime was

$$\begin{aligned}\hat{a}(t) &= \hat{a}(0)e^{(s_0-i\omega)t} + \frac{1}{i(\omega_c+\nu_0-2\omega)+\gamma_c+\gamma} \\ &\quad\times \int_0^t e^{(s_0-i\omega)(t-t')}\left\{\{i(\nu_0-\omega)+\gamma\}\hat{\Gamma}_f(t') - i\sum_m \kappa_m\hat{\Gamma}_m\right\} dt'\end{aligned} \tag{9.110}$$

Multiplying by $i(\hbar\omega_c/2)^{1/2}U_c(z) = i(\hbar\omega_c/\varepsilon_1 d)^{1/2}\sin\{(\omega_c/c_1)(z+d)\}$ (see Equations 3.1 and 2.19a) and using Equation 3.22b for κ_m, we rewrite the quantum part as

$$\begin{aligned}\hat{E}^{(+)}(z,t) &= \frac{\sin\{(\omega_k/c_1)(z+d)\}}{i(\omega_c+\nu_0-2\omega)+\gamma_c+\gamma}\, i\frac{p_a\omega_c}{\varepsilon_1 d}\int_0^t e^{(s_0-i\omega)(t-t')} \\ &\quad\times\sum_m \sin\{(\omega_k/c_1)(z_m+d)\}\hat{\Gamma}_m(t')\,dt'\end{aligned} \tag{9.111}$$

We compare it with the quantum part in Equation 9.64:

$$\begin{aligned}\hat{E}^{(+)}(z,t) &= \frac{\sin\Omega_c(z+d)/c_1}{\gamma+\gamma_c+i(\nu_0+\omega_c-2\omega)}\frac{i\nu_0 p_a}{\varepsilon_1 d}\sum_m \sin\{\Omega_c(z_m+d)/c_1\} \\ &\quad\times\int_0^t \hat{\Gamma}_m(\tau)\exp[(s_0-i\omega)(t-\tau)]d\tau\end{aligned} \tag{9.112}$$

So, the major difference is that the quasimode function $\sin\{(\omega_k/c_1)(z+d)\}$ replaces the complex cavity mode function $\sin\Omega_c(z+d)/c_1$. If we repeat the square of the integral in Equation 9.109 with the quasimode function, we have

$$\left(\int_{-d}^{0} dz\, |\sin\{\omega_c(z+d)/c_1\}|^2\right)^2 = \left(\frac{d}{2}\right)^2 \tag{9.113}$$

Thus

$$\frac{\left(\int_{-d}^{0} dz\,|\sin\{\Omega_c(z+d)/c_1\}|^2\right)^2}{\left(\int_{-d}^{0} dz\,|\sin\{\omega_c(z+d)/c_1\}|^2\right)^2} = \left(\frac{\beta_c}{\gamma_c}\right)^2 = K_L \tag{9.114}$$

Therefore, we may conclude that the linewidth correction factor $K_L = (\beta_c/\gamma_c)^2$ comes from the use of the proper cavity resonant mode function that reflects the output coupling, at least for the quantum noise. The physical interpretation of the correction factor, the excess noise factor, will be discussed and a more general derivation of the factor will be given in Chapter 14. In particular, the general derivation scheme will show that the quantity to be compared with that in Equation 9.109 is the squared modulus of the integrated squared mode function instead of that in Equation 9.113 due to the quasimode function (see Equation 14.46). This is related to the projection of the noise function onto the adjoint mode function, as was mentioned below Equation 9.44.

References

1 Ujihara, K. (**1977**) *Phys. Rev. A, 16,* 652–658.

2 Haken, H. (**1970**) Laser theory, in *Licht und Materie, IC,* Handbuch der Physik, vol. *XXV/2c* (eds. S. Flügge and L. Genzel), Springer, Berlin.

10
A One-Dimensional Laser with Output Coupling: Quantum Nonlinear Gain Analysis

In this chapter, we solve the laser equation of motion (Equation 5.33a) quantum mechanically for the one-sided cavity model described in Sections 1.3 and 1.4. We take into account the gain saturation behavior in the atomic motion: the atomic inversion is dependent on the field strength at the location of the atom. Because of the output coupling, the field distribution is not uniform. So we need to find consistent distributions of the atomic inversion and the field strength. Because of this nonlinear nature of the problem, it is difficult to solve the time-varying behavior of the laser. We concentrate on steady-state operation, assuming the presence of a time-independent field amplitude that still depends on the location. The field phase is, however, allowed to diffuse under the action of the noise forces. The steady-state, time-independent field distribution in the presence of gain saturation was found in Chapter 8, ignoring the noise. Thus, in this chapter, the main problem is to find the degree of phase diffusion, which determines the laser linewidth. The resultant expression for the laser linewidth will contain correction factors compared with the conventional formula when expressed in terms of the inverse output power. One of the correction factors, $(\beta_c/\gamma_c)^2$, is the same as that in the quantum linear gain analysis in the previous chapter. The other factor results in a non-power-reciprocal part in the linewidth formula. The essence of the contents of this chapter was published in Ref. [1].

10.1
The Equation for the Quantum Nonlinear Gain Analysis

From Equation 5.33, for the entire region $-d < z < L$, the equation to be solved reads

$$\hat{E}^{(+)}(z,t) = \hat{F}_t(z,t) + \hat{F}_q(z,t) + \sum_m \left[\frac{|p_m|^2 \nu_m^2}{2\hbar\omega} \int_0^t \sum_j U_j(z)\,U_j(z_m)\,e^{-i\omega_j(t-t')} \times \int_0^{t'} e^{-(i\nu_m+\gamma_m)(t'-t'')}\,\hat{E}^{(+)}(z_m,t'')\hat{\sigma}_m(t'')dt''dt' \right] \tag{10.1}$$

where

Output Coupling in Optical Cavities and Lasers: A Quantum Theoretical Approach
Kikuo Ujihara

ISBN: 978-3-527-40763-7

$$\hat{F}_t(z,t) = i\sum_j \sqrt{\frac{\hbar\omega_j}{2}}\, U_j(z)\hat{a}_j(0)e^{-i\omega_j t} \tag{10.2}$$

and

$$\hat{F}_q(z,t) = \sum_m \left[\frac{ip_m v_m}{2}\int_0^t \sum_j U_j(z)\,U_j(z_m)\,e^{-i\omega_j(t-t')}\int_0^{t'} e^{-(i v_m+\gamma_m)(t'-t'')}\,\hat{\Gamma}_m(t'')dt''dt'\right] \tag{10.3}$$

Assuming a single-frequency oscillation, we truncate the sinusoidal motion at the center angular frequency ω. We write

$$\hat{E}^{(+)}(z,t) = \tilde{E}^{(+)}(z,t)e^{-i\omega t}, \quad \hat{F}_t(z,t) = \tilde{F}_t(z,t)e^{-i\omega t}, \quad \hat{F}_q(z,t) = \tilde{F}_q(z,t)e^{-i\omega t} \tag{10.4}$$

Then, for $-d < z < L$, we have

$$\begin{aligned}\tilde{E}^{(+)}(z,t) = \sum_m &\left[\frac{|p_m|^2 v_m^2}{2\hbar\omega}\int_0^t \sum_j U_j(z)\,U_j(z_m)\,e^{i(\omega-\omega_j)(t-t')}\right.\\ &\times\left.\int_0^{t'} e^{\{i(\omega-v_m)-\gamma_m\}(t'-t'')}\,\tilde{E}^{(+)}(z_m,t'')\hat{\sigma}_m(t'')dt''dt'\right]\\ &+\tilde{F}_t(z,t)+\tilde{F}_q(z,t)\end{aligned} \tag{10.5}$$

We will first seek the differential equation for $\tilde{E}^{(+)}(z,t)$ with respect to time. Next, we will Laplace-transform the differential equation. Then we will look for the steady-state field distribution in the transformed domain. We will finally look for the phase diffusion in the inverse transformed domain, that is, in the time domain. The correlation for the noise forces was discussed in the previous chapter. These correlation characteristics determine the degree of phase diffusion.

The summation over j present on the right-hand side (RHS) was evaluated in Equation 9.8 for inside the cavity:

$$\begin{aligned}&\sum_j \left\{U_j(z)\,U_j(z_m)e^{-i(\omega_j-\omega)(t-t')}\right\}\\ &=\sum_j \frac{2}{\varepsilon_1 L}\frac{1}{1-K\sin^2 k_{1j}d}\sin k_{1j}(z+d)\,\sin k_{1j}(z_m+d)\,e^{-i\omega_j(t-t')}\\ &=\frac{1}{\varepsilon_1 c_1}\sum_{n=0}^{\infty}\frac{1}{1+\delta_{0,n}}(-r)^n\sum_{\rho=1}^{4}\alpha_\rho\left\{e^{i\tau_{\rho n}\omega}\delta(-\tau_{\rho n}+t-t')+e^{-i\tau_{\rho n}\omega}\delta(\tau_{\rho n}+t-t')\right\}\end{aligned} \tag{10.6}$$

where the factors $\alpha_1 = \alpha_2 = 1$ and $\alpha_3 = \alpha_4 = -1$. The delay times are

$$\tau_{1n} = \frac{2nd + z - z_m}{c_1}, \qquad \tau_{2n} = \frac{2nd - z + z_m}{c_1}$$
$$\tau_{3n} = \frac{2nd + 2d + z + z_m}{c_1}, \qquad \tau_{4n} = \frac{2nd - 2d - z - z_m}{c_1} \tag{10.7}$$

For outside the cavity, the summation was given by Equation 9.66 (ω is absent here):

$$\begin{aligned}
&\sum_j \left\{ U_j(z) U_j(z_m) e^{-i\omega_j(t-t')} \right\} \\
&= \sum_j \frac{2}{\varepsilon_1 L} \frac{1}{1 - K\sin^2 k_{1j}d} \left(\frac{k_{1j}}{k_{0j}} \cos k_{1j}d \sin k_{0j}z + \sin k_{1j}d \cos k_{0j}z \right) \\
&\times \sin k_{1j}(z_m + d) e^{-i\omega_j(t-t')} \\
&= \frac{1}{\varepsilon_1 c_1} \frac{2c_0}{(c_1 + c_0)} \\
&\times \sum_{n=0}^{\infty} (-r)^n \{\delta(t - t' + \tau_{5n}) + \delta(t - t' - \tau_{5n}) - \delta(t - t' + \tau_{6n}) - \delta(t - t' - \tau_{6n})\}
\end{aligned} \tag{10.8}$$

where

$$\tau_{5n} = \frac{z}{c_0} + \frac{2nd - z_m}{c_1}, \qquad \tau_{6n} = \frac{z}{c_0} + \frac{2nd + 2d + z_m}{c_1} \tag{10.9}$$

First, we consider the field inside the cavity. Using Equation 10.6 in Equation 10.5, we have (see Equation 9.10)

$$\begin{aligned}
\tilde{E}(z,t) &= \tilde{F}_t(z,t) + \tilde{F}_q(z,t) \\
&+ \sum_m g_m \left[\int_0^{t-|z-z_m|/c_1} e^{\{-i(\nu_m-\omega)-\gamma_m\}(t-t')+(i\nu_m+\gamma_m)|z-z_m|/c_1} \hat{\sigma}_m(t')\tilde{E}(z_m,t')dt' \right. \\
&- \int_0^{t-(2d+z+z_m)/c_1} e^{\{-i(\nu_m-\omega)-\gamma_m\}(t-t')+(i\nu_m+\gamma_m)(2d+z+z_m)/c_1} \hat{\sigma}_m(t')\tilde{E}(z_m,t')dt' \\
&\left. + \sum_{n=1}^{n_M} (-r)^n (I_{1n} + I_{2n} - I_{3n} - I_{4n}) \right]
\end{aligned} \tag{10.10a}$$

where

$$I_{\rho n} = \int_0^{t-\tau_{\rho n}} \exp\left[\{-i(\nu_m - \omega) - \gamma_m\}(t-t') + (i\nu_m + \gamma_m)\tau_{\rho n}\right] \hat{\sigma}_m(t')\tilde{E}(z_m,t')dt' \tag{10.10b}$$

and

$$g_m = \frac{|p_m|^2 \nu_m^2}{2\hbar\omega\varepsilon_1 c_1} \tag{10.10c}$$

10.2 Homogeneously Broadened Atoms and Uniform Pumping

Here we go to the case of homogeneous broadening, that is

$$\nu_m = \nu_0, \qquad p_m = p_a, \qquad \gamma_m = \gamma \tag{10.11}$$

$$g_m = g = \frac{\nu_0^2 |p_a|^2}{2\hbar\omega\varepsilon_1 c_1} \tag{10.12}$$

and to uniform pumping and uniform unsaturated atomic inversion, that is

$$\Gamma_{mp} = \Gamma_p, \qquad \hat{\sigma}_m^0 = \hat{\sigma}^0 \tag{10.13}$$

Then, by differentiation with respect to time t, the integral equation is converted to a simplified differential equation:

$$\begin{aligned}
&\frac{\partial}{\partial t}\left\{\tilde{E}(z,t) - \tilde{F}_t(z,t) - \tilde{F}_q(z,t)\right\} \\
&= \{-i(\nu_0 - \omega) - \gamma\}\left\{\tilde{E}(z,t) - \tilde{F}_t(z,t) - \tilde{F}_q(z,t)\right\} \\
&+ \sum_m g\left[\exp\left(i\omega\frac{|z-z_m|}{c_1}\right)\tilde{E}\left(z_m,\ t - \frac{|z-z_m|}{c_1}\right)\hat{\sigma}_m\left(t - \frac{|z-z_m|}{c_1}\right)\right. \\
&- \exp\left(i\omega\frac{2d+z+z_m}{c_1}\right)\tilde{E}\left(z_m,\ t - \frac{2d+z+z_m}{c_1}\right)\hat{\sigma}_m\left(t - \frac{2d+z+z_m}{c_1}\right) \\
&\left. + \sum_{n=1}^{n_M}(-r)^n\left\{\sum_{\rho=1}^{4}\alpha_\rho \exp(i\omega\tau_{\rho n})\tilde{E}(z_m,\ t-\tau_{\rho n})\,\hat{\sigma}_m(t-\tau_{\rho n})\right\}\right]
\end{aligned} \tag{10.14}$$

To go further, we divide the field waves and the driving noise forces into right- and left-going waves, respectively:

$$\begin{aligned}
\tilde{E}(z,t) &= \hat{e}^+(z,t)\exp\{+i\omega(z+d)/c_1\} + \hat{e}^-(z,t)\exp\{-i\omega(z+d)/c_1\} \\
\tilde{F}_{t,q}(z,t) &= \hat{f}_{t,q}^+(z,t)\exp\{+i\omega(z+d)/c_1\} + \hat{f}_{t,q}^-(z,t)\exp\{-i\omega(z+d)/c_1\}
\end{aligned} \tag{10.15}$$

where the suffices t and q signify thermal and quantum noise, respectively.

Then, comparing the terms of the right- and left-traveling waves, and ignoring those terms that are oscillating rapidly with z_m, having a factor $\exp(\pm 2i\omega z_m/c_1)$, we have (see Equations 9.16a and 9.16b):

$$\begin{aligned}
&\left(\frac{\partial}{\partial t} + \gamma'\right)\left\{\hat{e}^+(z,t) - \hat{\upsilon}^+(z,t)\right\} \\
&= \sum_m g\left[H(z-z_m)\hat{e}^+\left(z_m,\ t - \frac{z-z_m}{c_1}\right)\hat{\sigma}_m\left(t - \frac{z-z_m}{c_1}\right)\right. \\
&- \hat{e}^-\left(z_m,\ t - \frac{2d+z+z_m}{c_1}\right)\hat{\sigma}_m\left(t - \frac{2d+z+z_m}{c_1}\right)
\end{aligned}$$

$$+\sum_{n=1}^{n_M}(r')^n\left\{\hat{e}^+\left(z_m, t-\frac{2nd+z-z_m}{c_1}\right)\hat{\sigma}_m\left(t-\frac{2nd+z-z_m}{c_1}\right)\right.$$
$$\left.\left.-\hat{e}^-\left(z_m, t-\frac{2nd+2d+z+z_m}{c_1}\right)\hat{\sigma}_m\left(t-\frac{2nd+2d+z+z_m}{c_1}\right)\right\}\right] \tag{10.16a}$$

and

$$\left(\frac{\partial}{\partial t}+\gamma'\right)\{\hat{e}^-(z,t)-\hat{v}^-(z,t)\}$$
$$=\sum_m g\left[H(z_m-z)\,\hat{e}^-\left(z_m, t-\frac{-z+z_m}{c_1}\right)\hat{\sigma}_m\left(t-\frac{-z+z_m}{c_1}\right)\right.$$
$$+\sum_{n=1}^{n_M}(r')^n\left\{\hat{e}^-\left(z_m, t-\frac{2nd-z+z_m}{c_1}\right)\hat{\sigma}_m\left(t-\frac{2nd-z+z_m}{c_1}\right)\right.$$
$$\left.\left.-\hat{e}^+\left(z_m, \frac{2nd-2d-z-z_m}{c_1}\right)\hat{\sigma}_m\left(t-\frac{2nd-2d-z-z_m}{c_1}\right)\right\}\right] \tag{10.16b}$$

where

$$\hat{v}^+(z,t)=\hat{f}_t^+(z,t)+\hat{f}_q^+(z,t), \qquad \hat{v}^-(z,t)=\hat{f}_t^-(z,t)+\hat{f}_q^-(z,t) \tag{10.17}$$

and

$$\gamma'=\gamma+i(\nu_0-\omega) \tag{10.18}$$

$$r'=-r\exp(2id\omega/c_1) \tag{10.19}$$

Here the unit step function has been denoted as H.

10.3 The Steady-State and Laplace-Transformed Equations

We go to the steady state, which here means that the field amplitude fluctuates negligibly but the field phase diffuses freely. The amplitude is stabilized by the gain saturation effect but the noise sources cause a random walk of the phase. We make the assumption on the atomic inversion that the inversion keeps a constant value in time, the value being given by the steady-state ensemble average of the quantum-mechanical expectation value with respect to the reservoirs for the atoms and to the free thermal field. Accordingly, we write the saturated inversion as

$$\hat{\sigma}_m(t)=\langle\sigma_m\rangle \tag{10.20}$$

where the saturation property is given by Equation 8.16:

$$\langle\sigma_m\rangle=\frac{\sigma^0}{1+\bar{E}_m^2/|E_s|^2} \tag{10.21a}$$

where

$$\bar{E}_m = \sqrt{\left\langle \left| \tilde{E}^{(+)}(z_m) \right|^2 \right\rangle} \tag{10.21b}$$

is the reservoir-averaged quantum-mechanical expectation value of the local field amplitude. The operator sign drops from the atomic inversion.

In order to solve the coupled equations 10.16a and 10.16b involving space variable z and time variable t, we Laplace-transform the field operator and the noise operators with respect to time, as in Chapter 9 (we will concentrate on the spatial region for the time being):

$$\begin{aligned} \hat{e}^+(z,t) &\to \hat{L}^+(z,s) \\ \hat{e}^-(z,t) &\to \hat{L}^-(z,s) \\ \hat{\upsilon}^+(z,t) &\to \hat{V}^+(z,s) \\ \hat{\upsilon}^-(z,t) &\to \hat{V}^-(z,s) \end{aligned} \tag{10.22}$$

As the Laplace transform of $\hat{e}^+(z_m, t - \tau_{pn})$ is $\exp(-\tau_{\rho n} s)\hat{L}^+(z_m, s)$, the summations over n in Equations 10.16a and 10.16b reduce to geometric progressions, which can be easily evaluated. Replacing the summation over m by integration with the assumed uniform density of atoms N, we have (see Equations 9.21a and 9.21b):

$$\begin{aligned} &(s+\gamma')\left[\hat{L}^+(z,s) - \hat{V}^+(z,s)\right] \\ &= gN\left\{\int_{-d}^{z} \exp\left(-\frac{z-z_m}{c_1}s\right)\hat{L}^+(z_m,s)\langle\sigma_m\rangle dz_m \right. \\ &\quad - \frac{1}{1-r''(s)}\int_{-d}^{0} \exp\left(-\frac{2d+z+z_m}{c_1}s\right)\hat{L}^-(z_m,s)\langle\sigma_m\rangle dz_m \\ &\quad \left. + \frac{r''(s)}{1-r''(s)}\int_{-d}^{0} \exp\left(\frac{z_m-z}{c_1}s\right)\hat{L}^+(z_m,s)\langle\sigma_m\rangle dz_m \right\} \end{aligned} \tag{10.23a}$$

and

$$\begin{aligned} &(s+\gamma')\left[\hat{L}^-(z,s) - \hat{V}^-(z,s)\right] \\ &= gN\left\{\int_{z}^{0} \exp\left(\frac{z-z_m}{c_1}s\right)\hat{L}^-(z_m,s)\langle\sigma_m\rangle dz_m \right. \\ &\quad + \frac{r''(s)}{1-r''(s)}\int_{-d}^{0} \exp\left(\frac{z-z_m}{c_1}s\right)\hat{L}^-(z_m,s)\langle\sigma_m\rangle dz_m \\ &\quad \left. - \frac{r''(s)}{1-r''(s)}\int_{-d}^{0} \exp\left(\frac{2d+z+z_m}{c_1}s\right)\hat{L}^+(z_m,s)\langle\sigma_m\rangle dz_m \right\} \end{aligned} \tag{10.23b}$$

where

$$r''(s) = r' \exp(-2ds/c_1) = -r \exp\{(i\omega - s)2d/c_1\} \tag{10.24}$$

The initial values $\hat{e}^{\pm}(z,0) - \hat{v}^{\pm}(z,0)$ associated with the Laplace transform vanish, as can be shown by setting $t = 0$ in Equation 10.5 with the aid of Equation 10.15. Differentiation with respect to z yields the following coupled differential equations:

$$\frac{d}{dz}\hat{L}^{+}(z,s) = \left\{-\frac{s}{c_1} + \frac{gN\langle\sigma_z\rangle}{s+\gamma'}\right\}\hat{L}^{+}(z,s) + \left\{\frac{d}{dz} + \frac{s}{c_1}\right\}\hat{V}^{+}(z,s) \tag{10.25a}$$

$$\frac{d}{dz}\hat{L}^{-}(z,s) = \left\{\frac{s}{c_1} - \frac{gN\langle\sigma_z\rangle}{s+\gamma'}\right\}\hat{L}^{-}(z,s) + \left\{\frac{d}{dz} - \frac{s}{c_1}\right\}\hat{V}^{-}(z,s) \tag{10.25b}$$

where we have stressed that $\langle\sigma_m\rangle$ depends on the location z. Integrating, we obtain

$$\begin{aligned} L^{+}(z,s) = &\int_{-d}^{z} \exp\left[\int_{z'}^{z} dz''\left\{-\frac{s}{c_1} + \frac{gN\langle\sigma_{z''}\rangle}{s+\gamma'}\right\}\right]\left(\frac{d}{dz'} + \frac{s}{c_1}\right)V^{+}(z',s)dz' \\ &+ \exp\left[\int_{-d}^{z} dz'\left\{-\frac{s}{c_1} + \frac{gN\langle\sigma_{z'}\rangle}{s+\gamma'}\right\}\right]L^{+}(-d,s) \end{aligned} \tag{10.26a}$$

$$\begin{aligned} L^{-}(z,s) = &\int_{-d}^{z} \exp\left[\int_{z'}^{z} dz''\left\{\frac{s}{c_1} - \frac{gN\langle\sigma_{z''}\rangle}{s+\gamma'}\right\}\right]\left(\frac{d}{dz'} - \frac{s}{c_1}\right)V^{-}(z',s)dz' \\ &+ \exp\left[\int_{-d}^{z} dz'\left\{\frac{s}{c_1} - \frac{gN\langle\sigma_{z'}\rangle}{s+\gamma'}\right\}\right]L^{-}(-d,s) \end{aligned} \tag{10.26b}$$

Because of the nonlinear atomic inversion factors, it will be difficult to solve the coupled equations for a general location z. Thus we look for the boundary values $L^{\pm}(-d,s)$ and $L^{\pm}(0,s)$. From Equations 10.23a and 10.23b we have for $z = -d$

$$\hat{L}^{+}(-d,s) - \hat{V}^{+}(-d,s) = -\{\hat{L}^{-}(-d,s) - \hat{V}^{-}(-d,s)\} \tag{10.27}$$

But from Equations 10.2 and 10.3 $\hat{F}_t(-d,t) = \hat{F}_q(-d,t) = 0$ since $U_j(-d) = 0$ because of the vanishing boundary condition (see Equation 1.41b). Thus from Equation 10.15 we have

$$\hat{V}^{+}(-d,s) + \hat{V}^{-}(-d,s) = 0 \tag{10.28}$$

so that

$$\hat{L}^{+}(-d,s) = -\hat{L}^{-}(-d,s) \tag{10.29}$$

This is a statement that the right-going wave at $z=-d$ is just the left-going wave reflected at the perfectly conducting boundary. Also from Equations 10.23a and 10.23b we have for $z = 0$

$$(s+\gamma')\left[\hat{L}^{+}(0,s)-\hat{V}^{+}(0,s)\right] = gN\left\{-\frac{1}{1-r''(s)}\int_{-d}^{0}\exp\left(-\frac{2d+z_m}{c_1}s\right)\hat{L}^{-}(z_m,s)\langle\sigma_m\rangle dz_m + \frac{1}{1-r''(s)}\int_{-d}^{0}\exp\left(\frac{z_m}{c_1}s\right)\hat{L}^{+}(z_m,s)\langle\sigma_m\rangle dz_m\right\} \tag{10.30a}$$

and

$$(s+\gamma')\left[\hat{L}^{-}(0,s)-\hat{V}^{-}(0,s)\right] = gN\left\{\int_{-d}^{0}\frac{r''(s)}{1-r''(s)}\exp\left(\frac{-z_m}{c_1}s\right)\hat{L}^{-}(z_m,s)\langle\sigma_m\rangle dz_m - \int_{-d}^{0}\frac{r''(s)}{1-r''(s)}\exp\left(\frac{2d+z_m}{c_1}s\right)\hat{L}^{+}(z_m,s)\langle\sigma_m\rangle dz_m\right\} \tag{10.30b}$$

Comparing these two equations we obtain

$$-r''(s)e^{(2d/c_1)s}\left\{\hat{L}^{+}(0,s)-\hat{V}^{+}(0,s)\right\} = \hat{L}^{-}(0,s)-\hat{V}^{-}(0,s)$$

or by Equation 10.24

$$-r'\left\{\hat{L}^{+}(0,s)-\hat{V}^{+}(0,s)\right\} = \hat{L}^{-}(0,s)-\hat{V}^{-}(0,s) \tag{10.31}$$

Next, from Equations 10.26a and 10.26b we have for $z = 0$

$$\hat{L}^{+}(0,s) = \int_{-d}^{0}\exp\left[\int_{z'}^{0}dz''\left\{-\frac{s}{c_1}+\frac{gN\langle\sigma_{z''}\rangle}{s+\gamma'}\right\}\right]\left(\frac{d}{dz'}+\frac{s}{c_1}\right)\hat{V}^{+}(z',s)dz' + \exp\left[\int_{-d}^{0}dz'\left\{-\frac{s}{c_1}+\frac{gN\langle\sigma_{z'}\rangle}{s+\gamma'}\right\}\right]\hat{L}^{+}(-d,s) \tag{10.32a}$$

and

$$\hat{L}^{-}(0,s) = \int_{-d}^{0}\exp\left[\int_{z'}^{0}dz''\left\{\frac{s}{c_1}-\frac{gN\langle\sigma_{z''}\rangle}{s+\gamma'}\right\}\right]\left(\frac{d}{dz'}-\frac{s}{c_1}\right)\hat{V}^{-}(z',s)dz' + \exp\left[\int_{-d}^{0}dz'\left\{\frac{s}{c_1}-\frac{gN\langle\sigma_{z'}\rangle}{s+\gamma'}\right\}\right]\hat{L}^{-}(-d,s) \tag{10.32b}$$

Provided that the quantities $\hat{V}^{\pm}(z,s)$ are known, Equations 10.29, 10.31, 10.32a, and 10.32b constitute coupled equations for $\hat{L}^{\pm}(-d,s)$ and $\hat{L}^{\pm}(0,s)$. Anticipating the need for $\hat{e}^+(-0,t)$ in evaluation of the output field, we solve these coupled equations for $\hat{L}^+(0,s)$. First, we rewrite Equations 10.32a and 10.32b as

$$\begin{aligned}\hat{L}^+(-d,s) &= \exp\left[\int_{-d}^{0} dz'\left\{\frac{s}{c_1}-\frac{gN\langle\sigma_{z'}\rangle}{s+\gamma'}\right\}\right]\hat{L}^+(0,s)\\ &\quad -\int_{-d}^{0}\exp\left[\int_{-d}^{z'} dz''\left\{\frac{s}{c_1}-\frac{gN\langle\sigma_{z''}\rangle}{s+\gamma'}\right\}\right]\left(\frac{d}{dz'}+\frac{s}{c_1}\right)\hat{V}^+(z',s)dz'\end{aligned} \tag{10.33a}$$

and

$$\begin{aligned}\hat{L}^-(-d,s) &= \exp\left[\int_{-d}^{0} dz'\left\{-\frac{s}{c_1}+\frac{gN\langle\sigma_{z'}\rangle}{s+\gamma'}\right\}\right]\hat{L}^-(0,s)\\ &\quad -\int_{-d}^{0}\exp\left[-\int_{-d}^{z'} dz''\left\{\frac{s}{c_1}-\frac{gN\langle\sigma_{z''}\rangle}{s+\gamma'}\right\}\right]\left(\frac{d}{dz'}-\frac{s}{c_1}\right)\hat{V}^-(z',s)dz'\end{aligned} \tag{10.33b}$$

The sum of the RHS members of these two equations vanish because of Equation 10.29. Then using Equation 10.31 we eliminate $L^-(0,s)$ to obtain

$$\begin{aligned}&\left[\exp\left\{\int_{-d}^{0} dz'\left(\frac{s}{c_1}-\frac{gN\langle\sigma_{z'}\rangle}{s+\gamma'}\right)\right\}-r'\exp\left\{\int_{-d}^{0} dz'\left(-\frac{s}{c_1}+\frac{gN\langle\sigma_{z'}\rangle}{s+\gamma'}\right)\right\}\right]\hat{L}^+(0,s)\\ &=\int_{-d}^{0}\exp\left[\int_{-d}^{z'} dz''\left\{\frac{s}{c_1}-\frac{gN\langle\sigma_{z''}\rangle}{s+\gamma'}\right\}\right]\left(\frac{d}{dz'}+\frac{s}{c_1}\right)\hat{V}^+(z',s)dz'\\ &\quad+\int_{-d}^{0}\exp\left[-\int_{-d}^{z'} dz''\left\{\frac{s}{c_1}-\frac{gN\langle\sigma_{z''}\rangle}{s+\gamma'}\right\}\right]\left(\frac{d}{dz'}-\frac{s}{c_1}\right)\hat{V}^-(z',s)dz'\\ &\quad-\exp\left[\int_{-d}^{0} dz'\left\{-\frac{s}{c_1}+\frac{gN\langle\sigma_{z'}\rangle}{s+\gamma'}\right\}\right]\left\{r'\hat{V}^+(0,s)+\hat{V}^-(0,s)\right\}\end{aligned} \tag{10.34}$$

or

$$\begin{aligned}&\left[1-r'\exp\left\{2\int_{-d}^{0} dz'\left(-\frac{s}{c_1}+\frac{gN\langle\sigma_{z'}\rangle}{s+\gamma'}\right)\right\}\right]\hat{L}^+(0,s)\\ &=\int_{-d}^{0}\exp\left[-\int_{z'}^{0} dz''\left\{\frac{s}{c_1}-\frac{gN\langle\sigma_{z''}\rangle}{s+\gamma'}\right\}\right]\left(\frac{d}{dz'}+\frac{s}{c_1}\right)\hat{V}^+(z',s)dz'\\ &\quad+\exp\left\{-2\int_{-d}^{0} dz'\left(\frac{s}{c_1}-\frac{gN\langle\sigma_{z'}\rangle}{s+\gamma'}\right)\right\}\\ &\quad\times\int_{-d}^{0}\exp\left[\int_{z'}^{0} dz''\left\{\frac{s}{c_1}-\frac{gN\langle\sigma_{z''}\rangle}{s+\gamma'}\right\}\right]\left(\frac{d}{dz'}-\frac{s}{c_1}\right)\hat{V}^-(z',s)dz'\\ &\quad-\exp\left[2\int_{-d}^{0} dz'\left\{-\frac{s}{c_1}+\frac{gN\langle\sigma_{z'}\rangle}{s+\gamma'}\right\}\right]\left\{r'\hat{V}^+(0,s)+\hat{V}^-(0,s)\right\}\end{aligned} \tag{10.35}$$

This last equation has simple interpretations. The left-hand side (LHS) is the difference of the amplitudes $\hat{e}^+(t,0)$ and $\hat{e}^+(t-2d/c_1,0)$, where the latter is multiplied by the net gain and the reflection coefficients at both end surfaces with phase shift associated with one round trip in the cavity. Note that $r' = -r\exp(2id\omega/c_1)$ by Equation 10.19. The first term on the RHS is the contribution to the difference from the right-going components of the noise forces associated with amplification and the proper time delay during the path from the location z' of the noise source and the output end $z = 0$ of the cavity. The noise source extends from $z = -d$ to $z = 0$. The sum of (d/dz') and (s/c_1) represents the total derivative $(\partial/\partial z) + (\partial/\partial t)(\partial t/\partial z)$ for the right-going wave along the path. The second term comes from the left-going component of the noise forces with path length $2d-|z'|$ for the noise to reach $z = 0$ from z'. This term is also associated with the proper net gain and phase shift as well as the proper time delay. The sum of (d/dz') and $-(s/c_1)$ represents the total derivative for the left-going wave multiplied by the phase shift of π at the perfect reflector, that is $-\{-(\partial/\partial z) + (\partial/\partial t)(\partial t/\partial z)\}$. The third term is the contribution of the noise forces that existed at $z = 0$ one round-trip time before. Note that the right-going component in the third term is associated with amplification and reflection at both end surfaces, whereas the left-going component is amplified and phase-shifted by π at the left end surface but has not reflected at the output surface.

10.4 The Lowest-Order Solution

We solve Equation 10.35 in a perturbative manner. For the lowest order, we solve the equation that is ensemble averaged with respect to the noise sources. We assume that the ensemble averages of the thermal and quantum noise sources vanish, that is,

$$\begin{aligned} \langle \hat{a}_j(0) \rangle &= 0 \\ \langle \hat{\Gamma}_m(t) \rangle &= 0 \end{aligned} \tag{10.36}$$

so that we have

$$\left\langle \hat{V}^{\pm}(z,s) \right\rangle = 0 \tag{10.37}$$

Then from Equation 10.35 we have

$$\left[1 - r'\exp\left\{2\int_{-d}^{0} dz'\left(-\frac{s}{c_1} + \frac{gN\langle\sigma_{z'}\rangle}{s+\gamma'}\right)\right\}\right]\hat{L}^+(0,s) = 0 \tag{10.38}$$

The inverse Laplace transform reads, using Equation 10.19,

$$\hat{e}^+(0,t) - \hat{e}^+(0,t-2d/c_1)\{-r\exp(2id\omega/c_1)\}\exp\int_{-d}^{0} \frac{2gN\langle\sigma_{z'}\rangle}{\gamma'}dz' = 0 \tag{10.39}$$

where we have ignored s as compared to γ'. This equation shows that, in the steady state, the right-going wave at the location $z = 0$ at time t is equal to the wave at $t{-}2d/c_1$ times the one round-trip net gain with associated phase shift. The net gain means double integrated gain times the product of r and –1, the reflection coefficients at both end surfaces. Note that the factor $gN\langle\sigma_z\rangle/\gamma \simeq \mathcal{g}N\langle\sigma_z\rangle/c_1|_{\omega_c=\nu_0}$ is the unsaturated gain per unit length of the laser medium for zero detuning, where $\mathcal{g}$, the gain per atom per unit time, was defined in Equation 4.14. The appearance of the frequency difference in $\gamma' = \gamma + i(\nu_0 - \omega)$ represents the dispersion of the amplifying medium. In the steady state, we have $\hat{e}^+(0,t) = \hat{e}^+(0, t - 2d/c_1)$. Therefore, for a non-trivial solution for $\hat{e}^+(0,t)$ to exist, we should have

$$1 + re^{2i\omega d/c_1} \exp\left\{2\int_{-d}^{0} dz' \left(\frac{gN\langle\sigma_{z'}\rangle}{\gamma'}\right)\right\} = 0 \tag{10.40}$$

Rewriting $\exp(2id\omega/c_1)$ as $\exp(2ikd)$ we have

$$\exp\int_{-d}^{0} dz' \left(\frac{2gN\langle\sigma_{z'}\rangle}{\gamma'}\right) = \frac{-1}{r\exp(2ikd)} \tag{10.41}$$

The integral on the LHS reads, by Equation 10.21a,

$$\int_{-d}^{0} dz' \left(\frac{2gN\langle\sigma_{z'}\rangle}{\gamma'}\right) = \frac{2gN\sigma^0}{\gamma'}\int_{-d}^{0} \frac{dz'}{1 + \bar{E}_{z'}^2/|E_s|^2} = 2\alpha^0 I \tag{10.42}$$

where the quantity I was defined in Equation 8.28b as

$$I \equiv \int_{-d}^{0} \frac{dz'}{1 + \bar{E}_{z'}^2/|E_s|^2} \tag{10.43a}$$

and the amplitude gain per unit length is

$$\alpha^0 = \frac{gN\sigma^0}{\gamma'} = \frac{gN\sigma^0}{\gamma + i(\nu_0 - \omega)} \tag{10.43b}$$

Therefore, Equation 10.41 reads

$$\exp(2\alpha^0 I) = \frac{-1}{r\exp(2ikd)} \tag{10.44}$$

This is exactly what appeared in Equation 8.29 in the semiclassical, nonlinear gain analysis of the same laser as in this chapter. Equation 8.29 describes the relation between the integrated local gain and the boundary conditions at both ends of the cavity. Since the analysis in Chapter 8 was done ignoring the noise sources, the situation there is the same as the situation here. Thus we can use all the results in Section 8.4 for the present, ensemble-averaged, and steady-state analysis. As we saw in Equation 7.33, $-r^{-1}\exp(-2ikd)$ is equal to $\exp\{(2d/c_1)(\gamma_c - i\omega + i\omega_c)\}$. Therefore

$$\frac{2g\sigma^0 N}{\gamma'}\int_{-d}^{0} \frac{dz'}{1 + \bar{E}_{z'}^2/|E_s|^2} = \frac{2d}{c_1}(\gamma_c - i\omega + i\omega_c) \tag{10.45}$$

From this equation, comparing the real and imaginary parts, we have the oscillation frequency and the space-averaged gain, which is equal to the cavity loss (see Equations 8.32 and 8.33)

$$\omega = \frac{\gamma\omega_c + \gamma_c \nu_0}{\gamma + \gamma_c} \tag{10.46}$$

$$\frac{g\sigma^0 N c_1}{\gamma}\frac{1}{1+\delta^2}\frac{1}{d}\int_{-d}^{0}\frac{dz'}{1+\bar{E}_{z'}^2/|E_s|^2} = \gamma_c \tag{10.47a}$$

where

$$\delta^2 = \left(\frac{\omega_c - \nu_0}{\gamma + \gamma_c}\right)^2 \tag{10.47b}$$

Note that the factor $g\sigma^0 N c_1/\{\gamma(1+\delta^2)\} = \mathcal{g}N\sigma^0$ is the unsaturated amplitude gain per unit time of the laser medium. The threshold atomic inversion, and the space-averaged, steady-state atomic inversion are (see Equations 8.48 and 8.53)

$$\sigma_{th}^0 = \bar{\sigma}_{ss} = \frac{2\hbar\omega\varepsilon_1}{\nu_0^2|p_a|^2 N}\gamma\gamma_c(1+\delta^2) = \frac{\gamma_c}{\mathcal{g}N} \tag{10.48}$$

Also, the field amplitude at the output end of the cavity is, from Equation 8.47,

$$\begin{aligned}\left\langle |e^+(0)|^2 \right\rangle &= \frac{|E_s|^2}{1-r^2}\left\{\frac{\nu_0^2|p|^2 N\sigma^0 d}{\varepsilon_1 c_1 \hbar\omega}\frac{1}{\gamma(1+\delta^2)} - \ln(1/r)\right\} \\ &= \frac{|E_s|^2}{1-r^2}\left\{\frac{2d}{c_1}\mathcal{g}N\sigma^0 - \ln(1/r)\right\}\end{aligned} \tag{10.49}$$

The quantity in the curly brackets in the second line is the difference between the gain and the loss for the field amplitude for one round trip in the cavity. As was stated in Chapter 8, we can in principle determine the amplitudes of the right- and the left-going waves, $e^{\pm}(z)$, except for undetermined phases.

Equation 10.47a also shows that the space-averaged, saturated gain is equal to the cavity loss.

10.5
The First-Order Solution: Temporal Evolution

10.5.1
The Formal Temporal Differential Equation

Next we consider Equation 10.35 in first order in the noise forces and in the parameter s. If we use the notation in Equation 10.43a, we have

$$\begin{aligned} 1 - r' \exp\left\{2\int_{-d}^{0} dz'\left(-\frac{s}{c_1}+\frac{gN\langle\sigma_{z'}\rangle}{s+\gamma'}\right)\right\} \\ = 1 - r' \exp\left\{-\frac{2ds}{c_1}+\frac{2GNI}{s+\gamma'}\right\} \end{aligned} \quad (10.50a)$$

where we have used the relation $g\sigma^0 = G$ (see Equations 7.16 and 8.10a). Here we refer to Equation 9.38. As a function of s, Equation 10.50a is in the same form as the LHS of Equation 9.38 multiplied by $\exp\{GNd/(s+\gamma')\}$. Therefore, the expansion on the RHS of Equation 9.38 can be used also in this nonlinear case. Thus we have

$$\begin{aligned} 1 - r' \exp\left\{2\int_{-d}^{0} dz'\left(-\frac{s}{c_1}+\frac{gN\langle\sigma_{z'}\rangle}{s+\gamma'}\right)\right\} \\ = \frac{2d/c_1}{s_0+\gamma'}\{\gamma+\gamma_c+i(\nu_0+\omega_c-2\omega)\}(s-s_0) \end{aligned} \quad (10.50b)$$

where s_0 is the pole given by Equation 10.50a obtained as in Equation 7.35:

$$s_0 = -\frac{\gamma\gamma_c+(\nu_0-\omega)(\omega-\omega_c)-(c_1/d)\int_{-d}^{0} gN\langle\sigma_{z'}\rangle dz' - i\{\gamma(\omega-\omega_c)+\gamma_c(\omega-\nu_0)\}}{\gamma+\gamma_c-i(\nu_0+\omega_c-2\omega)} \quad (10.50c)$$

Hereafter we set $s_0 = 0$ since the steady state analyzed using Equation 10.40 corresponded to this situation and the lowest-order results in Equations 10.46 and 10.47 can be obtained by setting $s_0=0$ in Equation 10.50c. Thus the first-order equation derived from Equation 10.35 reads, if we use Equation 10.40 on the RHS,

$$\begin{aligned} &\frac{2d}{c_1}\frac{\gamma'+\gamma_c'}{\gamma'} s\hat{L}^+(0,s) \\ &= \int_{-d}^{0} \exp\left[-\int_{z'}^{0} dz''\left\{\frac{s}{c_1}-\frac{gN\langle\sigma_{z''}\rangle}{s+\gamma'}\right\}\right]\left(\frac{d}{dz'}+\frac{s}{c_1}\right)\hat{V}^+(z',s)dz' \\ &\quad + e^{-2ds/c_1}\frac{1}{r'}\int_{-d}^{0} \exp\left[\int_{z'}^{0} dz''\left\{\frac{s}{c_1}-\frac{gN\langle\sigma_{z''}\rangle}{s+\gamma'}\right\}\right]\left(\frac{d}{dz'}-\frac{s}{c_1}\right)\hat{V}^-(z',s)dz' \\ &\quad - e^{-2ds/c_1}\hat{V}^+(0,s) - e^{-2ds/c_1}\frac{1}{r'}\hat{V}^-(0,s) \end{aligned} \quad (10.51)$$

where

$$\gamma' = \gamma + i(\nu_0 - \omega) \quad (10.52a)$$

$$\gamma_c' = \gamma_c + i(\omega_c - \omega) \quad (10.52b)$$

$$r' = -r\exp(2ikd) \quad (10.52c)$$

Inverse transformation of Equation 10.51 yields

$$
\begin{aligned}
&\frac{2d}{c_1}\frac{\gamma' + \gamma_c'}{\gamma'}\left\{\frac{d}{dt}\hat{e}^{+}(0,t) + \mathscr{L}^{-1}\hat{e}^{+}(0,0)\right\} \\
&= \int_{-d}^{0} \exp\left[\int_{z'}^{0} dz'' \left\{\frac{gN\langle\sigma_{z''}\rangle}{\gamma'}\right\}\right] \left\{\left(\frac{\partial}{\partial z'} + \frac{\partial}{c_1\partial t}\right)\hat{\upsilon}^{+}(z',t)\right\}_{t+z'/c_1} dz' \\
&+ \exp\left[\int_{-d}^{0} dz'' \left\{\frac{gN\langle\sigma_{z''}\rangle}{\gamma'}\right\}\right] \\
&\times \int_{-d}^{0} \exp\left[\int_{-d}^{z'} dz'' \left\{\frac{gN\langle\sigma_{z''}\rangle}{\gamma'}\right\}\right] \left\{\left(\frac{\partial}{\partial z'} - \frac{\partial}{c_1\partial t}\right)\hat{\upsilon}^{-}(z',t)\right\}_{t-(2d+z')/c_1} dz' \\
&+ \mathscr{L}^{-1}\int_{-d}^{0} \exp\left[\int_{z'}^{0} dz'' \left\{-\frac{s}{c_1} + \frac{gN\langle\sigma_{z''}\rangle}{\gamma'}\right\}\right] \left\{\frac{\hat{\upsilon}^{+}(z',0)}{c_1}\right\} dz' \\
&- \mathscr{L}^{-1}\exp\left[\int_{-d}^{0} dz'' \left\{\frac{gN\langle\sigma_{z''}\rangle}{\gamma'}\right\}\right] \\
&\times \int_{-d}^{0} \exp\left(-\frac{2d+z'}{c_1}s\right) \exp\left[\int_{-d}^{z'} dz'' \left\{\frac{gN\langle\sigma_{z''}\rangle}{\gamma'}\right\}\right] \left\{\frac{\hat{\upsilon}^{-}(z',0)}{c_1}\right\} dz' \\
&- \hat{\upsilon}^{+}(0, t - 2d/c_1) + \frac{1}{r\exp(2ikd)}\hat{\upsilon}^{-}(0, t-2d/c_1)
\end{aligned}
\tag{10.53}
$$

Here we have ignored s in the factor $s + \gamma'$, assuming much slower variation of the field envelope function than the dipolar relaxation. In the second term of $\hat{\upsilon}^{-}$, we have eliminated r' using Equation 10.40. The two terms with $\mathscr{L}^{-1}$ on the RHS come from the initial values associated with the Laplace transforms of time derivatives. These terms are proportional to $\delta(t - |z'|/c_1)$ and $\delta\{t - (2d + z')/c_1\}$, respectively, and important only at times $t \leq 2d/c_1$. These can be ignored for the steady state and will be neglected. Also, the inverse Laplace transform on the LHS is proportional to $\delta(t)$ and will be neglected. There appear retardation times that correctly represent the time required for a noise occurring at z' to reach $z = 0$. In the following, we shall be interested in the time variation of the field amplitude, which is much slower than the cavity decay rate and the reciprocal cavity round-trip time, so that our linewidth $\Delta\omega$ and the time difference of interest Δt should satisfy

$$
\begin{aligned}
\Delta\omega &\ll \gamma_c, \qquad c_1/2d \\
\Delta t &\gg \gamma_c^{-1}, \qquad 2d/c_1
\end{aligned}
\tag{10.54}
$$

The retardation times of the order $2d/c_1$ will accordingly be ignored. This assumption is in accordance with the approximation involved in going from Equation 10.35 to Equation 10.51: the LHS of Equation 10.35 may be written, for $s_0 = 0$, as $[1 - \exp\{\xi(-2d/c_1)s\}]L^{+}(0,s)$, where $\xi = (\gamma' + \gamma_c')/\gamma'$ (see Equation 10.50b). Therefore, Equation 10.51 is under the approximation $|s| \ll |\xi|^{-1}(c_1/2d)$, which is equivalent to Equation 10.54 except for the factor $|\xi|^{-1}$, which is usually of the order of unity. (Despite the approximation concerning the delay times in the following equation 10.55, we can show that the main results below can be derived without this approximation, that is, using Equation 10.53.) Thus Equation 10.53 becomes, moving the boundary values to the top,

$$\begin{aligned}\frac{d}{dt}\hat{e}^{+}(0,t) = &\frac{c_1}{2d}\frac{\gamma'}{\gamma'+\gamma'_c}\left[-\hat{\upsilon}^{+}(0,t)+\frac{1}{r\exp(2ikd)}\hat{\upsilon}^{-}(0,t)\right.\\&+\int_{-d}^{0}\exp\left\{\int_{z'}^{0}dz''\left(\frac{gN\langle\sigma_{z''}\rangle}{\gamma'}\right)\right\}\left\{\left(\frac{\partial}{\partial z'}+\frac{\partial}{c_1\partial t}\right)\hat{\upsilon}^{+}(z',t)\right\}dz'\\&+\exp\left\{\int_{-d}^{0}dz''\left(\frac{gN\langle\sigma_{z''}\rangle}{\gamma'}\right)\right\}\\&\left.\times\int_{-d}^{0}\exp\left\{\int_{-d}^{z'}dz''\left(\frac{gN\langle\sigma_{z''}\rangle}{\gamma'}\right)\right\}\left\{\left(\frac{\partial}{\partial z'}-\frac{\partial}{c_1\partial t}\right)\hat{\upsilon}^{-}(z',t)\right\}dz'\right]\end{aligned} \tag{10.55}$$

Here we give the expressions for the noise forces. For the thermal noise we have from Equation 9.30

$$\hat{f}_t^{\pm}(z,t) = \pm\sum_j\frac{1}{2}\sqrt{\frac{\hbar\omega_j}{\varepsilon_1 L(1-K\sin^2 k_{1j}^2 d)}}e^{\pm i(k_{1j}-k)(z+d)}\hat{a}_j(0)e^{-i(\omega_j-\omega)t} \tag{10.56}$$

For the quantum noise we have from Equations 9.34a and 9.34b

$$\begin{aligned}&\hat{f}_q^{+}(z,t)\exp\{i\omega(z+d)/c_1\}\\&=h\sum_m e^{-\gamma' t}\left[H(z-z_m)\exp\{(i\nu_0+\gamma)(z-z_m)/c_1\}\int_0^{t-(z-z_m)/c_1}\exp(\gamma' t')\tilde{\Gamma}_m(t')dt'\right.\\&-\exp\left\{(i\nu_0+\gamma)\frac{2d+z+z_m}{c_1}\right\}\int_0^{t-(2d+z+z_m)/c_1}\exp(\gamma' t')\tilde{\Gamma}_m(t')dt'\\&+\sum_{n=1}^{\infty}(-r)^n\left\{\exp\{(i\nu_0+\gamma)\tau_{1n}\}\int_0^{t-\tau_{1n}}\exp(\gamma' t')\tilde{\Gamma}_m(t')dt'\right.\\&\left.\left.-\exp\{(i\nu_0+\gamma)\tau_{3n}\}\int_0^{t-\tau_{3n}}\exp(\gamma' t')\tilde{\Gamma}_m(t')dt'\right\}\right]\end{aligned} \tag{10.57a}$$

and

$$\begin{aligned}&\hat{f}_q^{-}(z,t)\exp\{-i\omega(z+d)/c_1\}\\&=h\sum_m e^{-\gamma' t}\left[H(z_m-z)\exp\{(i\nu_0+\gamma)(z_m-z)/c_1\}\int_0^{t-(z_m-z)/c_1}\exp(\gamma' t')\tilde{\Gamma}_m(t')dt'\right.\\&+\sum_{n=1}^{\infty}(-r)^n\left\{\exp\{(i\nu_0+\gamma)\tau_{2n}\}\int_0^{t-\tau_{2n}}\exp(\gamma' t')\tilde{\Gamma}_m(t')dt'\right.\\&\left.\left.-\exp\{(i\nu_0+\gamma)\tau_{4n}\}\int_0^{t-\tau_{4n}}\exp(\gamma' t')\tilde{\Gamma}_m(t')dt'\right\}\right]\end{aligned} \tag{10.57b}$$

where the constant h was defined in Equation 9.33c as

$$h = \frac{i p_a v_0}{2\varepsilon_1 c_1} \tag{10.57c}$$

10.5.2 Thermal Noise

Now we examine the noise terms in Equation 10.55. Let us first examine the contribution from the thermal noise in the second and fourth lines in Equation 10.55. We note that, for $\hat{f}_t^+(z,t)$,

$$\begin{aligned}
&\left(\frac{\partial}{\partial z} + \frac{\partial}{c_1 \partial t}\right)\hat{f}_t^+(z,t) \\
&= \left(\frac{\partial}{\partial z} + \frac{\partial}{c_1 \partial t}\right)\sum_j \frac{1}{2}\sqrt{\frac{\hbar\omega_j}{\varepsilon_1 L(1 - K\sin^2 k_{1j}^2 d)}}\, e^{i(k_{1j}-k)(z+d)}\hat{a}_j(0)e^{-i(\omega_j-\omega)t} \qquad (10.58a) \\
&= 0
\end{aligned}$$

Similarly, the contribution from $\hat{f}_t^-(z,t)$ also vanishes:

$$\begin{aligned}
&\left(\frac{\partial}{\partial z} - \frac{\partial}{c_1 \partial t}\right)\hat{f}_t^-(z,t) \\
&= \left(\frac{\partial}{\partial z} - \frac{\partial}{c_1 \partial t}\right)\sum_j \frac{1}{2}\sqrt{\frac{\hbar\omega_j}{\varepsilon_1 L(1 - K\sin^2 k_{1j}^2 d)}}\, e^{-i(k_{1j}-k)(z+d)}\hat{a}_j(0)e^{-i(\omega_j-\omega)t} \qquad (10.58b) \\
&= 0
\end{aligned}$$

Thus the contribution from the thermal noise comes only from the quantity in the first line in Equation 10.55.

10.5.3 Quantum Noise

Next, let us examine the contribution from quantum noise to the first line in Equation 10.55. We have from Equations 10.57a and 10.57b, for $z = 0$,

$$\begin{aligned}
&\hat{f}_q^+(0,t)\exp\{i\omega d/c_1\} \\
&= h\sum_m e^{-\gamma' t}\Bigg[\exp\{-(iv_0+\gamma)z_m/c_1\}\int_0^{t+z_m/c_1}\exp(\gamma' t')\tilde{\Gamma}_m(t')dt' \\
&\quad - \exp\left\{(iv_0+\gamma)\frac{2d+z_m}{c_1}\right\}\int_0^{t-(2d+z_m)/c_1}\exp(\gamma' t')\tilde{\Gamma}_m(t')dt' \qquad (10.59a) \\
&\quad + \sum_{n=1}^{\infty}(-r)^n\left\{\exp\{(iv_0+\gamma)\tau_{1n}\}\int_0^{t-\tau_{1n}}\exp(\gamma' t')\tilde{\Gamma}_m(t')dt'\right. \\
&\quad \left.- \exp\{(iv_0+\gamma)\tau_{3n}\}\int_0^{t-\tau_{3n}}\exp(\gamma' t')\tilde{\Gamma}_m(t')dt'\right\}\Bigg]
\end{aligned}$$

and

$$\hat{f}_q^-(0,t)\exp\{-i\omega d/c_1\}$$

$$= h\sum_m e^{-\gamma' t}\left[\sum_{n=1}^{\infty}(-r)^n\left\{\exp\{(i\nu_0+\gamma)\tau_{2n}\}\int_0^{t-\tau_{2n}}\exp(\gamma' t')\tilde{\Gamma}_m(t')dt'\right.\right.$$
$$\left.\left.-\exp\{(i\nu_0+\gamma)\tau_{4n}\}\int_0^{t-\tau_{4n}}\exp(\gamma' t')\tilde{\Gamma}_m(t')dt'\right\}\right] \tag{10.59b}$$

with

$$\tau_{1n}=\frac{2nd-z_m}{c_1},\qquad \tau_{2n}=\frac{2nd+z_m}{c_1}$$
$$\tau_{3n}=\frac{2nd+2d+z_m}{c_1},\qquad \tau_{4n}=\frac{2nd-2d-z_m}{c_1} \tag{10.59c}$$

Equation 10.59a becomes

$$\hat{f}_q^+(0,t)\exp\{i\omega d/c_1\}$$

$$= h\sum_m e^{-\gamma' t}\left[\sum_{n=0}^{\infty}(-r)^n\left\{\exp\{(i\nu_0+\gamma)\tau_{1n}\}\int_0^{t-\tau_{1n}}\exp(\gamma' t')\tilde{\Gamma}_m(t')dt'\right.\right.$$
$$\left.\left.-\exp\{(i\nu_0+\gamma)\tau_{3n}\}\int_0^{t-\tau_{3n}}\exp(\gamma' t')\tilde{\Gamma}_m(t')dt'\right\}\right] \tag{10.60a}$$

Also, by setting $n \to n+1$ ($\tau_{2n} \to \tau_{3n}$, $\tau_{4n} \to \tau_{1n}$), Equation 10.59b may be rewritten as

$$\hat{f}_q^-(0,t)\exp\{-i\omega d/c_1\}$$

$$= h\sum_m e^{-\gamma' t}\left[\sum_{n+1=1}^{\infty}(-r)^{n+1}\left\{\exp\{(i\nu_0+\gamma)\tau_{3n}\}\int_0^{t-\tau_{3n}}\exp(\gamma' t')\tilde{\Gamma}_m(t')dt'\right.\right.$$
$$\left.\left.-\exp\{(i\nu_0+\gamma)\tau_{1n}\}\int_0^{t-\tau_{1n}}\exp(\gamma' t')\tilde{\Gamma}_m(t')dt'\right\}\right] \tag{10.60b}$$

Thus we see that

$$\hat{f}_q^+(0,t)\exp\{i\omega d/c_1\} = -(-r)^{-1}\hat{f}_q^-(0,t)\exp\{-i\omega d/c_1\} \tag{10.61a}$$

or

$$-\hat{f}_q^+(0,t)+\frac{1}{r\exp(2ikd)}\hat{f}_q^-(0,t)=0 \tag{10.61b}$$

Thus the contribution from quantum noise to the first line in Equation 10.55 vanishes.

Next we examine the contribution from the quantum noise sources to the second and fourth lines in Equation 10.55. If we use Equations 10.57a and 10.57b, the results of $\partial/\partial z$ and $(1/c_1)\partial/\partial t$ almost cancel each other, but $(\partial/\partial z)H(z-z_m) = \delta(z-z_m)$ and $(\partial/\partial z)H(z_m-z) = -\delta(z-z_m)$ remain. This is a rather strange result. To confirm it, we return to the original form of the quantum noise in Equation 10.3 and go to the case of homogeneous broadening:

$$\hat{F}_q(z,t) = \sum_m \left[\frac{ip_a \nu_0}{2} \int_0^t \sum_j U_j(z) U_j(z_m) e^{-i\omega_j(t-t')} \int_0^{t'} e^{-(i\nu_0+\gamma)(t'-t'')} \hat{\Gamma}_m(t'') dt'' dt' \right] \quad (10.62)$$

Dividing the function $U_j(z)$ for inside the cavity described in Equation 1.62b as

$$\begin{aligned} U_j(z) &= \sqrt{\frac{2}{\varepsilon_1 L(1 - K\sin^2 k_{1j} d)}} \sin k_{1j}(z+d) \\ &= \sqrt{\frac{2}{\varepsilon_1 L(1 - K\sin^2 k_{1j} d)}} \frac{e^{ik_{1j}(z+d)} - e^{-ik_{1j}(z+d)}}{2i} \end{aligned} \quad (10.63)$$

we have

$$\begin{aligned} \hat{f}_q^+(z,t) &= \exp\{i\omega t - i\omega(z+d)/c_1\} \\ &\times \sum_m \left[\frac{ip_a \nu_0}{2} \int_0^t \sum_j \sqrt{\frac{2}{\varepsilon_1 L(1 - K\sin^2 k_{1j} d)}} \frac{e^{ik_{1j}(z+d)}}{2i} U_j(z_m) \right. \\ &\left. \times e^{-i\omega_j(t-t')} \int_0^{t'} e^{-(i\nu_0+\gamma)(t'-t'')} \hat{\Gamma}_m(t'') dt'' dt' \right] \end{aligned} \quad (10.64a)$$

and

$$\begin{aligned} \hat{f}_q^-(z,t) &= \exp\{i\omega t + i\omega(z+d)/c_1\} \\ &\times \sum_m \left[-\frac{ip_a \nu_0}{2} \int_0^t \sum_j \sqrt{\frac{2}{\varepsilon_1 L(1 - K\sin^2 k_{1j} d)}} \frac{e^{-ik_{1j}(z+d)}}{2i} U_j(z_m) \right. \\ &\left. \times e^{-i\omega_j(t-t')} \int_0^{t'} e^{-(i\nu_0+\gamma)(t'-t'')} \hat{\Gamma}_m(t'') dt'' dt' \right] \end{aligned} \quad (10.64b)$$

The quantity in the second line in Equation 10.55 is thus

$$\begin{aligned} &\left(\frac{\partial}{\partial z} + \frac{\partial}{c_1 \partial t} \right) \hat{f}_q^+(z,t) \\ &= \left(\frac{\partial}{\partial z} + \frac{\partial}{c_1 \partial t} \right) \exp\{i\omega t - i\omega(z+d)/c_1\} \\ &\quad \times \sum_m \frac{ip_a \nu_0}{2} \int_0^t \sum_j \sqrt{\frac{2}{\varepsilon_1 L(1 - K\sin^2 k_{1j} d)}} \frac{e^{ik_{1j}(z+d)}}{2i} U_j(z_m) \\ &\quad \times e^{-i\omega_j(t-t')} \int_0^{t'} e^{-(i\nu_0+\gamma)(t'-t'')} \hat{\Gamma}_m(t'') dt'' dt' \end{aligned}$$

$$
\begin{aligned}
&= \frac{1}{c_1}\exp\{i\omega t - i\omega(z+d)/c_1\} \\
&\quad \times \sum_m \frac{ip_a v_0}{2} \sum_j \sqrt{\frac{2}{\varepsilon_1 L(1 - K\sin^2 k_{1j} d)}} \frac{e^{ik_{1j}(z+d)}}{2i} U_j(z_m) \\
&\quad \times \int_0^t e^{-(iv_0+\gamma)(t-t'')} \hat{\Gamma}_m(t'')dt'' \qquad (10.65a) \\
&= \frac{1}{c_1}\sum_m \frac{ip_a v_0}{2} \sum_j \frac{2}{\varepsilon_1 L(1 - K\sin^2 k_{1j} d)} \frac{e^{i(k_{1j}-k)(z+d)}}{2i} \sin k_{1j}(z_m + d) \\
&\quad \times \int_0^t e^{\{i(\omega - v_0)-\gamma\}(t-t'')} \tilde{\Gamma}_m(t'')dt''
\end{aligned}
$$

Likewise, the quantity in the fourth line in Equation 10.55 is

$$
\begin{aligned}
&\left(\frac{\partial}{\partial z} - \frac{\partial}{c_1 \partial t}\right)\hat{f}_q^-(z,t) \\
&= \left(\frac{\partial}{\partial z} - \frac{\partial}{c_1 \partial t}\right)\exp\{i\omega t + i\omega(z+d)/c_1\} \\
&\quad \times \sum_m \left(-\frac{ip_a v_0}{2}\right) \int_0^t \sum_j \sqrt{\frac{2}{\varepsilon_1 L(1 - K\sin^2 k_{1j} d)}} \frac{e^{-ik_{1j}(z+d)}}{2i} U_j(z_m) \\
&\quad \times e^{-i\omega_j(t-t')} \int_0^{t'} e^{-(iv_0+\gamma)(t'-t'')} \hat{\Gamma}_m(t'')dt''dt' \\
&= \frac{-1}{c_1}\exp\{i\omega t + i\omega(z+d)/c_1\} \qquad (10.65b) \\
&\quad \times \sum_m \left(-\frac{ip_a v_0}{2}\right) \sum_j \sqrt{\frac{2}{\varepsilon_1 L(1 - K\sin^2 k_{1j} d)}} \frac{e^{-ik_{1j}(z+d)}}{2i} U_j(z_m) \\
&\quad \times \int_0^t e^{-(iv_0+\gamma)(t-t'')} \hat{\Gamma}_m(t'')dt'' \\
&= \frac{1}{c_1}\sum_m \frac{ip_a v_0}{2} \sum_j \frac{2}{\varepsilon_1 L(1 - K\sin^2 k_{1j} d)} \frac{e^{-i(k_{1j}-k)(z+d)}}{2i} \sin k_{1j}(z_m + d) \\
&\quad \times \int_0^t e^{\{i(\omega - v_0)-\gamma\}(t-t'')} \tilde{\Gamma}_m(t'')dt''
\end{aligned}
$$

where we have set, according to Equation 10.4,

$$\hat{\Gamma}_m(t) = \tilde{\Gamma}_m(t)e^{-i\omega t} \qquad (10.65c)$$

The good thing about Equations 10.65a and 10.65b is that the summation over j does not contain time t, and the evaluation of the sum is similar to the proof of the completeness of the universal mode functions described in Equations 1.75–1.78.

In particular, using the expansion of the squared normalization constant described in the first line in Equation 1.70a and using the rule (set $k_{1j} - k = x$)

$$\int_0^{\infty} d\omega_j\, e^{\pm i(k_{1j}-k)(z+d)+ik_{1j}y} \simeq \int_{-\infty}^{\infty} c_1 dx\, e^{\pm ix(z+d)+i(x+k)y}$$
$$= 2\pi c_1 e^{iky}\delta(z+d\pm y) \tag{10.66}$$

we have, after some minor algebra, and noting that $\delta\{2nd \pm (z \pm z_m)\} = 0$ for $n \neq 0$,

$$\sum_j \frac{2}{\varepsilon_1 L(1-K\sin^2 k_{1j}d)} \frac{e^{\pm i(k_{1j}-k)(z+d)}}{2i} \sin k_{1j}(z_m+d)$$
$$= \pm\frac{e^{\mp ik(z_m+d)}}{\varepsilon_1}\delta(z-z_m) \tag{10.67}$$

Thus we have, as was expected,

$$\left(\frac{\partial}{\partial z} + \frac{\partial}{c_1 \partial t}\right)\hat{f}_q^{+}(z,t) = \frac{ip_a \nu_0}{2c_1\varepsilon_1}\sum_m e^{-ik(z_m+d)}\delta(z-z_m)$$
$$\times \int_0^t e^{\{i(\omega-\nu_0)-\gamma\}(t-t'')}\tilde{\Gamma}_m(t'')dt'' \tag{10.68a}$$

and

$$\left(\frac{\partial}{\partial z} - \frac{\partial}{c_1 \partial t}\right)\hat{f}_q^{-}(z,t) = -\frac{ip_a \nu_0}{2c_1\varepsilon_1}\sum_m e^{ik(z_m+d)}\delta(z-z_m)$$
$$\times \int_0^t e^{\{i(\omega-\nu_0)-\gamma\}(t-t'')}\tilde{\Gamma}_m(t'')dt'' \tag{10.68b}$$

10.5.4
The Temporal Differential Equation

Thus substituting Equations 10.58a, 10.58b, 10.61b, 10.68a, and 10.68b into Equation 10.55 we obtain

$$\frac{d\hat{e}^{+}(0,t)}{dt} = \frac{c_1}{2d}\frac{\gamma'}{\gamma' + \gamma'_c}\left[-\hat{f}_t^{\,+}(0,t) + \frac{1}{r\exp(2ikd)}\hat{f}_t^{\,-}(0,t)\right.$$
$$+\frac{ip_a\nu_0}{2c_1\varepsilon_1}e^{-ikd}\sum_m e^{-ikz_m}\exp\left\{\int_{z_m}^{0} dz\left(\frac{gN\langle\sigma_z\rangle}{\gamma'}\right)\right\}$$
$$\times \int_0^t e^{\{i(\omega-\nu_0)-\gamma\}(t-t'')}\tilde{\Gamma}_m(t'')dt''$$

$$
\begin{aligned}
&-\frac{ip_a\nu_0}{2c_1\varepsilon_1}e^{-ikd}\exp\left\{\int_{-d}^{0}dz\left(\frac{gN\langle\sigma_z\rangle}{\gamma'}\right)\right\} \\
&\times\sum_m e^{ik(z_m+2d)}\exp\left\{\int_{-d}^{z_m}dz\left(\frac{gN\langle\sigma_z\rangle}{\gamma'}\right)\right\} \qquad (10.69)\\
&\times\left.\int_0^t e^{\{i(\omega-\nu_0)-\gamma\}(t-t'')}\tilde{\Gamma}_m(t'')dt''\right]
\end{aligned}
$$

This equation has simple interpretations. The time rate of change of the right-going wave at the inner surface of the coupling interface is proportional to the sum of (i) some thermal noise contributions, (ii) quantum noise propagated to the right to the coupling surface with corresponding amplification and phase shift, and (iii) quantum noise propagated first to the left and reflected by the perfect conductor mirror and then propagated to the coupling surface. The common phase factor e^{-ikd} for the quantum noise comes from $[\exp\{ik(z+d)\}_{z=0}]^{-1}$. For the thermal noise, the coefficients for $\hat{f}_t^{+}(0,t)$ and $\hat{f}_t^{-}(0,t)$ come from the net phase shift plus amplification during one round trip within the cavity. This is seen by Equation 10.40 for the cavity round-trip gain and phase shift. The right-going part will first be reflected at the coupling surface and then amplified during the round trip with an additional phase jump of π at the perfect conductor mirror. The left-going part is not reflected at the coupling surface during the round trip and therefore gets a net gain of $1/r$. The reason why there are no noise contributions in Equation 10.69 that were generated further in the past than one round trip is that any single round trip in the past experiences an amplification that is canceled by the cavity loss and thus does not contribute to the change in the field amplitude.

10.5.5
Penetration of Thermal Noise into the Cavity

For the thermal noise, a more concise interpretation of the thermal contribution is that the quantity in the curly bracket in the first line in Equation 10.69 is the thermal noise that penetrated into the cavity from outside and was amplified through one round trip in the cavity. To see this, we note from Equation 10.56 that

$$
\hat{f}_t^{\pm}(z,t) = \pm\sum_j \frac{1}{2}\sqrt{\frac{\hbar\omega_j}{\varepsilon_1 L(1-K\sin^2 k_{1j}d)}}\,e^{\pm i(k_{1j}-k)(z+d)}\hat{a}_j(0)e^{-i(\omega_j-\omega)t}
$$

and from Equation 10.132 below, or from Equations 10.2 and 1.62b, the left-going thermal noise outside the cavity is

$$
\begin{aligned}
\hat{f}_{ot}^{-}(z,t) &= i\sum_j\sqrt{\frac{\hbar\omega_j}{2}}\sqrt{\frac{2}{\varepsilon_1 L(1-K\sin^2 k_{1j}^2 d)}} \qquad (10.70)\\
&\times\left(-\frac{k_{1j}}{k_{0j}}\cos k_{1j}d\frac{e^{-ik_{0j}z}}{2i}+\sin k_{1j}d\frac{e^{-ik_{0j}z}}{2}\right)\hat{a}_j(0)e^{-i(\omega_j-\omega)t}
\end{aligned}
$$

It is easy to see that, except for the common factor of

$$\sum_j \frac{1}{2}\sqrt{\frac{\hbar\omega_j}{\varepsilon_1 L(1-K\sin^2 k_{1j}d)}}\hat{a}_j(0)e^{-i(\omega_j-\omega)t}$$

the quantity in question is

$$\begin{aligned}
&-\hat{f}_t^{\,+}(-0,t)+\frac{1}{r\exp(2ikd)}\hat{f}_t^{\,-}(-0,t)\\
&\rightarrow -e^{i(k_{1j}-k)d}-\frac{1}{r\exp(2ikd)}e^{-i(k_{1j}-k)d}\\
&=-e^{-ikd}\left\{\left(1+\frac{1}{r}\right)\cos k_{1j}d+i\left(1-\frac{1}{r}\right)\sin k_{1j}d\right\}\\
&=ie^{-ikd}\left(\frac{1-r}{r}\right)\left\{-\frac{1}{i}\left(\frac{1+r}{1-r}\right)\cos k_{1j}d+\sin k_{1j}d\right\}\\
&\rightarrow \frac{T'}{r}e^{-ikd}\hat{f}_{ot}^{\,-}(+0,t)
\end{aligned}\tag{10.71}$$

where $T' = 1-r$ is the transmission coefficient for a wave incident on the coupling surface from outside. Also, $(1+r)/(1-r) = c_0/c_1 = k_{1j}/k_{0j}$. This relation can be rephrased as follows: "The left-going thermal wave just inside the cavity is the sum of the wave transmitted from outside and the right-going wave reflected with reflection coefficient r." It is important to note that the final form of Equation 10.71 implies that, except for the phase factor, the thermal noise reaching $z = -0$ in Equation 10.69 is the noise that penetrates into the cavity at $z = 0$ with transmission coefficient T' and is amplified by $|1/r| = |\exp(2\alpha_0 l)|$ (see Equation 10.44) during one round trip. In the next chapter, we use this property of the thermal noise, namely that it penetrates from outside and is amplified by the proper rate during one round trip.

10.6
Phase Diffusion and the Laser Linewidth

Here, as preparation for the subsequent analysis on the laser linewidth, we discuss the phase diffusion and its relation to the linewidth for the case of a laser operating above threshold with well-stabilized amplitude. We regard the amplitude $\hat{e}^+(0,t)$ as an essentially classical quantity that has well-stabilized real amplitude e_0 and a fluctuating phase $\phi(0,t)$ that is also real. Thus

$$\hat{e}^+(0,t) = e_0\exp[i\phi(0,t)] \tag{10.72}$$

where e_0 is given by the square root of Equation 10.49 for the case of the nonlinear laser discussed in this chapter.

The line profile or the power spectrum is obtained as the Fourier transform of the field correlation function. In the case of the field in the form of Equation 10.72, the correlation function has the form

$$\begin{aligned}\left\langle \hat{e}^{+\dagger}(0,\ t+\Delta t)\hat{e}^{+}(0,t)\right\rangle &= e_0^2\langle \exp[-i\{\phi(0,\ t+\Delta t)-\phi(0,t)\}]\rangle \\ &= e_0^2\langle \exp[-i\Delta\phi]\rangle \\ &\approx e_0^2\left\langle 1-i\Delta\phi+\tfrac{1}{2}(-i\Delta\phi)^2\right\rangle \\ &= e_0^2\left\{1-i\langle\Delta\phi\rangle-\tfrac{1}{2}\left\langle(\Delta\phi)^2\right\rangle\right\} \\ &= e_0^2\left\{1-\tfrac{1}{2}\left\langle(\Delta\phi)^2\right\rangle\right\} \\ &\approx e_0^2\exp\left\{-\tfrac{1}{2}\left\langle(\Delta\phi)^2\right\rangle\right\}\end{aligned} \tag{10.73a}$$

where

$$\Delta\phi \equiv \phi(0,\ t+\Delta t)-\phi(0,t) \tag{10.73b}$$

and we have assumed that the phase change $\Delta\phi$ in a time Δt is small compared to unity. The averaged phase change $\langle\Delta\phi\rangle$ vanishes because of the random nature of the phase change. Because of the delta-correlated nature of the noise sources, we anticipate that the ensemble-averaged, squared phase change is proportional to Δt. That is, we anticipate that

$$\left\langle(\Delta\phi)^2\right\rangle = B|\Delta t| \tag{10.74}$$

Then, writing the center frequency of oscillation as ω_0, we have the power spectrum:

$$\begin{aligned}I(\omega) &= \int_{-\infty}^{\infty} e^{-i\omega\Delta t}\left\langle \hat{E}^{(-)}(0,\ t+\Delta t)\hat{E}^{(+)}(0,t)\right\rangle d\Delta t \\ &= e_0^2\int_{-\infty}^{\infty} e^{-i\omega\Delta t}\langle \exp[-i\{\phi(0,\ t+\Delta t)-\phi(0,t)\}]\rangle e^{i\omega_o\Delta t}d\Delta t \\ &\approx e_0^2\int_{-\infty}^{\infty} e^{-i\omega\Delta t}\exp\left\{-\tfrac{1}{2}\left\langle(\Delta\phi)^2\right\rangle\right\}e^{i\omega_o\Delta t}d\Delta t \\ &= e_0^2\int_{-\infty}^{\infty} e^{-i\omega\Delta t}\exp\left\{-\tfrac{1}{2}B|\Delta t|\right\}e^{i\omega_o\Delta t}d\Delta t \\ &= e_0^2\frac{B}{(\omega-\omega_o)^2+(B/2)^2}\end{aligned} \tag{10.75a}$$

Therefore, the full width at half-maximum (FWHM) of the spectrum $\Delta\omega$ is B, the average squared phase change per unit time:

$$\Delta\omega = B \tag{10.75b}$$

Thus, using Equation 10.69 we seek the value of B. If we substitute Equation 10.72 and its complex conjugate into Equation 10.69, we have formally

$$\frac{d\phi(0,t)}{dt} = K'(t), \qquad \frac{d\phi(0,t)}{dt} = K'^{*}(t) \tag{10.76a}$$

where $K'(t)$ is the RHS of Equation 10.69 divided by $\{ie_0 \exp(i\phi)\}$. Since the phase ϕ is real, we have

$$\frac{d\phi(0,t)}{dt} = \mathrm{Re}K'(t) \equiv K(t) \tag{10.76b}$$

Here we have defined a real function $K(t)$. We have

$$\phi(0,t) = \int_{-\infty}^{t} K(t')dt' \tag{10.76c}$$

and the phase diffusion during t to $t + \Delta t$, measured as the ensemble average of the squared phase shift during t to $t + \Delta t$, is

$$\begin{aligned}\left\langle \{\Delta\phi(0,t)\}^2 \right\rangle &= \left\langle \int_t^{t+\Delta t} K(t')dt' \int_t^{t+\Delta t} K(t'')dt'' \right\rangle \\ &= \int_t^{t+\Delta t}\int_t^{t+\Delta t} \langle K(t')K(t'')\rangle dt'dt''\end{aligned} \tag{10.77}$$

Thus, if $K(t)$ is delta correlated, that is, if

$$\langle K(t')K(t'')\rangle = B\delta(t' - t'') \tag{10.78}$$

we have

$$\int_t^{t+\Delta t}\int_t^{t+\Delta t} \langle K(t')K(t'')\rangle dt'dt'' = B|\Delta t| \tag{10.79}$$

and we arrive at Equation 10.74.

10.7 Phase Diffusion in the Nonlinear Gain Regime

10.7.1 Phase Diffusion

We substitute Equation 10.72 into Equation 10.69 and multiply both sides by $\{ie_0 \exp(i\phi)\}^{-1}$ to obtain $d\phi(0,t)/dt$. Because it is not guaranteed that the quantity thus obtained is real, we have to add its Hermitian conjugate and divide by 2 to obtain a real phase, as in Equation 10.76b:

$$
\begin{aligned}
\frac{d\phi(0,t)}{dt} &= \frac{1}{2ie_0\exp[i\phi(0,t)]}\frac{c_1}{2d}\frac{\gamma'}{\gamma'+\gamma'_c}\left[-\hat{f}_t^{\,+}(0,t)+\frac{1}{r\exp(2ikd)}\hat{f}_t^{\,-}(0,t)\right.\\
&\quad+\frac{ip_a\nu_0}{2c_1\varepsilon_1}\sum_m e^{-ik(z_m+d)}\exp\left\{\int_{z_m}^{0}dz\left(\frac{gN\langle\sigma_z\rangle}{\gamma'}\right)\right\}\\
&\quad\times\int_0^t e^{\{i(\omega-\nu_0)-\gamma\}(t-t'')}\tilde{\Gamma}_m(t'')dt''\\
&\quad-\frac{ip_a\nu_0}{2c_1\varepsilon_1}\exp\left\{\int_{-d}^{0}dz\left(\frac{gN\langle\sigma_z\rangle}{\gamma'}\right)\right\}\\
&\quad\times\sum_m e^{ik(z_m+d)}\exp\left\{\int_{-d}^{z_m}dz\left(\frac{gN\langle\sigma_z\rangle}{\gamma'}\right)\right\}\\
&\quad\times\left.\int_0^t e^{\{i(\omega-\nu_0)-\gamma\}(t-t'')}\tilde{\Gamma}_m(t'')dt''\right]+\text{H.C.}
\end{aligned}
\tag{10.80}
$$

We assume that the phase diffusion is slow and the factor $\exp[-i\phi(0,t)]$ does not change much during the impulsive actions of a noise source. For this reason, this factor can safely be absorbed into the noise terms, and we will ignore this factor from now on. Since the function $K(t)$ defined by the RHS of Equation 10.76b and given by Equation 10.80 contains the operators $\hat{a}(0)$, $\hat{a}^\dagger(0)$, $\hat{\Gamma}_m(t)$, and $\hat{\Gamma}_m^\dagger(t)$, the LHS of Equation 10.79 contains $\left\langle\hat{a}(0)\hat{a}^\dagger(0)\right\rangle$, $\left\langle\hat{a}^\dagger(0)\hat{a}(0)\right\rangle$, $\left\langle\hat{\Gamma}_m(t)\hat{\Gamma}_m^\dagger(t)\right\rangle$, and $\left\langle\hat{\Gamma}_m^\dagger(t)\hat{\Gamma}_m(t)\right\rangle$ as non-vanishing noise correlations. Taking these into account and noting Equations 10.2, 10.3, and 10.15 for the definition of the noise terms, we construct the LHS of Equation 10.79 term by term.

First, we examine the terms of the form $\left\langle\hat{a}^\dagger(0)\hat{a}(0)\right\rangle$. The explicit expressions for $\hat{f}_t^{\pm}(z,t)$ are obtained from Equation 10.56. We have

$$
\begin{aligned}
&\left\langle K_{a^\dagger}(t')K_a(t'')\right\rangle\\
&=\frac{1}{4e_0^2}\left(\frac{c_1}{2d}\right)^2\left|\frac{\gamma'}{\gamma'+\gamma'_c}\right|^2\\
&\quad\times\left\langle\sum_i\frac{1}{2}\sqrt{\frac{\hbar\omega_i}{\varepsilon_1 L(1-K\sin^2 k_{1i}^2 d)}}\,\hat{a}_i^\dagger(0)e^{+i(\omega_i-\omega)t'}\right.\\
&\quad\times\left\{-e^{-i(k_{1i}-k)d}-\frac{1}{r\exp(-2ikd)}e^{+i(k_{1i}-k)d}\right\}\\
&\quad\times\sum_j\frac{1}{2}\sqrt{\frac{\hbar\omega_j}{\varepsilon_1 L(1-K\sin^2 k_{1j}^2 d)}}\,\hat{a}_j(0)e^{-i(\omega_j-\omega)t''}\\
&\quad\times\left.\left\{-e^{i(k_{1j}-k)d}-\frac{1}{r\exp(2ikd)}e^{-i(k_{1j}-k)d}\right\}\right\rangle
\end{aligned}
$$

$$= \frac{1}{4e_0^2}\left(\frac{c_1}{2d}\right)^2 \left|\frac{\gamma'}{\gamma' + \gamma'_c}\right|^2 \frac{1}{4}\sum_j \frac{\hbar\omega_j}{\varepsilon_1 L(1 - K\sin^2 k_{1j}^2 d)}$$

$$\times \left| e^{i(k_{1j}-k)d} + \frac{1}{r\exp(2ikd)} e^{-i(k_{1j}-k)d}\right|^2 \langle n_j\rangle e^{+i(\omega_j-\omega)(t'-t'')} \tag{10.81}$$

$$= \frac{1}{4e_0^2}\left(\frac{c_1}{2d}\right)^2 \left|\frac{\gamma'}{\gamma' + \gamma'_c}\right|^2 \frac{1}{4}\sum_j \frac{\hbar\omega_j}{\varepsilon_1 L}\left(\frac{1+r}{r}\right)^2 \langle n_j\rangle e^{+i(\omega_j-\omega)(t'-t'')}$$

where we have used Equation 9.4 for the correlation of the photon creation and annihilation operators. The normalization factor for the universal mode function has been canceled by the quantity in the absolute square with a residual factor $(1+r)^2/r^2$. Now the number of thermal photons $\langle n_j\rangle$ is a slowly varying function of the universal mode frequency ω_j and may be taken out of the summation together with ω_j itself. Then the summation over j can be approximated as

$$\sum_j \frac{\hbar\omega_j}{\varepsilon_1 L}\left(\frac{1+r}{r}\right)^2 \langle n_j\rangle e^{+i(\omega_j-\omega)(t'-t'')}$$

$$\begin{aligned} &= \frac{\hbar\omega}{\varepsilon_1 L}\left(\frac{1+r}{r}\right)^2 \langle n_\omega\rangle \int_0^\infty \frac{L}{c_0\pi} e^{+i(\omega_j-\omega)(t'-t'')} d\omega_j \\ &\simeq \frac{\hbar\omega}{\varepsilon_1 c_0 \pi}\left(\frac{1+r}{r}\right)^2 \langle n_\omega\rangle \int_{-\infty}^\infty e^{+ix(t'-t'')}dx \\ &= \frac{\hbar\omega}{\varepsilon_1 c_0 \pi}\left(\frac{1+r}{r}\right)^2 \langle n_\omega\rangle 2\pi\delta(t'-t'') \end{aligned} \tag{10.82}$$

where we have replaced ω_j by ω, the center frequency of oscillation, and $\langle n_j\rangle$ by $\langle n_\omega\rangle$. We have also set $\omega_j - \omega = x$ and extended the lower limit of integration to $-\infty$ on the grounds that the important contributions come from the universal modes that are within the constraint expressed by Equation 10.54 around the cavity resonant frequency. Thus we have

$$\left\langle K_{a^\dagger}(t')K_a(t'')\right\rangle = \left[\frac{1}{8e_0^2}\left(\frac{c_1}{2d}\right)^2 \left|\frac{\gamma'}{\gamma' + \gamma'_c}\right|^2 \frac{\hbar\omega}{\varepsilon_1 c_0}\left(\frac{1+r}{r}\right)^2 \langle n_\omega\rangle\right]\delta(t'-t'') \tag{10.83a}$$

By a similar calculation we obtain, using Equation 9.4,

$$\left\langle K_a(t')K_{a^\dagger}(t'')\right\rangle = \left[\frac{1}{8e_0^2}\left(\frac{c_1}{2d}\right)^2 \left|\frac{\gamma'}{\gamma' + \gamma'_c}\right|^2 \frac{\hbar\omega}{\varepsilon_1 c_0}\left(\frac{1+r}{r}\right)^2 \langle n_\omega + 1\rangle\right]\delta(t'-t'') \tag{10.83b}$$

These correlation functions satisfy the property in Equation 10.78. Adding these two for the thermal noise, we have

$$\begin{aligned}\left\langle \{\Delta\phi(0,t)\}^2 \right\rangle_t &= \int_t^{t+\Delta t}\int_t^{t+\Delta t} \langle K_t(t')K_t(t'')\rangle dt'dt'' \\ &= \frac{1}{4e_0^2}\left(\frac{c_1}{2d}\right)^2 \left|\frac{\gamma'}{\gamma'+\gamma'_c}\right|^2 \frac{\hbar\omega}{\varepsilon_1 c_0}\left(\frac{1+r}{r}\right)^2 \left(\langle n_\omega\rangle + \tfrac{1}{2}\right)|\Delta t|\end{aligned} \tag{10.83c}$$

The quantum noise part described by the second through sixth lines in Equation 10.80 looks complicated. Thus we use the following abbreviations:

$$A_m = \frac{ip_a v_0}{2c_1\varepsilon_1} e^{-ik(z_m+d)} \exp\int_{z_m}^0 dz\left(\frac{gN\langle\sigma_z\rangle}{\gamma'}\right) \tag{10.84a}$$

$$B_m = -\frac{ip_a v_0}{2c_1\varepsilon_1}\exp\left\{\int_{-d}^0 dz\left(\frac{gN\langle\sigma_z\rangle}{\gamma'}\right)\right\} e^{ik(z_m+d)} \exp\left\{\int_{-d}^{z_m} dz\left(\frac{gN\langle\sigma_z\rangle}{\gamma'}\right)\right\} \tag{10.84b}$$

We see that the constants A_m and B_m are functions of z_m and they are rapidly oscillating in space with the exponential factors $e^{\mp ik(z_m+d)}$. Then the quantum noise part of the phase change reads, from Equation 10.80,

$$\begin{aligned}\frac{d\phi_q(0,t)}{dt} &= \frac{1}{2ie_0}\frac{c_1}{2d}\frac{\gamma'}{\gamma'+\gamma'_c}\left\{\sum_m A_m\int_0^t e^{\{i(\omega-v_0)-\gamma\}(t-t'')}\tilde{\Gamma}_m(t'')dt''\right. \\ &\left.+\sum_m B_m\int_0^t e^{\{i(\omega-v_0)-\gamma\}(t-t'')}\tilde{\Gamma}_m(t'')dt''\right\} + \text{H.C.}\end{aligned} \tag{10.85}$$

Then the function for quantum noise $K_q(t)$ defined in Equation 10.77 is the RHS of this equation and the correlation function in Equation 10.79 becomes

$$\begin{aligned}\langle K_q(t')K_q(t)\rangle &= \frac{1}{4e_0^2}\left(\frac{c_1}{2d}\right)^2\left|\frac{\gamma'}{\gamma'+\gamma'_c}\right|^2 \\ &\times\left\langle\left[\left\{\sum_{m'} A^*_{m'}\int_0^{t'} e^{\{-i(\omega-v_0)-\gamma\}(t'-t''')}\tilde{\Gamma}^\dagger_{m'}(t''')dt'''\right.\right.\right. \\ &\left.\left.+\sum_{m'} B^*_{m'}\int_0^{t'} e^{\{-i(\omega-v_0)-\gamma\}(t'-t''')}\tilde{\Gamma}^\dagger_{m'}(t''')dt'''\right\} + \text{H.C.}\right] \\ &\times\left[\left\{\sum_m A_m\int_0^{t''} e^{\{i(\omega-v_0)-\gamma\}(t''-t'''')}\tilde{\Gamma}_m(t'''')dt''''\right.\right. \\ &\left.\left.\left.+\sum_m B_m\int_0^{t} e^{\{i(\omega-v_0)-\gamma\}(t''-t'''')}\tilde{\Gamma}_m(t'''')dt''''\right\} + \text{H.C.}\right]\right\rangle\end{aligned} \tag{10.86}$$

Because of the correlation properties (Equations 9.5a and 9.5b) for the quantum Langevin forces, the double sum reduces to a single sum and the double time integral reduces to a single time integral. For example, the portion that includes $A^*_{m'}A_m\left\langle\hat{\Gamma}^\dagger_{m'}\hat{\Gamma}_m\right\rangle$ reads, with suffix $\Gamma^\dagger\Gamma A$,

$$
\begin{aligned}
\langle K_q(t')K_q(t'')\rangle_{\Gamma^\dagger\Gamma A} =& \frac{1}{4e_0^2}\left(\frac{c_1}{2d}\right)^2\left|\frac{\gamma'}{\gamma'+\gamma'_c}\right|^2 \\
&\times\sum_m |A_m|^2 G^m_{21,12} e^{\{-i(\omega-\nu_0)(t'-t'')-\gamma(t'+t'')\}}\begin{cases}(e^{2\gamma t''}-1)/2\gamma, & t'\geq t''\\ (e^{2\gamma t'}-1)/2\gamma, & t'<t''\end{cases}\\
=& \frac{1}{4e_0^2}\left(\frac{c_1}{2d}\right)^2\left|\frac{\gamma'}{\gamma'+\gamma'_c}\right|^2\sum_m |A_m|^2 G^m_{21,12}e^{\{-i(\omega-\nu_0)(t'-t'')\}}\\
&\times\begin{cases}(e^{-\gamma(t'-t'')}-e^{-\gamma(t'+t'')})/2\gamma, & t'\geq t''\\ (e^{\gamma(t'-t'')}-e^{-\gamma(t'+t'')})/2\gamma, & t'<t''\end{cases}
\end{aligned}
\tag{10.87}
$$

The terms of $\exp\{-\gamma(t'+t'')\}$ decay after some time and are not important in a steady state, and so will be neglected. Similar results are obtained for $\langle K_q(t')K_q(t'')\rangle_{\Gamma^\dagger\Gamma B}$, $\langle K_q(t')K_q(t'')\rangle_{\Gamma\Gamma^\dagger A}$, and $\langle K_q(t')K_q(t'')\rangle_{\Gamma\Gamma^\dagger B}$. For the last two correlation functions, the noise correlation constants are $G^m_{12,21}$ that come from the correlation $\left\langle\hat{\Gamma}_m(t')\hat{\Gamma}^\dagger_m(t'')\right\rangle$. The cross-terms involving A^*_m and B_m or A_m and B^*_m contain space integrals of rapidly oscillating functions and have no contributions. Thus, summarizing the results, we have

$$
\begin{aligned}
\langle K_q(t')K_q(t'')\rangle =& \frac{1}{4e_0^2}\left(\frac{c_1}{2d}\right)^2\left|\frac{\gamma'}{\gamma'+\gamma'_c}\right|^2\sum_m\left(|A_m|^2+|B_m|^2\right)\\
&\times\left(G^m_{21,12}+G^m_{12,21}\right)e^{\{-i(\omega-\nu_0)(t'-t'')\}}\begin{cases}e^{-\gamma(t'-t'')}/2\gamma, & t'\geq t''\\ e^{\gamma(t'-t'')}/2\gamma, & t'<t''\end{cases}
\end{aligned}
\tag{10.88}
$$

Substituting this expression into Equation 10.77 we have for $\Delta t > 0$

$$
\begin{aligned}
&\int_t^{t+\Delta t}\int_t^{t+\Delta t}\langle K_q(t')K_q(t'')\rangle dt'dt''\\
&=\frac{1}{4e_0^2}\left(\frac{c_1}{2d}\right)^2\left|\frac{\gamma'}{\gamma'+\gamma'_c}\right|^2\sum_m\left(|A_m|^2+|B_m|^2\right)\left(G^m_{21,12}+G^m_{12,21}\right)\\
&\times\frac{1}{2\gamma}\int_t^{t+\Delta t}dt'\left\{\int_{t'}^{t+\Delta t}dt''\,e^{\{-i(\omega-\nu_0)(t'-t'')\}}e^{\gamma(t'-t'')}+\int_t^{t'}dt''\,e^{\{-i(\omega-\nu_0)(t'-t'')\}}e^{-\gamma(t'-t'')}\right\}\\
&=\frac{1}{4e_0^2}\left(\frac{c_1}{2d}\right)^2\left|\frac{\gamma'}{\gamma'+\gamma'_c}\right|^2\sum_m\left(|A_m|^2+|B_m|^2\right)\left(G^m_{21,12}+G^m_{12,21}\right)\\
&\times\frac{1}{2\gamma}\int_t^{t+\Delta t}dt'\left\{\frac{e^{\{-i(\omega-\nu_0)(t'-t-\Delta t)\}}e^{\gamma(t'-t-\Delta t)}-1}{i(\omega-\nu_0)-\gamma}+\frac{1-e^{\{-i(\omega-\nu_0)(t'-t)\}}e^{-\gamma(t'-t)}}{i(\omega-\nu_0)+\gamma}\right\}\\
&=\frac{1}{4e_0^2}\left(\frac{c_1}{2d}\right)^2\left|\frac{\gamma'}{\gamma'+\gamma'_c}\right|^2\sum_m\left(|A_m|^2+|B_m|^2\right)\left(G^m_{21,12}+G^m_{12,21}\right)\\
&\times\frac{1}{2\gamma}\left\{\frac{1-e^{i(\omega-\nu_0)\Delta t}e^{-\gamma\Delta t}}{-\{i(\omega-\nu_0)-\gamma\}^2}-\frac{\Delta t}{i(\omega-\nu_0)-\gamma}+\frac{\Delta t}{i(\omega-\nu_0)+\gamma}-\frac{e^{\{-i(\omega-\nu_0)\Delta t\}}e^{-\gamma\Delta t}-1}{-\{i(\omega-\nu_0)+\gamma\}^2}\right\}
\end{aligned}
\tag{10.89a}
$$

If $\Delta t < 0$ we have

$$\begin{aligned}
&\int_t^{t+\Delta t}\int_t^{t+\Delta t}\langle K_q(t')K_q(t'')\rangle dt' dt''\\
&=\frac{1}{4e_0^2}\left(\frac{c_1}{2d}\right)^2\left|\frac{\gamma'}{\gamma'+\gamma_c'}\right|^2\sum_m\left(|A_m|^2+|B_m|^2\right)\left(G_{21,12}^m+G_{12,21}^m\right)\\
&\times\frac{1}{2\gamma}\int_t^{t+\Delta t}dt'\left\{\int_{t'}^{t+\Delta t}dt''\,e^{\{-i(\omega-\nu_0)(t'-t'')\}}e^{-\gamma(t'-t'')}+\int_t^{t'}dt''\,e^{\{-i(\omega-\nu_0)(t'-t'')\}}e^{\gamma(t'-t'')}\right\}\\
&=\frac{1}{4e_0^2}\left(\frac{c_1}{2d}\right)^2\left|\frac{\gamma'}{\gamma'+\gamma_c'}\right|^2\sum_m\left(|A_m|^2+|B_m|^2\right)\left(G_{21,12}^m+G_{12,21}^m\right)\\
&\times\frac{1}{2\gamma}\int_t^{t+\Delta t}dt'\left\{\frac{e^{\{-i(\omega-\nu_0)(t'-t-\Delta t)\}}e^{-\gamma(t'-t-\Delta t)}-1}{i(\omega-\nu_0)+\gamma}+\frac{1-e^{\{-i(\omega-\nu_0)(t'-t)\}}e^{\gamma(t'-t)}}{i(\omega-\nu_0)-\gamma}\right\}\\
&=\frac{1}{4e_0^2}\left(\frac{c_1}{2d}\right)^2\left|\frac{\gamma'}{\gamma'+\gamma_c'}\right|^2\sum_m\left(|A_m|^2+|B_m|^2\right)\left(G_{21,12}^m+G_{12,21}^m\right)\\
&\times\frac{1}{2\gamma}\left\{\frac{1-e^{i(\omega-\nu_0)\Delta t}e^{\gamma\Delta t}}{-\{i(\omega-\nu_0)+\gamma\}^2}-\frac{\Delta t}{i(\omega-\nu_0)+\gamma}+\frac{\Delta t}{i(\omega-\nu_0)-\gamma}-\frac{e^{\{-i(\omega-\nu_0)\Delta t\}}e^{\gamma\Delta t}-1}{-\{i(\omega-\nu_0)-\gamma\}^2}\right\}
\end{aligned}\tag{10.89b}$$

Now we are assuming that the time variation of the field amplitude is much slower than the dipole relaxation. Remember that we neglected s compared with γ' in Equation 10.53, implying that the time derivative can be ignored compared to γ'. In another words, the time scale of interest is much larger than the reciprocal dipole damping rate. Thus we have

$$\gamma\,|\Delta t| \gg 1 \tag{10.90}$$

Under this approximation only the terms of Δt in the curly brackets in Equations 10.89a and 10.89b remain. Thus

$$\begin{aligned}
\left\langle\{\Delta\phi(0,t)\}^2\right\rangle_q &=\int_t^{t+\Delta t}\int_t^{t+\Delta t}\langle K_q(t')K_q(t'')\rangle dt' dt''\\
&=\frac{1}{4e_0^2}\left(\frac{c_1}{2d}\right)^2\left|\frac{\gamma'}{\gamma'+\gamma_c'}\right|^2\sum_m\left(|A_m|^2+|B_m|^2\right)\\
&\times\left(G_{21,12}^m+G_{12,21}^m\right)\frac{|\Delta t|}{\gamma^2+(\omega-\nu_0)^2}
\end{aligned}\tag{10.91}$$

Summarizing, the contribution to the laser linewidth comes from the thermal noise in the form of Equation 10.83c and from the quantum noise in the form of Equation 10.91. The remaining task is to evaluate the sum over m in Equation 10.91, where the coefficients A_m and B_m are defined in Equations 10.84a and 10.84b. For homogeneous broadening of the atoms and homogeneous pumping, the coefficients $G_{21,12}^m$ and $G_{12,21}^m$ are constants that are independent of the suffix m and, from Equations 9.5a and 9.5b, the sum is simply

$$G_{21,12}^m+G_{12,21}^m=2\gamma \tag{10.92}$$

Therefore we examine the sum

$$\sum_m \left(|A_m|^2 + |B_m|^2\right) = \sum_m \left(\frac{\rho a \nu_0}{2c_1\varepsilon_1}\right)^2 \left[\exp\int_{z_m}^{0} dz\left(\frac{2\gamma g N\langle\sigma_z\rangle}{\gamma^2+(\omega-\nu_0)^2}\right) + \exp\left\{\int_{-d}^{0} dz\left(\frac{2\gamma g N\langle\sigma_z\rangle}{\gamma^2+(\omega-\nu_0)^2}\right)\right\} \times \exp\left\{\int_{-d}^{z_m} dz\left(\frac{2\gamma g N\langle\sigma_z\rangle}{\gamma^2+(\omega-\nu_0)^2}\right)\right\}\right] \tag{10.93}$$

10.7.2
Evaluation of the Sum $\sum_m \left(|A_m|^2 + |B_m|^2\right)$

The evaluation of the above sum can be done by consulting the results of Chapter 8, where the same laser as here was analyzed semiclassically in the gain saturated regime. Note that $\langle\sigma_z\rangle$ in Equation 10.93 depends on the field amplitude through Equation 10.21a, which, in turn, depends on the location z. First, for the spatial differential equation for the absolute square of the right- and left-traveling field amplitudes, we again cite Equations 8.35a and 8.35b, which are, respectively,

$$(d/dz)|e^+(z)|^2 = \frac{\alpha^0+\alpha^{0*}}{1+|E_{z/s}|^2}|e^+(z)|^2$$

$$(d/dz)|e^-(z)|^2 = \frac{-(\alpha^0+\alpha^{0*})}{1+|E_{z/s}|^2}|e^-(z)|^2$$

Because we are now assuming that the magnitude of the field amplitude is constant in time, allowing only phase diffusion, we can use these equations here. These are integrated as

$$|e^+(z)|^2 = |e^+(z_0)|^2 \exp\left(\int_{z_0}^{z} \frac{\alpha^0+\alpha^{0*}}{1+|E_{z'/s}|^2}dz'\right) \tag{10.94a}$$

$$|e^-(z)|^2 = |e^-(z_0)|^2 \exp\left(-\int_{z_0}^{z} \frac{\alpha^0+\alpha^{0*}}{1+|E_{z'/s}|^2}dz'\right) \tag{10.94b}$$

where, from Equation 8.23c, the power gain per unit length is

$$\alpha^0+\alpha^{0*} = \frac{g}{\gamma'}\sigma^0 N + \text{C.C.} = \frac{2\gamma g N\sigma^0}{(\nu_0-\omega)^2+\gamma^2} = \frac{2\mathcal{g}N\sigma^0}{c_1} \tag{10.94c}$$

Comparing these with Equation 10.21a, that is,

$$\langle \sigma_m \rangle = \frac{\sigma^0}{1 + \bar{E}_m^2/|E_s|^2} \tag{10.95}$$

we see that the integrand in Equation 10.93 is equal to the integrand on the RHS of Equation 10.94a:

$$\frac{2\gamma g N\langle \sigma_z \rangle}{\gamma^2 + (\omega - \nu_0)^2} = \frac{\alpha^0 + \alpha^{0*}}{1 + |E_{z/s}|^2} \tag{10.96}$$

Thus we have

$$\exp \int_{z_m}^{0} dz \left(\frac{2\gamma g N\langle \sigma_z \rangle}{\gamma^2 + (\omega - \nu_0)^2} \right) = \frac{|e^+(0)|^2}{|e^+(z_m)|^2} \tag{10.97a}$$

and

$$\exp \int_{-d}^{z_m} dz \left(\frac{2\gamma g N\langle \sigma_z \rangle}{\gamma^2 + (\omega - \nu_0)^2} \right) = \frac{|e^+(z_m)|^2}{|e^+(-d)|^2} \tag{10.97b}$$

Also, we have

$$\exp\left\{ \int_{-d}^{0} dz \left(\frac{2\gamma g N\langle \sigma_z \rangle}{\gamma^2 + (\omega - \nu_0)^2} \right) \right\} = \frac{|e^+(0)|^2}{|e^+(-d)|^2} \tag{10.97c}$$

Now, as can be seen from Equations 10.94a and 10.94b, the product of $|e^+(z)|^2$ and $|e^-(z)|^2$ is a constant:

$$|e^+(z)|^2 |e^-(z)|^2 = C \tag{10.98a}$$

Since the field vanishes at the perfect conductor surface at $z = -d$, Equation 10.15 shows that

$$e^+(-d) = -e^-(-d) \tag{10.98b}$$

Therefore, we can set

$$C = |e^+(-d)|^2 |e^-(-d)|^2 = |e^+(-d)|^4 \tag{10.98c}$$

Thus we have

$$|e^+(0)|^2 \Big/ |e^+(z_m)|^2 = |e^+(0)|^2 |e^-(z_m)|^2 \Big/ |e^+(-d)|^4 \tag{10.99}$$

Therefore Equation 10.93 becomes

$$\sum_m \left(|A_m|^2 + |B_m|^2\right) = N\left(\frac{|p_a|\nu_0}{2c_1\varepsilon_1}\right)^2 \frac{|e^+(0)|^2}{|e^+(-d)|^2} \times \int_{-d}^{0} dz_m \left(\frac{|e^-(z_m)|^2 + |e^+(z_m)|^2}{|e^+(-d)|^2}\right) \tag{10.100}$$

In Appendix E we evaluate the integral in Equation 10.100. The result is

$$\int_{-d}^{0} dz_m \left(\frac{|e^-(z_m)|^2 + |e^+(z_m)|^2}{|e^+(-d)|^2}\right) = 2d\frac{\beta_c}{\gamma_c}\left\{1 + \frac{\Delta}{1+\Delta}g(r)\right\} \tag{10.101}$$

where the function

$$\begin{aligned} g(r) &= \frac{1}{2}\left(\frac{\gamma_c}{\beta_c}\right)^2 + \frac{1+r^2}{4r}\frac{\gamma_c}{\beta_c} - 1 \\ &= 2\left\{\frac{\ln(1/r)}{(1-r^2)/r}\right\}^2 + \frac{\frac{1}{2}\ln(1/r)\{(1-r^4)/r^2\}}{\{(1-r^2)/r\}^2} - 1 \end{aligned} \tag{10.102}$$

is monotonically decreasing from $+\infty$ to 0 as r goes from 0 to 1, and

$$\Delta = \frac{\sigma^0 - \sigma^0_{th}}{\sigma^0_{th}} \tag{10.103}$$

is the fractional excess atomic inversion. The factor $\beta_c = (c_1/2d)(1-r^2)/2r$ was introduced in Equation 6.35 and appeared also in Equations 9.92 and 9.108 concerning the integrated, absolute squared field strength of the cavity resonant mode. We stress here that the integral would be simply $2d$ if the field distribution were uniform, as in the quasimode analysis. For $r \to 1$ we can show that $\gamma_c/\beta_c \to 1$ and $g(r) \to 0$.

Now Equation 10.100 reads, owing to Equation E.10,

$$\sum_m \left(|A_m|^2 + |B_m|^2\right) = N\left(\frac{|p_a|\nu_0}{2c_1\varepsilon_1}\right)^2 \frac{2d}{r}\frac{\beta_c}{\gamma_c}\left\{1 + \frac{\Delta}{1+\Delta}g(r)\right\} \tag{10.104}$$

Returning now to Equation 10.91 and using Equation 10.92, we obtain

$$\begin{aligned} &\int_t^{t+\Delta t}\int_t^{t+\Delta t} \langle K_q(t')K_q(t'')\rangle dt'dt'' \\ &= \left[\frac{1}{4e_0^2}\left(\frac{c_1}{2d}\right)^2 \left|\frac{\gamma'}{\gamma' + \gamma'_c}\right|^2 \frac{2\gamma N}{\gamma^2 + (\omega - \nu_0)^2}\left(\frac{|p_a|\nu_0}{2c_1\varepsilon_1}\right)^2 \frac{2d}{r}\frac{\beta_c}{\gamma_c}\left\{1 + \frac{\Delta}{1+\Delta}g(r)\right\}\right]|\Delta t| \end{aligned} \tag{10.105}$$

Thus the contribution from the quantum noise to the linewidth (FWHM) is the quantity in the square bracket. We rewrite this equation using the expression for the threshold atomic inversion in Equation 10.48 and the expression for β_c:

$$\left\langle \{\Delta\phi(0,t)\}^2 \right\rangle_q = \int_t^{t+\Delta t}\int_t^{t+\Delta t} \langle K_q(t')K_q(t'')\rangle dt'dt'' \\ = \left[\frac{1}{4e_0^2}\left(\frac{c_1}{2d}\right)^2\left|\frac{\gamma'}{\gamma'+\gamma'_c}\right|^2 \frac{\hbar\omega}{c_1\varepsilon_1}\frac{1-r^2}{r^2}\frac{N}{2N\sigma_{th}^0}\left\{1+\frac{\Delta}{1+\Delta}g(r)\right\}\right]|\Delta t| \tag{10.106}$$

10.7.3
The Linewidth and the Correction Factors

Adding the contributions from the thermal noise, Equation 10.83c, and from the quantum noise, Equation 10.106, we have the total diffusion and the linewidth (FWHM):

$$\left\langle \{\Delta\phi(0,t)\}^2 \right\rangle = \frac{1}{4e_0^2}\left(\frac{c_1}{2d}\right)^2\left|\frac{\gamma'}{\gamma'+\gamma'_c}\right|^2 \left[\frac{\hbar\omega}{\varepsilon_1 c_0}\left(\frac{1+r}{r}\right)^2\left(\langle n_\omega\rangle+\frac{1}{2}\right)\right. \\ \left. +\frac{\hbar\omega}{c_1\varepsilon_1}\frac{1-r^2}{r^2}\frac{N}{2N\sigma_{th}^0}\left\{1+\frac{\Delta}{1+\Delta}g(r)\right\}\right]|\Delta t| \tag{10.107a}$$

Thus

$$\Delta\omega = \frac{1}{4e_0^2}\left(\frac{c_1}{2d}\right)^2\left|\frac{\gamma'}{\gamma'+\gamma'_c}\right|^2 \left[\frac{\hbar\omega}{\varepsilon_1}\frac{1}{c_0}\left(\frac{1+r}{r}\right)^2\left(\langle n_\omega\rangle+\frac{1}{2}\right)\right. \\ \left. +\frac{\hbar\omega}{c_1\varepsilon_1}\frac{1-r^2}{r^2}\frac{N}{2N\sigma_{th}^0}\left\{1+\frac{\Delta}{1+\Delta}g(r)\right\}\right] \\ = \frac{1}{4e_0^2}\left(\frac{c_1}{2d}\right)^2\left|\frac{\gamma'}{\gamma'+\gamma'_c}\right|^2 \frac{\hbar\omega}{\varepsilon_1}\frac{1}{c_0}\left(\frac{1+r}{r}\right)^2 \\ \times\left[\left(\langle n_\omega\rangle+\tfrac{1}{2}\right)+\frac{N}{2N\sigma_{th}^0}\left\{1+\frac{\Delta}{1+\Delta}g(r)\right\}\right] \tag{10.107b}$$

where in the second equality we have used the relation $c_0/c_1 = (1+r)/(1-r)$. Here the real field amplitude e_0 is given by the square root of Equation 10.49.

Now we try to express the linewidth in terms of the power output. In order to obtain the output power dependence on the linewidth, we require the relation between the amplitude e_0 and the power output P. The power output per unit cross-sectional area ρ may be related to the real amplitude e_0 of the right-traveling wave at the inner surface of the coupling interface by

$$P = 2\varepsilon_0 c_0 |Te_0|^2 = 2\varepsilon_0 c_0 (1+r)^2 |e_0|^2 \tag{10.108}$$

where we have put the absolute sign on the real amplitude for later comparison. In the next section we will examine how the use of the transmission coefficient T in

Equation 10.108 can be justified. Using this expression and noting that $\varepsilon_0/\varepsilon_1 = c_1^2/c_0^2 = (1-r)^2/(1+r)^2$, we have

$$\Delta\omega = \frac{2\hbar\omega}{P}\left(\frac{c_1}{2d}\right)^2\left|\frac{\gamma'}{\gamma'+\gamma'_c}\right|^2\left(\frac{1-r^2}{2r}\right)^2 \times \left[\left(\langle n_\omega\rangle + \tfrac{1}{2}\right) + \frac{N}{2N\sigma_{th}^0}\left\{1+\frac{\Delta}{1+\Delta}g(r)\right\}\right] \tag{10.109a}$$

Finally, using the definition $\beta_c = (c_1/2d)(1-r^2)/2r$ (Equation 6.35) and Equations 10.52a and 10.52b for γ' and γ'_c together with the expression (Equation 10.46) for the steady-state oscillation angular frequency, we have

$$\Delta\omega = \frac{2\hbar\omega\gamma_c^2}{P}\left(\frac{\beta_c}{\gamma_c}\right)^2\frac{\gamma^2(1+\delta^2)}{(\gamma+\gamma_c)^2+\delta^2(\gamma-\gamma_c)^2} \times \left[\left(\langle n_\omega\rangle + \tfrac{1}{2}\right) + \frac{N_2+N_1}{2N\sigma_{th}^0}\left\{1+\frac{\Delta}{1+\Delta}g(r)\right\}\right] \tag{10.109b}$$

We use Equations 10.48 and 10.49 to relate the power output and the fractional excess atomic inversion as

$$P = 2\varepsilon_0 c_0 T^2\left\langle\left|e^+(0)\right|^2\right\rangle = 2\varepsilon_1 c_1|E_s|^2\ln\left(\frac{1}{r}\right)\left\{\frac{\sigma^0}{\sigma_{th}^0}-1\right\} = P_s\ln\left(\frac{1}{r}\right)\Delta \tag{10.110}$$

where $P_s = 2\varepsilon_1 c_1|E_s|^2$ is the saturation power. Therefore, the fractional excess inversion is proportional to the output power, and Equation 10.109b can be rewritten as

$$\Delta\omega = \frac{2\hbar\omega\gamma_c^2\gamma^2(1+\delta^2)}{(\gamma+\gamma_c)^2+\delta^2(\gamma-\gamma_c)^2}\left(\frac{\beta_c}{\gamma_c}\right)^2 \times \left[\frac{1}{P}\left\{\left(\langle n_\omega\rangle+\tfrac{1}{2}\right)+\frac{N_2+N_1}{2N\sigma_{th}^0}\right\} + \frac{N_2+N_1}{2N\sigma_{th}^0}\frac{g(r)}{P+P_s\ln(1/r)}\right] \tag{10.111}$$

This expression has two corrections compared to the conventional formula obtained in Equation 4.82 and in Refs. [2] and [3]. One correction is the factor

$$K_L = \left(\frac{\beta_c}{\gamma_c}\right)^2 = \left\{\frac{(1-r^2)/2r}{\ln(1/r)}\right\}^2 \tag{10.112}$$

which also appeared in the previous chapter in the quantum linear gain analysis.

The other is the newly added term that is proportional to $g(r)$. This term originates in the quantum noise contribution. Figures 10.1 and 10.2 depict the factor $K_L = (\beta_c/\gamma_c)^2$ and the function $g(r)$, respectively, as functions of the reflection coefficient r. The first correction factor is a decreasing function of the reflection coefficient r: it decreases from $+\infty$ to 1 as r goes from 0 to 1, and thus is

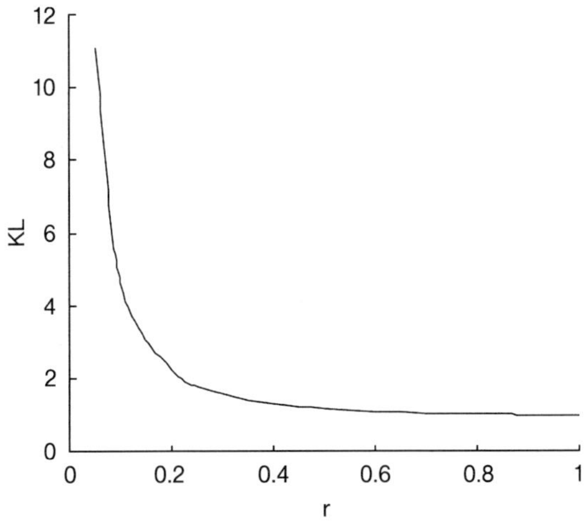

Figure 10.1 The longitudinal excess noise factor $K_L = (\beta_c/\gamma_c)^2$ as a function of the amplitude reflectivity *r*.

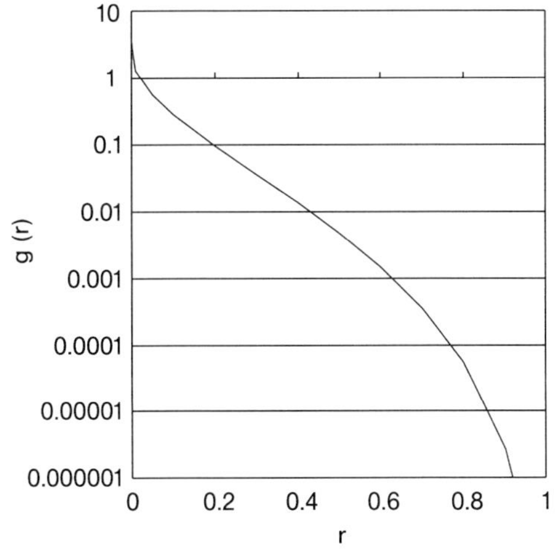

Figure 10.2 The function $g(r)$.

important for small r. The second correction is important for large fractional excess atomic inversion Δ, or small saturation power, and for small r. The function $g(r)$ decreases monotonically from $+\infty$ to 0 as r goes from 0 to 1 as stated earlier. The second correction brings to the laser linewidth a non-power-reciprocal part of the linewidth. For the region of small power output such that $P \ll P_s \ln(1/r)$, there appears a power-independent part of the linewidth as noticed by Prasad [4] and by Van Exter *et al.* [5]. The quantity $P_s \ln(1/r)$ corresponds to the power output for the case where the field amplitude squared is equal to the saturation parameter $|E_s|^2$, whence $\sigma_{ss} = \sigma_{th} = \sigma^0/2$ and $\Delta = 1$ (see Equation 10.95).

The form of the noise $\langle n_\omega \rangle + \frac{1}{2} + (N_2 + N_1)/(2N\sigma^0_{th})$ in Equation 10.111 looks different from that in Equation 9.105, $\{\sigma^2/(\sigma_{th}\sigma_{th0})\}\langle n_\omega \rangle + N_2/(N\sigma_{th})$, obtained for the linear gain analysis. In the case of Equation 9.105, the above form appeared directly from the normally ordered correlation functions in Equations 9.4a and 9.5a. In the case of Equation 10.111, the factors $\langle n_c \rangle + \frac{1}{2}$ and $N/(2N\sigma_{th})$ appeared because of the symmetrically ordered correlation functions used for the evaluation of the real phase of the field. In particular, the symmetric ordering appeared in Equations 10.83a, 10.83b and 10.88 because of the Hermitian conjugate terms. So, in this case of nonlinear gain analysis, the anti-normally ordered correlation functions in Equations 9.4b and 9.5b were also taken into account. Except for the factor $\sigma^2/(\sigma_{th}\sigma_{th0})$, the above two forms are the same since $\frac{1}{2} + (N_2 + N_1)/(2N\sigma_{th}) = N_2/(N\sigma_{th})$. It should be noted that different orderings of the noise operators lead to almost the same form of the noise contributions.

10.8
The Field Outside the Cavity

Up to now we have considered the linewidth for the field at the output end of the cavity. The output power dependence of the linewidth was derived through the *ad hoc* Equation 10.108, expressing the assumed relation between the output power and the field strength at the inner surface of the coupling interface. We will now calculate the output field for homogeneously broadened atoms and for uniform pumping, and examine how the *ad hoc* equation can be justified. Equations 10.1 and 10.8 give, in a similar manner as in Chapter 9,

$$\begin{aligned}
\hat{E}^{(+)}(z,t) &= \hat{F}_t(z,t) + \hat{F}_q(z,t) \\
&\quad + \sum_m \frac{|p_a|^2 v_0^2}{2\hbar\omega} \frac{1}{\varepsilon_1 c_1} \frac{2c_0}{(c_1 + c_0)} \sum_{n=0}^{\infty} (-r)^n \int_0^t \{\delta(t - t' + \tau_{5n}) + \delta(t - t' - \tau_{5n}) \\
&\quad - \delta(t - t' + \tau_{6n}) - \delta(t - t' - \tau_{6n})\} \int_0^{t'} e^{-(i\nu_0+\gamma)(t'-t'')} \hat{E}^{(+)}(z_m, t'')\hat{\sigma}_m(t)dt''dt' \\
&= \hat{F}_t(z,t) + \hat{F}_q(z,t) \\
&\quad + \sum_m \frac{|p_a|^2 v_0^2}{2\hbar\omega\varepsilon_1 c_1} \frac{2c_0}{(c_1 + c_0)} \sum_{n=0}^{\infty} (-r)^n \\
&\quad \times \left\{ \int_0^{t-\tau_{5n}} e^{-(i\nu_0+\gamma)(t-\tau_{5n}-t'')} \hat{E}^{(+)}(z_m, t'')\hat{\sigma}_m(t'')dt'' \right. \\
&\quad \left. - \int_0^{t-\tau_{6n}} e^{-(i\nu_0+\gamma)(t-\tau_{6n}-t'')} \hat{E}^{(+)}(z_m, t'')\hat{\sigma}_m(t'')dt'' \right\}
\end{aligned} \tag{10.113}$$

where, from Equation 10.9,

$$\tau_{5n} = \frac{z}{c_0} + \frac{2nd - z_m}{c_1}, \qquad \tau_{6n} = \frac{z}{c_0} + \frac{2nd + 2d + z_m}{c_1} \tag{10.114}$$

Truncating the rapid oscillation with frequency ω as in Equation 10.4,

$$\hat{E}^{(+)}(z,t) = \tilde{E}^{(+)}(z,t)e^{-i\omega t}, \quad \hat{F}_t(z,t) = \tilde{F}_t(z,t)e^{-i\omega t},$$
$$\hat{F}_q(z,t) = \tilde{F}_q(z,t)e^{-i\omega t} \tag{10.115}$$

we have

$$\tilde{E}^{(+)}(z,t) = \tilde{F}_t(z,t) + \tilde{F}_q(z,t) + \sum_m \frac{|p_a|^2 \nu_0^2}{2\hbar\omega\varepsilon_1 c_1} \frac{2c_0}{(c_1 + c_0)}$$
$$\times \sum_{n=0}^{\infty} (-r)^n \left\{ \int_0^{t-\tau_{5n}} e^{-\{i(\nu_0-\omega)+\gamma\}(t-\tau_{5n}-t'')} e^{i\omega\tau_{5n}} \tilde{E}^{(+)}(z_m, t'')\hat{\sigma}_m(t'')dt'' \right.$$
$$\left. - \int_0^{t-\tau_{6n}} e^{-\{i(\nu_0-\omega)+\gamma\}(t-\tau_{6n}-t'')} e^{i\omega\tau_{6n}} \tilde{E}^{(+)}(z_m, t'')\hat{\sigma}_m(t'')dt'' \right\} \tag{10.116}$$

Differentiating with respect to time t, we obtain

$$\left\{ \frac{\partial}{\partial t} + \gamma' \right\} \left\{ \tilde{E}^{(+)}(z,t) - \tilde{F}_t(z,t) - \tilde{F}_q(z,t) \right\}$$
$$= Tg \sum_m \sum_{n=0}^{\infty} (-r)^n \left\{ e^{i\omega\tau_{5n}} \tilde{E}^{(+)}(z_m, t-\tau_{5n})\hat{\sigma}_m(t-\tau_{5n}) \right. \tag{10.117}$$
$$\left. - e^{i\omega\tau_{6n}} \tilde{E}^{(+)}(z_m, t-\tau_{6n})\hat{\sigma}_m(t-\tau_{6n}) \right\}$$

where $\gamma' = \gamma + i(\nu_0 - \omega)$, the transmission coefficient $T = 2c_0/(c_0 + c_1) = 1 + r$, and the constant g is defined in Equation 10.12. We decompose the field into right- and left-going waves, as in Equation 10.15, but with the wave constant $k_0 = \omega/c_0$:

$$\tilde{E}_o(z,t) = \hat{e}_o^+(z,t)\exp\{+i\omega z/c_0\} + \hat{e}_o^-(z,t)\exp\{-i\omega z/c_0\}$$
$$\tilde{F}_{oq,t}(z,t) = \hat{f}_{oq,t}^+(z,t)\exp\{+i\omega z/c_0\} + \hat{f}_{oq,t}^-(z,t)\exp\{-i\omega z/c_0\} \tag{10.118}$$

where the suffix o has been added to signify a quantity existing outside the cavity. We note from Equation 10.114 that the right-hand member of Equation 10.117 yields only right-going waves. We have

$$\left\{ \frac{\partial}{\partial t} + \gamma' \right\} \left\{ \hat{e}_o^+(z,t) - \hat{f}_{ot}^+(z,t) - \hat{f}_{oq}^+(z,t) \right\}$$
$$= Tg \sum_m \sum_{n=0}^{\infty} (-r)^n \left\{ \exp\left(i\omega \frac{2nd - z_m}{c_1} \right) \tilde{E}^{(+)}\left(z_m, t - \frac{z}{c_0} - \frac{2nd - z_m}{c_1} \right) \langle \hat{\sigma}_m \rangle \right. \tag{10.119}$$
$$\left. - \exp\left(i\omega \frac{2nd + 2d + z_m}{c_1} \right) \tilde{E}^{(+)}\left(z_m, t - \frac{z}{c_0} - \frac{2nd + 2d + z_m}{c_1} \right) \langle \hat{\sigma}_m \rangle \right\}$$

For the left-going waves we have

$$\left\{\frac{\partial}{\partial t}+\gamma'\right\}\left\{\hat{e}_o^-(z,t)-\hat{f}_{ot}^-(z,t)-\hat{f}_{oq}^-(z,t)\right\}=0 \tag{10.120}$$

We Laplace-transform these by the correspondences

$$\begin{aligned}
\hat{e}_o^+(z,t) &\rightarrow \hat{L}_o^+(z,s)\\
\hat{e}_o^-(z,t) &\rightarrow \hat{L}_o^-(z,s)\\
\hat{v}_o^+(z,t) &\rightarrow \hat{V}_o^+(z,s)\\
\hat{v}_o^-(z,t) &\rightarrow \hat{V}_o^-(z,s)
\end{aligned} \tag{10.121a}$$

where

$$\hat{v}_o^{\pm}(z,t)=\hat{f}_{ot}^{\pm}+\hat{f}_{oq}^{\pm} \tag{10.121b}$$

We have from Equation 10.119

$$\begin{aligned}
&\{s+\gamma'\}\{\hat{L}_o^+(z,s)-\hat{V}_o^+(z,s)\}\\
&=Tg\sum_m\sum_{n=0}^{\infty}(-r)^n\left[\exp\left\{i\omega\left(\frac{2nd-z_m}{c_1}\right)\right\}\exp\left(-\frac{z}{c_0}s-\frac{2nd-z_m}{c_1}s\right)\right.\\
&\quad\times\left\{\hat{L}^+(z_m,s)e^{ik(z_m+d)}+\hat{L}^-(z_m,s)e^{-ik(z_m+d)}\right\}\langle\hat{\sigma}_m\rangle\\
&\quad-\exp\left\{i\omega\left(\frac{2nd+2d+z_m}{c_1}\right)\right\}\exp\left(-\frac{z}{c_0}s-\frac{2nd+2d+z_m}{c_1}s\right)\\
&\quad\left.\times\left\{\hat{L}^+(z_m,s)e^{ik(z_m+d)}+\hat{L}^-(z_m,s)e^{-ik(z_m+d)}\right\}\langle\hat{\sigma}_m\rangle\right]\\
&=TgN\int_{-d}^{0}dz_m\frac{1}{1-r''(s)}\left[\exp\left\{-\frac{z}{c_0}s-(i\omega-s)\left(\frac{z_m}{c_1}\right)\right\}\hat{L}^+(z_m,s)e^{ik(z_m+d)}\langle\hat{\sigma}_m\rangle\right.\\
&\quad\left.-\exp\left\{-\frac{z}{c_0}s+(i\omega-s)\left(\frac{2d+z_m}{c_1}\right)\right\}\hat{L}^-(z_m,s)e^{-ik(z_m+d)}\langle\hat{\sigma}_m\rangle\right]
\end{aligned} \tag{10.122a}$$

where the factor $r''(s)=-re^{2d(i\omega-s)/c_1}$ was defined in Equation 10.24. The initial value $\hat{e}_o^+(z,0)-\hat{v}_o^+(z,0)$ vanishes, as can be seen by examining Equations 10.113 and 10.118 for $t=0$. In the second expression, we have converted the summation over m into an integration over z_m. We have ignored the terms in the integration that are rapidly oscillating with z_m. We further rewrite Equation 10.122a as

$$\begin{aligned}
&\{s+\gamma'\}\{\hat{L}_o^+(z,s)-\hat{V}_o^+(z,s)\}\\
&=TgN\exp\left(-\frac{z}{c_0}s\right)e^{ikd}\int_{-d}^{0}dz_m\frac{1}{1-r''(s)}\\
&\quad\times\left[\exp\left(\frac{z_m}{c_1}s\right)\hat{L}^+(z_m,s)\langle\hat{\sigma}_m\rangle-\exp\left(-\frac{2d+z_m}{c_1}s\right)\hat{L}^-(z_m,s)\langle\hat{\sigma}_m\rangle\right]
\end{aligned} \tag{10.122b}$$

We compare this equation with the right-going wave in Equation 10.23a for inside the cavity at the coupling surface $z = -0$:

$$(s+\gamma')\left[\hat{L}^{+}(0,s) - \hat{V}^{+}(0,s)\right]$$
$$= gN\left\{-\frac{1}{1-r''(s)}\int_{-d}^{0}\exp\left(-\frac{2d+z_m}{c_1}s\right)\hat{L}^{-}(z_m,s)\langle\sigma_m\rangle dz_m \right. \tag{10.123}$$
$$\left. + \frac{1}{1-r''(s)}\int_{-d}^{0}\exp\left(\frac{z_m}{c_1}s\right)\hat{L}^{+}(z_m,s)\langle\sigma_m\rangle dz_m\right\}$$

Thus we have

$$\left\{\hat{L}_o^{+}(z,s) - \hat{V}_o^{+}(z,s)\right\} = T\exp\left(-\frac{z}{c_0}s\right)e^{ikd}\left[\hat{L}^{+}(0,s) - \hat{V}^{+}(0,s)\right] \tag{10.124}$$

From Equation 10.120 we have

$$(s+\gamma')\left[\hat{L}_o^{-}(z,s) - \hat{V}_o^{-}(z,s)\right] = 0$$

or

$$\hat{L}_o^{-}(z,s) - \hat{V}_o^{-}(z,s) = 0 \tag{10.125}$$

We inverse Laplace-transform Equations 10.124 and 10.125 to obtain

$$\hat{e}_o^{+}(z,t) = \hat{f}_o^{+}(z,t) + Te^{ikd}\hat{e}^{+}\left(-0,\ t-\frac{z}{c_0}\right) - Te^{ikd}\hat{f}^{+}\left(-0,\ t-\frac{z}{c_0}\right) \tag{10.126}$$

$$\hat{e}_o^{-}(z,t) = \hat{f}_o^{-}(z,t) \tag{10.127}$$

Here we show that, outside the cavity, the only relevant noise source is the thermal noise. In fact, if we use the expansion in Equations 10.8 and 10.9 for outside the cavity, we have from Equation 10.62

$$\begin{aligned}
&\hat{F}_q(z,t) \\
&= \frac{ip_a\nu_0}{2}\frac{1}{\varepsilon_1 c_1}\frac{2c_0}{(c_1+c_0)} \\
&\times\sum_m\sum_{n=0}^{\infty}(-r)^n\left\{\int_0^{t-\tau_{5n}} e^{-(i\nu_0+\gamma)(t-\tau_{5n}-t'')}\hat{\Gamma}_m(t'')dt'' - \int_0^{t-\tau_{6n}} e^{-(i\nu_0+\gamma)(t-\tau_{6n}-t'')}\hat{\Gamma}_m(t'')dt''\right\} \\
&= T\frac{ip_a\nu_0}{2\varepsilon_1 c_1}\sum_m\sum_{n=0}^{\infty}(-r)^n\left\{\int_0^{t-\frac{z}{c_0}-\frac{2nd-z_m}{c_1}}\exp\left[-(i\nu_0+\gamma)\left(t-\frac{z}{c_0}-\frac{2nd-z_m}{c_1}-t''\right)\right]\hat{\Gamma}_m(t'')dt''\right. \\
&\left.-\int_0^{t-\frac{z}{c_0}-\frac{2nd+2d+z_m}{c_1}}\exp\left[-(i\nu_0+\gamma)\left(t-\frac{z}{c_0}-\frac{2nd+2d+z_m}{c_1}-t''\right)\right]\hat{\Gamma}_m(t'')dt''\right\} \\
&= \hat{f}_{oq}^{+}(z,t)e^{-i\omega(t-z/c_0)}
\end{aligned} \tag{10.128}$$

This contains only right-going waves and has no left-going component, as we have indicated in the last line.

Next, we examine the quantum noise that is transmitted from inside to outside the cavity. From Equation 10.60a the right-going wave inside the cavity at the coupling surface is, recovering the oscillation at the angular frequency ω,

$$\begin{aligned}
&\hat{f}_q^{+}(-0,t)\exp\{i\omega d/c_1\}\exp(-i\omega t)\\
&= h\sum_m e^{-\gamma' t}e^{-i\omega t}\left[\sum_{n=0}^{\infty}(-r)^n\left\{\exp\{(i\nu_0+\gamma)\tau_{1n}\}\int_0^{t-\tau_{1n}}\exp(\gamma' t')\tilde{\Gamma}_m(t')dt'\right.\right.\\
&\quad\left.\left.-\exp\{(i\nu_0+\gamma)\tau_{3n}\}\int_0^{t-\tau_{3n}}\exp(\gamma' t')\tilde{\Gamma}_m(t')dt'\right\}\right]\\
&= \frac{ip_a\nu_0}{2\varepsilon_1 c_1}\sum_m e^{-(\gamma+i\nu_0)t}\left[\sum_{n=0}^{\infty}(-r)^n\left\{\exp\left((i\nu_0+\gamma)\frac{2nd-z_m}{c_1}\right)\right.\right.\\
&\quad\times\int_0^{t-\frac{2nd-z_m}{c_1}} e^{\{\gamma+i(\nu_0-\omega)\}t'}\tilde{\Gamma}_m(t')dt'\\
&\quad\left.\left.-\exp\left((i\nu_0+\gamma)\frac{2nd+2d+z_m}{c_1}\right)\int_0^{t-\frac{2nd+2d+z_m}{c_1}} e^{\{\gamma+i(\nu_0-\omega)\}t'}\tilde{\Gamma}_m(t')dt'\right\}\right]
\end{aligned}\tag{10.129}$$

where Equations 10.57c and 10.52a were used for the constant h and γ'. The values of τ_{1n} and τ_{3n} were substituted from Equation 10.7. Comparison of Equations 10.128 and 10.129 shows that

$$\hat{f}_{oq}^{+}(z,t) = T\hat{f}_q^{+}(-0,t-z/c_0)\exp\{i\omega d/c_1\} \tag{10.130}$$

Therefore, concerning the quantum noise, the first and the last terms in Equation 10.126 cancel. Thus the "raw" quantum noise disappears outside the cavity – it appears only as an amplified noise that constitutes a part of transmitted internal light field.

Next, we examine the case of thermal noise in Equation 10.126. From Equation 10.56 the right- and left-going waves inside the cavity are

$$\hat{f}_t^{\pm}(z,t) = \pm\sum_j \frac{1}{2}\sqrt{\frac{\hbar\omega_j}{\varepsilon_1 L(1-K\sin^2 k_{1j}^2 d)}}\,e^{\pm i(k_{1j}-k)(z+d)}\hat{a}_j(0)e^{-i(\omega_j-\omega)t} \tag{10.131}$$

while the right- and left-going waves outside the cavity are, using Equations 10.2 and 1.62b,

$$\begin{aligned}
\hat{f}_{ot}^{\pm}(z,t) &= i\sum_j\sqrt{\frac{\hbar\omega_j}{2}}\sqrt{\frac{2}{\varepsilon_1 L(1-K\sin^2 k_1^2 d)}}\\
&\quad\times\left(\pm\frac{k_{1j}}{k_{0j}}\cos k_{1j}d\,\frac{e^{\pm ik_{0j}z}}{2i}+\sin k_{1j}d\,\frac{e^{\pm ik_{0j}z}}{2}\right)\hat{a}_j(0)e^{-i(\omega_j-\omega)t}
\end{aligned}\tag{10.132}$$

The third term in Equation 10.126 is, using $\hat{f}_t^+$ in Equation 10.131,

$$-Te^{ikd}\hat{f}_t^+(-0,\ t-z/c_0) = -Te^{ikd}\sum_j \frac{1}{2}\sqrt{\frac{\hbar\omega_j}{\varepsilon_1 L(1-K\sin^2 k_{1j}^2 d)}} \times e^{i(k_{1j}-k)d}\hat{a}_j(0)e^{-i(\omega_j-\omega)(t-z/c_0)} \tag{10.133}$$

Thus, concerning the thermal noise, the sum of the first and the third terms in Equation 10.126 is

$$\begin{aligned}
&\hat{f}_{ot}^+(z,t) - Te^{ikd}\hat{f}_t^+(-0,\ t-z/c_0) \\
&= i\sum_j \sqrt{\frac{\hbar\omega_j}{2}}\sqrt{\frac{2}{\varepsilon_1 L(1-K\sin^2 k_{1j}d)}}\left(\frac{k_{1j}}{k_{0j}}\cos k_{1j}d\frac{e^{ik_{0j}z}}{2i} + \sin k_{1j}d\frac{e^{ik_{0j}z}}{2}\right)\hat{a}_j(0) \\
&\quad \times e^{-i(\omega_j-\omega)t}e^{-ik_0 z} - Te^{ikd}\sum_j \frac{1}{2}\sqrt{\frac{\hbar\omega_j}{\varepsilon_1 L(1-K\sin^2 k_{1j}d)}} \\
&\quad \times e^{i(k_{1j}-k)d}\hat{a}_j(0)e^{-i(\omega_j-\omega)(t-z/c_0)} \\
&= \sum_j \sqrt{\frac{\hbar\omega_j}{2}}\sqrt{\frac{2}{\varepsilon_1 L(1-K\sin^2 k_{1j}d)}}\hat{a}_j(0)e^{-i(\omega_j-\omega)(t-z/c_0)} \\
&\quad \times \left(\frac{k_{1j}}{k_{0j}}\cos k_{1j}d\frac{1}{2} + i\sin k_{1j}d\frac{1}{2} - T\frac{1}{2}e^{i(k_{1j})d}\right)
\end{aligned} \tag{10.134}$$

Then, noting that $T = 1 + r$ and that $k_{1j}/k_{0j} = c_0/c_1 = (1+r)/(1-r)$, it can be shown that the quantity in the large round bracket is

$$\begin{aligned}
&\left(\frac{k_{1j}}{k_{0j}}\cos k_{1j}d\frac{1}{2} + i\sin k_{1j}d\frac{1}{2} - T\frac{1}{2}e^{i(k_{1j})d}\right) \\
&= -\frac{k_{1j}}{k_{0j}}\cos k_{1j}d \times \frac{r'}{2} + i\sin k_{1j}d \times \frac{r'}{2}
\end{aligned} \tag{10.135}$$

where $r' = -r$. Thus

$$\begin{aligned}
&\hat{f}_{ot}^+(z,t) - Te^{ikd}\hat{f}_t^+(-0,\ t-z/c_0) \\
&= r'\sum_j \sqrt{\frac{\hbar\omega_j}{2}}\sqrt{\frac{2}{\varepsilon_1 L(1-K\sin^2 k_1^2 d)}}\hat{a}_j(0)e^{-i(\omega_j-\omega)(t-z/c_0)} \\
&\quad \times \left(-\frac{k_{1j}}{k_{0j}}\cos k_{1j}d \times \frac{1}{2} + i\sin k_{1j}d \times \frac{1}{2}\right)
\end{aligned} \tag{10.136}$$

From Equation 10.132 this quantity describes a right-going wave that is just the left-going wave outside the cavity $\hat{f}_{ot}^-(z,t)$ reflected at the coupling surface with the reflection coefficient $r' = -r$ for the wave incident on the coupling surface from outside. Thus the net result for Equation 10.126 is

$$\hat{e}_o^+(z,t) = Te^{ikd}\hat{e}^+\left(-0,\ t-\frac{z}{c_0}\right) + r'\hat{f}_{ot}^-\left(+0,\ t-\frac{z}{c_0}\right) \tag{10.137}$$

This shows that, also for thermal noise, the "raw" thermal noise inside the cavity does not appear outside the cavity. Consequently, Equation 10.137 together with Equation 10.127 gives the final result for the expression for the laser field outside the cavity. Since the second term in Equation 10.137 represents the ambient thermal noise existing outside the cavity, the relevant output field is given by the first term in the equation and is a copy of the field at the inner surface of the coupling interface at $z=0$:

$$\hat{e}_o^+(z,t) = Te^{ikd}\hat{e}^+\left(-0,\ t-\frac{z}{c_0}\right) \tag{10.138}$$

Thus the phase fluctuation and the linewidth should be the same as was given for the field at the inner surface of the coupling interface. The formal expression for the latter field is given by Equation 10.69 and the linewidth due to phase diffusion for this field is given by Equations 10.107b and 10.109a. In going from Equation 10.107b to 10.109a we have expressed in Equation 10.108 the field amplitude outside the cavity as $e_o = Te_0$, where the real amplitude e_0 was defined by Equation 10.72 as $\hat{e}^+(0,t) = e_0 \exp[i\phi(0,t)]$. On the other hand, Equation 10.138 gives, except for the fluctuating phase, the constant part of the amplitude outside the cavity as

$$e_o = Te^{ikd}e_0 \tag{10.139}$$

This does not affect the evaluation of the output power in Equation 10.108. Therefore, the output power dependence of the linewidth of the output field is the same as that in Equations 10.109b and 10.111, as expected.

We note, however, that this equivalence is not true if the output field is not a laser field but a minute field from inside the cavity, because this time the thermal field expressed by the second term in Equation 10.137 cannot be ignored. An explicit consequence of including this second term will be described in Section 15.1.1.1.

We note that for the expression in Equation 10.138 for the field outside the cavity, the phase diffusion along the distance on the laser axis can likewise be evaluated as the temporal diffusion. It is easy to see that the general expression for the diffusion in this case reads (see Equation 10.109b)

$$\begin{aligned}\left\langle\{\Delta\phi(z,t)\}^2\right\rangle =& \frac{2\hbar\omega_c\beta_c^2}{P}\frac{\gamma^2(1+\delta^2)}{(\gamma+\gamma_c)^2+\delta^2(\gamma-\gamma_c)^2}\\ &\times\left[\left\langle n_\omega+\tfrac{1}{2}\right\rangle + \frac{N_2+N_1}{2N\sigma_{th}^0}\left\{1+\frac{\Delta}{1+\Delta}g(r)\right\}\right]\left|\Delta t-\frac{\Delta z}{c_0}\right|\end{aligned} \tag{10.140}$$

References

1 Ujihara, K. (**1984**) *Phys. Rev. A*, *29*, 3253–3263.

2 Haken, H. (**1970**) *Laser theory*, in *Licht und Materie, IC*, Handbuch der Physik, vol. *XXV/2c* (eds. S. Flügge and L. Genzel), Springer, Berlin.

3 Sargent, M., III, Scully, M.O., and Lamb, W.E., Jr. (**1974**) *Laser Physics*, Addison-Wesley, Reading, MA.

4 Prasad, S. (**1992**) *Phys. Rev. A*, *46*, 1540–1559.

5 van Exter, M.P., Kuppens, S.J.M., and Woerdman, J.P. (**1995**) *Phys. Rev. A*, *51*, 809–816.

11
Analysis of a One-Dimensional Laser with Two-Side Output Coupling: The Propagation Method

In the previous chapter, we have extensively developed the rigorous method to evaluate the phase diffusion of a one-dimensional laser with output coupling starting from the full coupled equations of motion for the field and for the atomic dipoles. The atomic inversion was assumed to be constant in time but was dependent on the field strength. The detailed analysis in the previous chapter indicates the existence of a simplified, *ad hoc* method that relies on the existence of two counter-propagating waves inside the cavity and on the optical rules applied to them at the boundaries. We name this method the propagation method or propagation theory. This method also relies on the simplified correlation functions for the noise sources. By limiting the number of traveling waves to two, and allowing for their reflection and transmission rules at the two boundaries, one can follow the development of the field during one round-trip time. On the basis of this time development, one will obtain a diffusion equation for the phase of the field, which can be evaluated rather easily by use of simplified models of the noise sources. The essence of the contents of this chapter was published in Ref. [1] using a simpler cavity model.

11.1
Model of the Laser and the Noise Sources

The model cavity is depicted in Figure 11.1. The cavity extends from $z=0$ to $z=d$. Infinitely thin mirrors M1 and M2 are attached to the cavity ends. The dielectric constant and the velocity of light inside the cavity are, respectively, ε_1 and c_1. The amplitude reflection coefficients of mirrors M1 and M2 are r_1 and r_2, respectively, for the wave incident from inside the cavity. The amplitude transmission coefficients for mirrors M1 and M2 for the waves incident from inside are, respectively, T_1 and T_2. The amplitude transmission coefficients for the waves incident from outside are T_1' and T_2', respectively. The outside regions $z < 0$ and $d < z$ are vacuum. The dielectric constant and the velocity of light for the outside regions are ε_0 and c_0, respectively. A gain medium made up of gain atoms of uniform

Output Coupling in Optical Cavities and Lasers: A Quantum Theoretical Approach
Kikuo Ujihara

ISBN: 978-3-527-40763-7

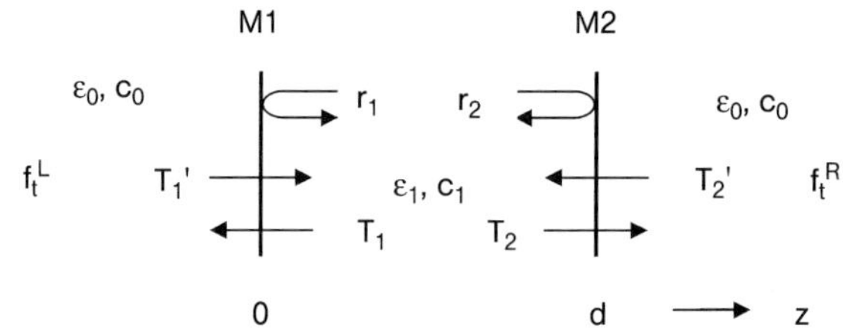

Figure 11.1 The model of an asymmetric two-sided cavity.

density N exists in the region $0 < z < d$ and the medium is uniformly pumped. The atoms are two-level atoms with angular transition frequency ν_0 and homogeneous full width at half-maximum (FWHM) 2γ.

We recall from Equation 10.69 that the slowly varying field amplitude at the output surface of the one-sided cavity has its time derivative in the form of a sum of the thermal and quantum contributions. A concise interpretation of the thermal contribution was that the quantity in the curly bracket in the first line in Equation 10.69 is the thermal noise that penetrated into the cavity from outside and amplified through one round trip in the cavity.

For the quantum part of the time derivative of the slowly varying field amplitude at the inner surface of the coupling interface, there was a sum of the right-going and left-going quantum noise fields, that is, the quantum noise propagated to the right to the coupling surface with corresponding amplification and phase shift, and the quantum noise propagated first to the left and reflected by the perfect conductor mirror and then propagated to the coupling surface. The quantum noise field is generated within the cavity from the laser active atoms and amplified before it reaches the output mirror.

Here we are considering an asymmetric two-sided cavity with mirrors M1 and M2, in contrast to the one-sided cavity considered in Chapter 10. We assume that the same interpretation for the time derivative of the fields at the respective ends of the cavity is applicable in principle, except that the thermal noise penetrates into the cavity from both sides of the cavity instead of from the single side. Thus, according to the arguments leading to Equation 10.71, we assume that two thermal noise sources of amplitudes $f_t^{R,L}$ are penetrating from outside to inside the cavity. The waves f_t^R and f_t^L come from the right-hand and left-hand free space (vacuum), respectively. The thermal part of the time derivative of the slowly varying field amplitude at the inner surface of mirror M1 (M2) will be the sum of the thermal fields penetrating into the cavity from both sides and reaching the mirror M1 (M2) with amplification during one round trip. The wave $f_t^R(+0, t)$ corresponds to $\hat{f}_{ot}^-(+0, t)$ and its correlation properties can be obtained by use of the expression for $\hat{f}_{ot}^-(z, t)$ in Equation 10.70. For the quantum noise, the interpretation of the contributions from right- and left-going waves remains the same, except that the perfect conductor mirror is replaced by a mirror of finite reflectivity.

With the above considerations in mind, we construct the noise models. The thermal noise has the correlation properties as obtained using Equation 10.70.

Noting the fact that the absolute square of the quantity in the large round bracket in Equation 10.70 for $z=0$ is equal to $1 - K\sin^2 k_{1j}d$ times $n^2/4$, where n is the refractive index, converting the summation over j to an integration with the density of modes $L/(c_0\pi)$, extending the lower limit of integration from $-\omega$ to $-\infty$, and finally using Equation 2.8, we obtain, for example,

$$\begin{aligned}\left\langle \hat{f}_t^{R\dagger}(t)\hat{f}_t^{R}(t')\right\rangle &= \sum_j\sum_i \frac{\hbar\sqrt{\omega_j\omega_i}}{2}\frac{2}{\varepsilon_1 L}\frac{n^2}{4}\left\langle \hat{a}_j^\dagger(0)\hat{a}_i(0)\right\rangle e^{i(\omega_j-\omega)t-i(\omega_i-\omega)t'} \\ &= \sum_j \frac{\hbar\omega_j}{2}\frac{2}{\varepsilon_1 L}\frac{n^2}{4}\langle n_j\rangle e^{i(\omega_j-\omega)(t-t')} \\ &= \int_{-\infty}^{\infty} d\omega_j \frac{L}{c_0\pi}\frac{\hbar\omega_j}{2}\frac{2}{\varepsilon_1 L}\frac{n^2}{4}\langle n_j\rangle e^{i(\omega_j-\omega)(t-t')} \\ &\simeq \frac{\hbar\omega}{4\pi\varepsilon_0 c_0} n_\omega 2\pi\delta(t-t')\end{aligned} \tag{11.1a}$$

where we have put $\omega_j\langle n_j\rangle$ outside the integral sign as ωn_ω, noting that the important contributions come from around ω. Thus, with similar considerations, we have

$$\begin{aligned}2\varepsilon_0 c_0\left\langle \hat{f}_t^{R\dagger}(t)\hat{f}_t^{R}(t')\right\rangle &= n_\omega\hbar\omega\delta(t-t') \\ 2\varepsilon_0 c_0\left\langle \hat{f}_t^{R}(t)\hat{f}_t^{R\dagger}(t')\right\rangle &= (n_\omega+1)\hbar\omega\delta(t-t')\end{aligned} \tag{11.1b}$$

$$\begin{aligned}2\varepsilon_0 c_0\left\langle \hat{f}_t^{L\dagger}(t)\hat{f}_t^{L}(t')\right\rangle &= n_\omega\hbar\omega\delta(t-t') \\ 2\varepsilon_0 c_0\left\langle \hat{f}_t^{L}(t)\hat{f}_t^{L\dagger}(t')\right\rangle &= (n_\omega+1)\hbar\omega\delta(t-t')\end{aligned} \tag{11.1c}$$

Here we have used the thermal field function for the one-sided cavity and applied the resulting properties to the thermal noise on both sides of the two-sided cavity. This may be justified because of the quite general structure of the correlation functions, which do not depend on the structure of the cavity. (The above derivation of Equations 11.1b and 11.1c can also be deduced if we follow the calculations from Equations 10.80 to 10.83, since the thermal noise in Equation 10.80 (the first large curly bracket) is proportional to $\hat{f}_{ot}^{-}(+0,t)$ by Equation 10.71, which corresponds to $f_t^R(+0,t)$.)

The factor 1 on the right-hand side (RHS) of Equations 11.1b and 11.1c, associated with the anti-normally ordered products of the noise operators, represents the vacuum fluctuations. The coefficient n_ω is the thermal photon number per mode of free vacuum at the angular frequency ω. Equations 11.1b and 11.1c imply that the delta-correlated normally ordered noise power in the free field is equal to the energy of photons present multiplied by a delta function of time, which has the dimension of "per second."

As for the quantum noise sources, comparison of Equations 10.89a and 10.89b to Equation 10.91 shows that we may have delta-correlated noise forces under the condition of Equation 10.90, that is, the noise is delta correlated at a time scale that is larger than the reciprocal atomic linewidth. In Equation 10.69, the exponential factors with integral of the atomic inversion are the amplification factors for the noise fields. In the same equation, the integral containing the quantum noise $\tilde{\Gamma}_m$ may be evaluated, under the above condition expressed by Equation 10.90, as

$$\int_0^t e^{\{i(\omega-\nu_0)-\gamma\}(t-t'')}\tilde{\Gamma}_m(t)dt'' = \tilde{\Gamma}_m(t)\frac{1}{\gamma - i(\omega-\nu_0)} \tag{11.2}$$

Taking this into account, we assume that the quantum noise field $\hat{f}_m(t)$ associated with atom m has the form

$$\hat{f}_m(t) = \frac{ip_a\nu_0}{2c_1\varepsilon_1}\frac{\tilde{\Gamma}_m(t)}{\gamma - i(\omega-\nu_0)} \tag{11.3}$$

which has the property (see Equations 9.5a, 9.5b, and 10.92)

$$2\varepsilon_1 c_1\left\{\left\langle \hat{f}_m^\dagger(t)\hat{f}_{m'}(t')\right\rangle + \left\langle \hat{f}_{m'}(t)\hat{f}_m^\dagger(t')\right\rangle\right\} = (2g\hbar\omega/c_1)\delta_{mm'}\delta(t-t') \tag{11.4}$$

where $2g = \gamma|p_a|^2\nu_0^2/\left[\varepsilon_1\hbar\omega\left\{\gamma^2+(\nu_0-\omega)^2\right\}\right]$ is the stimulated transition rate per atom per unit density of photons, that is, the spontaneous emission rate per atom (see Equation 4.14). The RHS of Equation 11.4 may be interpreted as the delta-correlated noise intensity with instantaneous intensity $2g\hbar\omega/c_1$.

11.2
The Steady State and the Threshold Condition

In the previous section, we have defined the noise sources and their correlation functions. Before we proceed to evaluate the effect of the accumulated noise on the field amplitude, we examine the steady state of the two-sided cavity laser in the saturated gain regime ignoring noise. In this section, we ignore the operator aspect of the waves and treat them as classical quantities. We assume an above-threshold oscillation with well-stabilized amplitude. If we write the slowly varying amplitudes of the right- and left-going waves inside the cavity as $e^+(z)$ and $e^-(z)$, respectively, they satisfy the differential equations (see Equations 8.35a and 8.35b as well as Equation 10.94c)

$$\begin{aligned}\frac{d}{dz}|e^+(z)|^2 &= 2\mathrm{Re}\{\alpha(z)\}|e^+(z)|^2\\ \frac{d}{dz}|e^-(z)|^2 &= -2\mathrm{Re}\{\alpha(z)\}|e^-(z)|^2\end{aligned} \tag{11.5}$$

with the boundary conditions

$$\begin{aligned} |e^{+}(0)|^{2} &= |r_1|^{2}|e^{-}(0)|^{2} \\ |e^{-}(d)|^{2} &= |r_2|^{2}|e^{+}(d)|^{2} \end{aligned} \tag{11.6}$$

The amplitude gain per unit length is given as

$$\alpha(z) = \frac{g}{\gamma'} N\sigma(z) \tag{11.7}$$

where g is given by Equation 10.12 and (see Equations 8.16 and 8.22)

$$\sigma(z) = \frac{\sigma^0}{1 + \{|e^{+}(z)|^2 + |e^{-}(z)|^2\}/|E_s|^2} \tag{11.8}$$

Also, we have the rule

$$|e^{+}(z)|^{2}|e^{-}(z)|^{2} = \text{const} = C \tag{11.9}$$

For the steady state, the round-trip gain is compensated for by the mirror losses. We write the steady-state condition as

$$|r_1|^2|r_2|^2 \exp \int_0^d 4\text{Re}\{\alpha(z)\}dz = 1 \tag{11.10}$$

Some simple rules concerning the absolute squared field amplitudes $|e^{\pm}(z)|^2$ will be given in Appendix F.

We define a neutral point z_c where

$$|e^{+}(z_c)|^2 = |e^{-}(z_c)|^2 = \sqrt{C} \tag{11.11}$$

As in Chapter 10 we will later need the integrated local intensity, which is the sum of the absolute squared $e^{+}(z)$ and $e^{-}(z)$. That is, we explore

$$I = \int_0^d \left\{|e^{+}(z)|^2 + |e^{-}(z)|^2\right\}dz \tag{11.12}$$

We write

$$|e^{+}(z)|^2 + |e^{-}(z)|^2 = X \tag{11.13}$$

Then, because of Equation 11.9, we have

$$|e^{+}(z)|^2 - |e^{-}(z)|^2 = \pm\sqrt{X^2 - 4C} \tag{11.14}$$

The plus sign applies for $z > z_c$ and the minus sign for $z < z_c$. Then Equation 11.5 yields

$$\begin{aligned} \frac{d}{dz}X &= \pm 2\text{Re}\{\alpha(z)\}\sqrt{X - 4C} \\ &= \pm 2\text{Re}\{\alpha^0\}\frac{\sqrt{X - 4C}}{1 + (X/|E_s|^2)} \end{aligned} \tag{11.15}$$

where the linear, unsaturated amplitude gain per unit length is

$$\alpha^0 = \frac{gN\sigma^0}{\gamma'} \tag{11.16}$$

In Appendix F, the integral I in Equation 11.12 is evaluated using Equation 11.15. Here we determine the constant C by integrating Equation 11.15. From Equation 11.15 we have

$$2\mathrm{Re}\{\alpha^0\}dz = \pm\frac{\{1 + (X/|E_s|^2)\}dX}{\sqrt{X^2 - 4C}} \tag{11.17}$$

Integrating from $z = 0$ to $z = d$, we have

$$\begin{aligned}
2\mathrm{Re}\{\alpha^0\}d &= -\int_{X(0)}^{X(z_c)} \frac{\{1 + (X/|E_s|^2)\}}{\sqrt{X^2 - 4C}}dX + \int_{X(z_c)}^{X(d)} \frac{\{1 + (X/|E_s|^2)\}}{\sqrt{X^2 - 4C}}dX \\
&= -\ln\left|X + \sqrt{X^2 - 4C}\right|\Bigg|_{X(0)}^{X(z_c)} - \frac{\sqrt{X^2 - 4C}}{|E_s|^2}\Bigg|_{X(0)}^{X(z_c)} \\
&\quad + \ln\left|X + \sqrt{X^2 - 4C}\right|\Bigg|_{X(z_c)}^{X(d)} + \frac{\sqrt{X^2 - 4C}}{|E_s|^2}\Bigg|_{X(z_c)}^{X(d)} \\
&= \ln\frac{\{X(0) + \sqrt{X^2(0) - 4C}\}\{X(d) + \sqrt{X^2(d) - 4C}\}}{\{X(z_c) + \sqrt{X^2(z_c) - 4C}\}^2} \\
&\quad + \frac{1}{|E_s|^2}\left\{\sqrt{X^2(0) - 4C} + \sqrt{X^2(d) - 4C} - 2\sqrt{X^2(z_c) - 4C}\right\}
\end{aligned} \tag{11.18}$$

We note that

$$\begin{aligned}
\sqrt{X^2(0) - 4C} &= |e^-(0)|^2 - |e^+(0)|^2 \\
\sqrt{X^2(d) - 4C} &= |e^+(d)|^2 - |e^-(d)|^2 \\
\sqrt{X^2(z_c) - 4C} &= |e^+(z_c)|^2 - |e^-(z_c)|^2 = 0
\end{aligned} \tag{11.19}$$

Thus we have

$$\begin{aligned}
2\mathrm{Re}\{\alpha^0\}d &= \ln\frac{2|e^-(0)|^2 \times 2|e^+(d)|^2}{4C} \\
&\quad + \frac{1}{|E_s|^2}\left\{|e^-(0)|^2 - |e^+(0)|^2 + |e^+(d)|^2 - |e^-(d)|^2\right\}
\end{aligned} \tag{11.20}$$

Substituting Equations F.6 and F.7 in Appendix F into Equation 11.20, we have

$$2\text{Re}\{\alpha^0\}d = \ln\frac{1}{|r_1||r_2|} + \frac{\sqrt{C}}{|E_s|^2}\left\{\frac{1}{|r_1|} - |r_1| + \frac{1}{|r_2|} - |r_2|\right\} \tag{11.21}$$

Therefore, we have

$$\begin{aligned}\sqrt{C} = {} & |e^{\pm}(z_c)|^2 \\ = {} & |E_s|^2\left(2\text{Re}\{\alpha^0\}d - \ln\frac{1}{|r_1||r_2|}\right)\frac{|r_1||r_2|}{(|r_1| + |r_2|)(1 - |r_1||r_2|)}\end{aligned} \tag{11.22}$$

The threshold condition is obtained by setting $C = 0$. Thus

$$\text{Re}\{\alpha^0_{th}\} = \frac{1}{2d}\ln\frac{1}{|r_1||r_2|} \tag{11.23}$$

or, by Equation 11.16,

$$\frac{\gamma g N \sigma^0_{th}}{\gamma^2 + (\nu_0 - \omega)^2} = \frac{1}{2d}\ln\frac{1}{|r_1||r_2|} \tag{11.24}$$

If we define the cavity decay constant γ_c for the present cavity model as

$$\gamma_c = \frac{c_1}{2d}\ln\frac{1}{|r_1||r_2|} \tag{11.25}$$

and assume the usual frequency pulling described by Equation 10.46, then Equation 11.24 reduces to the form of Equation 10.48 (note that $g = \nu_0^2|p_a|^2/(2\hbar\omega\varepsilon_1 c_1)$ from Equation 10.12):

$$\sigma^0_{th} = \bar{\sigma}_{ss} = \frac{2\hbar\omega\varepsilon_1}{\nu_0^2|p_a|^2 N}\gamma\gamma_c(1 + \delta^2) \tag{11.26a}$$

This equation can be recast in the form

$$N\mathcal{g}\sigma^0_{th} = \gamma_c \tag{11.26b}$$

That the threshold atomic inversion σ^0_{th} in Equation 11.26a is equal to the space-averaged steady-state inversion $\bar{\sigma}_{ss} = (1/d)\int_0^d \sigma(z)dz$ can be shown as follows. We use Equation F.1a with $z' = d$ and $z = 0$ to get $|e^+(d)|^2 = |e^+(0)|^2 \exp\int_0^d 2\text{Re}\,\alpha(z)dz$ and $|e^-(d)|^2 = |e^-(0)|^2 \exp\int_0^d\{-2\text{Re}\,\alpha(z)\}dz$. These two equations and Equation F.5 yield $\exp\int_0^d 2\text{Re}\,\alpha(z)dz = 1/(|r_1||r_2|)$, from which the second equality in Equation 11.26a follows if we use Equation 11.7, the expression for g given above Equation 11.26a, and that for γ_c in Equation 11.25.

The cavity resonant frequency ω_c for the present cavity model will be discussed below.

11.3 The Time Rate of the Amplitude Variation

Next, we consider the time rate of change of the field amplitude corresponding to Equation 10.69 used for the case of a single-sided cavity. Here, we cite Equation 10.69 again to derive assistance for further consideration:

$$\begin{aligned}\frac{d\hat{e}^{+}(0,t)}{dt} &= \frac{c_1}{2d}\frac{\gamma'}{\gamma'+\gamma'_c}\left[-\hat{f}_t^{+}(0,t)+\frac{1}{r\exp(2ikd)}\hat{f}_t^{-}(0,t)\right.\\ &+\frac{ip_a\nu_0}{2c_1\varepsilon_1}e^{-ikd}\sum_m e^{-ikz_m}\exp\left\{\int_{z_m}^{0}dz\left(\frac{gN\langle\sigma_z\rangle}{\gamma'}\right)\right\}\int_0^t e^{\{i(\omega-\nu_0)-\gamma\}(t-t'')}\\ &\times\tilde{\Gamma}_m(t'')dt''-\frac{ip_a\nu_0}{2c_1\varepsilon_1}e^{-ikd}\exp\left\{\int_{-d}^{0}dz\left(\frac{gN\langle\sigma_z\rangle}{\gamma'}\right)\right\}\\ &\left.\times\sum_m e^{ik(z_m+2d)}\exp\left\{\int_{-d}^{z_m}dz\left(\frac{gN\langle\sigma_z\rangle}{\gamma'}\right)\right\}\int_0^t e^{\{i(\omega-\nu_0)-\gamma\}(t-t'')}\tilde{\Gamma}_m(t'')dt''\right]\end{aligned}$$

The field amplitude is now an operator. We have already commented upon the amplified thermal and quantum noise. Here we note the two factors at the front of the right-hand member of this equation. The first factor is $c_1/(2d)$. This is the reciprocal round-trip time. So the quantity in the large square bracket represents the contributions to the amplitude change that occur in one round-trip time. This factor originates in Equation 10.50b. Here, we cite Equation 10.50b again for convenience:

$$\begin{aligned}1-r'\exp&\left\{2\int_{-d}^{0}dz'\left(-\frac{s}{c_1}+\frac{gN\langle\sigma_{z'}\rangle}{s+\gamma'}\right)\right\}\\ &=\frac{2d/c_1}{s_0+\gamma'}\{\gamma+\gamma_c+i(\nu_0+\omega_c-2\omega)\}(s-s_0)\end{aligned}$$

where $r'=-r\exp(2ikd)$ in the single-sided model. The second factor is $\gamma'/(\gamma'+\gamma'_c)$. This factor also originates in the above equation (Equation 10.50b) under the assumption of a slower variation of the field envelope function than the natural dipolar decay, that is, under the assumption that $|\partial/\partial t|\ll\gamma$. Thus, Equation 10.50b determines the pole in the s-plane.

We want to rewrite Equation 10.50b so as to take into account the two-sided feature of the present cavity model. We remember that the factor $-r$ represents the product of r and -1, the amplitude reflection coefficients at the right and left end surfaces of the single-sided cavity, respectively. Thus we may replace r' in Equation 10.50b by $r_1r_2\exp(2ikd)$. Also, the range of spatial integration should be the region $0\le z\le d$ instead of $-d\le z\le 0$ of the previous chapter. Thus, for a pole $s=s_0$, we should have

$$1 - r_1 r_2 \exp(2ikd) \exp\left\{2\int_0^d dz'\left(-\frac{s}{c_1} + \frac{gN\langle\sigma_{z'}\rangle}{s+\gamma'}\right)\right\} = 0 \tag{11.27}$$

This can be rewritten as

$$1 - \exp\left\{-2mi\pi + \ln(r_1 r_2) + 2ikd - \frac{2sd}{c_1} + 2\frac{gN}{s+\gamma'}\int_0^d \langle\sigma_{z'}\rangle dz'\right\} = 0 \tag{11.28}$$

Now if we define the cavity decay constant γ_c as in Equation 11.25 and newly define the cavity resonant angular frequency ω_c as

$$\omega_c = \frac{c_1}{2d}\{2m\pi - \arg(r_1 r_2)\} \tag{11.29}$$

then Equation 11.28 may be rewritten as

$$1 - \exp\left[-\frac{2d}{c_1}\{s + \gamma_c + i(\omega_c - \omega)\} + \frac{2gN}{s+\gamma+i(\nu_0-\omega)}\int_0^d \langle\sigma_{z'}\rangle dz'\right] = 0 \tag{11.30}$$

where we have used Equation 10.52a for γ'. The pole is the value of s that makes the quantity in the square bracket null. Thus

$$\{s_0 + \gamma_c + i(\omega_c - \omega)\}\{s_0 + \gamma + i(\nu_0 - \omega)\} - \frac{c_1}{2d}2gN\int_0^d \langle\sigma_{z'}\rangle dz' = 0 \tag{11.31}$$

Ignoring the square of s_0, we have

$$s_0 = -\frac{\gamma\gamma_c + (\nu_0 - \omega)(\omega - \omega_c) - (c_1/d)\int_0^d gN\langle\sigma_{z'}\rangle dz' - i\{\gamma(\omega - \omega_c) + \gamma_c(\omega - \nu_0)\}}{\gamma + \gamma_c - i(\nu_0 + \omega_c - 2\omega)} \tag{11.32}$$

just as in Equation 10.50c. Then the left-hand side (LHS) of Equation 11.30 can be approximated as

$$\begin{aligned} &1 - \exp\left[-\frac{2d}{c_1}\{s + \gamma_c + i(\omega_c - \omega)\} + \frac{2gN}{s+\gamma+i(\nu_0-\omega)}\int_0^d \langle\sigma_{z'}\rangle dz'\right] \\ &= \left(\frac{2d}{c_1}\frac{\gamma' + \gamma'_c}{s+\gamma'}\right)(s - s_0) \simeq \left\{\frac{2d}{c_1}\frac{\gamma' + \gamma'_c}{\gamma'}\right\}(s - s_0) \end{aligned} \tag{11.33}$$

where

$$\gamma' = \gamma + i(\nu_0 - \omega), \qquad \gamma'_c = \gamma_c + i(\omega_c - \omega) \tag{11.34}$$

Thus, just as in Equation 10.69, the reciprocal of the quantity $(\gamma' + \gamma'_c)/\gamma'$ appears in the time derivative of the slowly varying field amplitude. The physical meaning of this quantity, a bad cavity effect, will be discussed in the next chapter. Note that the cavity decay constant γ_c and the cavity resonant angular frequency ω_c have been newly defined in Equations 11.25 and 11.29, respectively.

We evaluate the accumulated noise at the inner surface of the coupling mirror M2. That for mirror M1 can be treated similarly. For the slowly varying amplitude of the right-traveling wave just inside the mirror M2, the time rate of change will be given as

$$\frac{d\hat{e}^{+}(d-0,t)}{dt} = \frac{c_1}{2d}\frac{\gamma'}{\gamma' + \gamma'_c}\left(\hat{F}_t + \hat{F}_q\right) \tag{11.35}$$

where $\hat{F}_t$ and $\hat{F}_q$ are the sums of the thermal and quantum noise field operators, respectively, which have emerged during the last round-trip time and reached the inner surface of the mirror M2. The reason for the absence in Equation 11.35 of noise contributions that were generated in the past older than one round trip is that any one round trip in the past yields a net amplification (amplification plus cavity loss) of unity and does not contribute to the change in the field amplitude. As we saw in the last chapter, the noise sources are amplified basically with the same rate as for the coherent laser field under the assumption of time-independent atomic inversion. Now the amplified thermal field is given by

$$\hat{F}_t = \left\{ T'_2 r_1 G_{tR} e^{2ikd} \hat{f}_t^R \left(d+0,\ t-\frac{2d}{c_1}\right) + T'_1 G_{tL} e^{ikd} \hat{f}_t^L \left(-0,\ t-\frac{d}{c_1}\right)\right\} \tag{11.36}$$

Here e^{2ikd} and e^{ikd} are phase shifts associated with respective propagations. The amplifying constants G_{tR} and G_{tL} are determined as follows. The steady-state amplitude for the oscillating laser field is maintained under the condition given by Equation 11.27 with $s=0$. Thus

$$1 - r_1 r_2 \exp(2ikd)\exp\left\{2\int_0^d dz' \left(\frac{gN\langle\sigma_{z'}\rangle}{\gamma'}\right)\right\} = 0 \tag{11.37}$$

We define the single path gain G_s as

$$G_s \equiv \exp\left\{\int_0^d \alpha(z')dz'\right\} \tag{11.38}$$

where $\alpha(z)$ is given by Equation 11.7. This gain satisfies

$$1 - r_1 r_2 \exp(2ikd) G_s^2 = 0 \tag{11.39}$$

Then we have

$$\begin{aligned} G_{tR} &= G_s^2 = \exp(-2ikd)/(r_1 r_2) \\ G_{tL} &= G_s = \exp(-ikd)/\sqrt{r_1 r_2} \end{aligned} \tag{11.40}$$

For the amplified quantum noise we have

$$\hat{F}_q = \sum_m \left\{ \hat{f}_m \left(t - \frac{d - z_m}{c_1} \right) g_{mR} e^{ik(d - z_m)} + \hat{f}_m \left(t - \frac{d + z_m}{c_1} \right) r_1 g_{mL} e^{ik(d + z_m)} \right\} \tag{11.41}$$

where g_{mR} is the amplification associated with the path from $z = z_m$ to $z = d$ and g_{mL} stands for the amplification associated with the path from $z = z_m$ to $z = 0$ and then to $z = d$:

$$g_{mR} = \exp \int_{z_m}^{d} \alpha(z) dz$$
$$g_{mL} = G_s \exp \int_{0}^{z_m} \alpha(z) dz \tag{11.42}$$

The factors $e^{ik(d \pm z_m)}$ represent the phase shifts associated with respective propagations. The quantum noise $\hat{F}_q$ may have a constant phase factor corresponding to e^{-ikd} in Equation 10.69 depending on the definition of $\hat{e}^+(z, t)$ compared to $\hat{E}^{(+)}(z, t)$, but it does not contribute to the phase diffusion and will be ignored.

11.4
The Phase Diffusion of the Output Field

The output field $\hat{e}^+_{0,2}(z, t)$ coming from $\hat{e}^+(d - 0, t)$ and coupled out of the mirror M2 is

$$\hat{e}^+_{0,2}(z, t) = T_2 \hat{e}^+ \left(d - 0,\ t - \frac{z - d}{c_0} \right) \tag{11.43}$$

As in the previous chapter, we assume a well-stabilized field amplitude $e_{0,2}$ and examine the diffusion of the phase of the field $\hat{e}^+_{0,2}(z, t)$. We set

$$\hat{e}^+_{0,2}(z, t) = e_{0,2} \exp\{i\phi_2(z, t)\} \tag{11.44}$$

Then we have

$$\frac{\partial}{\partial t} \phi_2(z, t) = -\frac{ie^{-i\phi_2}}{2e_{0,2}} \frac{\partial}{\partial t} \hat{e}^+_{0,2}(z, t) + \text{H.C.} \tag{11.45}$$

Using Equation 11.35 we obtain

$$\frac{\partial}{\partial t} \phi_2(z, t) = -\frac{ie^{-i\phi_2}}{2e_{0,2}} \frac{c_1}{2d} \frac{\gamma'}{\gamma' + \gamma'_c} T_2 \left(\hat{F}_t + \hat{F}_q \right) + \text{H.C.} \tag{11.46}$$

where $\hat{F}_t$ and $\hat{F}_q$ are given by Equations 11.36 and 11.41, respectively, with the time being replaced by $t - (z - d)/c_0$. Now the phase change $\Delta\phi_2$ during time t to $t + \Delta t$ is given by the integral over this time region of the RHS of Equation 11.46. As in the previous chapter, we assume that the phase ϕ_2 is slowly changing on the time scale of the correlation times of $\hat{F}_t$ and $\hat{F}_q$, so that the factor $e^{-i\phi_2}$ can be absorbed into the noise forces without affecting their delta-correlated characteristics described in Equations 11.1b and 11.4. Thus we see that the ensemble-averaged value of $\Delta\phi_2$ squared is proportional to Δt. Using Equations 11.36 and 11.41 together with Equations 11.1b and 11.4 we have

$$\left\langle \{\Delta\phi_2(z,t)\}^2 \right\rangle = \left[-\frac{i}{2e_{0,2}} \frac{c_1}{2d} \frac{\gamma'}{\gamma' + \gamma'_c} T_2 \left\{ \hat{F}_t(t'') + \hat{F}_q(t'') \right\} + \text{H.C.} \right] \Big\rangle dt' dt''$$
$$= \frac{1}{4|e_{0,2}|^2} \left(\frac{c_1}{2d}\right)^2 \left| \frac{\gamma'}{\gamma' + \gamma'_c} \right|^2 |T_2|^2 \left[\left(|T'_2 r_1 G_{tR}|^2 + |T'_1 G_{tL}|^2 \right) \frac{\hbar\omega(2n_\omega + 1)}{2\varepsilon_0 c_0} + \sum_m \left(|g_{mR}|^2 + |r_1 g_{mL}|^2 \right) \frac{2g\hbar\omega}{2\varepsilon_1 c_1^2} \right] |\Delta t| \tag{11.47}$$

Here we have assumed that $\hat{f}_t^R$ and $\hat{f}_t^L$ have no mutual correlation. Also, we have ignored the cross-terms of g_{mR} and g_{mL} because the phase factors $e^{ik(d-z_m)} e^{-ik(d+z_m)}$ and $e^{ik(d+z_m)} e^{-ik(d-z_m)}$ associated with these terms will yield vanishing results when summed over m.

The summation over m in Equation 11.47 can be converted to the spatially integrated field intensity as follows. Now by Equations 11.42, F.1a, 11.9, and F.7

$$|g_{mR}|^2 = \exp \int_{z_m}^{d} 2\text{Re}\{\alpha(z)\} dz = \frac{|e^+(d-0)|^2}{|e^+(z_m)|^2} = \frac{|e^+(d-0)|^2}{C} |e^-(z_m)|^2 = \frac{|e^-(z_m)|^2}{|r_2|\sqrt{C}} \tag{11.48}$$

Also, using Equations 11.42, F.1a, F.6, and 11.40, we have

$$|r_1 g_{mL}|^2 = |r_1 G_s|^2 \exp \int_0^{z_m} 2\text{Re}\{\alpha(z)\} dz = |r_1 G_s|^2 \frac{|e^+(z_m)|^2}{|e^+(0)|^2} = |r_1 G_s|^2 \frac{|e^+(z_m)|^2}{|r_1|\sqrt{C}} = \frac{|e^+(z_m)|^2}{|r_2|\sqrt{C}} \tag{11.49}$$

Combining Equations 11.48 and 11.49 we have

$$\sum_m \left(|g_{mR}|^2 + |r_1 g_{mL}|^2 \right) = \frac{1}{|r_2|} \sum_m \frac{1}{\sqrt{C}} \left(|e^+(z_m)|^2 + |e^-(z_m)|^2 \right) = \frac{N}{|r_2|} \frac{I}{\sqrt{C}} \tag{11.50}$$

where, in third term, we have rewritten the sum using the definition of the integrated intensity I in Equation 11.12. The quantity $I/\sqrt{C}$ is evaluated in Appendix F.

Using Equation 11.40 in the quantity in the round bracket for the thermal noise in Equation 11.47 we obtain

$$\left|T_2' r_1 G_{tR}\right|^2 + \left|T_1' G_{tL}\right|^2 = \frac{1}{|r_2|}\left(\frac{|T_2'|^2}{|r_2|} + \frac{|T_1'|^2}{|r_1|}\right) \tag{11.51}$$

In Appendix G we shall discuss a general multilayered mirror and show that

$$\frac{\left|T_{1,2}'\right|^2}{|r_{1,2}|} = \frac{1}{n}\left(\frac{1}{|r_{1,2}|} - |r_{1,2}|\right) \tag{11.52a}$$

and that

$$\frac{|T_{1,2}|^2}{|r_{1,2}|} = n\left(\frac{1}{|r_{1,2}|} - |r_{1,2}|\right) \tag{11.52b}$$

where n ($= \sqrt{\varepsilon_1/\varepsilon_0}$) is the refractive index of the material inside the cavity.

11.5
The Linewidth for the Nonlinear Gain Regime

Thus the linewidth becomes, from Equation 11.47,

$$\begin{aligned}
(\Delta\omega)_2 &= \frac{\left\langle \{\Delta\phi_2(z,t)\}^2 \right\rangle}{|\Delta t|} \\
&= \frac{1}{4|e_{0,2}|^2}\left(\frac{c_1}{2d}\right)^2 \left|\frac{\gamma'}{\gamma' + \gamma_c'}\right|^2 n\left(\frac{1}{|r_2|} - |r_2|\right)\left(\frac{1}{|r_1|} - |r_1| + \frac{1}{|r_2|} - |r_2|\right) \\
&\quad \times \left[\frac{1}{n}\frac{\hbar\omega(2n_\omega + 1)}{2\varepsilon_0 c_0} + \frac{Nd}{\ln(1/|r_1||r_2|)}\left\{1 + \frac{\Delta}{1+\Delta} g(|r_1|,|r_2|)\right\}\frac{2g\hbar\omega}{2\varepsilon_1 c_1^2}\right] \\
&= \frac{1}{4|e_{0,2}|^2}\frac{\hbar\omega}{\varepsilon_0 c_0}\left(\frac{c_1}{2d}\right)^2 \left|\frac{\gamma'}{\gamma' + \gamma_c'}\right|^2 \left(\frac{1}{|r_2|} - |r_2|\right)\frac{(|r_1| + |r_2|)(1 - |r_1||r_2|)}{|r_1||r_2|} \\
&\quad \times \left[\left(n_\omega + \frac{1}{2}\right) + \frac{N}{2N\sigma_{th}^0}\left\{1 + \frac{\Delta}{1+\Delta} g(|r_1|,|r_2|)\right\}\right]
\end{aligned} \tag{11.53}$$

where we have used Equation F.18 for $I/\sqrt{C}$. Also, we have used Equations 11.25 and 11.26b in the last equality. In addition, we have used the relation $n/(\varepsilon_1 c_1) = 1/(\varepsilon_0 c_0)$. Here $g(|r_1|,|r_2|)$ is given as

$$\begin{aligned}
&g(|r_1|,|r_2|) \\
&= \frac{2[\ln(1/|r_1||r_2|)]^2 + \frac{1}{2}[\ln(1/|r_1||r_2|)]\left(|r_1|^2 + |r_2|^2\right)\left(1 - |r_1|^2|r_2|^2\right)/|r_1|^2|r_2|^2}{[(|r_1| + |r_2|)(1 - |r_1||r_2|)/|r_1||r_2|]^2} - 1
\end{aligned} \tag{11.54}$$

This is symmetric with respect to $|r_1|$ and $|r_2|$. If we had calculated the linewidth from the phase diffusion of the output field from mirror M1, by symmetry we would have obtained

$$(\Delta\omega)_1 = \frac{\left\langle \{\Delta\phi_1(z,t)\}^2 \right\rangle}{|\Delta t|}$$

$$= \frac{1}{4|e_{0,1}|^2}\frac{\hbar\omega}{\varepsilon_0 c_0}\left(\frac{c_1}{2d}\right)^2\left|\frac{\gamma'}{\gamma'+\gamma_c'}\right|^2\left(\frac{1}{|r_1|}-|r_1|\right)\frac{(|r_1|+|r_2|)(1-|r_1||r_2|)}{|r_1||r_2|} \tag{11.55}$$

$$\times\left[\left(n_\omega+\frac{1}{2}\right)+\frac{N}{2N\sigma_{th}^0}\left\{1+\frac{\Delta}{1+\Delta}g(|r_1|,|r_2|)\right\}\right]$$

Now the two expressions for the linewidth in fact give the same width. We note from Equations 11.43 and 11.44 that

$$|e_{0,2}|^2 = |T_2 e^+(d-0,\ t)|^2 \tag{11.56}$$

But by Equations F.7 and 11.52b we have

$$|e_{0,2}|^2 = \frac{|T_2|^2}{|r_2|}\sqrt{C} = n\left(\frac{1}{|r_2|}-|r_2|\right)\sqrt{C} \tag{11.57}$$

where $\sqrt{C} = |e^{\pm}(z_c)|^2$ is the field intensity at the neutral point as defined in Equation 11.9 and is given by Equation 11.22. Thus

$$\frac{(1/|r_2|)-|r_2|}{|e_{0,2}|^2} = \frac{1}{n\sqrt{C}} \tag{11.58a}$$

Similarly, we have

$$\frac{(1/|r_1|)-|r_1|}{|e_{0,1}|^2} = \frac{1}{n\sqrt{C}} \tag{11.58b}$$

Thus we have

$$(\Delta\omega) = (\Delta\omega)_1 = (\Delta\omega)_2$$

$$= \frac{1}{4n\sqrt{C}}\frac{\hbar\omega}{\varepsilon_0 c_0}\left(\frac{c_1}{2d}\right)^2\left|\frac{\gamma'}{\gamma'+\gamma_c'}\right|^2\frac{(|r_1|+|r_2|)(1-|r_1||r_2|)}{|r_1||r_2|} \tag{11.59}$$

$$\times\left[\left(n_\omega+\frac{1}{2}\right)+\frac{N}{2N\sigma_{th}^0}\left\{1+\frac{\Delta}{1+\Delta}g(|r_1|,|r_2|)\right\}\right]$$

Now, the fractional excess atomic inversion Δ is related to $\sqrt{C}$ by Equations 11.22 and 11.23 as

$$\sqrt{C} = |E_s|^2 \frac{|r_1||r_2| \ln(1/|r_1||r_2|)}{(|r_1| + |r_2|)(1 - |r_1||r_2|)} \Delta \tag{11.60}$$

Therefore, we have

$$\begin{aligned}
(\Delta\omega) &= \frac{1}{4n\sqrt{C}} \frac{\hbar\omega}{\varepsilon_0 c_0} \left(\frac{c_1}{2d}\right)^2 \left|\frac{\gamma'}{\gamma' + \gamma_c'}\right|^2 \frac{(|r_1| + |r_2|)(1 - |r_1||r_2|)}{|r_1||r_2|} \\
&\times \left[\left(n_\omega + \frac{1}{2}\right) + \frac{N}{2N\sigma_{th}^0}\left\{1 + \frac{\sqrt{C}}{\sqrt{C} + |E_s|^2 h(|r_1|, |r_2|)} g(|r_1|, |r_2|)\right\}\right] \\
&= \frac{1}{4n} \frac{\hbar\omega}{\varepsilon_0 c_0} \left(\frac{c_1}{2d}\right)^2 \left|\frac{\gamma'}{\gamma' + \gamma_c'}\right|^2 \frac{(|r_1| + |r_2|)(1 - |r_1||r_2|)}{|r_1||r_2|} \\
&\times \left[\frac{1}{\sqrt{C}}\left\{\left(n_\omega + \frac{1}{2}\right) + \frac{N}{2N\sigma_{th}^0}\right\} + \frac{N}{2N\sigma_{th}^0} \frac{g(|r_1|, |r_2|)}{\sqrt{C} + |E_s|^2 h(|r_1|, |r_2|)}\right]
\end{aligned} \tag{11.61}$$

where

$$h(|r_1|, |r_2|) = \frac{|r_1||r_2| \ln(1/|r_1||r_2|)}{(|r_1| + |r_2|)(1 - |r_1||r_2|)} \tag{11.62}$$

Now let us consider the dependence of the laser linewidth on the total output power. From Equation 11.57 the output power measured outside the mirror M2 is

$$P_2 = 2c_0\varepsilon_0 |e_{0,2}|^2 = 2c_0\varepsilon_0 n\left(\frac{1}{|r_2|} - |r_2|\right)\sqrt{C} = \left(\frac{1}{|r_2|} - |r_2|\right) P_c \tag{11.63}$$

Similarly, the output power measured outside the mirror M1 is

$$P_1 = 2c_0\varepsilon_0 |e_{0,1}|^2 = 2c_0\varepsilon_0 n\left(\frac{1}{|r_1|} - |r_1|\right)\sqrt{C} = \left(\frac{1}{|r_1|} - |r_1|\right) P_c \tag{11.64}$$

The total output power P_t is

$$\begin{aligned}
P_t = P_1 + P_2 &= 2c_0\varepsilon_0 n\left(\frac{1}{|r_1|} - |r_1| + \frac{1}{|r_2|} - |r_2|\right)\sqrt{C} \\
&= 2c_0\varepsilon_0 n \frac{(|r_1| + |r_2|)(1 - |r_1||r_2|)}{|r_1||r_2|}\sqrt{C} \\
&= \frac{(|r_1| + |r_2|)(1 - |r_1||r_2|)}{|r_1||r_2|} P_c
\end{aligned} \tag{11.65}$$

where $P_c = 2c_0\varepsilon_0 n\sqrt{C}$ is the power at z_c of the right- or left-going wave. Therefore, in terms of the output powers, we have

$$\begin{aligned}(\Delta\omega) &= \frac{\hbar\omega}{2}\left(\frac{c_1}{2d}\right)^2\left|\frac{\gamma'}{\gamma'+\gamma'_c}\right|^2\left(\frac{1}{|r_1|}-|r_1|\right)\frac{(|r_1|+|r_2|)(1-|r_1||r_2|)}{|r_1||r_2|} \\ &\times\left[\frac{1}{P_1}\left\{\left(n_\omega+\frac{1}{2}\right)+\frac{N}{2N\sigma_{th}^0}\right\}+\frac{N}{2N\sigma_{th}^0}\frac{g(|r_1|,|r_2|)}{P_1+P_s(1/|r_1|-|r_1|)h(|r_1|,|r_2|)}\right] \\ &= \frac{\hbar\omega}{2}\left(\frac{c_1}{2d}\right)^2\left|\frac{\gamma'}{\gamma'+\gamma'_c}\right|^2\left(\frac{1}{|r_2|}-|r_2|\right)\frac{(|r_1|+|r_2|)(1-|r_1||r_2|)}{|r_1||r_2|} \\ &\times\left[\frac{1}{P_2}\left\{\left(n_\omega+\frac{1}{2}\right)+\frac{N}{2N\sigma_{th}^0}\right\}+\frac{N}{2N\sigma_{th}^0}\frac{g(|r_1|,|r_2|)}{P_2+P_s(1/|r_2|-|r_2|)h(|r_1|,|r_2|)}\right] \\ &= \frac{\hbar\omega}{2}\left(\frac{c_1}{2d}\right)^2\left|\frac{\gamma'}{\gamma'+\gamma'_c}\right|^2\left\{\frac{(|r_1|+|r_2|)(1-|r_1||r_2|)}{|r_1||r_2|}\right\}^2 \\ &\times\left[\frac{1}{P_t}\left\{\left(n_\omega+\frac{1}{2}\right)+\frac{N}{2N\sigma_{th}^0}\right\}+\frac{N}{2N\sigma_{th}^0}\frac{g(|r_1|,|r_2|)}{P_t+P_s\ln(1/|r_1||r_2|)}\right]\end{aligned} \tag{11.66}$$

where

$$P_s = 2\varepsilon_0 c_0 n|E_s|^2 = 2\varepsilon_1 c_1|E_s|^2 \tag{11.67}$$

is the saturation power. The factor $|\gamma'/(\gamma'+\gamma'_c)|^2$ is evaluated as

$$\begin{aligned}\left|\frac{\gamma'}{\gamma'+\gamma'_c}\right|^2 &= \left|\frac{\gamma+i(\nu_0-\omega)}{\gamma+i(\nu_0-\omega)+\gamma_c+i(\omega_c-\omega)}\right|^2 \\ &= \frac{\gamma^2(1+\delta^2)}{(\gamma+\gamma_c)^2+(\gamma-\gamma_c)^2\delta^2}\end{aligned} \tag{11.68}$$

with

$$\delta^2 = \frac{(\nu_0-\omega_c)^2}{(\gamma+\gamma_c)^2} \tag{11.69}$$

where we have used the linear pulling relation obtained from Equation 11.32 for the steady state ($s_0 = 0$):

$$\omega = \frac{\gamma\omega_c+\gamma_c\nu_0}{\gamma+\gamma_c} \tag{11.70}$$

Henry [2] obtained the same $r_{1,2}$ dependence as the one in the first line of Equation 11.66 in his linear gain analysis based on the Green's function method. The Green's function method will be described in Section 14.2 in Chapter 14.

The result obtained in Equation 11.66 as compared to the quasimode theoretical result in Equation 4.82 has two corrections. The expression for the linewidth (FWHM) in angular frequency containing the total output power, the last expression in Equation 11.66, may be rewritten in the form

$$\Delta\omega = \frac{2\hbar\omega K_L \gamma_c^2 \gamma^2 (1+\delta^2)}{(\gamma+\gamma_c)^2 + (\gamma-\gamma_c)^2\delta^2} \left[\frac{1}{P_t} \left\{ \langle n_\omega \rangle + \frac{1}{2} + \frac{N}{2N\sigma_{th}^0} \right\} + \frac{N}{2N\sigma_{th}^0} C_q \right] \quad (11.71)$$

where

$$\begin{aligned} K_L &= \frac{\{(|r_1|+|r_2|)(1-|r_1||r_2|)/(2|r_1||r_2|)\}^2}{[\ln(1/|r_1||r_2|)]^2} \\ &= \frac{\{(1-|r_1|^2)/2|r_1| + (1-|r_2|^2)/2|r_2|\}^2}{[\ln(1/|r_1|) + \ln(1/|r_2|)]^2} \end{aligned} \quad (11.72)$$

$$C_q = \frac{1}{P_t + P_s \ln(1/|r_1||r_2|)} g(|r_1|, |r_2|) \quad (11.73)$$

The correction factor K_L is the generalization of the longitudinal excess noise factor that appeared in Equation 10.112 for the one-sided laser model. Note that below Equation 7.37 we discussed the replacement of the cavity decay constant in the quasimode analysis by that of the new cavity model based on the equivalence of the two in the decay equations for the field amplitude. Likewise, the cavity decay constant γ_c in Equation 11.71 is equivalent to that of the quasimode cavity model that appears in Equation 4.82. Thus the correction to the conventional formula is given rightly by Equation 11.72. The factor C_q is the generalization of the factor $g(r)/\{P + P_s \ln(1/r)\}$ in Equation 10.111. When P_s is large compared to P, or, roughly, when $\varepsilon_0 c_0 |e_{0,1,2}|^2 \ll \varepsilon_1 c_1 |E_s|^2$, this yields a power-independent part of the linewidth. That is, it will yield a contribution that is independent of P.

It is easy to see that the K_L in Equation 11.72 reduces to that in Equation 10.112 if we set, for example, $r_1 \to r$ and $r_2 \to -1$. Also, for the same settings, C_q reduces to $g(r)/\{P + P_s \ln(1/r)\}$ in Equation 10.111. Note that $g(|r_1|, |r_2|)$ is given by Equation 11.54. The expression for the function $g(r)$ is found in Equation 10.102 or in Equation E.24. For a symmetric cavity with $|r_1| = |r_2| = r$, we have

$$K_L = \frac{[(1-r^2)/r]^2}{[2\ln(1/r)]^2} \quad (11.74)$$

which is the same as that for the one-sided cavity obtained in Equation 10.112. Also, for this case, it can be shown that $g(|r_1|, |r_2|) = g(r)$. For this case of $|r_1| = |r_2| = r$, C_q should read

$$C_q = \frac{1}{P_t + 2P_s \ln(1/r)} g(r) \quad (11.75)$$

The generalized longitudinal excess noise factor K_L is large for an asymmetric cavity, as will be discussed in Section 12.4.1. For a fixed value of $|r_2|$, the generalized longitudinal excess noise factor K_L diverges for small $|r_1|$ as

$$\lim_{|r_1|\to 0} K_L = \left[\frac{1/(2|r_1|)}{\ln(1/|r_1|)}\right]^2 \tag{11.76}$$

A different treatment of the C_q term will be given in Section 12.11.

11.6
The Linewidth for the Linear Gain Regime

Up to now we have considered the phase diffusion above threshold using the simplified method. In this section, we briefly discuss, using the same simplified method, the linewidth below threshold, where a linear gain model is appropriate. Now, because the net gain for the coherent oscillation is negative, we have no steady amplitude but instead have decaying amplitude. The steady state in the field power is maintained by the added noise components, which compensate for the negative net loss. Thus Equation 11.35 for the right-going wave just inside the coupling mirror M2 will read

$$\frac{d\hat{e}^{+}(d-0,t)}{dt} = s_0\hat{e}^{+}(d-0,t) + \frac{c_1}{2d}\frac{\gamma'}{\gamma'+\gamma_c'}\left(\hat{F}_t + \hat{F}_q\right) \tag{11.77}$$

where s_0 is given by Equation 11.32 with relatively small linear gain. Here Re $s_0 < 0$, so that the coherent amplitude always decays. The thermal noise source $\hat{F}_t$ and the quantum noise source $\hat{F}_q$ are given, respectively, by Equations 11.36 and 11.41. Integrating Equation 11.77 we have

$$\begin{aligned}\hat{e}^{+}(d-0,t) &= \frac{c_1}{2d}\frac{\gamma'}{\gamma'+\gamma_c'}\int_0^t e^{s_0(t-t')} \\ &\quad\times\left\{\hat{F}_t(t') + \hat{F}_q(t')\right\}dt' + \hat{e}^{+}(d-0,\ 0)e^{s_0 t}\end{aligned} \tag{11.78}$$

We will calculate the correlation function for this field amplitude and Fourier-transform the correlation function to obtain the line profile. Thus

$$\begin{aligned}&\left\langle \hat{e}^{+\dagger}(d-0,\ t+\tau)\,\hat{e}^{+}(d-0,t)\right\rangle \\ &= \left(\frac{c_1}{2d}\right)^2\left|\frac{\gamma'}{\gamma'+\gamma_c'}\right|^2\int_0^{t+\tau}dt''\int_0^t dt'\,e^{(s_0+s_0^*)t - s_0^* t'' - s_0 t' + s_0^*\tau} \\ &\quad\times\left\{\left\langle \hat{F}_t^\dagger(t'')\hat{F}_t(t')\right\rangle + \left\langle \hat{F}_q^\dagger(t'')\hat{F}_q(t')\right\rangle\right\}\end{aligned} \tag{11.79}$$

Here we have ignored the contributions from the initial value, which decay with time or vanish because $\left\langle \hat{F}_{t,q}(t)\right\rangle = \left\langle \hat{F}_{t,q}^\dagger(t)\right\rangle = 0$. Taking into account the relations in the first lines in Equations 11.1b and 11.1c, we write

$$\left\langle \hat{F}_t^\dagger(t'')\hat{F}_t(t')\right\rangle = D_{tt}\delta(t''-t') \tag{11.80}$$

where D_{tt} will be determined later. For the quantum noise, since we are using normally ordered correlations, we use, instead of Equation 11.4, the relation in

Equation 4.50 or 9.73, which is applicable to the case of unsaturated, linear gain. Thus we write (see Equation 11.3 and the expression for $2\mathcal{g}$ below Equation 11.4)

$$2\varepsilon_1 c_1 \left\langle \hat{f}_m^{\dagger}(t)\hat{f}_{m'}(t')\right\rangle = (2\mathcal{g}\hbar\omega/c_1)\{(1+\sigma)/2\}\delta_{mm'}\delta(t-t') \tag{11.81}$$

and

$$\left\langle \hat{F}_q^{\dagger}(t'')\hat{F}_q(t')\right\rangle = D_{qq}\delta(t''-t') \tag{11.82}$$

where D_{qq} will be determined later. Then, Equation 11.79 can be integrated to yield

$$\begin{aligned}&\langle \hat{e}^{+\dagger}(d-0,\ t+\tau)\,\hat{e}^{+}(d-0,\ t)\rangle\\ &= \left(\frac{c_1}{2d}\right)^2\left|\frac{\gamma'}{\gamma'+\gamma'_c}\right|^2(D_{tt}+D_{qq})\begin{cases}\dfrac{e^{s_0^*\tau}-e^{(s_0+s_0^*)t+s_0^*\tau}}{-(s_0+s_0^*)}, & \tau>0\\[2ex] \dfrac{e^{-s_0\tau}-e^{(s_0+s_0^*)t+s_0^*\tau}}{-(s_0+s_0^*)}, & \tau<0\end{cases}\end{aligned} \tag{11.83}$$

Discarding the terms that decay for $t\to\infty$ and noting that $\mathrm{Re}s_0<0$, we have

$$\begin{aligned}&\langle \hat{e}^{+\dagger}(d-0,\ t+\tau)\,\hat{e}^{+}(d-0,\ t)\rangle\\ &= \left(\frac{c_1}{2d}\right)^2\left|\frac{\gamma'}{\gamma'+\gamma'_c}\right|^2\frac{D_{tt}+D_{qq}}{2|\mathrm{Re}s_0|}\begin{cases}e^{s_0^*\tau}, & \tau>0\\ e^{-s_0\tau}, & \tau<0\end{cases}\end{aligned} \tag{11.84}$$

Fourier-transforming Equation 11.84, we have the power spectrum, similarly to Equation 9.99,

$$\begin{aligned}I(\omega) &= \int_{-\infty}^{+\infty}\langle \hat{e}^{+\dagger}(d-0,\ t+\tau)\,\hat{e}^{+}(d-0,\ t)\rangle e^{-i\omega\tau}d\tau\\ &\propto \int_{-\infty}^{0} e^{-s_0\tau-i\omega\tau}d\tau+\int_{0}^{+\infty}e^{s_0^*\tau-i\omega\tau}d\tau\\ &= \frac{-2\mathrm{Re}\,s_0}{(\omega+\mathrm{Im}\,s_0)^2+(\mathrm{Re}\,s_0)^2}\end{aligned} \tag{11.85}$$

Thus the angular FWHM is

$$\Delta\omega = 2|\mathrm{Re}\,s_0| \tag{11.86}$$

The power output through mirror M2 is

$$\begin{aligned}P_2 &= 2\varepsilon_0 c_0|T_2|^2\langle \hat{e}^{+\dagger}(d-0,\ t)\,\hat{e}^{+}(d-0,\ t)\rangle\\ &= 2\varepsilon_0 c_0|T_2|^2\left(\frac{c_1}{2d}\right)^2\left|\frac{\gamma'}{\gamma'+\gamma'_c}\right|^2\frac{D_{tt}+D_{qq}}{2|\mathrm{Re}\,s_0|}\end{aligned} \tag{11.87}$$

where we have used Equation 11.84 with $\tau=0$. Combining Equations 11.86 and 11.87 we have the formal expression for the linewidth in terms of the power output:

$$\Delta\omega = \frac{2\varepsilon_0 c_0 |T_2|^2}{P_2}\left(\frac{c_1}{2d}\right)^2 \left|\frac{\gamma'}{\gamma' + \gamma'_c}\right|^2 (D_{tt} + D_{qq}) \tag{11.88}$$

The remaining task is to evaluate $D_{tt} + D_{qq}$. First, we consider D_{tt}. Now Equation 11.80 becomes, by Equation 11.36,

$$\begin{aligned} D_{tt}\delta(t'-t) = \Bigg\langle \Bigg\{ & \left(T'_2 r_1 G_{tR} e^{2ikd}\right)^* \hat{f}_t^{R\dagger}\left(d+0,\ t' - \frac{2d}{c_1}\right) \\ & + \left(T'_1 G_{tL} e^{ikd}\right)^* \hat{f}_t^{L\dagger}\left(-d-0,\ t' - \frac{d}{c_1}\right)\Bigg\} \\ & \times \Bigg\{ T'_2 r_1 G_{tR} e^{2ikd} \hat{f}_t^{R}\left(d+0,\ t - \frac{2d}{c_1}\right) \\ & + T'_1 G_{tL} e^{ikd} \hat{f}_t^{L}\left(-d-0,\ t - \frac{d}{c_1}\right)\Bigg\}\Bigg\rangle \end{aligned} \tag{11.89}$$

Using Equations 11.1b and 11.1c we have

$$D_{tt} = \frac{n_\omega \hbar\omega}{2\varepsilon_0 c_0}\left(\left|T'_2 r_1 G_{tR}\right|^2 + \left|T'_1 G_{tL}\right|^2\right) \tag{11.90}$$

Now the amplifying constants G_{tR} and G_{tL} may be given approximately by Equation 11.40 also in this linear gain regime as long as the operation is not far below threshold. Thus remembering Equation 11.52a we have

$$D_{tt} = \frac{n_\omega \hbar\omega}{2\varepsilon_0 c_0}\left(\left|\frac{T'_2}{r_2}\right|^2 + \left|\frac{T'_1}{\sqrt{r_1 r_2}}\right|^2\right) = \frac{n_\omega \hbar\omega}{2\varepsilon_0 c_0}\frac{1}{n|r_2|}\left(\frac{1}{|r_2|} - |r_2| + \frac{1}{|r_1|} - |r_1|\right) \tag{11.91}$$

Next, from Equation 11.41 we have

$$\begin{aligned} D_{qq}\delta(t'-t) = \Bigg\langle \sum_m \Bigg\{ & \hat{f}_m^\dagger\left(t' - \frac{d-z_m}{c_1}\right) g_{mR}^* e^{-ik(d-z_m)} \\ & + \hat{f}_m^\dagger\left(t' - \frac{d+z_m}{c_1}\right)(r_1 g_{mL})^* e^{-ik(d+z_m)}\Bigg\} \\ & \times \sum_{m'} \Bigg\{ \hat{f}_{m'}\left(t - \frac{d-z_{m'}}{c_1}\right) g_{m'R} e^{ik(d-z_{m'})} \\ & + \hat{f}_{m'}\left(t - \frac{d+z_{m'}}{c_1}\right) r_1 g_{m'L} e^{ik(d+z_{m'})}\Bigg\}\Bigg\rangle \end{aligned} \tag{11.92}$$

Using Equation 11.81 we have

$$D_{qq} = \frac{2g\hbar\omega(1+\sigma)}{4\varepsilon_1 c_1^2}\sum_m \left(|g_{mR}|^2 + |r_1 g_{mL}|^2\right) \tag{11.93}$$

where we have ignored the cross-terms of g_{mR} and g_{mL}, which vanish on taking the summation over m because of the rapidly oscillating functions of z_m. Now the amplification constants g_{mR} and g_{mL} are defined in Equation 11.42, which, in the linear gain regime, are

$$g_{mR} = \exp\int_{z_m}^{d} \alpha^0 dz = \exp\{\alpha^0 (d - z_m)\}$$
$$g_{mL} = G_s \exp\int_{0}^{z_m} \alpha^0 dz = G_s \exp(\alpha^0 z_m) \tag{11.94}$$

where, from Equation 11.7,

$$\alpha^0 = \frac{gN\sigma^0}{\gamma'} \tag{11.95}$$

and from Equation 11.38

$$G_s \equiv \exp\int_0^d \alpha^0 dz' = \exp(\alpha^0 d) \tag{11.96}$$

Thus we have

$$\sum_m \left(|g_{mR}|^2 + |r_1 g_{mL}|^2\right) = N\int_0^d \left(e^{2\mathrm{Re}\,\alpha^0(d-z_m)} + |r_1 G_s|^2 e^{2\mathrm{Re}\,\alpha^0 z_m}\right) dz_m$$
$$= \frac{N}{2\mathrm{Re}\,\alpha^0}\left\{e^{2\mathrm{Re}\,\alpha^0 d} - 1 + |r_1|^2 e^{2\mathrm{Re}\,\alpha^0 d}\left(e^{2\mathrm{Re}\alpha^0 d} - 1\right)\right\} \tag{11.97}$$

But as we have approximately

$$|r_1 r_2| e^{2\mathrm{Re}\,\alpha^0 d} = 1 \tag{11.98}$$

then Equation 11.97 becomes

$$\sum_m \left(|g_{mR}|^2 + |r_1 g_{mL}|^2\right)$$
$$= \frac{Nd}{\ln(1/|r_1||r_2|)}\left\{\frac{1}{|r_1||r_2|} - 1 + |r_1|^2\left(\frac{1}{|r_1|^2|r_2|^2} - \frac{1}{|r_1||r_2|}\right)\right\} \tag{11.99}$$
$$= \frac{Nd}{\ln(1/|r_1||r_2|)}\frac{1}{|r_2|}\left(\frac{1}{|r_1|} - |r_2| + \frac{1}{|r_2|} - |r_1|\right)$$

(For somewhat below threshold, the LHS member of Equation 11.98 is smaller than unity, and the above sum becomes smaller than this expression.) So we have

$$D_{qq} = \frac{2g\hbar\omega(1+\sigma)}{4\varepsilon_1 c_1^2}\frac{Nd}{\ln(1/|r_1||r_2|)}\frac{1}{|r_2|}\left(\frac{1}{|r_1|} - |r_1| + \frac{1}{|r_2|} - |r_2|\right)$$
$$= \frac{\hbar\omega}{2\varepsilon_1 c_1}\frac{N_2}{N\sigma_{th}^0}\frac{1}{|r_2|}\left(\frac{1}{|r_1|} - |r_1| + \frac{1}{|r_2|} - |r_2|\right) \tag{11.100}$$

where we have used Equations 11.25 and 11.26b as well as the relation $(1+\sigma)N = 2N_2$ in the last line. As a result, the linewidth (FWHM) in angular frequency becomes, by Equations 11.88, 11.91, and 11.100,

$$\begin{aligned}\Delta\omega &= \frac{\hbar\omega}{P_2}\left(\frac{c_1}{2d}\right)^2\left|\frac{\gamma'}{\gamma'+\gamma'_c}\right|^2\frac{|T_2|^2}{n|r_2|}\left(\frac{1}{|r_2|}-|r_2|+\frac{1}{|r_1|}-|r_1|\right)\left(n_\omega+\frac{N_2}{N\sigma^0_{th}}\right)\\ &= \frac{\hbar\omega}{P_2}\left(\frac{c_1}{2d}\right)^2\left|\frac{\gamma'}{\gamma'+\gamma'_c}\right|^2\left(\frac{1}{|r_2|}-|r_2|\right)\frac{(|r_1|+|r_2|)(1-|r_1||r_2|)}{|r_1||r_2|}\left(n_\omega+\frac{N_2}{N\sigma^0_{th}}\right)\end{aligned} \tag{11.101}$$

We have used Equation 11.52b in the second line. Now, let us think of the neutral point z_c inside the cavity, where $|e^+(z_c)| = |e^-(z_c)|$ holds. If the amplitude gain from z_c to the output port $z=d$ is G_{cd}, we have $\left|r_2 G_{cd}^2\right| \simeq 1$. So, we have

$$P_2 n|r_2|/|T_2|^2 = 2\varepsilon_0 c_0 |e^+(d-0)|^2 n|r_2| = 2\varepsilon_1 c_1 \left|e^\pm(z_c)\right|^2 = P_c \tag{11.102}$$

where P_c is the power associated with $e^\pm(z_c)$. Thus we have

$$\Delta\omega = \frac{\hbar\omega}{P_c}\left(\frac{c_1}{2d}\right)^2\left|\frac{\gamma'}{\gamma'+\gamma'_c}\right|^2\left(\frac{1}{|r_2|}-|r_2|+\frac{1}{|r_1|}-|r_1|\right)\left(n_\omega+\frac{N_2}{N\sigma^0_{th}}\right) \tag{11.103}$$

This expression is independent of the choice of the output port. By symmetry we have

$$\begin{aligned}\Delta\omega &= \frac{\hbar\omega}{P_1}\left(\frac{c_1}{2d}\right)^2\left|\frac{\gamma'}{\gamma'+\gamma'_c}\right|^2\left(\frac{1}{|r_1|}-|r_1|\right)\\ &\quad\times\frac{(|r_1|+|r_2|)(1-|r_1||r_2|)}{|r_1||r_2|}\left(n_\omega+\frac{N_2}{N\sigma^0_{th}}\right)\end{aligned} \tag{11.104}$$

Thus, for the total output power $P_t = P_1 + P_2$ we have

$$\Delta\omega = \frac{\hbar\omega}{P_t}\left(\frac{c_1}{2d}\right)^2\left|\frac{\gamma'}{\gamma'+\gamma'_c}\right|^2\left\{\frac{(|r_1|+|r_2|)(1-|r_1||r_2|)}{|r_1||r_2|}\right\}^2\left(n_\omega+\frac{N_2}{N\sigma^0_{th}}\right) \tag{11.105}$$

with (see Equation 11.68)

$$\left|\frac{\gamma'}{\gamma'+\gamma'_c}\right|^2 = \frac{\gamma^2(1+\delta^2)}{(\gamma+\gamma_c)^2+(\gamma-\gamma_c)^2\delta^2} \tag{11.106}$$

This is just twice the last line in Equation 11.66 except for the term of $g(|r_1|,|r_2|)$ that appears because of the saturated atomic inversion. The factor of 2 can be traced back to the difference between Equation 11.46 for the phase diffusion with constant amplitude and Equation 11.77 for the field driven by the noise sources, including the amplitude noise. This point will be discussed further in Section 12.5. Comparison with the standard result for a quasimode laser in Equation 4.62 yields the longitudinal excess noise factor:

$$\begin{aligned}K_L &= \frac{(c_1/2d)^2\{(|r_1|+|r_2|)(1-|r_1||r_2|)/|r_1||r_2|\}^2}{4\gamma_c^2}\\ &= \left\{\frac{(|r_1|+|r_2|)(1-|r_1||r_2|)/(2|r_1||r_2|)}{\ln(1/|r_1||r_2|)}\right\}^2\end{aligned} \tag{11.107}$$

This is the same as the factor K_L in Equation 11.72 obtained for the nonlinear gain regime. This is the generalization of the longitudinal excess noise factor in

Equation 9.106, which was obtained for the one-sided cavity laser. Note that this factor approaches unity in the good cavity limit $|r_1||r_2| \to 1$.

In Equation 11.105 the thermal noise term of n_ω and the quantum noise term of $N_2/N\sigma_{th}^0$ have appeared through normally ordered noise operators, while in Equation 11.66 the terms of $n_\omega + \frac{1}{2}$ and $N/(2N\sigma_{th}^0)$ have appeared through symmetrically ordered noise operators. The sum $\frac{1}{2} + N/(2N\sigma_{th}^0)$ makes $N_2/N\sigma_{th}^0$.

The simplified *ad hoc* method of the present chapter, which we have named the propagation method or propagation theory, was proposed by Ujihara [1] and by Goldberg *et al.* [3], who considered spatial hole burning effects on the linewidth. Prasad [4] also used the same method and obtained similar results, including the power-independent part in the linewidth. The results obtained in this chapter were also derived by Henry [2] for the linear gain regime and by van Exter *et al.* [5] for the linear and nonlinear gain regime by the Green's function method, which will be described in Chapter 14.

Finally, we note that the present propagation method directly shows that the excess noise factor is the result of noise amplification during the one round trip analyzed in this chapter. We consider the linear gain regime for simplicity. If amplification is absent but a steady state is still required, we should require that the reflectivity of the two mirrors be unity ($|r_{1,2}| \to 1$) by Equation 11.98. Then, noting that

$$|r_{1,2}|^{-1} - |r_{1,2}| = (1 + |r_{1,2}|)(1 - |r_{1,2}|)/|r_{1,2}| \to 2(1 - |r_{1,2}|)$$

the diffusion constant D_{tt} in Equation 11.91 becomes

$$D_{tt} \to \frac{n_\omega \hbar\omega}{2\varepsilon_0 c_0} \frac{1}{n} 2(2 - |r_1| - |r_2|) = \frac{n_\omega \hbar\omega}{\varepsilon_0 c_0} \frac{1}{n} \frac{2d}{c_1} \gamma_c \qquad (11.108)$$

where we have used the relation

$$\gamma_c = \left(\frac{c_1}{2d}\right) \ln \frac{1}{|r_1||r_2|} \to \left(\frac{c_1}{2d}\right)(2 - |r_1| - |r_2|) \qquad (11.109)$$

which holds in this limit. The sum in Equation 11.99 merely becomes

$$\sum_m \left(|g_{mR}|^2 + |r_1 g_{mL}|^2\right) = 2Nd \qquad (11.110)$$

Thus the diffusion constant D_{qq} in Equation 11.100 becomes, simply,

$$D_{qq} = \frac{g\hbar\omega(1+\sigma)Nd}{\varepsilon_1 c_1^2} \qquad (11.111)$$

Thus the linewidth in terms of the output power P_2 in Equation 11.88 becomes

$$\Delta\omega = \frac{2\varepsilon_0 c_0 |T_2|^2}{P_2} \left(\frac{c_1}{2d}\right)^2 \left|\frac{\gamma'}{\gamma' + \gamma_c'}\right|^2 \left(\frac{n_\omega \hbar\omega}{\varepsilon_0 c_0 n} \frac{2d}{c_1} \gamma_c + \frac{g\hbar\omega(1+\sigma)Nd}{\varepsilon_1 c_1^2}\right) \qquad (11.112)$$

We will have a similar expression in terms of the output power P_1 with $|T_2|^2$ replaced by $|T_1|^2$. Thus in terms of the total output power P_t we have

$$\Delta\omega = \frac{2\varepsilon_0 c_0\left(|T_1|^2+|T_2|^2\right)}{P_t}\left(\frac{c_1}{2d}\right)^2\left|\frac{\gamma'}{\gamma'+\gamma'_c}\right|^2 \times\left(\frac{n_\omega\hbar\omega\,2d}{\varepsilon_0 c_0 n\; c_1}\gamma_c+\frac{g\hbar\omega(1+\sigma)Nd}{\varepsilon_1 c_1^2}\right) \tag{11.113}$$

We use Equation G.23 in Appendix G to obtain $|T_{1,2}|^2 \to 2n(1-|r_{1,2}|)$, and we eliminate the constant g using the threshold relation (Equation 11.26b) $gN\sigma_{th} = \gamma_c$ as in the previous evaluations of the linewidth. Then noting that $\varepsilon_1 c_1 = (\varepsilon_0 n^2)(c_0/n)$, it is easy to see that the linewidth reduces to the standard result in Equation 4.62a for the linear gain regime, which lacks the excess noise factor:

$$\Delta\omega = \frac{4\hbar\omega\gamma_c^2}{P_t}\left|\frac{\gamma'}{\gamma'+\gamma'_c}\right|^2\left(n_\omega+\frac{N_2}{N\sigma_{th}}\right) \tag{11.114}$$

We have shown that, if the two mirrors are nearly perfectly reflecting and the gain of the laser medium is unity, we will have no excess noise factor. This shows directly that the excess noise factor originates from the finite mirror transmissions as well as finite amplification of the thermal and quantum noise during one round trip in the cavity. A similar argument may be given for the nonlinear, saturated gain medium.

References

1 Ujihara, K. (1984) *IEEE J. Quantum Electron.*, *QE-20*, 814–818.

2 Henry, C.H. (1986) *J. Lightwave Technol.*, *LT-4*, 288–297.

3 Goldberg, P., Milonni, P.W., and Sundaram, B. (1991) *Phys. Rev. A*, *44*, 1969–1985.

4 Prasad, S. (1992) *Phys. Rev. A*, *46*, 1540–1559.

5 van Exter, M.P., Kuppens, S.J.M., and Woerdman, J.P. (1995) *Phys. Rev. A*, *51*, 809–816.

12
A One-Dimensional Laser with Output Coupling: Summary and Interpretation of the Results

Starting with the basic equations of motion for a laser having output coupling derived in Chapter 5, we have analyzed the equations with the thermal and quantum noise sources taken into account. We have analyzed the equations by the use of a contour integral method and by the use of the Fourier series expansion of the normalization factor of the mode of the "universe." The former method described in Chapter 6 was effective only for the linear gain regime. The latter method described in Chapters 7 through 10 was correct, but it took laborious efforts to solve the resulting equations. Chapters 7 and 8 were devoted to semiclassical analyses and were intended as preparation for the quantum-mechanical analyses developed in Chapters 9 and 10. Chapter 9 analyzed the correlation of the field driven by the noise sources. Chapter 10 investigated phase diffusion in the steady state. Both chapters had to deal with the complexity brought in by the presence of the output coupling. In Chapter 11 we have presented a simplified, *ad hoc*, quantum-mechanical model for the laser with output coupling that could be analyzed with less effort. We have named this method the propagation method or propagation theory. Except for the standard results for laser operation, we have obtained the longitudinal excess noise factor in the expression for the laser linewidth by taking the output coupling into account. Also, we have encountered the power-independent part of the linewidth. The extensive calculations needed may have left the reader adrift from the physics involved. In this chapter we retrace the calculations in Chapters 7 through 11 and discuss the physical aspects of the results.

12.1
Models of the Quasimode Laser and Continuous Mode Laser

First of all we discuss the difference between the quantum-mechanical laser models used in the quasimode theory in Chapter 4 and in the multimode theory in Chapters 9 and 10.

Output Coupling in Optical Cavities and Lasers: A Quantum Theoretical Approach
Kikuo Ujihara

ISBN: 978-3-527-40763-7

Figure 12.1 depicts the thermodynamic models of (a) the quasimode laser and (b) the continuous mode laser. In the quasimode laser model, the atoms couple with the (possibly single) cavity mode. The atoms are coupled with the pumping and the damping reservoirs, while the cavity mode is coupled with the loss reservoir. In the continuous mode laser model, the atoms are coupled with the pumping and the damping reservoir as in the quasimode laser model. But the atoms are coupled with the continuous, "universal" field modes. Some of the continuous modes make up the relevant cavity mode, which has no explicit loss reservoir. These continuous modes act as the resonant mode of the cavity as well as the loss reservoir. The exact treatment of the output coupling is secured by this model, provided the rigorous forms of the universal mode functions are used. This treatment has led to the appearance of correction factors in the expression for the laser linewidth compared with the conventional formula obtained by use of the quasimode theory. These are the excess noise factor for the laser linewidth and another factor for the "power-independent part" of the linewidth.

12.2 Noise Sources

12.2.1 Thermal Noise and Vacuum Fluctuation as Input Noise

One cause of laser linewidth is the ambient field fluctuation. Mathematically, this was introduced as the fluctuating field $\hat{F}_t(z,t)$ coming from the initial values, $\hat{a}_j(0)$, of the field as in Equation 5.33b. In the quantum linear gain analysis in Chapter 9 it yielded a thermal contribution proportional to $\langle n_\omega \rangle$ in the linewidth formula (Equation 9.105). This originated in Equation 9.72 or in the first line of Equation 5.36. In Chapter 9 the line profile was derived as the Fourier transform of the field correlation function in the time region, and the correlation function was defined as the ensemble average of the normally ordered field operators. The appearance

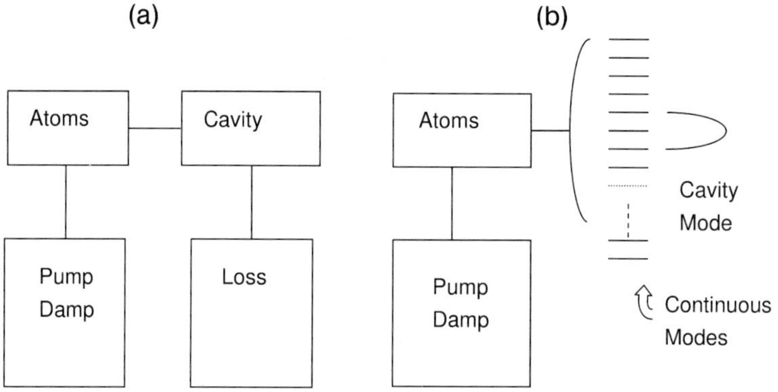

Figure 12.1 The models of (a) the quasimode laser and (b) the continuous mode laser.

of $\langle n_\omega \rangle$ was determined by this ordering. If the ambient temperature was zero, then $\langle n_\omega \rangle = 0$ and no thermal contribution appears.

The causes of the linewidth for the nonlinear, above-threshold operation obtained in Chapter 10 have somewhat different origins. The form of the linewidth formula in Equation 10.111 suggests that the ambient field resulted in the factor $\langle n_\omega \rangle + \frac{1}{2}$ in contrast to the case in Chapter 9. In Chapter 10 the linewidth was obtained through calculations of the phase diffusion, where the phase was a real quantity. For this requirement, the phase was evaluated from a sum of an operator quantity proportional to the field and its Hermitian conjugate. The field correlation function was thus in a symmetrically ordered form. The term $\frac{1}{2}$ above came from the 1 in the second line in Equation 5.36. This factor would not vanish even though the ambient temperature is zero and $\langle n_\omega \rangle = 0$. This contribution from the ambient field should be interpreted as coming from the vacuum field fluctuation. It is to be noted that in Chapter 10 the ambient field including the vacuum fluctuation was shown in Equation 10.71 to come from outside the cavity.

In contrast, the thermal or vacuum part of the noise in the quasimode laser considered in Chapter 4 came from the "artificial" Langevin force $\hat{\Gamma}_f(t)$ introduced in Equations 3.35, 3.36–3.37 for the field decay in the cavity. The thermal noise in the propagation method in Chapter 11 was described as coming in to the cavity from outside. The correlation functions in Equations 11.1b and 11.1c of this noise were deduced by referring to Equations 10.69 to 10.71, which resulted in a reasonable physical interpretation in terms of the thermal noise incident onto the cavity. The resultant thermal noise parts of the linewidth were generalizations of the results in Chapters 9 and 10. Note that Equations 10.69 to 10.71 are based on the quantum-mechanically correct continuous mode theory.

12.2.2 Quantum Noise

The quantum noise, on the other hand, originated in the Langevin force $\hat{\Gamma}_m(t)$ introduced in conjunction with the damping term in the equation of motion (Equation 3.45) for the atomic dipole. It appeared in Equation 9.105 in a form proportional to N_2, which came from the normally ordered correlation function of the quantum Langevin noise in Equation 9.73. This implies that the noise is proportional to the upper-level population, which is directly responsible for spontaneous emission. Thus in this case the quantum noise is interpreted as coming from spontaneous emission by upper-level atoms.

The quantum noise part in Equation 10.111, however, is proportional to $N = N_1 + N_2$. Thus the quantum noise coming from the atoms is not merely from the inverted population but also from the non-inverted population. The interpretation may be that not only spontaneous emission events proportional to N_2 in number but also absorption events proportional to N_1 disturb the field phase. The sum of $\frac{1}{2}$ and $(N_1 + N_2)/\{2(N_2 - N_1)\}$ is $N_2/(N_2 - N_1)$, and thus Equation 10.111 yields the same form of the linewidth as Equation 9.105 except for the over all factor of $\frac{1}{2}$ and for the added term for the quantum noise containing the saturation power P_s.

For the case of the quasimode laser analyzed in Chapter 4, the situation is almost the same except for the absence of the excess noise factor. The quantum noise source for the propagation method described in Chapter 11 was a modified version of those used in Chapters 9 and 10 or in Chapter 4. The force $\hat{f}_m(t)$ in Equation 11.3 was defined on the assumption that we are interested in the field variation on a time scale that is larger than the reciprocal atomic linewidth. The pertinent correlation function had an interpretation in terms of the power emitted in a spontaneous emission event. Spontaneous emission is examined in Section 12.12 and in Chapter 13 below.

12.3 Operator Orderings

In optics, normally ordered correlation functions or intensities are preferentially used because the usual wide-band optical detector that uses absorption as the detection mechanism responds to the normally ordered quantities [1]. The forms of noise $\langle n_c\rangle$ versus $N_2/(N_2-N_1)$ in the expression (Equation 4.82) for the linewidth of a quasimode laser in the nonlinear gain analysis are the same as in Equation 4.62a obtained for the linear gain analysis. In the case of Equation 4.62a, these forms appeared directly from the normally ordered correlation functions in Equations 3.36 and 4.50. However, in the case of Equation 4.82, these factors originally appeared in the forms of $\langle n_c\rangle+\frac{1}{2}$ and $N/(2N\sigma_{th})$, respectively, as seen from Equation 4.81. These forms appeared from the symmetrically ordered correlation functions used for the evaluation of the real phase of the field. The symmetric ordering appeared in Equation 4.76 because of the Hermitian conjugate terms. We note that the anti-normally ordered contributions from the thermal noise and quantum noise are $\langle n_c\rangle+1$ and $N_1/(N_2-N_1)$, respectively. It is interesting to note that the sum of the normally ordered contributions $\langle n_c\rangle+N_2/(N_2-N_1)$ is equal to that of the anti-normally ordered contributions $\langle n_c\rangle+1+N_1/(N_2-N_1)$. It should be noted that different orderings of the noise operators lead to the same form of the noise contributions.

A similar situation obtains for the continuous mode analysis of the one-sided cavity laser. The form of the noise $\langle n_\omega\rangle+\frac{1}{2}+(N_2+N_1)/(2N\sigma_{th}^0)$ in Equation 10.111 obtained in the nonlinear gain analysis looks different from that in Equation 9.105, $\{\sigma^2/(\sigma_{th}\sigma_{th0})\}\langle n_\omega\rangle+N_2/(N\sigma_{th})$, obtained for the linear gain analysis. In the case of Equation 9.105, the above forms appeared directly from the normally ordered correlation functions in Equations 9.4a and 9.5a. In the case of Equation 10.111, the factors $\langle n_\omega\rangle+\frac{1}{2}$ and $N/(2N\sigma_{th})$ appeared because of the symmetrically ordered correlation functions used for the evaluation of the real phase of the field. In particular, the symmetric ordering appeared in Equations 10.83a, 10.83b and 10.88 because of the Hermitian conjugate terms. Except for the factor $\sigma^2/(\sigma_{th}\sigma_{th0})$, the above two forms are the same since $\frac{1}{2}+(N_2+N_1)/(2N\sigma_{th}^0)=N_2/(N\sigma_{th}^0)$. Also in this case different orderings of the noise operators lead to almost the same form of the noise contributions.

Similarly, for the generalized two-sided cavity lasers, similar arguments can be given for the forms in Equation 11.105, $n_\omega + (N_2/N\sigma^0_{th})$, for the linear gain analysis, and in Equation 11.71, $\langle n_\omega \rangle + \frac{1}{2} + (N/2N\sigma^0_{th})$, for the nonlinear gain analysis. Goldberg *et al.* [2] discussed extensively the relation between the operator ordering and the physical cause of the noise.

12.4 Longitudinal Excess Noise Factor

An important result that arises from using the continuous mode analysis is the appearance of the longitudinal excess noise factor in the expression for the laser linewidth. The longitudinal excess noise factor was found in Chapters 9 and 10 both for the linear gain analysis and for the saturated, nonlinear gain analysis for the one-sided cavity laser model. A somewhat different result was found in the contour integral method in Chapter 6. The result was generalized to the case of a general asymmetric cavity in Chapter 11. This factor was defined as the ratio of the linewidth obtained in the continuous mode analysis to that obtained in the standard quasimode theory. More precisely, the factor is the ratio of the linewidth in a theory taking into account the local output coupling at the mirrors to that in a quasimode theory where the coupling loss is not localized. In other words, the difference originates in the assumed field distribution in the cavity: non-uniform or uniform.

12.4.1 Longitudinal Excess Noise Factor Below Threshold

In the linear gain analysis applicable for a below-threshold operation, we obtained Equation 9.106 for the one-sided cavity laser:

$$K_L = \left\{ \frac{(1 - r^2)/2r}{\ln(1/r)} \right\}^2 \tag{12.1a}$$

For a two-sided cavity laser we obtained Equation 11.107:

$$K_L = \left\{ \frac{(|r_1| + |r_2|)(1 - |r_1||r_2|)/2|r_1||r_2|}{\ln(1/|r_1||r_2|)} \right\}^2 \tag{12.1b}$$

This generalizes Equation 12.1a, and reduces to the form in Equation 12.1a for the case of a one-sided cavity with $|r_1| = 1$ and $|r_2| = r$ or for the case of a symmetric cavity with $|r_1| = |r_2| = r$.

First of all we note that the factor K_L is unity in the good cavity limit $|r_1||r_2| \to 1$. In this limit, the amplification coefficients for the last one round trip or the shorter trips that appeared in Equations 11.36 and 11.41 for the two-sided cavity laser (or Equation 10.69 for the one-sided cavity laser) are all unity, and the noise is not amplified during the last one round trip or in the respective shorter trips. This shows that finite amplification of the noise and, consequently, the non-uniformity of the field distribution is the physical origin of the excess noise factor.

The longitudinal excess noise factor is large for an asymmetric cavity: one can show that, for a given value ρ of the product $|r_1||r_2|$, that is, for a given value of the cavity decay rate, K_L is maximum when $|r_1| = 1$ and $|r_2| = \rho$ or when $|r_1| = \rho$ and $|r_2| = 1$. On the other hand, K_L is at its minimum when $|r_1| = |r_2| = \sqrt{\rho}$. This asymmetry effect is pronounced when $\rho \ll 1$. When the mirror asymmetry is pronounced, the field distribution inside the cavity is strongly non-uniform and the factor K_L is large. Hamel and Woerdman [3] verified this asymmetry effect experimentally by measuring the laser linewidth of semiconductor lasers with various combinations of facet mirror reflectivity.

Also, in Equation 9.114, we have indicated that the form of Equation 12.1a may come from the ratio

$$\frac{\left(\int_{-d}^{0} dz\, |\sin\{\Omega_c(z+d)/c_1\}|^2\right)^2}{\left(\int_{-d}^{0} dz\, |\sin\{\omega_c(z+d)/c_1\}|^2\right)^2} = \left(\frac{\beta_c}{\gamma_c}\right)^2 = K_L \tag{12.2}$$

where Ω_c is the complex cavity frequency in Equation 1.18b for a one-sided cavity and ω_c is the cavity frequency of a quasimode cavity, which is equal to one of the ω_k values in Equation 3.2. Improvement on this derivation of K_L will be discussed in Chapter 14, which will discuss the physical origin of the longitudinal excess noise factor.

12.4.2
Longitudinal Excess Noise Factor Above Threshold

We have from Equation 10.112 for the one-sided cavity laser

$$K_L = \left(\frac{\beta_c}{\gamma_c}\right)^2 = \left\{\frac{(1-r^2)/2r}{\ln(1/r)}\right\}^2 \tag{12.3}$$

For a general asymmetric cavity laser we may use the result of Chapter 11 expressed by Equation 11.72:

$$K_L = \frac{\{(|r_1|+|r_2|)(1-|r_1||r_2|)/2|r_1||r_2|\}^2}{\{\ln(1/|r_1||r_2|)\}^2} \tag{12.4}$$

Except for those factors that are common to linewidths for below-threshold operation, we have found additional terms for the linewidth for above-threshold operation. These are associated with the quantum noise contributions. We cite Equation 10.111 for the one-sided laser:

$$\Delta\omega = \frac{2\hbar\omega\gamma_c^2\gamma^2(1+\delta^2)}{(\gamma+\gamma_c)^2+\delta^2(\gamma-\gamma_c)^2}\left(\frac{\beta_c^2}{\gamma_c^2}\right) \times \left[\frac{1}{P}\left\{\left\langle n_\omega+\frac{1}{2}\right\rangle+\frac{N_2+N_1}{2N\sigma_{th}^0}\right\}+\frac{N_2+N_1}{2N\sigma_{th}^0}\frac{g(r)}{P+P_s\ln(1/r)}\right] \quad (12.5)$$

If the factor $g(r)/\{P+P_s\ln(1/r)\}$ is not small compared with $1/P$, the correction is appreciable. Similarly, for the generalized cavity model in Chapter 11, we had Equation 11.71:

$$\Delta\omega = \frac{2\hbar\omega K_L\gamma_c^2\gamma^2(1+\delta^2)}{(\gamma+\gamma_c)^2+(\gamma-\gamma_c)^2\delta^2}\left[\frac{1}{P_t}\left\{\langle n_\omega\rangle+\frac{1}{2}+\frac{N}{2N\sigma_{th}}\right\}+\frac{N}{2N\sigma_{th}}C_q\right] \quad (12.6)$$

where

$$C_q = \frac{1}{P_t+P_s\ln(1/|r_1||r_2|)}g(|r_1|,|r_2|) \quad (12.7)$$

If the factor C_q is not small compared with $1/P_t$, the correction is appreciable.

By retracing the calculations leading to the longitudinal excess noise factor and the additional correction associated with the quantum noise, we see that these arise from the amplification of the noise along the amplifying medium and local dumping at the coupling surfaces. For a quasimode laser the noise is amplified and dumped with average, mean-field rates and no local effect is involved. The corrections stem from the local aspects of the noise amplification. Existing theories for the physical interpretation of the longitudinal excess noise factor will be surveyed in Chapter 14.

12.5 Mathematical Relation between Below-Threshold and Above-Threshold Linewidths

As we have shown several times, the laser linewidth above threshold is just half that below threshold, except for the additional term for above threshold coming from gain saturation. This is interpreted as the result of suppression of the amplitude noise due to the gain saturation. That is, the gain reacts so as to cancel the amplitude variation of the field. Here we discuss briefly the mathematical origin of the factor $\frac{1}{2}$ in reducing the linewidth on going from below to above threshold.

The linewidth below threshold was calculated by noting the temporal decay of the field amplitude $\hat{e}(t)$, which is compensated for by the noise $\hat{F}(t)$ to maintain the average number of photons in the oscillating mode. The decay rate is given by the difference between the cavity loss rate and the gain $\gamma_c - g_a \equiv -s_0$. Thus we have, approximately,

$$\frac{d}{dt}\hat{e}(t) = s_0\hat{e}(t) + \hat{F}(t) \tag{12.8}$$

Thus we have, except for the term ê(0) that decays in time,

$$\hat{e}(t) = \int_0^t e^{s_0(t-t')}\hat{F}(t')dt' \tag{12.9}$$

and the correlation function is

$$\langle \hat{e}^\dagger(t+\tau)\hat{e}(t)\rangle = \left\langle \int_0^{t+\tau} e^{s_0^*(t+\tau-t')}\hat{F}^\dagger(t')dt' \int_0^t e^{s_0(t-t'')}\hat{F}(t'')dt''\right\rangle$$
$$\simeq D_{F^\dagger F} \times \begin{cases} \dfrac{e^{s_0^*\tau}}{2|\mathrm{Re}\, s_0|}, & \tau > 0 \\ \dfrac{e^{-s_0\tau}}{2|\mathrm{Re}\, s_0|}, & \tau < 0 \end{cases} \tag{12.10}$$

where we have set

$$\left\langle \hat{F}^\dagger(t')\hat{F}(t)\right\rangle = D_{F^\dagger F}\delta(t'-t) \tag{12.11}$$

By the Fourier transform of the correlation function, we know that the laser linewidth (full width at half-maximum) $\Delta\omega$ is

$$\Delta\omega = 2|\mathrm{Re}\, s_0| \tag{12.12}$$

Assuming a conversion coefficient S, the power output P is

$$P = S\langle \hat{e}^\dagger(t)\hat{e}(t)\rangle = \frac{SD_{F^\dagger F}}{2|\mathrm{Re}\, s_0|} \tag{12.13}$$

Thus we have

$$\Delta\omega = 2|\mathrm{Re}\, s_0| = \frac{SD_{F^\dagger F}}{P} \tag{12.14}$$

For an above-threshold operation we assume a stable amplitude e_0 and a diffusing phase $\phi(t)$. In this case we assume that the net gain $\gamma_c - g_a \equiv -s_0$ is exactly 0. Then Equation 12.8 becomes

$$\frac{d}{dt}\phi(t) = -i\frac{\hat{F}(t)}{e_0} \tag{12.15}$$

where we have assumed that the phase factor $e^{-i\phi(t)}$ is slowly varying and can be absorbed into the noise $\hat{F}(t)$. The crucial point here is that the phase should be a real quantity, and we warrant this claim by writing

$$\frac{d}{dt}\phi(t) = -i\frac{\hat{F}(t)}{2e_0} + i\frac{\hat{F}^\dagger(t)}{2e_0^*} \tag{12.16}$$

We evaluate the expected value of the squared phase change $\left\langle \{\Delta\phi(t)\}^2 \right\rangle$ during time Δt. Then it can be shown, by using the Fourier transform argument, that the linewidth is (see Section 10.6)

$$\Delta\omega = \frac{\left\langle \{\Delta\phi(t)\}^2 \right\rangle}{\Delta t} \tag{12.17}$$

Now $\left\langle \{\Delta\phi(t)\}^2 \right\rangle$ is evaluated as

$$\begin{aligned} &\left\langle \{\Delta\phi(t)\}^2 \right\rangle \\ &= \left\langle \int_t^{t+\Delta t} \int_t^{t+\Delta t} \left\{ -i\frac{\hat{F}(t')}{2e_0} + i\frac{\hat{F}^\dagger(t')}{2e_0^*} \right\} \left\{ -i\frac{\hat{F}(t'')}{2e_0} + i\frac{\hat{F}^\dagger(t'')}{2e_0^*} \right\} dt' dt'' \right\rangle \end{aligned} \tag{12.18}$$

Assuming that

$$\begin{aligned} &\left\langle \hat{F}(t')\hat{F}^\dagger(t) \right\rangle = D_{FF^\dagger}\delta(t'-t) \qquad \text{and} \\ &\left\langle \hat{F}(t')\hat{F}(t) \right\rangle = \left\langle \hat{F}^\dagger(t')\hat{F}^\dagger(t) \right\rangle = 0 \end{aligned} \tag{12.19}$$

we have

$$\left\langle \{\Delta\phi(t)\}^2 \right\rangle = \frac{D_{F^\dagger F} + D_{FF^\dagger}}{4|e_0|^2}\Delta t \tag{12.20}$$

Therefore, we have

$$\Delta\omega = \frac{\left\langle \{\Delta\phi(t)\}^2 \right\rangle}{\Delta t} = \frac{S(D_{F^\dagger F} + D_{FF^\dagger})}{4P} \tag{12.21}$$

Now, we have, formally,

$$(D_{F^\dagger F} + D_{FF^\dagger})_{\text{above threshold}} = 2(D_{F^\dagger F})_{\text{below threshold}} \tag{12.22}$$

as long as $\hat{F}(t)$ stands for the sum of the thermal noise and the quantum noise (see Section 12.3 above). Thus Equation 12.21 for the linewidth above threshold gives, formally, just half the linewidth for below threshold given by Equation 12.14. Note that Equation 12.22 describes a formal equivalence and that it does not mean $D_{F^\dagger F} = D_{FF^\dagger}$.

12.6 Detuning Effects

First of all we stress that, in the starting equations in Chapter 3 for the quasimode laser or in Chapter 5 for a one-sided cavity laser, we have assumed a temporarily constant atomic inversion and otherwise made no limiting assumptions on the relative magnitude of the decay rates of the atomic polarization and cavity field nor on the atomic or cavity resonance frequencies. The well-known consequence of the presence of the detuning between the cavity resonance frequency ω_c and the atomic transition frequency ν_0 is the linear pulling effect that appears on the oscillation frequency of a laser. This has appeared in all the laser models considered up to now. In Equations 4.12 and 4.46 we had

$$\omega_{th} = \frac{\gamma\omega_c + \gamma_c\nu_0}{\gamma + \gamma_c} \tag{12.23a}$$

where the subscript *th* stands for the threshold condition. This applies for the extreme case of the threshold condition in the linear gain analysis. We have obtained the same form in Chapters 6, 7, and 9. Also, Equation 11.32 will yield the same result if we set $s_0 = 0$ in this equation. In Equations 4.33 and 4.66 we had the angular oscillation frequency

$$\omega = \frac{\gamma\omega_c + \gamma_c\nu_0}{\gamma + \gamma_c} \tag{12.23b}$$

for the nonlinear, saturated gain regime. We have obtained the same result in Chapters 8, 10, and 11.

The laser model used in Chapter 4 is the quasimode cavity model, that in Chapters 7–10 is the one-sided cavity laser model, and the one in Chapter 11 is the generalized two-sided cavity laser model. For these different models, the expressions for the cavity decay constants and for the cavity resonant frequencies are different. In spite of these differences, the expressions for the oscillation or threshold frequencies are the same. This reflects a general rule for a pair of oscillators oscillating at a single frequency as a whole. In this case the frequency is pulled towards the oscillation frequency of the oscillator with the higher quality factor. Thus in Equations 12.23a and 12.23b, if the cavity has a higher Q value or sharper width $2\gamma_c$ than the atom of natural width 2γ, then ω tends to go to ω_c. This is the usual situation in a single-frequency laser. If, on the other hand, the atom has a sharper width, $2\gamma \ll 2\gamma_c$, the oscillation occurs close to ν_0.

A qualitative argument of the linear pulling based on the dispersion of the atomic medium can be given as follows. The amplifying medium of inverted atoms has a dispersion that gives a positive increase in the refractive index for $\omega > \nu_0$ and vice versa. Therefore, if the cavity resonance becomes higher in frequency than the atom, the cavity field sees a longer cavity length because of the positive refractive index increase. Then, the cavity resonant wavelength tends to increase, that is, the cavity resonant frequency decreases compared to the bare cavity case. If, on the other hand, the cavity frequency is smaller than the atomic frequency, the dispersion gives a decrease in the refractive index. Thus the cavity becomes effectively shorter, thus increasing the effective cavity frequency. This is the linear pulling.

Another effect of detuning between the cavity resonance frequency ω_c and the atomic transition frequency ν_0 appears on the laser linewidth as the factor in Equation 12.6:

$$F = \frac{\gamma_c^2 \gamma^2 (1 + \delta^2)}{(\gamma + \gamma_c)^2 + (\gamma - \gamma_c)^2 \delta^2}$$
$$\delta^2 = \frac{(\nu_0 - \omega_c)^2}{(\gamma + \gamma_c)^2} \tag{12.24}$$

When $\gamma \gg \gamma_c$, we have $F \approx \gamma_c^2$, with no detuning effect. Similarly, for $\gamma \gg \gamma_c$, we have $F \approx \gamma^2$, with no detuning effect. But when $\gamma \sim \gamma_c$, then $F \propto 1 + \delta^2$, which means that any detuning results in a broadening of the laser line. This may be understood as the result of the broad total response of the atom–cavity system, which appears when two equally broadened oscillators with different center frequencies cooperate.

12.7
Bad Cavity Effect

When the detuning is small, the factor F in Equation 12.24 reads

$$F = \frac{\gamma_c^2 \gamma^2}{(\gamma + \gamma_c)^2} \tag{12.25a}$$

A cavity is said to be a *good cavity* when the cavity bandwidth $2\gamma_c$ is much smaller than the atomic width 2γ. In this case the factor

$$F = \gamma_c^2 \tag{12.25b}$$

This is the standard form for F obtained for example by Schawlow and Townes [4], Haken [5], and Sargent *et al.* [6], assuming a good cavity.

A cavity is called a *bad cavity* when $2\gamma_c$ is not smaller than 2γ. The factor F in Equation 12.24 can be rewritten in the form

$$F = \frac{\gamma_c^2}{1 + (\gamma_c/\gamma)^2} \tag{12.25c}$$

Thus for a bad cavity, an appreciable reduction of the linewidth as compared to the standard form arises. Van Exter *et al.* [7] gave a physical interpretation of the reduction in terms of the effective elongation of the cavity due to the dispersion of the gain medium. They argue that the important light velocity is the group velocity rather than the phase velocity when the light burst from the noise sources travel within the cavity. They showed that the group refractive index in the presence of the Lorentzian gain is approximately $1 + (\mathcal{g} N \sigma_{ss}/\gamma)^2$ because of the atomic dispersion. This is equal to the factor $1 + (\gamma_c/\gamma)^2$ that we found in Equation 12.25c. Thus the cavity length is effectively elongated, which reduces the cavity decay rate.

Note, however, that the factor F and thus the laser linewidth is an increasing function of the cavity decay rate γ_c. Prasad [8] showed that, in the extreme case of very large γ_c, the linewidth below threshold tends to be that of the natural linewidth of the atoms. The linewidth formula for below threshold, Equation 9.100, for $\gamma_c \gg \gamma$ and for far below threshold, $\sigma \ll \sigma_{th}$, also yields this result:

$$\Delta\omega = \frac{2(\gamma+\gamma_c)\gamma\gamma_c(1+\delta^2)[1-\sigma/\sigma_{th}]}{(\gamma+\gamma_c)^2+\delta^2(\gamma-\gamma_c)^2} \to 2\gamma \tag{12.26}$$

Also, Equation 11.32 with $\Delta\omega = 2|\mathrm{Re}\, s_0|$ (Equation 11.86) yields this result. Thus a bad cavity laser has in general a broadened linewidth. Note that this result is obtained as a result of our neutral treatment of the cavity and the atomic bandwidths. We have treated the constants γ and γ_c as symmetrically as possible. Sometimes in the literature it is assumed from the outset that the cavity bandwidth is much smaller than the atomic width. In such a treatment, the bad cavity effect cannot be derived.

12.8
Incomplete Inversion and Level Schemes

In the expressions for the laser linewidth, Equations 9.105, 10.111, 11.105, and 11.71, we had the quantum noise factor for the linear gain cases,

$$\frac{N_2}{N\sigma_{th}} = \frac{N_2}{(N_2-N_1)_{th}} \tag{12.27}$$

or for the saturated gain cases,

$$\frac{1}{2}+\frac{N}{2N\sigma^0_{th}} = \frac{N_{2th}}{(N_2-N_1)_{th}} \tag{12.28}$$

In both of these equations, N_2 or N_{2th} are steady-state values of the upper-level population. The appearance of N_1, the lower-level population, in the denominator makes these factors larger than those obtained for $N_1=0$. This makes the linewidth broader than those obtained for $N_1=0$. This effect is called the incomplete inversion effect and appears in lasers other than an ideal four-level atom laser. In an ideal four-level laser the lower-level population $N_1=0$ and the spontaneous emission noise is proportional to N_2. When the lower-level population exists, the absorption events by the lower-level atoms also disturb the field coherence and lead to a larger noise than is given by N_2.

In our laser model we have employed the two-level atoms, with the upper level 2 and the lower level 1, as the model atoms. The pumping process is described as (see Equation 3.46)

$$(d/dt)\hat{\sigma}_m(t) = -\Gamma_{mp}\{\hat{\sigma}_m(t)-\sigma^0_m\} \tag{12.29a}$$

where we have ignored the noise term associated with the relaxation of the atomic inversion. The unsaturated atomic inversion σ^0_m and the relaxation constant Γ_{mp} are given by Equation 3.51 as

$$\sigma_m^0 = \frac{w_{m12} - w_{m21}}{w_{m12} + w_{m21}}$$
$$\Gamma_{mp} = w_{m12} + w_{m21} \tag{12.29b}$$

where w_{m12} and w_{m21} are, respectively, the upward and downward incoherent transition rates including the pumping and the natural relaxation rates. The minimum downward transition rate W_{m21} is given by the spontaneous emission rate. The unsaturated atomic inversion may be taken arbitrarily close to unity if we could make the pumping rate Γ_{mp} much larger than W_{m21}, whence $\Gamma_{mp} \approx w_{m12}$ and $\sigma_m^0 \simeq 1$. But this is usually not easy to realize in practical systems.

In this connection, the parameter N in Equations 12.27 and 12.28 and in previous chapters should be taken as the sum of the upper- and lower-level population densities of the laser,

$$N = N_1 + N_2 \tag{12.30}$$

not the total atomic density, which may include populations of levels other than the lasing levels. For example, a three-level system, where the uppermost level acts only as the intermediate level for pumping and has a very large relaxation rate, can be accurately simulated by a two-level atom. The possible existence of other levels and their effects are somehow squeezed into the equations for the atomic inversion, Equations 3.46 and 5.27. These lead to the steady-state, saturated atomic inversion in Equations 4.29, 4.37, 8.16, and 10.21a. In Chapter 11 we have assumed a similar form of saturated inversion in Equation 11.8.

A general four-level model was analyzed by Van Exter *et al.* [7] in relation to the longitudinal excess noise factor.

12.9 The Constants of Output Coupling

The constant $2\gamma_c$ is usually taken as the ratio of the output power and the energy stored in the cavity. We will show that this is not the case in the nonlinear gain regime of a laser with finite end mirror coupling.

For a quasimode laser the damping of the stored energy is governed by the field decay described by Equation 4.1:

$$\frac{d}{dt}\hat{a}(t) = -i\omega_c\hat{a}(t) - \gamma_c\hat{a}(t) - i\sum_m \kappa_m(\hat{b}_{m1}^\dagger\hat{b}_{m2})(t) \tag{12.31}$$

Ignoring the last term, which describes the energy flow to or from the atoms, we see that the energy flow to the cavity is governed by

$$\frac{d}{dt}\hat{a}(t) = -i\omega_c\hat{a}(t) - \gamma_c\hat{a}(t) \tag{12.32}$$

Taking the Hermitian conjugate we have

$$\frac{d}{dt}\hat{a}^{\dagger}(t) = i\omega_c\hat{a}^{\dagger}(t) - \gamma_c\hat{a}^{\dagger}(t) \tag{12.33}$$

Multiplying Equation 12.32 by $\hat{a}^{\dagger}(t)$ from the left and multiplying Equation 12.33 by $\hat{a}(t)$ from the right and adding, we obtain

$$\frac{d}{dt}\{\hat{a}^{\dagger}(t)\hat{a}(t)\} = -2\gamma_c\{\hat{a}^{\dagger}(t)\hat{a}(t)\} \tag{12.34}$$

As was described in Equation 4.60 the stored energy is proportional to (the ensemble average of) $\{\hat{a}^{\dagger}(t)\hat{a}(t)\}$. Thus $2\gamma_c$ is the correct energy damping rate of the quasimode cavity. This was used in both the linear and the nonlinear gain regimes as in Equations 4.61 and 4.80.

For the one-sided cavity model, let us first consider the case of the linear gain regime considered in Chapter 9. In this case the starting equation for the field amplitude is Equation 9.1, which has no explicit cavity decay constant. One of the original equations leading to Equation 9.1 is Equation 5.25 for the jth mode of the universe, which also has no explicit decay constant. Therefore, it is not obvious from these equations whether the stored energy decays with a certain decay constant. The results in Equation 9.108 show, however, that the constant $2\gamma_c = 2(c_1/2d)\ln(1/r)$ is the correct damping rate. Note that the expression $\gamma_c = (c_1/2d)\ln(1/r)$ was derived in Equation 1.18 as the natural decay constant for the one-sided cavity model.

Next we consider the nonlinear gain regime discussed in Chapter 10. We have the expression for the power output in Equation 10.110:

$$P = 2\varepsilon_1 c_1 |E_s|^2 \ln\left(\frac{1}{r}\right)\left\{\frac{\sigma^0}{\sigma^0_{th}} - 1\right\} \tag{12.35}$$

On the other hand, we have the integrated intensity in the cavity as in Equation 10.101:

$$\int_{-d}^{0} dz_m \left(\frac{|e^-(z_m)|^2 + |e^+(z_m)|^2}{|e^+(-d)|^2}\right) = 2d\frac{\beta_c}{\gamma_c}\left\{1 + \frac{\Delta}{1+\Delta}g(r)\right\} \tag{12.36}$$

from which we derive the stored energy W as

$$\begin{aligned} W &= 2\varepsilon_1 \int_{-d}^{0} |e^-(z_m)|^2 + |e^+(z_m)|^2 dz_m \\ &= 4\varepsilon_1 d\frac{\beta_c}{\gamma_c}\left\{1 + \frac{\Delta}{1+\Delta}g(r)\right\}|e^+(-d)|^2 \end{aligned} \tag{12.37}$$

where the factor $|e^+(-d)|^2$ can be found in Equation E.12 in Appendix E as

$$\begin{aligned} |e^+(-d)|^2 &= |E_s|^2\left(\frac{1}{r} - r\right)^{-1}\left\{2\mathrm{Re}\,\alpha^0 d - \ln\left(\frac{1}{r}\right)\right\} \\ &= |E_s|^2\left(\frac{1}{r} - r\right)^{-1}\ln\left(\frac{1}{r}\right)\left(\frac{\sigma^0}{\sigma^0_{th}} - 1\right) \end{aligned} \tag{12.38}$$

where we have used Equations 10.94c and 10.48 to go to the second line. Therefore we have

$$\begin{aligned}\frac{P}{W} &= \frac{c_1\gamma_c(1/r - r)}{2d\beta_c\{1 + [\Delta/(1+\Delta)]g(r)\}} = \frac{2\gamma_c}{1 + [\Delta/(1+\Delta)]g(r)} \\ &= \frac{2\gamma_c}{1 + g(r)/[1 + (P_s/P)\ln(1/r)]}\end{aligned} \tag{12.39}$$

where Equation 10.110 has been used in the last equality. This shows that $2\gamma_c$ is not the correct damping factor in this case of the nonlinear gain regime. The output coupling is more or less reduced depending on the relative excess atomic inversion $\Delta = (\sigma^0 - \sigma^0_{th})/\sigma^0_{th}$ or on the relative magnitude of the saturation power and the output power. The reduction is pronounced when the reflection coefficient r is small so that the function $g(r)$ is large and the output power P is appreciably larger than the saturation power P_s. This is a consequence of the gain saturation, which brings the laser out of the linear operation condition and deforms the field distribution from that of a natural resonant field distribution of the cavity.

A similar result for the case of a general two-sided cavity laser is anticipated. The reader may show that the ratio of the total power output to the stored energy is

$$\begin{aligned}\frac{P_t}{W} &= \frac{2\gamma_c}{1 + [\Delta/(1+\Delta)]g(|r_1|, |r_2|)} \\ &= \frac{2\gamma_c}{1 + g(|r_1|, |r_2|)/[1 + (P_s/P_t)\ln(1/|r_1||r_2|)]}\end{aligned} \tag{12.40}$$

where

$$\gamma_c = \frac{c_1}{2d}\ln\frac{1}{|r_1||r_2|} \tag{12.41}$$

The factor Δ vanishes for the linear gain regime and the ratio reduces to the usual $2\gamma_c$. But for a strongly nonlinear gain regime, where $P_t \gg P_s$, the ratio is reduced if $g(|r_1|, |r_2|)$ is large. Van Exter *et al.* [7] derived a different expression for the damping constant for the nonlinear gain regime that is also smaller than $2\gamma_c$.

12.10 Threshold Atomic Inversion and Steady-State Atomic Inversion

The threshold atomic inversion is the minimum value of the atomic inversion to maintain laser oscillation. The threshold is reached when the gain due to the inversion equals the cavity loss. The threshold inversion is equal to the steady-state value in a steady-state oscillation above threshold.

In the case of the quasimode laser in Chapter 4, the threshold atomic inversion was given in Equation 4.13b as

$$\sigma_{th} = \frac{2\hbar\varepsilon_1\gamma\gamma_c}{|p_a|^2 \nu_0 N}(1+\delta^2) \tag{12.42a}$$

which can be rewritten as

$$g N \sigma_{th} = \gamma_c \tag{12.42b}$$

where

$$g = \frac{|p_a|^2 \nu_0}{2\varepsilon_1 \hbar\gamma(1+\delta^2)}, \qquad \delta^2 = \frac{(\nu_0 - \omega_c)^2}{(\gamma+\gamma_c)^2} \tag{12.43}$$

Here the uniform density of atoms N is assumed to be large enough that many atoms exist in a region the length of an optical wavelength. If the density was sparse, the expression may depend on the degree of the overlap of the distributed atoms and the standing mode field. The threshold inversion is smaller for larger atomic density, larger electric dipole matrix element, smaller cavity loss, smaller atomic width, and smaller detuning between the cavity and the atomic resonances. Also, it was shown that the steady-state atomic inversion is the same as the threshold inversion:

$$\sigma_{ss} = \sigma_{th} \tag{12.44}$$

where the steady-state inversion is the saturated value according to

$$\sigma_{ss} = \frac{\sigma^0}{1 + \left|\tilde{E}^{(+)}\right|^2 / |E_s|^2} \tag{12.45}$$

Here $\left|\tilde{E}^{(+)}\right|^2$ is the squared oscillation amplitude suitably averaged over the region internal to the cavity and $|E_s|^2$ is the saturation parameter.

For the case of the one-sided cavity laser, the contour integral method for the linear gain regime in Chapter 6 gives the same result for the threshold inversion (see Equation 6.12)

$$\sigma_{th} = \frac{2\hbar\varepsilon_1\gamma\gamma_c}{|p_a|^2 \nu_0 N}(1+\delta^2) \tag{12.46}$$

Note that the cavity decay constant here has the explicit expression given by Equation 1.18a as compared to the abstract decay constant in Equation 3.35 for the quasimode cavity laser. The semiclassical and the quantum linear gain analyses based on the Laplace transform in Chapters 7 and 9 give the threshold inversion as (see Equation 7.44a)

$$\left\{ \frac{|p_a|^2 \nu_0^2}{2\hbar\omega\varepsilon_1\gamma(1+\delta^2)} \right\} N\sigma_{th} = \gamma_c \tag{12.47}$$

The content of this equation is essentially the same as Equation 12.42a. This equation can be recast in the form

$$g N \sigma_{th} = \gamma_c \tag{12.48}$$

which is the same as Equation 12.42b. Here g is the amplitude gain per unit density of inverted atoms per unit time. Equation 12.48 states that, at threshold, the amplitude gain per unit time is equal to the cavity loss rate.

The semiclassical and the quantum nonlinear, saturated gain analyses in Chapters 8 and 10 give the same threshold inversion as for the linear gain analyses (see Equation 8.48):

$$N\sigma_{th}^{0} = \frac{2\varepsilon_1 \hbar\omega\gamma\gamma_c}{\nu_0^2|p_a|^2}(1+\delta^2) \tag{12.49}$$

This can be rewritten in the form

$$g N\sigma_{th}^{0} = \gamma_c \tag{12.50}$$

Here the superscript 0 on the inversion σ denotes the unsaturated value. The steady-state inversion averaged over the length of the cavity is equal to the threshold value (see Equations 8.49 and 8.53)

$$\bar{\sigma}_{ss} = \sigma_{th} \tag{12.51}$$

where

$$\bar{\sigma}_{ss} \equiv \frac{1}{d}\int_{-d}^{0} \sigma_m dz_m \tag{12.52}$$

For the case of the generalized two-sided cavity laser analyzed in Chapter 11 we had formally the same results as for the quasimode cavity laser and for the one-sided cavity laser (see Equation 11.26a):

$$\sigma_{th}^{0} = \bar{\sigma}_{ss} = \frac{2\hbar\omega\varepsilon_1}{\nu_0^2|p_a|^2 N}\gamma\gamma_c(1+\delta^2) \tag{12.53}$$

The expression for the cavity decay rate γ_c is now given by Equation 11.25. This equation can also be recast in the form

$$g N\sigma_{th}^{0} = \gamma_c \tag{12.54}$$

12.11 The Power-Independent Part of the Linewidth

In the previous chapters we have expressed the laser linewidth in terms of the threshold atomic inversion σ_{th}^{0} in conjunction with the reciprocal output power. In our two-level atom model the former is a constant independent of the pump level and thus of the power output (see Equation 12.54 above). In the expressions for the linewidth for above threshold, we had the "power-independent part." In the literature, some authors [7, 8] prefer to express the linewidth in terms of the unsaturated atomic inversion σ_0. Note that this factor depends on the pump level and thus is related to the power output. The merit of these latter forms of expression is that they are a little more compact than our previous expressions. Let us see how they look.

For the saturated gain regime of the quasimode cavity laser, Equation 4.82 may read

$$\Delta\omega = \frac{2\hbar\omega_c\gamma_c^2}{P}\frac{\gamma^2(1+\delta^2)}{(\gamma+\gamma_c)^2+(\gamma-\gamma_c)^2\delta^2}\left(\langle n_c\rangle + \frac{N_2}{N\sigma_{th}^{0}}\right) \tag{12.55}$$

Here N_2 is the steady-state value, that is, the saturated value of N_2. We note that

$$\frac{1}{\sigma^0_{th}} = \frac{1}{\sigma^0} + \frac{1}{\sigma^0}\Delta \tag{12.56}$$

where $\Delta = (\sigma^0 - \sigma^0_{th})/\sigma^0_{th}$ is the fractional excess atomic inversion, and that

$$P = \frac{2d}{c_1}\gamma_c P_s \Delta \tag{12.57}$$

which can be derived from Equation 4.67 noting that $P = 4\gamma_c \varepsilon_1 d|\tilde{E}^{(+)}|^2$ and $P_s = 2\varepsilon_1 c_1 |E_s|^2$. We also note that the N_2 term in Equation 12.55 came from $\frac{1}{2} + N/(2N\sigma^0_{th})$ (see Equation 4.81). Thus we have

$$\Delta\omega = \frac{2\hbar\omega_c \gamma_c^2 \gamma^2 (1+\delta^2)}{(\gamma+\gamma_c)^2 + (\gamma-\gamma_c)^2\delta^2} \times \left\{ \frac{1}{P}\left(\langle n_c \rangle + \frac{N_2^0}{N\sigma^0} \right) + \frac{1}{P_s(2d/c_1)\gamma_c} \frac{N}{2N\sigma^0} \right\} \tag{12.58}$$

where N_2^0 is the unsaturated value of N_2. In this form we have a constant, power-independent part of the laser linewidth in the second term in the curly bracket. We note that the linewidth below threshold in Equation 4.62a cannot be rewritten in a similar form to have a power-independent part because Equation 12.57 is meaningless below threshold.

For the case of the one-sided cavity laser above threshold, we have (see Equation 10.111)

$$\Delta\omega = \frac{2\hbar\omega \beta_c^2 \gamma^2 (1+\delta^2)}{(\gamma+\gamma_c)^2 + \delta^2(\gamma-\gamma_c)^2} \times \left[\frac{1}{P}\left\{ \left\langle n_\omega + \frac{1}{2} \right\rangle + \frac{N_2+N_1}{2N\sigma^0_{th}} \right\} + \frac{N_2+N_1}{2N\sigma^0_{th}} \frac{g(r)}{P + P_s \ln(1/r)} \right] \tag{12.59}$$

This can be rewritten, using Equation 10.110 instead of Equation 12.57, as

$$\begin{aligned} \Delta\omega &= \frac{2\hbar\omega \beta_c^2 \gamma^2 (1+\delta^2)}{(\gamma+\gamma_c)^2 + \delta^2(\gamma-\gamma_c)^2} \times \left[\frac{1}{P}\left\{ \left\langle n_\omega + \frac{1}{2} \right\rangle + \frac{N_2+N_1}{2N\sigma^0} \right\} + \frac{N_2+N_1}{2N\sigma^0} \frac{\{g(r)+1\}}{P_s \ln(1/r)} \right] \\ &= \frac{2\hbar\omega \gamma^2 (1+\delta^2)}{(\gamma+\gamma_c)^2 + \delta^2(\gamma-\gamma_c)^2} \times \left[\frac{\beta_c^2}{P}\left\{ \left\langle n_\omega + \frac{1}{2} \right\rangle + \frac{N_2+N_1}{2N\sigma^0} \right\} + \frac{N_2+N_1}{2N\sigma^0} \frac{f(r)}{P_s} \right] \end{aligned} \tag{12.60}$$

where, by Equation 10.102,

$$f(r) = \left(\frac{c_1}{2d}\right)^2 \left\{ \frac{1}{2}\ln\left(\frac{1}{r}\right) + \frac{1-r^4}{8r^2} \right\} \tag{12.61}$$

Here we have used the relation $\beta_c = \{c_1/(2d)\}(1 - r^2)/(2r)$ (Equation 6.35).

For the case of the two-sided cavity laser above threshold, we have from Equation 11.71

$$\Delta\omega = \frac{2\hbar\omega K_L \gamma_c^2 \gamma^2 (1+\delta^2)}{(\gamma+\gamma_c)^2 + (\gamma-\gamma_c)^2\delta^2} \times \left[\frac{1}{P_t}\left\{\langle n_\omega\rangle + \frac{1}{2} + \frac{N}{2N\sigma_{th}^0}\right\} + \frac{N}{2N\sigma_{th}^0}\frac{1}{P_t + P_s \ln(1/|r_1||r_2|)} g(|r_1|,|r_2|)\right] \tag{12.62}$$

We can similarly rewrite this equation as

$$\Delta\omega = \frac{2\hbar\omega\gamma^2(1+\delta^2)}{(\gamma+\gamma_c)^2 + (\gamma-\gamma_c)^2\delta^2} \times \left[\frac{K_L\gamma_c^2}{P_t}\left\{\langle n_\omega\rangle + \frac{1}{2} + \frac{N}{2N\sigma^0}\right\} + \frac{N}{2N\sigma^0}\frac{f(|r_1|,|r_2|)}{P_s}\right] \tag{12.63}$$

where

$$\begin{aligned} f(|r_1|,|r_2|) &= K_L\gamma_c^2 \frac{g(|r_1|,|r_2|)+1}{\ln(1/|r_1||r_2|)} \\ &= \left(\frac{c_1}{2d}\right)^2 \left\{\frac{1}{2}\ln\left(\frac{1}{|r_1||r_2|}\right) + \frac{(|r_1|^2+|r_2|^2)(1-|r_1|^2|r_2|^2)}{8|r_1|^2|r_2|^2}\right\} \end{aligned} \tag{12.64}$$

We again note that the unsaturated atomic inversion σ^0 is not independent of the power output, as seen from Equation 12.57. Van Exter *et al.* [7] obtained, by the Green's function method of Tromborg *et al.* [9], the same result as Equation 12.63 except that the factor before $f(|r_1|,|r_2|)/P_s$ is $N_2/(N\sigma^0)$, in our notation, instead of $N/(2N\sigma^0)$. This difference comes from their neglect of the vacuum fluctuation or their reliance on the normally ordered correlation functions. Their results also contain a factor concerning the degree of incomplete inversion coming from the laser level scheme. Prasad [8] found, through a method similar to that in Chapter 11, similar results for a one-sided cavity laser. His vacuum fluctuation term is multiplied by the refractive index of the laser medium. He found a factor proportional to the function $f(r)$ in Equation 12.61 for the power-independent part.

12.12
Linewidth and Spontaneous Emission Rate

In Equation 4.13d and in other equations relating to the threshold condition, there appeared the constant $\wp$, which was interpreted as half the coefficient of the stimulated emission rate. This rate should be related to the spontaneous emission rate. As the laser linewidth is sometimes interpreted as resulting from spontaneous emission, it will be instructive to derive the expression for the spontaneous emission rate and its relation to the laser linewidth. Also, this discussion will become necessary in Chapter 14 where theories of the excess noise factor are reviewed.

12.12.1
Spontaneous Emission in the Quasimode Laser

We first consider the quasimode cavity laser. For the below-threshold case, we start with the time derivative of Equation 4.43 for the field amplitude and its Hermitian adjoint (we ignore the initial value term, which will decay eventually)

$$\begin{aligned}\frac{d}{dt}\tilde{a}(t) &= s_0\tilde{a}(t) + \frac{\{i(\nu_0-\omega)+\gamma\}\tilde{\Gamma}_f(t) - i\sum_m \kappa_m\tilde{\Gamma}_m(t)}{i(\omega_c+\nu_0-2\omega)+\gamma_c+\gamma} \\ \frac{d}{dt}\tilde{a}^\dagger(t) &= s_0^*\tilde{a}^\dagger(t) + \frac{\{-i(\nu_0-\omega)+\gamma\}\tilde{\Gamma}_f^\dagger(t) + i\sum_m \kappa_m^*\tilde{\Gamma}_m^\dagger(t)}{-i(\omega_c+\nu_0-2\omega)+\gamma_c+\gamma}\end{aligned} \qquad (12.65)$$

Multiplying the first equation by $\tilde{a}^\dagger(t)$ from the left and the second equation by $\tilde{a}(t)$ from the right and adding, we obtain

$$\begin{aligned}\frac{d}{dt}\{\tilde{a}^\dagger(t)\tilde{a}(t)\} &= (s_0+s_0^*)\tilde{a}^\dagger(t)\tilde{a}(t) + \tilde{a}^\dagger(t) \\ &\quad \frac{\times\{i(\nu_0-\omega)+\gamma\}\tilde{\Gamma}_f(t) - i\sum_m \kappa_m\tilde{\Gamma}_m(t)}{i(\omega_c+\nu_0-2\omega)+\gamma_c+\gamma} \\ &\quad + \frac{\{-i(\nu_0-\omega)+\gamma\}\tilde{\Gamma}_f^\dagger(t) + i\sum_m \kappa_m^*\tilde{\Gamma}_m^\dagger(t)}{-i(\omega_c+\nu_0-2\omega)+\gamma_c+\gamma}\tilde{a}(t)\end{aligned} \qquad (12.66)$$

We use Equation 4.43 and its Hermitian adjoint on the right-hand side and take the reservoir average. Using the delta-correlated natures of the noise forces (see Equations 3.36, 3.37, and 9.5c), we have for the reservoir average of the photon number

$$\begin{aligned}\frac{d}{dt}\langle\tilde{a}^\dagger(t)\tilde{a}(t)\rangle &= (s_0+s_0^*)\langle\tilde{a}^\dagger(t)\tilde{a}(t)\rangle \\ &\quad + \left\langle \frac{\int_0^t e^{s^*(t-t')}\left\{-i(\nu_0-\omega)+\gamma\right\}\tilde{\Gamma}_f^\dagger(t') + i\sum_m \kappa_m^*\tilde{\Gamma}_m^\dagger(t')\right\}dt'}{-i(\omega_c+\nu_0-2\omega)+\gamma_c+\gamma} \right. \\ &\quad \left. \times \frac{\{i(\nu_0-\omega)+\gamma\}\tilde{\Gamma}_f(t) - i\sum_m \kappa_m\tilde{\Gamma}_m(t)}{i(\omega_c+\nu_0-2\omega)+\gamma_c+\gamma} \right\rangle \\ &\quad + \left\langle \frac{\left\{-i(\nu_0-\omega)+\gamma\right\}\tilde{\Gamma}_f^\dagger(t)\right\} + i\sum_m \kappa_m^*\tilde{\Gamma}_m^\dagger(t)}{-i(\omega_c+\nu_0-2\omega)+\gamma_c+\gamma} \right. \\ &\quad \left. \times \frac{\int_0^t e^{s(t-t')}\{i(\nu_0-\omega)+\gamma\}\tilde{\Gamma}_f(t') - i\sum_m \kappa_m\tilde{\Gamma}_m(t')\}dt'}{i(\omega_c+\nu_0-2\omega)+\gamma_c+\gamma} \right\rangle \\ &= (s_0+s_0^*)\langle\tilde{a}^\dagger(t)\tilde{a}(t)\rangle \\ &\quad + 2\frac{\{(\nu_0-\omega)^2+\gamma^2\}\frac{1}{2}2\gamma_c\langle n_c\rangle + \sum_m |\kappa_m|^2\frac{1}{2}\gamma(1+\sigma)}{(\omega_c+\nu_0-2\omega)^2+(\gamma_c+\gamma)^2}\end{aligned} \qquad (12.67a)$$

Here, we have used the integral $\int_0^t \delta(t-t')dt' = 1/2$. Thus we have

$$\frac{d}{dt}\langle\tilde{a}^\dagger(t)\tilde{a}(t)\rangle = (s_0 + s_0^*)\langle\tilde{a}^\dagger(t)\tilde{a}(t)\rangle + R_t + R_{sp} \tag{12.67b}$$

with

$$R_t = \left|\frac{\gamma'}{\gamma' + \gamma_c'}\right|^2 2\gamma_c\langle n_c\rangle \tag{12.68}$$

$$R_{sp} = \left|\frac{\gamma'}{\gamma' + \gamma_c'}\right|^2 2\mathcal{g}N_2 \tag{12.69}$$

where Equation 3.26 has been used for κ_m, the summation over m of $U_c^2(z_m)$ has been evaluated as N/ε_1, the relation $|\gamma'|^2 = \gamma^2(1+\delta^2)$ has been used, and the factor $\mathcal{g}$ is given in Equation 12.43.

In Equation 12.67b the first term on the right-hand side gives the decay rate of the photon number, which is the difference between the stimulated emission rate and the cavity loss rate (see Equation 4.53). In Equations 12.68 and 12.69 the absolute squared factor is given in Equation 11.68 and represents the detuning and the bad cavity effects. Except for this factor, the rate R_t is interpreted as the injection rate of thermal photons and the rate R_{sp} is the *total spontaneous emission rate* within the cavity. Below Equation 4.14 the constant $\mathcal{g}$ was interpreted as "half the stimulated transition rate per atom per unit density of photons." Therefore, the quantity 2g is the stimulated transition rate per atom per unit density of photons. For a free space mode, this rate is equal to the spontaneous emission rate per atom into the mode. However, in this case, the spontaneous emission rate into the cavity mode per atom is multiplied by the absolute squared factor because of the above-mentioned effects in the quasimode cavity. Hereafter, we use the term "total spontaneous emission rate" for the spontaneous emission by all the atoms in an active cavity to distinguish it from the spontaneous emission by individual atoms, for which we retain the term "spontaneous emission rate."

This total spontaneous emission rate is related to the laser linewidth in Equation 4.62a. Noting that the output power $P = 2\gamma_c\hbar\omega\langle\hat{a}^\dagger\hat{a}\rangle$ and that $\mathcal{g}N\sigma_{th} = \gamma_c$ (Equation 4.47), we can show that

$$\Delta\omega = \frac{R_t + R_{sp}}{\langle\hat{a}^\dagger\hat{a}\rangle} \tag{12.70}$$

for below-threshold operation. For above-threshold operation, we have from Equation 4.82

$$\Delta\omega = \frac{R_t + R_{sp}}{2\langle\hat{a}^\dagger\hat{a}\rangle} \tag{12.71}$$

Thus the linewidth in angular frequency above threshold is, except for the thermal injection rate, the spontaneous emission rate divided by twice the photon number in the cavity, as noted by Henry [10]. (In Henry's formula, this linewidth is multiplied by $1+\alpha^2$, the Henry factor – see Section 12.13.3.)

12.12.2
Spontaneous Emission in the One-Sided Cavity Laser

Next we consider the one-sided cavity laser below threshold analyzed in Chapter 9. We have from Equation 9.64 (for $-d \leq z \leq 0$)

$$\hat{E}^{(+)}(z,t) = \frac{\sin\Omega_c(z+d)/c_1}{\gamma+\gamma_c+i(\nu_0+\omega_c-2\omega)} \times \left[\sum_j C_j\hat{a}_j(0)\int_0^t e^{-i\omega_j\tau}\exp[(s_0-i\omega)(t-\tau)]d\tau + \frac{i\nu_0 p_a}{\varepsilon_1 d}\right.$$
$$\left.\times\sum_m \sin\{\Omega_c(z_m+d)/c_1\}\int_0^t \hat{\Gamma}_m(\tau)\exp[(s_0-i\omega)(t-\tau)]d\tau\right] \tag{9.64}$$

Differentiation with respect to time yields

$$\frac{d}{dt}\hat{E}^{(+)}(z,t) = s_0\hat{E}^{(+)}(z,t) + \frac{\sin\Omega_c(z+d)/c_1}{\gamma+\gamma_c+i(\nu_0+\omega_c-2\omega)} \times\left[\sum_j C_j\hat{a}_j(0)\,e^{-i\omega_j t} + \frac{i\nu_0 p_a}{\varepsilon_1 d}\sum_m \sin\{\Omega_c(z_m+d)/c_1\}\,\hat{\Gamma}_m(t)\right] \tag{12.72}$$

As in Equation 12.67a we have

$$\frac{d}{dt}\left\langle \hat{E}^{(-)}(z,t)\hat{E}^{(+)}(z,t)\right\rangle = (s_0+s_0^*)\left\langle \hat{E}^{(-)}(z,t)\hat{E}^{(+)}(z,t)\right\rangle + \left|\frac{\sin\Omega_c(z+d)/c_1}{\gamma'+\gamma'_c}\right|^2 \times\left[D + \left|\frac{\nu_0 p_a}{\varepsilon_1 d}\right|^2\sum_m |\sin\{\Omega_c(z_m+d)/c_1\}|^2\gamma(1+\sigma)\right] \tag{12.73}$$

where D is given in Equation 9.81b. Here $\hat{E}^{(-)}(z,t)$ is the Hermitian adjoint of $\hat{E}^{(+)}(z,t)$ (see Equation 2.19b). In this case $\hat{E}^{(+)}(z,t)$ is not the photon annihilation operator of the cavity mode. In view of the spatial dependence of Equation 9.64 we assume the form

$$\hat{E}^{(+)}(z,t) = B\hat{a}(t)\sin\{\Omega_c(z+d)/c_1\} \tag{12.74}$$

where B is the normalization constant to be determined and $\langle\hat{a}^\dagger(t)\hat{a}(t)\rangle$ describes the photon number in the cavity. Since the field energy stored inside the cavity divided by $\hbar\omega$ is equal to the photon number, we have

$$2\varepsilon_1\int_{-d}^0 dz\left\langle \hat{E}^{(-)}(z,t)\hat{E}^{(+)}(z,t)\right\rangle = \hbar\omega\langle\hat{a}^\dagger(t)\hat{a}(t)\rangle \tag{12.75}$$

or

$$2\varepsilon_1|B|^2\langle\hat{a}^\dagger(t)\hat{a}(t)\rangle\int_{-d}^0 |\sin\{\Omega_c(z+d)/c_1\}|^2dz = \hbar\omega\langle\hat{a}^\dagger(t)\hat{a}(t)\rangle \tag{12.76}$$

Thus we have

$$|B|^2 = \frac{\hbar\omega}{2\varepsilon_1} \frac{1}{\int_{-d}^{0} |\sin\{\Omega_c(z+d)/c_1\}|^2 dz} \tag{12.77}$$

Substituting Equation 12.74 into Equation 12.73 and using Equation 12.77 we have

$$\begin{aligned} \frac{d}{dt}\langle \hat{a}^\dagger(t)\hat{a}(t)\rangle &= (s_0 + s_0^*)\langle \hat{a}^\dagger(t)\hat{a}(t)\rangle + \frac{2\varepsilon_1}{\hbar\omega} \frac{\int_{-d}^{0} |\sin\{\Omega_c(z+d)/c_1\}|^2 dz}{|\sin\{\Omega_c(z+d)/c_1\}|^2} \\ &\times \left| \frac{\sin \Omega_c(z+d)/c_1}{\gamma' + \gamma_c'} \right|^2 \\ &\times \left[D + \left| \frac{\nu_0 p_a}{\varepsilon_1 d} \right|^2 \sum_m |\sin\{\Omega_c(z_m+d)/c_1\}|^2 \gamma(1+\sigma) \right] \end{aligned} \tag{12.78}$$

Referring to Equation 9.92 for the integral and to Equation 4.14 for the constant g, we have

$$\frac{d}{dt}\langle \hat{a}^\dagger(t)\hat{a}(t)\rangle = (s_0 + s_0^*)\langle \hat{a}^\dagger(t)\hat{a}(t)\rangle + R_t' + R_{sp} \tag{12.79}$$

where

$$\begin{aligned} R_t' &= R_t \frac{\sigma^2}{\sigma_{th}\sigma_{th}^0} = \left| \frac{\gamma'}{\gamma' + \gamma_c'} \right|^2 \left(\frac{\beta_c}{\gamma_c} \right)^2 \frac{\sigma^2}{\sigma_{th}\sigma_{th}^0} 2\gamma_c \langle n_\omega \rangle \\ R_{sp} &= \left| \frac{\gamma'}{\gamma' + \gamma_c'} \right|^2 \left(\frac{\beta_c}{\gamma_c} \right)^2 2g N_2 \end{aligned} \tag{12.80}$$

Here we have used Equation 9.81b for D and Equation 9.14 for G in the expression for D. Also we have used Equation 7.44a for σ_{th}. This time both the thermal photon injection rate and the total spontaneous emission rate in the cavity have a new factor, $(\beta_c/\gamma_c)^2$, which was identified as the longitudinal excess noise factor for the case of the one-sided cavity laser below threshold. Also, the factor $\sigma^2/(\sigma_{th}\sigma_{th}^0)$ appears here as in Equation 9.105. The laser linewidth in Equation 9.105 can be rewritten as

$$\Delta\omega = \frac{R_t' + R_{sp}}{\langle \hat{a}^\dagger \hat{a} \rangle} \tag{12.81}$$

For above-threshold operation, Equation 10.111 for the same laser reduces, except for the extra term coming from the gain saturation (the term containing $g(r)$), to

$$\Delta\omega = \frac{R_t + R_{sp}}{2\langle \hat{a}^\dagger \hat{a} \rangle} \tag{12.82}$$

Here R_t is defined in Equation 12.80. Comparison of Equation 12.81 or Equation 12.82 with Equation 12.70 or Equation 12.71 shows that, except for the thermal contribution and the term coming from the gain saturation, the linewidth is in

They showed that the parameter $\Omega = (\nu_0 - \omega_c)/\gamma_{net}$, where γ_{net} is the net damping rate of the field in the active cavity, has the same role as the parameter α and yields a linewidth enhancement factor $1+\Omega^2$. However, they showed numerically that, for a large output coupling, the detuning effect is smaller than the $1+\Omega^2$ factor.

In this context we have obtained in Section 12.6 a factor

$$1 + \delta^2 = 1 + \frac{(\nu_0 - \omega_c)^2}{(\gamma + \gamma_c)^2} \tag{12.84}$$

only for the case $\gamma \sim \gamma_c$. When either $\gamma \gg \gamma_c$ or $\gamma \ll \gamma_c$ we had no detuning effect.

12.13.4
Internal Loss

By the term *internal loss* we mean the optical losses other than that due to output coupling. One physical cause of internal loss is scattering of the laser field by impurities or optical imperfections. The other is absorption of the laser field by any absorber other than the lasing atoms or by the lasing atoms involving non-lasing levels.

The loss may be localized or distributed through the cavity. The internal loss merely consumes the laser field energy, while the output coupling loss leads to the laser output. As mentioned in Chapter 3 and in Appendix C, these losses are associated with fluctuating forces so as to maintain the quantum-mechanical consistency.

12.13.4.1 Internal Loss in a Quasimode Laser

In the case of the quasimode laser model, the addition of an internal loss mechanism does not lead to much difficulty. The internal loss adds a decay term with a decay constant γ_l in Equation 4.39a together with a fluctuating force term, for example, $\hat{\Gamma}_l(t)$. Here the subscript l signifies the internal loss. Thus

$$\frac{d}{dt}\hat{a} = -i\omega_c\hat{a} - \gamma_c\hat{a} - \gamma_l\hat{a} - i\sum_m \kappa_m(\hat{b}_{m1}^\dagger \hat{b}_{m2}) + \hat{\Gamma}_f(t) + \hat{\Gamma}_l(t) \tag{12.85}$$

The correlation function for $\hat{\Gamma}_l(t)$ is given analogously to Equations 3.36 and 3.37:

$$\begin{aligned} &\langle \hat{\Gamma}_l(t) \rangle = 0, && \left\langle \hat{\Gamma}_l^\dagger(t) \right\rangle = 0 \\ &\left\langle \hat{\Gamma}_l^\dagger(t)\hat{\Gamma}_l(t') \right\rangle = 2\gamma_l \langle n_c \rangle \delta(t-t'), && \left\langle \hat{\Gamma}_l(t)\hat{\Gamma}_l^\dagger(t') \right\rangle = 2\gamma_l(\langle n_c \rangle + 1)\delta(t-t') \end{aligned} \tag{12.86}$$

Therefore, the analysis for the quasimode laser in Chapter 4 does not change significantly: the cavity decay rate becomes $\gamma_c + \gamma_l$ instead of γ_c, and the normally or anti-normally ordered noise correlations yield (see Appendix C)

$$\left\langle \left\{ \hat{\Gamma}_f^{\dagger}(t) + \hat{\Gamma}_l^{\dagger}(t) \right\} \left\{ \hat{\Gamma}_f(t') + \hat{\Gamma}_l(t') \right\} \right\rangle = 2(\gamma_c + \gamma_l) n_\omega \delta(t - t')$$
$$\left\langle \left\{ \hat{\Gamma}_f(t) + \hat{\Gamma}_l(t) \right\} \left\{ \hat{\Gamma}_f^{\dagger}(t') + \hat{\Gamma}_l^{\dagger}(t') \right\} \right\rangle = 2(\gamma_c + \gamma_l)(n_\omega + 1)\delta(t - t') \quad (12.87)$$

where we have assumed the mutual independence of the two noise forces. The laser output power is still $2\gamma_c$ times the stored energy. The product of $2\gamma_l$ and the stored energy gives the power dissipation rate to the internal loss mechanism. So, most of the γ_c in Chapter 4 become $\gamma_c + \gamma_l$ except the one in Equation 4.80 for the power output and one of the γ_c in the numerator in the expression for the linewidth in Equations 4.81 and 4.82.

12.13.4.2 **Internal Loss in a Two-Sided Cavity Laser**

For the case of continuous mode analysis starting in Chapter 5, it is not obvious how to take the decay rate γ_l and the fluctuating force $\hat{\Gamma}_l(t)$ into account in the starting equation (Equation 5.25) because this deals with the individual universal mode but not the cavity mode as a whole. Therefore, it is not advisable to use the continuous mode analysis from the outset. Any internal loss may be artificially squeezed into the theory after we get the equation of motion for the total cavity field, for example, Equation 10.69 obtained for the nonlinear, saturated gain analysis of a one-sided cavity laser. Here the time rate of change of the internal field is related to the thermal and quantum noise forces.

The simplified, propagation method used for the analysis of a generalized two-sided cavity laser in Chapter 11 was based on this equation. Equations 11.35 and 11.77 for the nonlinear and linear gain regimes, respectively, were the disguised forms of Equation 10.69 with suitably dressed-up noise forces. Two coupling surfaces with respective arbitrary reflectivity, r_1 and r_2, were introduced instead of the two with r and -1 for the one-sided cavity. The noise forces associated with possible internal losses may be introduced into Equations 11.35 and 11.77. For generality, let us assume that we have local losses at the two mirrors with respective loss rate γ_{l1} and γ_{l2}, respectively, and a distributed loss with overall loss rate γ_{ld}. Then, the factor $\hat{F}_t + \hat{F}_q$ in Equations 11.35, for example, may have additional terms due to these internal losses:

$$\frac{d\hat{e}^+(d-0,\, t)}{dt} = \frac{c_1}{2d}\frac{\gamma'}{\gamma' + \gamma_c'}\left(\hat{F}_t + \hat{F}_q + \hat{F}_{l1} + \hat{F}_{l2} + \hat{F}_{ld}\right) \quad (12.88)$$

The noise from the mirrors may be treated just as the thermal noise entering from the two mirrors. We cite Equation 11.36:

$$\hat{F}_t = \left\{ T_2' r_1 G_{tR} e^{2ikd} \hat{f}_t^R\left(d + 0,\, t - \frac{2d}{c_1}\right) + T_1' G_{tL} e^{ikd} \hat{f}_t^L\left(-0,\, t - \frac{d}{c_1}\right) \right\} \quad (11.36)$$

Here the noise forces are multiplied by the net amplification plus phase shift for one round trip or a one-way trip to mirror M2. Inspection of the calculation procedure to go from this equation to the linewidth formula in Equation 11.66 shows that the resultant squared transmission coefficients $|T_{1,2}'|^2$ result in the factor $(|r_1| + |r_2|)(1 - |r_1||r_2|)/|r_1||r_2|$, which, together with the factor $(c_1/2d)$, reduces to $2\ln(1/|r_1||r_2|) = 2\gamma_c$ in the limit $|r_1|, |r_2| \to 1$. If we look at the above factor more precisely, it appeared in Equation 11.53 in the form $|r_1|^{-1} - |r_1|$ plus $|r_2|^{-1} - |r_2|$. They came from $|T_1'|^2$ and $|T_2'|^2$, respectively. In the limit $|r_1|, |r_2| \to 1$, they would lead to $2\gamma_{c1} = 2(c_1/2d)\ln(1/|r_1|)$ and $2\gamma_{c2} = 2(c_1/2d)\ln(1/|r_2|)$, respectively, where, of course, we have $\gamma_{c1} + \gamma_{c2} = \gamma_c$. Thus we see that the usual cavity decay rate was incorporated through the transmission coefficients. Therefore $\hat{F}_{l1,2}$ may be given as

$$\hat{F}_{l2} + \hat{F}_{l1} = T'_{l2} r_1 G_{tR} e^{2ikd} \hat{f}_t^R \left(t - \frac{2d}{c_1} \right) + T'_{l1} \, G_{tL} e^{ikd} \hat{f}_t^L \left(t - \frac{d}{c_1} \right) \tag{12.89}$$

where T'_{l1} and T'_{l2} should lead to their respective loss rates γ_{l1} and γ_{l2}. One prescription for this is to set

$$T'_{l1} = T'_1 \sqrt{\frac{\gamma_{l1}}{\gamma_{c1}}}, \qquad T'_{l2} = T'_2 \sqrt{\frac{\gamma_{l2}}{\gamma_{c2}}} \tag{12.90}$$

For the distributed loss, if it is uniformly distributed, we may have, as in Equation 11.41 for the quantum noise,

$$\hat{F}_{ld} = \sum_i \left\{ \hat{f}_i^{ld} \left(t - \frac{d - z_i}{c_1} \right) g_{mR} e^{ik(d - z_i)} + \hat{f}_i^{ld} \left(t - \frac{d + z_i}{c_1} \right) r_1 g_{mL} e^{ik(d + z_i)} \right\} \tag{12.91}$$

The factor $\hat{f}_i^{ld}$ is the noise amplitude emitted by the ith fragment of the distributed loss mechanism. The assembled noise

$$\hat{f}_{ld} = \sum_i \hat{f}_i^{ld} \tag{12.92}$$

should lead to the decay rate γ_{ld}. Therefore, this may be given, as in Equation 12.90, by

$$\hat{f}_{ld} = T'_{ld} \hat{f}_t^R, \qquad T'_{ld} = T'_2 \sqrt{\frac{\gamma_{ld}}{\gamma_{c2}}} \tag{12.93a}$$

or

$$\hat{f}_{ld} = T''_{ld} \hat{f}_t^L, \qquad T''_{ld} = T'_1 \sqrt{\frac{\gamma_{ld}}{\gamma_{c1}}} \tag{12.93b}$$

These two alternatives should lead to the same results. The individual noises $\hat{f}_i^{ld}$, which are mutually independent, may be determined so that

$$\hat{f}_{ld}^\dagger(t) \hat{f}_{ld}(t') = \sum_i \hat{f}_i^{ld\dagger}(t) \hat{f}_i^{ld}(t') \tag{12.94}$$

Finally, the noises $\hat{F}_{l1} + \hat{F}_{l2} + \hat{F}_{ld}$ should be determined so that, in the final expression for the linewidth and in a good cavity limit, every γ_c, for example in Equation 11.66, is replaced by $\gamma_c + \gamma_{l1} + \gamma_{l2} + \gamma_{ld}$ except for the one that is related directly to the power output.

12.13.4.3 Internal Loss and Optimum Output Coupling

A practical benefit of taking into account the internal loss is that this allows us to know the optimum output coupling or the optimum mirror reflectivity to get the largest power output for a given pumping strength. As seen from Equation 4.34, in the presence of an internal loss mechanism, the power output P is (see Equation 4.14)

$$P = 2\gamma_c \varepsilon_1 d |E_s|^2 \left[\frac{\mathcal{g} N \sigma^0}{\gamma_c + \gamma_l} - 1 \right] \tag{12.95}$$

Note that $\mathcal{g}$ does not contain γ_c nor γ_l (see Equation 4.14). We assume, for simplicity, that the γ_c (and γ_l) dependence of $|E_s|^2$ given in Equation 4.35 can be ignored. Then the power output P as a function of γ_c vanishes at $\gamma_c = 0$ and at $\gamma_c = \mathcal{g} N \sigma^0 - \gamma_l$ but has a maximum at

$$\gamma_c = \sqrt{\mathcal{g} N \sigma^0 \gamma_l} - \gamma_l \tag{12.96}$$

The maximum power is

$$P = 2\varepsilon_1 d |E_s|^2 \left(\sqrt{\mathcal{g} N \sigma^0} - \sqrt{\gamma_l} \right)^2 \tag{12.97}$$

If we had $\gamma_l = 0$, P would increase with decreasing γ_c and would be at its maximum at $\gamma_c = 0$. This is unphysical since $\gamma_c = 0$ means zero output power. In reality we cannot avoid some internal loss, which leads to a finite optimum coupling loss.

12.13.5 Spatial Hole Burning

In a laser where the laser active atoms cannot move freely in space, the gain saturation occurs selectively at the locations where the field intensity is large. In lasers with Fabry–Perot type cavities, there exist two counter-traveling waves that interfere to bring an interference pattern with a period of half the wavelength in the laser medium. If the pumping is uniform, the portion of bright interference becomes more saturated than the portion of dark interference. This is called spatial hole burning in contrast to the usual hole burning that occurs in the gain spectral region in an inhomogeneously broadened medium. Spatial hole burning leads to a quasi-periodically modulated gain distribution along the length of the laser medium with the period being a half-wavelength.

In previous chapters, this phenomenon was disregarded because an accurate inclusion of this effect demands more complicated mathematics. For example, in Chapter 4 for a quasimode laser, we have replaced $\sin^2(z_m + d)$ by its space average $\frac{1}{2}$ in going from Equation 4.29 to Equation 4.32. Similarly, in Chapter 8 devoted to the semiclassical analysis of the one-sided cavity laser, we have ignored, in Equation 8.22, the interference terms in the absolute square of the total electric field, which would lead to spatial holes. This approximation was carried over to Chapter 10 for the quantum, nonlinear gain analysis of the same laser.

Agrawal and Lax [14] developed a method to treat the spatial hole burning and showed that the gain is in general different for the right- and left-going waves in the presence of spatial holes. Goldberg *et al.* [13] extensively examined the effect of spatial holes on the longitudinal excess noise factor, extending the method of Agrawal and Lax to include asymmetric cavities, and showed that the spatial holes increase the linewidth because of the decreased output power and that, when the cavity is two-sided and asymmetric, the directional gain results in a further increase in the excess noise factor.

12.13.6
Transition From Below Threshold to Above Threshold

A laser is an oscillator in the optical frequency region. An oscillator has in general a clear threshold gain and its operation changes abruptly at threshold: from no coherent output power below threshold to finite coherent output power above threshold. A laser also has a clear threshold behavior: we experience an abrupt emergence of a bright light beam at threshold. Accordingly, the analyses presented so far have been divided into linear gain analysis for below-threshold operation, and nonlinear, saturated gain analysis for above-threshold operation. In particular, the expression for the laser linewidth had a decrease by a factor of 2 from below to above threshold.

A closer look at threshold, however, reveals a smooth transition from below threshold to above threshold of various quantities in a laser. Risken [15] calculated the smooth change of the linewidth, showing a decrease by a factor of 2 from below to far above threshold. He treated the classical field amplitude using a Fokker–Planck equation approach. The Fokker–Planck equation for a laser deals with the probability distribution of the field amplitude [5,6]. Another method suitable for analyzing the smooth change through threshold is the density matrix equation method, used by, among others, Scully and Lamb [16]. This method treats the photon number distribution and derives a smooth change in the photon number distribution: from that of a black-body radiation for far below threshold to a Poisson distribution for far above threshold.

References

1. Glauber, R.J. (**1963**) *Phys. Rev., 130,* 2529–2539.
2. Goldberg, P., Milonni, P.W., and Sundaram, B. (**1991**) *Phys. Rev. A, 44,* 1969–1985.
3. Hamel, W.A. and Woerdman, J.P. (**1990**) *Phys. Rev. Lett., 64,* 1506–1509.
4. Schawlow, A.L. and Townes, C.H. (**1958**) *Phys. Rev., 112,* 1940–1949.
5. Haken, H. (**1970**) *Laser theory*, in *Licht und Materie, IC,* Handbuch der Physik, vol. XXV/2c (eds. S. Flügge and L. Genzel), Springer, Berlin.
6. Sargent, M., III, Scully, M.O., and Lamb, W.E., Jr. (**1974**) *Laser Physics,* Addison-Wesley, Reading, MA.
7. van Exter, M.P., Kuppens, S.J.M., and Woerdman, J.P. (**1995**) *Phys. Rev. A, 51,* 809–816.
8. Prasad, S. (**1992**) *Phys. Rev. A, 46,* 1540–1559.
9. Tromborg, B., Olesen, H., and Pan, X. (**1991**) *IEEE J. Quantum Electron., 27,* 178–192.
10. Henry, C.H. (**1986**) *J. Lightwave Tech., LT-4,* 288–297.

11 Kuppens, S.J.M., van Exter, M.P., Woerdman, J.P., and Kolobov, M.I. (**1996**) *Opt. Commun.*, *126*, 79–84.
12 Henry, C.H. (**1982**) *J. Quantum Electron.*, *QE-18*, 259–264.
13 Goldberg, P., Milonni, P.W., and Sundaram, B. (**1991**) *Phys. Rev. A*, *44*, 4556–4563.
14 Agrawal, G.P. and Lax, M. (**1981**) *J. Opt. Soc. Am.*, *71*, 515–519.
15 Risken, H. (**1966**) *Z. Phys.*, *191*, 302–312.
16 Scully, M.O. and Lamb, W.E., Jr. (**1967**) *Phys. Rev.*, *159*, 208–226.

13
Spontaneous Emission in a One-Dimensional Optical Cavity with Output Coupling

In this chapter we analyze spontaneous emission from a single excited atom in a one-dimensional, symmetric two-sided cavity. Perturbative treatments and exact non-perturbative treatments are given. We show that the spontaneous emission rate can be enhanced, compared to the case in a free one-dimensional space, by the so-called Purcell factor, but not by the excess noise factor. Thus this analysis makes it clear that the excess noise factor of the previous chapters is not a result of the enhancement of the spontaneous emission rate of individual atoms by the excess noise factor. Parts of the analyses presented in this chapter are due to Feng and Ujihara [1] and Takahashi and Ujihara [2]. Extension to three dimensions is considered in the final section.

13.1
Equations Describing the Spontaneous Emission Process

We consider the spontaneous emission process by a two-level atom located in an optical cavity having output coupling. The atom is initially prepared in the upper state and the field modes are initially in the vacuum states. Here we use the symmetric two-sided cavity model of Section 1.3.2 where the cavity is composed of a dielectric slab extending in the region $-d \le z \le d$. The dielectric has dielectric constant ε_1 and refractive index n with the velocity of light inside the dielectric being c_1. The outside regions are vacuum with dielectric constant ε_0 and velocity of light c_0. We naturally have $c_1 = c_0/n$. For quantization, we have imposed on the field modes a cyclic boundary condition with period $L + 2d \approx L$. The mode functions of the universe are given in Equation 1.58, where we have antisymmetric and symmetric functions.

In this chapter we use the Schrödinger equation in contrast to the Heisenberg equation used in Chapters 3 through 11. The Hamiltonian describing the spontaneous emission process by a two-level atom is given as

Output Coupling in Optical Cavities and Lasers: A Quantum Theoretical Approach
Kikuo Ujihara

ISBN: 978-3-527-40763-7

$$\hat{H}_t = \hat{H}_f + \hat{H}_a + \hat{H}_{int} \tag{13.1}$$

Here the field Hamiltonian is

$$\hat{H}_f = \sum_j \hat{H}_j = \sum_j \hbar\omega_j \left(\hat{a}_j^\dagger \hat{a}_j + \frac{1}{2} \right) \tag{13.2}$$

where the suffix j denotes a "universal" mode. The atom with Hamiltonian $\hat{H}_a$ has two levels $|u\rangle$ and $|l\rangle$ with energy eigenvalues

$$\hat{H}_a|u\rangle = \hbar\omega_A|u\rangle, \qquad \hat{H}_a|l\rangle = 0|l\rangle = 0 \tag{13.3}$$

The interaction Hamiltonian is (cf. Equations 3.17 and 3.19)

$$\hat{H}_{int} = -i\sum_j \sqrt{\frac{\hbar\omega_j}{2}}\, U_j(z_A)\hat{\mu}(\hat{a}_j - \hat{a}_j^\dagger) \tag{13.4}$$

where $\hat{\mu} = e\hat{x}$ is the component of the atom's electric dipole operator $e\hat{\mathbf{r}}$ in the direction of the polarization of the electric field, which is assumed to be in the x-direction. The function $U_j(z)$ is the jth mode of the "universe" and z_A is the location of the atom.

Now the electric dipole is a constant operator that does not change with time. Also, the annihilation and creation operators of the modes of the "universe" are constant in time. What changes with time is now the wavefunction, for which we assume the form

$$|\varphi(t)\rangle = C_u(t)|u\rangle|0\rangle e^{-i\omega_A t} + \sum_j C_{lj}(t)|l\rangle|1_j\rangle e^{-i\omega_j t} \tag{13.5}$$

with the initial conditions

$$C_u(0) = 1 \qquad \text{and} \qquad C_{lj}(0) = 0 \tag{13.6}$$

Here the field state $|0\rangle$ denotes the state where no photon exists in any mode, while $|1_j\rangle$ denotes the state where one photon exists in the mode j but no photon exists in any of the other modes. The time-varying coefficients $C_u(t)$ and $C_{lj}(t)$ are the probability amplitudes for the combined states $|u\rangle|0\rangle$ and $|l\rangle|1_j\rangle$, respectively. In Equation 13.5 we are implicitly assuming that the system has an appreciable probability only for one excitation state, that is, the state with an excited atom and no photons, or the state with a de-excited atom and one photon. We assume that the probabilities for states with zero excitations and with two excitations or more can be ignored. This corresponds to the rotating-wave approximation where energy-conserving terms are selectively treated.

Now the Schrödinger equation

$$i\hbar\frac{\partial}{\partial t}|\varphi(t)\rangle = \hat{H}_t|\varphi(t)\rangle \tag{13.7a}$$

reads

$$
\begin{aligned}
& i\hbar\frac{\partial}{\partial t}\left\{C_u(t)|u\rangle|0\rangle e^{-i\omega_A t} + \sum_j C_{lj}(t)|l\rangle\left|1_j\right\rangle e^{-i\omega_j t}\right\} \\
& \quad = \left\{\sum_j \hbar\omega_j\left(\hat{a}_j^{\dagger}\hat{a}_j + \frac{1}{2}\right) + \hat{H}_a - i\sum_j\sqrt{\frac{\hbar\omega_j}{2}}\,U_j(z_A)\hat{\mu}(\hat{a}_j - \hat{a}_j^{\dagger})\right\} \\
& \quad \times\left\{C_u(t)|u\rangle|0\rangle e^{-i\omega_A t} + \sum_j C_{lj}(t)|l\rangle\left|1_j\right\rangle e^{-i\omega_j t}\right\}
\end{aligned}
\tag{13.7b}
$$

Thus the left-hand side (LHS) is

$$
\begin{aligned}
\mathrm{LHS} = {} & i\hbar\dot{C}_u(t)|u\rangle|0\rangle e^{-i\omega_A t} + \hbar\omega_A C_u(t)|u\rangle|0\rangle e^{-i\omega_A t} \\
& + i\hbar\sum_j \dot{C}_{lj}(t)|l\rangle\left|1_j\right\rangle e^{-i\omega_j t} + \hbar\sum_j \omega_j C_{lj}(t)|l\rangle\left|1_j\right\rangle e^{-i\omega_j t}
\end{aligned}
\tag{13.8}
$$

On the right-hand side (RHS) we use Equations 2.14, 2.15 and 13.3 and ignore states with two photons. Thus the RHS is

$$
\begin{aligned}
\mathrm{RHS} = {} & \sum_i \frac{\hbar\omega_i}{2}\left\{C_u(t)|u\rangle|0\rangle e^{-i\omega_A t} + \sum_j C_{lj}(t)|l\rangle\left|1_j\right\rangle e^{-i\omega_j t}\right\} \\
& + \sum_j \hbar\omega_j C_{lj}(t)|l\rangle\left|1_j\right\rangle e^{-i\omega_j t} + \hbar\omega_A C_u(t)|u\rangle|0\rangle e^{-i\omega_A t} \\
& - i\sum_j\sqrt{\frac{\hbar\omega_j}{2}}\,U_j(z_A)\hat{\mu}C_{lj}(t)|l\rangle|0\rangle e^{-i\omega_j t} \\
& + i\sum_j\sqrt{\frac{\hbar\omega_j}{2}}\,U_j(z_A)\hat{\mu}C_u(t)|u\rangle\left|1_j\right\rangle e^{-i\omega_A t}
\end{aligned}
\tag{13.9}
$$

Note that the second and fourth terms on the LHS cancel with the fourth and third terms on the RHS, respectively. Multiplying both sides by $\langle u|\langle 0|$ and by $\langle l|\langle 1_j|$, we have, respectively,

$$
\begin{aligned}
i\hbar\dot{C}_u(t) &= \sum_i \frac{\hbar\omega_i}{2}C_u(t) - i\sum_j\sqrt{\frac{\hbar\omega_j}{2}}\,U_j(z_A)\langle u|\hat{\mu}|l\rangle C_{lj}(t)\,e^{i(\omega_A-\omega_j)t} \\
i\hbar\dot{C}_{lj}(t) &= \sum_i \frac{\hbar\omega_i}{2}C_{lj}(t) + i\sqrt{\frac{\hbar\omega_j}{2}}\,U_j(z_A)\langle l|\hat{\mu}|u\rangle C_u(t)e^{-i(\omega_A-\omega_j)t}
\end{aligned}
\tag{13.10}
$$

The factor $\sum_i \hbar\omega_i/2$ can be shown to give C_u and C_{lj} the same phase factor $\exp\{i\sum_j(\omega_j/2)t\}$, which we ignore from now on. Thus we ignore the first terms on the RHS in both lines of Equations 13.10. We obtain

$$\begin{aligned}\dot{C}_u(t) &= -\sum_j \sqrt{\frac{\omega_j}{2\hbar}} U_j(z_A)\mu_A C_{lj}(t) e^{i(\omega_A-\omega_j)t} \\ \dot{C}_{lj}(t) &= \sqrt{\frac{\omega_j}{2\hbar}} U_j(z_A)\mu_A^* C_u(t) e^{-i(\omega_A-\omega_j)t}\end{aligned} \tag{13.11}$$

where we have written $\langle u|\hat{\mu}|l\rangle = \mu_A$.

13.2
The Perturbation Approximation

We first consider the process perturbatively where the probability amplitude of the initial state $|u\rangle|0\rangle$ does not change much so that $C_u(t) \approx 1$. Thus the second line of Equations 13.10 can be integrated to yield

$$C_{lj}(t) = \sqrt{\frac{\omega_j}{2\hbar}} U_j(z_A)\mu_A^* \frac{e^{-i(\omega_A-\omega_j)t} - 1}{-i(\omega_A - \omega_j)} \tag{13.12}$$

The total probability P of the atom being in the lower state is

$$P = \sum_j \left|C_{lj}(t)\right|^2 = \sum_j \frac{\omega_j}{2\hbar} U_j^2(z_A)|\mu_A|^2 \frac{4\sin^2\{(\omega_A - \omega_j)t/2\}}{(\omega_A - \omega_j)^2} \tag{13.13}$$

We take the limit $t \to \infty$ whence

$$\frac{4\sin^2\{(\omega_A - \omega_j)t/2\}}{(\omega_A - \omega_j)^2} \to 2\pi t\delta(\omega_A - \omega_j) \tag{13.14}$$

Using the density of modes $\rho(\omega)$ given by Equation 1.67 and the mode functions given in Equations 1.58 and 1.65, we have

$$P = t\frac{\pi\omega_A|\mu_A|^2}{\hbar}\left\{\rho^a(\omega_A)\left(U^a_{\omega_A}(z_A)\right)^2 + \rho^b(\omega_A)\left(U^b_{\omega_A}(z_A)\right)^2\right\} \tag{13.15}$$

Thus the spontaneous emission rate R is

$$\begin{aligned}R = \frac{d}{dt}P &= \frac{\pi\omega_A|\mu_A|^2}{\hbar}\left\{\rho^a(\omega_A)\left(U^a_{\omega_A}(z_A)\right)^2 + \rho^b(\omega_A)\left(U^b_{\omega_A}(z_A)\right)^2\right\} \\ &= A_0\left(\frac{c_1}{c_0}\right)\left(\frac{\sin^2 k_{1A}z_A}{1 - K\sin^2 k_{1A}d} + \frac{\cos^2 k_{1A}z_A}{1 - K\cos^2 k_{1A}d}\right)\end{aligned} \tag{13.16}$$

where

$$A_0 = \frac{\omega_A|\mu_A|^2}{\hbar c_1 \varepsilon_1} \tag{13.17}$$

is the spontaneous emission rate in a one-dimensional, unbounded dielectric of dielectric constant ε_1. This expression for A_0 can be obtained if we use the density of modes $L/(2c_1\pi)$ for both the mode functions $\sqrt{2/(\varepsilon_1 L)}\sin k_{1j}z$ and $\sqrt{2/(\varepsilon_1 L)}\cos k_{1j}z$.

We see that the spontaneous emission rate depends both on the atomic location and on the atomic frequency relative to the cavity resonant frequencies. Note that $k_{1A} = \omega_A/c_1$. The emission rate is large when either resonant condition $\sin^2 k_{1A}d = 1$ or $\cos^2 k_{1A}d = 1$ is satisfied. In the case of a single-sided cavity, where the second term in Equation 13.16 becomes the same as the first term, the emission rate dependences on the atomic location and on the atomic frequency relative to the cavity resonant frequencies are more pronounced. Since $C_u(t) \approx 1$ is assumed, this perturbation approximation is valid only for $e^{-Rt} \simeq 1$ or $t \ll 1/R$. Another restriction for time t comes from the width of the delta function in Equation 13.14 ($\approx 2\pi/t$), which should be narrower than the spectral width of the mode functions ($\approx 2\gamma_c$).

13.3 Wigner–Weisskopf Approximation

In the Wigner–Weisskopf approximation, an exponential decay of the upper level $|u\rangle$ is assumed:

$$C_u(t) = e^{-(\gamma/2)t} \tag{13.18}$$

where the decay constant γ should be determined in a consistent manner. This approximation is valid up to $t \sim 1/\gamma$. We use this in the second equation in Equation 13.11 to obtain

$$C_{lj}(t) = \sqrt{\frac{\omega_j}{2\hbar}}\, U_j(z_A)\mu_A^* \frac{e^{\{-(\gamma/2)-i(\omega_A-\omega_j)\}t} - 1}{-(\gamma/2) - i(\omega_A - \omega_j)} \tag{13.19}$$

Substituting Equations 13.18 and 13.19 into the first equation in Equation 13.11 we obtain

$$\begin{aligned} \gamma &= \sum_j \frac{\omega_j}{\hbar} U_j^2(z_A)|\mu_A|^2 \frac{1 - e^{i(\omega_A-\omega_j)t+(\gamma/2)t}}{-(\gamma/2) - i(\omega_A - \omega_j)} \\ &\simeq \sum_j \frac{\omega_j}{\hbar} U_j^2(z_A)|\mu_A|^2 \frac{1 - e^{i(\omega_A-\omega_j)t}}{-i(\omega_A - \omega_j)} \end{aligned} \tag{13.20}$$

where in the second line we have ignored $\gamma/2$ as small compared to ω_A. The factor $\{1 - e^{i(\omega_A-\omega_j)t}\}/(\omega_A - \omega_j)$ is $\zeta(\omega_A - \omega_j)$ for $t \to \infty$ where the zeta function is given by Equation 2.53b. Thus for $t \to \infty$

$$\begin{aligned} \mathrm{Re}\gamma &\simeq \sum_j \frac{\omega_j}{\hbar} U_j^2(z_A)|\mu_A|^2 \pi\delta(\omega_A - \omega_j) \\ &= \frac{\pi\omega_A|\mu_A|^2}{\hbar}\left\{\rho^a(\omega_A)\left(U^a_{\omega_A}(z_A)\right)^2 + \rho^b(\omega_A)\left(U^b_{\omega_A}(z_A)\right)^2\right\} \\ &= R \end{aligned} \tag{13.21}$$

Thus we have shown that the real part of the assumed decay constant is the same as the spontaneous emission rate in Equation 13.16 obtained by the lowest-order perturbation calculation. The imaginary part expressing the line shift can be shown to be small [3]. Because this solution is valid up to $t \sim 1/\gamma$, we can discuss the emission spectrum, which is determined after the atom has fully decayed. The absolute square of $C_{lj}(t)$ in Equation 13.19 with $\gamma t \gg 1$ multiplied by the density of modes $\rho(\omega_j)$ gives the emission spectrum $I(\omega_j)$:

$$I(\omega_j) = \rho(\omega_j)\left|C_{lj}(t)\right|^2 = \rho(\omega_j)\frac{\omega_j}{2\hbar}U_j^2(z_A)|\mu_A|^2\frac{1}{(\gamma/2)^2 + (\omega_A - \omega_j)^2} \tag{13.22}$$

where

$$\rho(\omega_j)U_j^2(z_A) = \frac{1}{\pi c_0 \varepsilon_1}\left(\frac{\sin^2 k_{1j}z_A}{1 - K\sin^2 k_{1j}d} + \frac{\cos^2 k_{1j}z_A}{1 - K\cos^2 k_{1j}d}\right) \tag{13.23}$$

If the mode functions were of a flat spectrum, we would simply have a Lorentzian spectrum with full width at half-maximum (FWHM) of γ. However, both the spatially antisymmetric and symmetric modes have peaks at $\omega_j = \omega_{cm}^a = (2m+1)(\pi c_1/2d)$ and $\omega_j = \omega_{cm}^b = 2m(\pi c_1/2d)$ (integer m), respectively, with FWHM of $2\gamma_c$, where $\gamma_c \equiv (c_1/2d)\ln(1/r)$and r is the amplitude reflectivity of the coupling surfaces for waves incident from the inside. Therefore, the transition from the first to the second equation in Equation 13.20 may be allowed only if $\gamma < 2\gamma_c$. Thus the present results under the Wigner–Weisskopf approximation are limited also by the above inequality. If the reverse inequality holds, the emitted photon energy will accumulate in the cavity and will be reabsorbed by the atom, leading to the damped Rabi oscillation discussed below.

13.4 The Delay Differential Equation

Integrating the second equation in Equation 13.11 and substituting the result into the first equation, we have

$$\dot{C}_u(t) = -\frac{|\mu_A|^2}{2\hbar}\int_0^t dt' \sum_j \omega_j U_j^2(z_A)C_u(t')e^{i(\omega_A - \omega_j)(t-t')} \tag{13.24}$$

Using the density of modes $\rho(\omega)$ we rewrite it in the form

$$\dot{C}_u(t) = -\frac{|\mu_A|^2\omega_A}{2\hbar}\int_0^t dt' \int_{-\infty}^{\infty} d\omega_j\, \rho(\omega_j)U_j^2(z_A)e^{i(\omega_A - \omega_j)(t-t')}C_u(t') \tag{13.25}$$

Here, assuming that the time variation of $C_u(t)$ is much slower than the optical frequency, and hence the actual integration range in frequency is small compared to the optical frequency, we have taken ω_j out of the integration, replacing it by ω_A, and extended the lower limit of the frequency integral to $-\infty$. Using the form of the product $\rho(\omega_j)U_j^2(z_A)$ in Equation 13.23 we have

$$\dot{C}_u(t) = -\frac{|\mu_A|^2\omega_A}{2\pi\hbar c_0\varepsilon_1}\int_0^t dt'\int_{-\infty}^{\infty} d\omega_j \left(\frac{\sin^2 k_{1j}z_A}{1-K\sin^2 k_{1j}d}+\frac{\cos^2 k_{1j}z_A}{1-K\cos^2 k_{1j}d}\right)e^{i(\omega_A-\omega_j)(t-t')}C_u(t') \tag{13.26}$$

Using the Fourier series expansion in Equation 1.70b we have

$$\frac{\sin^2 k_{1j}z_A}{1-K\sin^2 k_{1j}d}+\frac{\cos^2 k_{1j}z_A}{1-K\cos^2 k_{1j}d} = \frac{2c_0}{c_1}\sum_{n=0}^{\infty}\frac{r^n}{1+\delta_{0,n}}\cos(2nk_{1j}d)\times\begin{cases}1, & n \text{ even}\\ \cos(2k_{1j}z_A), & n \text{ odd}\end{cases} \tag{13.27}$$

By a similar procedure to Equations 7.5–7.10 we have

$$\begin{aligned}&\int_{-\infty}^{\infty} d\omega_j\left(\frac{\sin^2 k_{1j}z_A}{1-K\sin^2 k_{1j}d}+\frac{\cos^2 k_{1j}z_A}{1-K\cos^2 k_{1j}d}\right)e^{i(\omega_A-\omega_j)(t-t')}\\ &= \frac{2c_0}{c_1}\sum_{n=0}^{\infty} r^{2n}\int_{-\infty}^{\infty} d\omega_j\\ &\times\left[\frac{1}{1+\delta_{0,n}}\cos(\omega_j nt_r)+\frac{r}{2}\cos\{\omega_j(nt_r+t_1)\}+\frac{r}{2}\cos\{\omega_j(nt_r+t_2)\}\right]\\ &\times e^{i(\omega_A-\omega_j)(t-t')} = \frac{2c_0}{c_1}\pi\sum_{n=0}^{\infty} r^{2n}\\ &\times\left[\frac{1}{1+\delta_{0,n}}\{e^{i\omega_A nt_r}\delta(t'-t+nt_r)+e^{-i\omega_A nt_r}\delta(t'-t-nt_r)\}\right.\\ &+\frac{r}{2}\left\{e^{i\omega_A(nt_r+t_1)}\delta(t'-t+nt_r+t_1)+e^{-i\omega_A(nt_r+t_1)}\delta(t'-t-nt_r-t_1)\right\}\\ &\left.+\frac{r}{2}\left\{e^{i\omega_A(nt_r+t_2)}\delta(t'-t+nt_r+t_2)+e^{-i\omega_A(nt_r+t_2)}\delta(t'-t-nt_r-t_2)\right\}\right]\end{aligned} \tag{13.28}$$

where

$$t_r = 4d/c_1, \qquad t_1 = 2(d-z_A)/c_1, \qquad t_2 = 2(d+z_A)/c_1 \tag{13.29}$$

Here t_r is the cavity round-trip time. The delay time $t_1(t_2)$ is for a round trip between the atom and the right-hand (left-hand) mirror. Substituting Equation 13.28 into Equation 13.26 and performing the integration, we have the delay differential equation

$$\dot{C}_u(t) = -\frac{A_0}{2}\Bigg[C_u(t)H(t) + 2\sum_{n=1}^{\infty} r^{2n} e^{i\omega_A n t_r} C_u(t-nt_r)H(t-nt_r)$$
$$+\sum_{n=0}^{\infty} r^{2n+1} e^{i\omega_A(nt_r+t_1)} C_u(t-nt_r-t_1)H(t-nt_r-t_1) \tag{13.30}$$
$$+\sum_{n=0}^{\infty} r^{2n+1} e^{i\omega_A(nt_r+t_2)} C_u(t-nt_r-t_2)H(t-nt_r-t_2)\Bigg]$$

where $H(t)$ is the unit step function and A_0 is the spontaneous emission rate in a one-dimensional, unbounded dielectric of dielectric constant ε_1, which is given by Equation 13.17. The first term describes the natural decay process in a free dielectric. It lasts until $t = \min(t_1, t_2)$. For $t > \min(t_1, t_2)$ the decay process is affected by the fed-back "signal" with reduced magnitude determined by the mirror reflectivity. There are four kinds of routes for the "signal" to come back to the atom. For larger time t, more and more terms with smaller and smaller magnitudes come into play. According to Milonni and Knight [4], these terms can also be interpreted as coming from the mirror images of the atom decaying cooperatively. Equation 13.30 is the same as that obtained by Cook and Milonni [5] who used resonant mode functions of the cavity and introduced the mirror reflectivity phenomenologically.

In the special case where the atom is at the center of the symmetric two-sided cavity, $t_1 = t_2 = t_r/2$ and Equation 13.30 is written as

$$\dot{C}_u(t) = -\frac{A_0}{2} C_u(t)H(t) - A_0 \sum_{n=1}^{\infty} r^n e^{i\omega_A n(t_r/2)} \times C_u(t-n(t_r/2))H(t-n(t_r/2)) \tag{13.31a}$$

We further rewrite it as

$$\dot{C}_u(t) = -\frac{A_0}{2} C_u(t)H(t) + \sum_{n=1}^{\infty} q_n(\omega_A n t_h) C_u(t_n)H(t_n) \tag{13.31b}$$

where

$$q_n(x) = -A_0 r^n e^{ix}, \qquad t_n = t - nt_h, \qquad t_h = t_r/2 \tag{13.31c}$$

Then the solution is obtained as [6]

$$C_u(t) = \sum_{n=0}^{\infty}\left[\sum \frac{q_1^{a_1}(\omega_A t_h) q_2^{a_2}(2\omega_A t_h)\cdots q_n^{a_n}(n\omega_A t_h)}{a_1! a_2! \cdots a_n!} t_n^m \times \exp\left(-\frac{A_0}{2}t_n\right)\right] H(t_n) \tag{13.32a}$$

where the sum is over all non-negative integers a_i ($i = 1, 2, \ldots, n$) that satisfy

$$1a_1 + 2a_2 + \cdots + na_n = n \tag{13.32b}$$

and

$$m = a_1 + a_2 + \cdots + a_n \tag{13.32c}$$

For a free space ($r \to 0$) Equation 13.31a shows that the upper-state population decays as $|C_u(t)|^2 = \exp(-A_0 t)$, as expected.

In Figure 13.1, typical decay curves are presented (after [2]). Here, the cavity length is $2d = (5/2)\lambda$, $r = 0.6$, and $A_0 t_r = 0.5$. The curves are for the atom at an antinode (curve A) at $z_A = (1/5)d = (1/4)\lambda$, at a node (curve C) at $z_A = (2/5)d = (1/2)\lambda$, and between an antinode and a node (curve B) at $z_A = 0.35d \simeq 0.43\lambda$. Note that curve A shows weak Rabi oscillation (see next section). A curve for a natural decay in "free" space ($e^{-A_0 t}$) would come between the lowest and uppermost curves.

13.5 Expansion in Terms of Resonant Modes and Single Resonant Mode Limit

Using the second, resonant mode expansion of Equation 1.70b in Equation 13.26, we have

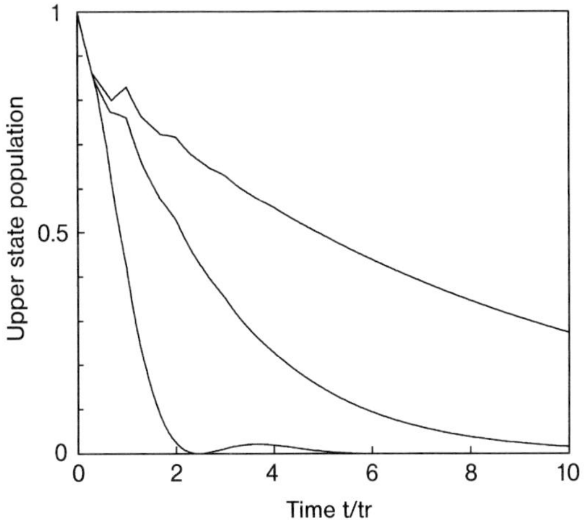

Figure 13.1 The time evolution of the upper-state population $|C_u(t)|^2$, for $2d = 2.5\lambda$, $r = 0.6$, and $A_0 t_r = 0.5$. The time is scaled to the round-trip time t_r. (lowest curve) A, $z_A / d = 0.2$ (antinode); (middle curve) B, $z_A / d = 0.35$ (between node and antinode); and (uppermost curve) C, $z_A / d = 0.4$ (node). After Ref. [2].

$$\begin{aligned}\dot{C}_u(t) &= -\frac{|\mu_A|^2\omega_A}{2\pi\hbar c_0\varepsilon_1}\int_0^t dt' \int_{-\infty}^{\infty} d\omega_j\, e^{i(\omega_A-\omega_j)(t-t')}C_u(t')\\ &\times \sum_{m=-\infty}^{\infty}\left\{\frac{c_0\gamma_c/d}{\gamma_c^2+(\omega_j-\omega_{cm}^a)^2}\sin^2 k_{1j}z_A + \frac{c_0\gamma_c/d}{\gamma_c^2+(\omega_j-\omega_{cm}^b)^2}\cos^2 k_{1j}z_A\right\}\\ &= -\frac{A_0}{\pi t_r}\int_0^t dt' \int_{-\infty}^{\infty} d\omega_j\, e^{i(\omega_A-\omega_j)(t-t')}C_u(t')\\ &\times \sum_{m=-\infty}^{\infty}\left\{\left(\frac{i}{\omega_j-\omega_{cm}^a+i\gamma_c}+\text{C.C.}\right)\right.\\ &\times \sin^2 k_{1j}z_A + \left.\left(\frac{i}{\omega_j-\omega_{cm}^b+i\gamma_c}+\text{C.C.}\right)\cos^2 k_{1j}z_A\right\}\end{aligned} \tag{13.33}$$

where C.C. denotes the complex conjugate. This equation contains contributions from all the cavity resonant modes. When the spectrum of $C_u(t)$ is limited, we may choose several cavity modes around the atomic frequency ω_A. Further, if the atom decays slowly during one round-trip time, and if for some ω_{cm} both the cavity half-width γ_c and the detuning $\omega_A - \omega_{cm}$ are small compared to the cavity mode spacing $\Delta\omega_c = \pi c_1/(2d)$, that is, if

$$A_0 \ll 1/t_r, \qquad \gamma_c \ll \Delta\omega_c, \qquad |\omega_A-\omega_{cm}| \ll \Delta\omega_c \tag{13.34}$$

then we may choose only the cavity mode ω_{cm}. For $\omega_{cm} = \omega_{cm}^a$, we have in the single resonant mode limit

$$\begin{aligned}\dot{C}_u(t) &= -\frac{A_0}{\pi t_r}\int_0^t dt'\, C_u(t')\int_{-\infty}^{\infty} d\omega_j\left(\frac{i}{\omega_j-\omega_{cm}^a+i\gamma_c}+\text{C.C.}\right)\\ &\times \sin^2 k_{1j}z_A e^{i(\omega_A-\omega_j)(t-t')}\end{aligned} \tag{13.35}$$

(For $\omega_{cm} = \omega_{cm}^b$ we have $\cos^2 k_{1j}z_A$ instead of $\sin^2 k_{1j}z_A$ in Equation 13.35.) In the integrand we have a pole at $\omega_j = \omega_{cm}^a - i\gamma_c \simeq \omega_{cm}^a$. Since we are not interested in the variation during a time of order $t_r \sim z_A/c_1$, we can take $\sin^2 k_{1j}z_A$ outside of the integral concerning ω_j and set it equal to $\sin^2(\omega_{cm}^a z_A/c_1)$. Thus, by the contour integral on the lower half-plane of ω_j and evaluation of the residue, we have

$$\dot{C}_u(t) = -(2A_0/t_r)\sin^2(\omega_{cm}^a z_A/c_1)\int_0^t dt'\, C_u(t')e^{i(\omega_A-\omega_{cm}^a+i\gamma_c)(t-t')} \tag{13.36}$$

Differentiation with respect to time t yields

$$\begin{aligned}&\ddot{C}_u(t) - i(\omega_A-\omega_{cm}^a+i\gamma_c)\dot{C}_u(t) + (2A_0/t_r)\\ &\quad\times \sin^2(\omega_{cm}^a z_A/c_1)C_u(t) = 0\end{aligned} \tag{13.37a}$$

which can be formally written as

$$\begin{aligned}
&\ddot{C}_u(t) + (\gamma_c + i\Delta)\dot{C}_u(t) + (\Omega_R/2)^2 C_u(t) = 0 \\
&\Delta = \omega_{cm}^a - \omega_A \\
&(\Omega_R/2)^2 = \begin{cases} (2A_0/t_r)\sin^2(\omega_{cm}^a z_A/c_1), & \omega_{cm} = \omega_{cm}^a \\ (2A_0/t_r)\cos^2(\omega_{cm}^b z_A/c_1), & \omega_{cm} = \omega_{cm}^b \end{cases}
\end{aligned} \tag{13.37b}$$

where we have added the result for $\omega_{cm} = \omega_{cm}^b$. From the initial condition, Equation 13.6, we have $C_u\ (0) = 1$ and $\dot{C}_u(0) = 0$ (see Equation 13.11), so that the solution to Equation 13.37b is

$$\begin{aligned}
C_u(t) &= \frac{\lambda_2 e^{\lambda_1 t} - \lambda_1 e^{\lambda_2 t}}{\lambda_2 - \lambda_1} \\
\lambda_{1,2} &= \frac{1}{2}\left\{-(\gamma_c + i\Delta) \pm \sqrt{(\gamma_c + i\Delta)^2 - \Omega_R^2}\right\}
\end{aligned} \tag{13.38}$$

where the + sign is for λ_1 and the – sign for λ_2.

We examine two cases. The first is the underdamped case where $\gamma_c < \Omega_R$. For simplicity we assume that $\Delta = 0$. Then we have

$$\begin{aligned}
&C_u(t) \\
&= \frac{-\left(\gamma_c + i\sqrt{\Omega_R^2 - \gamma_c^2}\right)e^{-\left\{\left(\gamma_c - i\sqrt{\Omega_R^2-\gamma_c^2}\right)/2\right\}t} + \left(\gamma_c - i\sqrt{\Omega_R^2 - \gamma_c^2}\right)e^{-\left\{\left(\gamma_c + i\sqrt{\Omega_R^2-\gamma_c^2}\right)/2\right\}t}}{-2i\sqrt{\Omega_R^2 - \gamma_c^2}} \\
&= \frac{e^{-\gamma_c t/2}}{-2i\sqrt{\Omega_R^2-\gamma_c^2}}\left\{-2i\sqrt{\Omega_R^2-\gamma_c^2}\cos\left(\frac{\sqrt{\Omega_R^2-\gamma_c^2}}{2}t\right) - 2i\gamma_c \sin\left(\frac{\sqrt{\Omega_R^2-\gamma_c^2}}{2}t\right)\right\}
\end{aligned} \tag{13.39}$$

In the limit $\gamma_c \ll \Omega_R$ we have a damped oscillation

$$|C_u(t)|^2 = e^{-\gamma_c t}\cos^2\left(\frac{1}{2}\Omega_R t\right) = \frac{1}{2}e^{-\gamma_c t}\{1 + \cos(\Omega_R t)\} \tag{13.40}$$

The oscillation frequency Ω_R is known as the Rabi frequency. Note that the decay rate is the same as the decay rate of the cavity field amplitude but is half that of the cavity field energy. The damped Rabi oscillation was derived by Sachdev [7] using a different method, a reservoir method in a single mode context. The regime $\gamma_c \ll \Omega_R$ is known as the strong coupling regime in the field of cavity quantum electrodynamics.

The second case is the overdamped case where $\gamma_c > \Omega_R$. Then we have, assuming again that $\Delta = 0$,

$$C_u(t) = \frac{-\left(\gamma_c+\sqrt{\gamma_c^2-\Omega_R^2}\right)e^{-\left\{\left(\gamma_c-\sqrt{\gamma_c^2-\Omega_R^2}\right)/2\right\}t}+\left(\gamma_c-\sqrt{\gamma_c^2-\Omega_R^2}\right)e^{-\left\{\left(\gamma_c+\sqrt{\gamma_c^2-\Omega_R^2}\right)/2\right\}t}}{-2\sqrt{\gamma_c^2-\Omega_R^2}} \quad (13.41)$$

In the limit $\gamma_c \gg \Omega_R$ we have

$$|C_u(t)|^2 = e^{-\left(\gamma_c-\sqrt{\gamma_c^2-\Omega_R^2}\right)t} = \exp\left(-\frac{\Omega_R^2}{2\gamma_c}t\right) = \begin{cases} \exp\left(-\frac{4A_0\sin^2(\omega_{cm}^a z_A/c_1)}{\gamma_c t_r}t\right), & \omega_{cm}=\omega_{cm}^a \\ \exp\left(-\frac{4A_0\cos^2(\omega_{cm}^a z_A/c_1)}{\gamma_c t_r}t\right), & \omega_{cm}=\omega_{cm}^b \end{cases} \quad (13.42)$$

At a node where $\sin^2(\omega_{cm}^a z_A/c_1)=1$ for $\omega_{cm}=\omega_{cm}^a$ or $\cos^2(\omega_{cm}^a z_A/c_1)=1$ for $\omega_{cm}=\omega_{cm}^b$, the spontaneous decay rate is enhanced by a factor

$$f = \frac{4}{\gamma_c t_r} = \frac{4}{\pi}\frac{\Delta\omega_c}{2\gamma_c} = \frac{4}{\pi}F \quad (13.43a)$$

where the cavity mode spacing $\Delta\omega_c = c_1\pi/(2d)$ and F is the finesse of the cavity. When the mode number m is 1, or the mode is the lowest resonant mode, we have

$$f = \frac{4}{\gamma_c t_r} = \frac{4}{\pi}Q \quad (13.43b)$$

where Q is the cavity quality factor. Thus in an overdamped cavity the spontaneous emission rate is enhanced by roughly the finesse or the cavity quality factor. The enhancement by the so-called Purcell factor, $f = 3Q\lambda^3/4\pi^2 V \sim Q$, where V is the volume of the cavity and λ the transition wavelength, was predicted by Purcell [8] for a cavity used in the radio frequency. The regime $\gamma_c \gg \Omega_R$ is known as the weak coupling regime.

Note that the enhancement factor in Equation 13.43a is different from the excess noise factor in Equation 12.1a applicable to the present symmetric cavity model, which reads

$$K_L = \left\{\frac{(1-r^2)/2r}{\ln(1/r)}\right\}^2 \quad (13.44)$$

If we have $\sin^2(\omega_{cm}^a z_A/c_1)=0$ $(\cos^2(\omega_{cm}^a z_A/c_1)=0)$ in Equation 13.42, which occurs for an atom at a node, the spontaneous emission is inhibited as the decay rate becomes 0. This occurs in the single-mode limit in general, as seen from Equation 13.36, where for $\omega_{cm}=\omega_{cm}^a$ we have $\dot{C}_u=0$ for $\sin^2(\omega_{cm}^a z_A/c_1)=0$. (This zero decay rate apparently contradicts the statement below the delay differential equation, Equation 13.30, that a natural decay lasts until $t=\min(t_1,t_2)$. Of course, a natural decay lasts until $t=\min(t_1,t_2)$. But after that, the alternate terms in Equation 13.30 destructively add to inhibit all-over decay when the zero-decay condition in this single-mode limit holds (see, for example, curve B in Figure 13.1). A different kind of inhibition of spontaneous emission, inhibition by a smaller cavity size than the radiation wavelength, was predicted by Kleppner [9].

13.6 Spontaneous Emission Spectrum Observed Outside the Cavity

We consider the observation of spontaneously emitted light at a detector outside the cavity. Regardless of the location of the detector, the detected intensity is proportional to [10]

$$I(z,t) = \langle \varphi(t)|\hat{E}^{(-)}(z)\hat{E}^{(+)}(z)|\varphi(t)\rangle \tag{13.45}$$

Substituting Equation 13.5 for the wavefunction, the expression from Equation 2.19a for the positive frequency part of the electric field operator,

$$\hat{E}^{(+)}(z) = i\sum_j \sqrt{\frac{\hbar\omega_j}{2}}\,U_j(z)\hat{a}_j$$

and its adjoint into Equation 13.45, we obtain

$$\begin{aligned} I(z,t) &= \left(C_u^*(t)\langle u|\langle 0|e^{i\omega_A t} + \sum_{i'} C_{li'}^*(t)\langle l|\langle 1_{i'}|\,e^{i\omega_{i'} t}\right) \sum_i \sqrt{\frac{\hbar\omega_i}{2}}\,U_i(z)\hat{a}_i^\dagger \\ &\times \sum_j \sqrt{\frac{\hbar\omega_j}{2}}\,U_j(z)\hat{a}_j \left(C_u(t)|u\rangle|0\rangle e^{-i\omega_A t} + \sum_{j'} C_{lj'}(t)|l\rangle\,|1_{j'}\rangle e^{-i\omega_{j'} t}\right) \\ &= \sum_i\sum_j \frac{\hbar}{2}\sqrt{\omega_i\omega_j}\,U_i(z)U_j(z)C_{li}^*(t)C_{lj}(t)e^{i(\omega_i-\omega_j)t} \\ &= \left|\sum_j \sqrt{\frac{\hbar\omega_j}{2}}\,U_j(z)C_{lj}(t)e^{-i\omega_j t}\right|^2 \end{aligned} \tag{13.46}$$

where $C_{lj}(t)$ is given by Equations 13.11 and 13.6 as

$$C_{lj}(t) = \sqrt{\frac{\omega_j}{2\hbar}}\,U_j(z_A)\mu_A^* \int_0^t dt'\, C_u(t')e^{-i(\omega_A-\omega_j)t'} \tag{13.47}$$

Thus the intensity at the observation point z_B ($>d$) is

$$\begin{aligned} I(z_B,t) &= \left|\sum_j \mu_A^* \frac{\omega_j}{2}\,U_j(z_B)U_j(z_A)\int_0^t dt'\, C_u(t')e^{i(\omega_j-\omega_A)(t'-t)}\right|^2 \\ &= \left|\frac{\mu_A^*}{2}\int_0^t dt'\, C_u(t')\sum_j \omega_j\left\{U_j^a(z_B)U_j^a(z_A) + U_j^b(z_B)U_j^b(z_A)\right\}e^{i(\omega_j-\omega_A)(t'-t)}\right|^2 \end{aligned} \tag{13.48}$$

We use Equations 1.58a and 1.58b with Equations 1.65a and 1.65b for the functions $U_j^a(z_B)$, $U_j^a(z_A)$, $U_j^b(z_B)$, and $U_j^b(z_A)$. We also use the Fourier series expansion of the normalization constants in Equation 1.70b. Then, by a similar procedure as in Equations 13.25–13.30, we have

$$I(z_B,t) = \frac{\omega_A^2|\mu_A|^2}{4\varepsilon_0\varepsilon_1 c_0 c_1}(1-r^2)|f(z_A,z_B,t)|^2 \tag{13.49a}$$

where

$$\begin{aligned} f(z_A,z_B,t)=e^{-i\omega_A t}\Bigg[&\sum_{n=0}^{\infty} r^{2n}e^{i\omega_A(nt_r+t_R)}C_u(t-nt_r-t_R)H(t-nt_r-t_R)\\ &+\sum_{n=0}^{\infty} r^{2n+1}e^{i\omega_A(nt_r+t_L)}C_u(t-nt_r-t_L)H(t-nt_r-t_L)\Bigg] \quad (z_B>d) \end{aligned} \tag{13.49b}$$

and

$$\begin{aligned} t_r &= \frac{4d}{c_1}\\ t_R &= \frac{d-z_A}{c_1}+\frac{z_B-d}{c_0} \qquad (z_B>d)\\ t_L &= \frac{t_r}{2}+\frac{d+z_A}{c_1}+\frac{z_B-d}{c_0} \end{aligned} \tag{13.49c}$$

The retardation time t_R is the time required for an optical wave to go directly from z_A to z_B. The time t_L is the time required to go from z_A to z_B after reflection at the left interface at $z=-d$. The equation for $z_B<-d$ can be obtained from Equations 13.49a and 13.49b by replacing z_A and z_B in Equation 13.49c by $-z_A$ and $-z_B$, respectively. The "intensity" in Equation 13.49a has a simple interpretation: the field amplitude at z_B and at t is made up of discrete contributions that were "emitted" with the strength of the probability amplitudes C_u at respective retarded times and underwent a phase shift as well as a reduction caused during the trip associated with multiple reflections. Note that the field amplitude at z_B and at t can be regarded as being proportional to $f(z_A,z_B,t)$ in Equation 13.49b. Then the observed spectrum can be obtained as the absolute square of the Fourier transform of the field amplitude. Thus we have the power spectrum $S(z_A,z_B,\omega)$

$$S(z_A,z_B,\omega) = \frac{\omega_A^2|\mu_A|^2}{4\varepsilon_0\varepsilon_1 c_0 c_1}(1-r^2)\left|\int_{-\infty}^{\infty} dt f(z_A,z_B,t)e^{i\omega t}\right|^2 \tag{13.50}$$

Here we define the time history of the emitter amplitude as

$$D(t) = e^{-i\omega_A t}C_u(t)H(t) \tag{13.51}$$

Then, the term of $e^{-i\omega_A t}r^{2n}e^{i\omega_A(nt_r+t_R)}C_u(t-nt_r-t_R)H(t-nt_r-t_R)$, for example, is transformed as

$$\begin{aligned}
&\int_{-\infty}^{\infty} dt\, e^{i\omega t} e^{-i\omega_A t} r^{2n} e^{i\omega_A(nt_r+t_R)} C_u(t-nt_r-t_R) H(t-nt_r-t_R) \\
&= r^{2n} e^{i\omega(nt_r+t_R)} \int_{-\infty}^{\infty} dt\, e^{i\omega(t-nt_r-t_R)} e^{-i\omega_A(t-nt_r-t_R)} C_u(t-nt_r-t_R) H(t-nt_r-t_R) \\
&= r^{2n} e^{i\omega(nt_r+t_R)} D(\omega)
\end{aligned} \tag{13.52}$$

where $D(\omega)$ is the Fourier transform of $D(t)$. Then, carrying out the summation over n, the resultant spectrum is obtained as

$$\begin{aligned}
S(z_A, z_B, \omega) &= \frac{\omega_A^2 |\mu_A|^2}{4\varepsilon_0 \varepsilon_1 c_0 c_1} T |D(\omega)|^2 \\
&\quad \times \left[\frac{1 + R + 2\sqrt{R}\cos\{2\omega(z_A+d)/c_1\}}{1 + R^2 - 2R\cos(4\omega d/c_1)} \right] \qquad (z_B > d)
\end{aligned} \tag{13.53}$$

where $R = r^2$ and $T = 1 - r^2$. For $z_B < -d$ the sum $z_A + d$ should be replaced by $d - z_A$.

Note that the spectrum is independent of the location z_B of the detector, as expected from the one-dimensional nature of the process being considered. Note also that the power spectrum is not determined simply by the spectrum of the "emitter" history $D(\omega)$. It depends also on the quantity in the square bracket, which is dependent on the cavity structure and the atomic location. This factor represents how the emitted field is transferred to the observation point. In fact, this factor is proportional to the absolute square of the response function defined in Equation 2.53a for the present case of the two-sided cavity with source point inside and observation point outside the cavity:

$$Y(z_B, z_A, \omega) = -\frac{1+r}{2\varepsilon_1 c_1} \left[\frac{1 + r\exp\{2i\omega(z_A+d)/c_1\}}{1 - r^2\exp(4i\omega d/c_1)} \right] e^{i\omega t_R} \qquad (z_B > d) \tag{13.54}$$

For $z_B < -d$, the sum $z_A + d$ should be replaced by $d - z_A$. This response function can be derived by use of the mode functions in Equations 1.58 and 1.65 and the Fourier series expansion of the normalization constants in Equation 1.70b and by performing principal part integrations. The evaluation can be done term by term. One has two geometric progressions, which can easily be summed. The response function in Equation 13.54 can also be obtained intuitively by a classical consideration, as follows. One assumes a current source $J\exp(-i\omega t)\delta(z - z_A)$ inside the two-sided cavity, then we have an induced electric field $E\exp(-i\omega t) = -JZ\exp(-i\omega t)$ at $z = z_A$ where $Z = \sqrt{\mu_0/\varepsilon_1} = 1/(\varepsilon_1 c_1)$ is the space impedance. Because the problem is one dimensional here, this electric field amplitude is transmitted to the two sides with half the magnitude, that is, with $-(JZ/2)\exp(-i\omega t)$. These waves are transmitted to z_B ($> d$) directly or after a single or multiple reflections at the coupling surfaces with respective phase changes and amplitude reductions. All the contributions have associated transmission coefficient $1 + r$ at the coupling surface at $z = d$. The two geometrical progressions thus obtained easily yield Equation 13.54 if we divide the resultant total field at z_B ($< d$) by $J\exp(-i\omega t)$.

Now using Equation 13.54 in Equation 13.53 and noting that $\varepsilon_1 c_1 = \varepsilon_0 c_0 n = \varepsilon_0 c_0 (1+r)/(1-r)$, we can write

$$S(z_A, z_B, \omega) = \omega_A^2 |\mu_A|^2 |D(\omega)|^2 |Y(z_B, z_A, \omega)|^2 \tag{13.55}$$

That this is a general formula not restricted to the present two-sided cavity can be shown by using the first equation in Equation 13.48. We show the derivation in Appendix H. Figure 13.2 shows an example of the calculated spectrum to be observed outside the cavity. Figure 13.2a is $|D(\omega)|^2$ multiplied by ω_c^2, Figure 13.2b is the absolute squared response function $|Y(z_B, z_A, \omega)|^2$ multiplied by $\varepsilon_0^2 c_0^2$, and Figure 13.2c is the intensity spectrum observed outside the cavity, in arbitrary units. The parameters are the same as for curve A in Figure 13.1. In Figure 13.2b,c the solid and dashed curves are for $z_B < d$ and $z_B < -d$, respectively. Two peaks corresponding to the Rabi oscillation with $\Omega_R = 0.32\Delta\omega_c$ are seen. In the figure ω_c stands for the intermode spacing $\Delta\omega_c$.

13.7 Extension to Three Dimensions

The three-dimensional analysis of a laser with spatially distributed active atoms is difficult in general, even if the cavity has a simple structure such as the planar ones considered so far in this book. This is because the propagation of the emitted field from one atom to the next involves three-dimensional effects concerning the direction of propagation and the direction of field polarization. The influence of the transmitted field on the receiving atom is dependent on the atom's polarization direction. Thus it is extremely difficult to write down consistent equations of motion for the field and the atoms when the atoms are distributed in three-dimensional space.

On the other hand, the process of spontaneous emission usually involves only a single atom. In this case a three-dimensional analysis in free space is well established. The spontaneous emission process in a cavity can also be analyzed three dimensionally if the cavity structure is simple and the relevant field mode functions are available. Here we briefly describe the case where the cavity is a dielectric slab, or a simple two-sided cavity, as was considered one dimensionally in Section 1.3.2 and in the previous sections in this chapter. The description will follow Ho and Ujihara [6].

The cavity is composed of a dielectric slab extending in the region $-d \le z \le d$. The dielectric has dielectric constant ε_1 and refractivity n, with the velocity of light inside the dielectric being c_1. The outside regions are vacuum with dielectric constant ε_0 and the velocity of light c_0. The slab is assumed to have infinite extents in the x- and y-directions, and the atom of transition frequency ω_A and dipole operator $\hat{\boldsymbol{\mu}}$ is positioned inside the cavity at $\mathbf{r}_A = (0,0,z_A)$. The mode functions of the three-dimensional "universe" $\mathbf{U}_j(\mathbf{r})$ are defined by imposing periodic boundary conditions in the x-, y-, and z-directions with periods L_x, L_y, and $L_z + 2d \approx L_z$, respectively. They are normalized so that

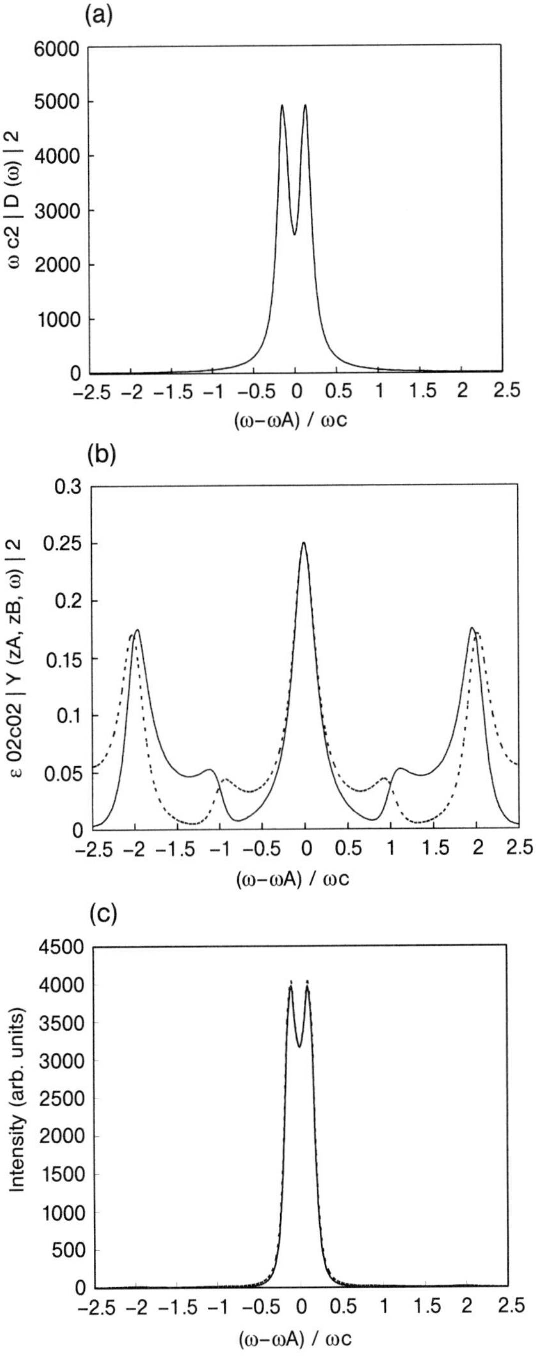

Figure 13.2 The spectra of (a) $D(t)$, (b) the absolute squared response function $|Y(z_B, z_A, \omega)|^2$, and (c) the emission spectrum $S(z_A, z_B, t)$ observed outside the cavity. After Ref. [2].

$$\int_V \varepsilon(\mathbf{r})\mathbf{U}_i(\mathbf{r})\mathbf{U}_j(\mathbf{r})d\mathbf{r} = \delta_{ij} \tag{13.56}$$

where $V = L_x L_y L_z$. As before, the mode functions inside the cavity and outside are denoted as $\mathbf{U}_{1j}$ and $\mathbf{U}_{0j}$, respectively, and will be given in terms of the relevant mode wavevectors $\mathbf{k}_{1,0} = (k_x, k_y, k_{1,0z})$ for inside and outside the cavity and their projections onto the x–y plane $\mathbf{k}_p = (k_x, k_y, 0)$ in addition to the mode index j. The mode functions are categorized into TE and TM mode functions of even and odd symmetries in the z-direction, and they are further classified into even and odd modes in the x–y direction.

The odd x–y TE mode functions are given, suppressing the mode index j, as

$$\mathbf{U}_{1,0}(\mathbf{r}) = \alpha(\hat{x}k_y - \hat{y}k_x)\sin(\mathbf{k}_p \cdot \mathbf{r})u_{1,0}(z) \tag{13.57}$$

where $u_{1,0}(z)$ and the normalization constant α for odd z TE modes are

$$\begin{aligned} u_1^{ozTE}(z) &= \sin k_{1z}z \\ u_0^{ozTE}(z) &= \sin k_{1z}d\cos k_{0z}(z-d) + \frac{k_{1z}}{k_{0z}}\cos k_{1z}d\sin k_{0z}(z-d) \end{aligned} \tag{13.58}$$

and

$$\begin{aligned} \alpha^{ozTE} &= \frac{1}{k_p}\frac{k_{0z}}{k_{1z}}\frac{2}{\sqrt{\varepsilon_0 V(1 - K\sin^2 k_{1z}d)}} \\ K &= 1 - \left(\frac{k_{0z}^2}{k_{1z}^2}\right) \end{aligned} \tag{13.59}$$

and for even z TE modes

$$\begin{aligned} u_1^{ezTE}(z) &= \cos k_{1z}z \\ u_0^{ezTE}(z) &= \cos k_{1z}d\cos k_{0z}(z-d) - \frac{k_{1z}}{k_{0z}}\sin k_{1z}d\sin k_{0z}(z-d) \end{aligned} \tag{13.60}$$

and

$$\begin{aligned} \alpha^{ezTE} &= \frac{1}{k_p}\frac{k_{0z}}{k_{1z}}\frac{2}{\sqrt{\varepsilon_0 V(1 - K\cos^2 k_{1z}d)}} \\ K &= 1 - \left(\frac{k_{0z}^2}{k_{1z}^2}\right) \end{aligned} \tag{13.61}$$

The even x–y TE mode functions are obtained by replacing $\sin(\mathbf{k}_p \cdot \mathbf{r})$ by $\cos(\mathbf{k}_p \cdot \mathbf{r})$ in Equation 13.57.

The odd x–y TM mode functions are given, again suppressing the mode index j, as

$$\mathbf{U}_{1,0}(\mathbf{r}) = \alpha\left[(\hat{x}k_x + \hat{y}k_y)\sin(\mathbf{k}_p\cdot\mathbf{r})\frac{d}{dz}u_{1,0}(z) - \hat{z}k_p^2\cos(\mathbf{k}_p\cdot\mathbf{r})u_{1,0}(z)\right] \tag{13.62}$$

where $u_{1,0}$ (z) and the normalization constant α for odd z TM modes are

$$\begin{aligned} u_1^{ozTM}(z) &= \sin k_{1z}z \\ u_0^{ozTM}(z) &= \sin k_{1z}d\cos k_{0z}(z-d) + \frac{k_{1z}}{k_{0z}}\frac{k_0^2}{k_1^2}\cos k_{1z}d\sin k_{0z}(z-d) \end{aligned} \tag{13.63}$$

and

$$\begin{aligned} \alpha^{ozTM} &= \frac{1}{k_pk_0}\frac{k_{0z}}{k_{1z}}\frac{k_1^2}{k_0^2}\frac{2}{\sqrt{\varepsilon_0 V(1-K'\sin^2 k_{1z}d)}} \\ K' &= 1-\left(\frac{k_{0z}^2}{k_{1z}^2}\frac{k_1^4}{k_0^4}\right) \end{aligned} \tag{13.64}$$

and for even z TM modes are

$$\begin{aligned} u_1^{ezTM}(z) &= \cos k_{1z}z \\ u_0^{ezTM}(z) &= \cos k_{1z}d\cos k_{0z}(z-d) - \frac{k_{1z}}{k_{0z}}\frac{k_0^2}{k_1^2}\sin k_{1z}d\sin k_{0z}(z-d) \end{aligned} \tag{13.65}$$

and

$$\begin{aligned} \alpha^{ezTM} &= \frac{1}{k_pk_0}\frac{k_{0z}}{k_{1z}}\frac{k_1^2}{k_0^2}\frac{2}{\sqrt{\varepsilon_0 V(1-K'\cos^2 k_{1z}d)}} \\ K' &= 1-\left(\frac{k_{0z}^2}{k_{1z}^2}\frac{k_1^4}{k_0^4}\right) \end{aligned} \tag{13.66}$$

The even x–y TM mode functions are obtained by replacing $\sin(\mathbf{k}_p\cdot\mathbf{r})$ in front of $(d/dz)u_{1,0}(z)$ by $\cos(\mathbf{k}_p\cdot\mathbf{r})$ and $-\cos(\mathbf{k}_p\cdot\mathbf{r})$ in front of $u_{1,0}$ (z) by $\sin(\mathbf{k}_p\cdot\mathbf{r})$ in Equation 13.62.

The reader may notice some similarities of the mode functions and the normalization constants in these equations and in those for the one-dimensional versions in Sections 1.3.2 and 1.4. The equations here reduce to those in Section 1.3.2 in the limit $k_p \to 0$, whence the distinction between the TE and TM modes disappears.

The normalization constants have similar Fourier series expansions as those in Equation 1.70b:

$$
\begin{aligned}
\left(\alpha^{ezTE}\right)^2 &= \frac{4}{\varepsilon_0 V}\frac{1}{k_p^2}\frac{k_{0z}}{k_{1z}}\left\{1+2\sum_{n=1}^{\infty} r^n \cos 2nk_{1z}d\right\} \\
r &= \frac{k_{1z}-k_{0z}}{k_{1z}+k_{0z}} \\
\left(\alpha^{ozTM}\right)^2 &= \frac{4}{\varepsilon_0 V}\frac{1}{k_p^2 k_0^2}\frac{k_{0z}}{k_{1z}}\frac{k_1^2}{k_0^2}\left\{1+2\sum_{n=1}^{\infty} (r')^n \cos 2nk_{1z}d\right\} \\
r' &= \frac{k_{0z}k_1^2-k_{1z}k_0^2}{k_{0z}k_1^2+k_{1z}k_0^2}
\end{aligned}
\qquad (13.67)
$$

Expansions for odd z TE and even z TM modes are obtained by replacing r and r' by $-r$ and $-r'$, respectively.

From now on, we assume for simplicity that the dielectric constant ε_1 of the slab is equal to the vacuum dielectric constant ε_0 but the reflectivities in the Fourier series expansions in Equation 13.67 are kept finite. Moreover, we assume that $r = r'$ and that they are independent of the **k**-vector orientation. This assumption makes the three-dimensional summation over mode j tractable.

Now that we have the relevant mode functions, we can formulate the spontaneous emission process as in Section 13.1 with the Hamiltonian and the wavefunction

$$
\begin{aligned}
\hat{H}_t &= \hat{H}_f + \hat{H}_a + \hat{H}_{int} \\
\hat{H}_{int} &= -i\sum_j \sqrt{\frac{\hbar\omega_j}{2}}\mathbf{U}_j(\mathbf{r}_A)\cdot\hat{\boldsymbol{\mu}}(\hat{a}_j-\hat{a}_j^{\dagger})
\end{aligned}
\qquad (13.68)
$$

$$
|\varphi(t)\rangle = C_u(t)|u\rangle|0\rangle e^{-i\omega_A t} + \sum_j C_{lj}(t)|l\rangle\left|1_j\right\rangle e^{-i\omega_j t} \qquad (13.69)
$$

where the interaction Hamiltonian is rewritten for the three-dimensional mode functions and a vector dipole moment operator.

With these formulations in hand, we can perform perturbative and non-perturbative analyses as before. For example, as in Section 13.4, one can derive the delay differential equation for the probability amplitude of the upper-state population:

$$
\begin{aligned}
\dot{C}_u(t) = &-\frac{A_0}{2}C_u(t)H(t) + \sum_{n=1}^{\infty} p_{2n}(\omega_A n t_r)C_u(t-nt_r)H(t-nt_r) \\
&+\frac{1}{2}\sum_{n=0}^{\infty}\left[p_{2n+1}\{\omega_A(nt_r+t_1)\}C_u(t-nt_r-t_1)H(t-nt_r-t_1).\right. \\
&\left.+p_{2n+1}\{\omega_A(nt_r+t_2)\}C_u(t-nt_r-t_2)H(t-nt_r-t_2)\right]
\end{aligned}
\qquad (13.70)
$$

where the spontaneous emission rate in a free vacuum is

$$A_0 = \frac{\omega_A^3 |\mu_A|^2}{3\pi\hbar c_0^3 \varepsilon_0} \tag{13.71}$$

and

$$\begin{aligned} p_n^{\|}(x) &= -\frac{3A_0}{2} r^n \left(\frac{1}{ix} + \frac{1}{x^2} - \frac{1}{ix^3} \right) e^{ix} \\ p_n^{\perp}(x) &= 3A_0(-r)^n \left(\frac{1}{x^2} - \frac{1}{ix^3} \right) e^{ix} \end{aligned} \tag{13.72}$$

The retardation times t_r, t_1, and t_2 are given by Equation 13.29 with c_1 replaced by c_0. The coefficient $p_n^{\|}(x)$ applies when the atomic dipole is oriented parallel to the x–y plane or to the mirror surfaces, and $p_n^{\perp}(x)$ applies when the dipole is perpendicular to the plane. For the derivation of the delay differential equation (Equation 13.70), the reader is referred to the paper by Ho and Ujihara [6]. If the atom is at the center of the cavity $z_A = 0$, we have

$$\dot{C}_u(t) = -\frac{A_0}{2} C_u(t) H(t) + \sum_{n=1}^{\infty} p_n(2k_A nd) C_u(t_n) H(t_n) \tag{13.73}$$

where

$$t_n = t - \frac{2nd}{c_0} \tag{13.74}$$

Then the solution is obtained as in Equation 13.32a

$$C_u(t) = \sum_{n=0}^{\infty} \left[\sum \frac{p_1^{a_1}(2k_A d) p_2^{a_2}(4k_A d) \cdots p_n^{a_n}(2nk_A d)}{a_1! a_2! \cdots a_n!} t_n^m \exp\left(-\frac{A_0}{2} t_n \right) \right] H(t_n) \tag{13.75a}$$

where the sum is over all non-negative integers a_i $(i = 1, 2, \ldots, n)$ that satisfy

$$1a_1 + 2a_2 + \cdots + na_n = n \tag{13.75b}$$

and

$$m = a_1 + a_2 + \cdots + a_n \tag{13.75c}$$

The non-perturbative results are obtained numerically using Equation 13.70 and assuming that the atomic dipole is oriented parallel to the x–y plane. One finds, for a very good cavity of length of $\lambda_A/2$, a Rabi-type oscillation in the upper atomic population. For longer cavities with lengths $2 \times (\lambda_A/2)$, $3 \times (\lambda_A/2)$, . . . , the spontaneous decay rate approaches that in free vacuum.

The perturbation approximation result is obtained by replacing the C_u on the RHS of Equation 13.70 by the initial value $C_u(0) = 1$. Then the spontaneous emission rate R can be obtained by using the relation

$$R = -\frac{d}{dt}|C_u|^2 \tag{13.76}$$

as

$$R = A_0 - 2\mathrm{Re}\sum_{n=1}^{\infty} p_{2n}(\omega_A n t_r) - \mathrm{Re}\sum_{n=0}^{\infty}\Big[p_{2n+1}\{\omega_A(nt_r + t_1)\}$$
$$+ p_{2n+1}\{\omega_A(nt_r + t_2)\}\Big] \tag{13.77}$$

The spontaneous emission rate based on the perturbation approximation can of course be obtained as in Equations 13.13–13.16:

$$R = \sum_j \frac{\pi\omega_j}{\hbar}\left|\mathbf{U}_j(\mathbf{r}_A)\cdot\hat{\boldsymbol{\mu}}\right|^2 \delta(\omega_j - \omega_A) \tag{13.78}$$

That this is equal to Equation 13.77 may be proved by going to summation over the categories and to three-dimensional integration over the $\mathbf{k}_j$-vector and by using the Fourier series expansion of the normalization constants.

Equation 13.77 shows directly that $R \to A_0$ as $r \to 0$, that is, as the reflectivity vanishes, the spontaneous emission rate becomes equal to that in free vacuum. The reader may show also that the rate in Equation 13.78 reduces to A_0 in the limit $r \to 0$.

A perturbation result based on Equation 13.78 was derived by De Martini *et al.* [11] by using traveling-wave mode functions associated with a cavity composed of two, parallel, infinitely thin mirrors of different complex reflectivities. They gave compact expressions for the spontaneous emission rate for high-Q cavities and detailed numerical results on decay rate dependence on various parameters. They also gave experimental results on inhibition and enhancement of spontaneous emission using a Fabry–Perot microcavity with europium atoms in dibenzoylmethane complex as the emitter.

The field intensity observed at an arbitrary location outside the cavity can be obtained as for the one-dimensional case in Section 13.6. One can examine the intensity at the observation point $\mathbf{r}_B$, $I(\mathbf{r}_B, t) = \langle\varphi(t)|\hat{E}^{(-)}(\mathbf{r}_B)\hat{E}^{(+)}(\mathbf{r}_B)|\varphi(t)\rangle$, just as in Equation 13.48, which can be evaluated analytically for the summation over the field modes j and numerically for $C_u(t)$ (see [6]).

One may wonder if the present cavity model, where the mirrors extend infinitely in transverse directions, can adequately simulate actual plane parallel cavities of finite transverse size. Specifically, one may wonder if the present model does not ignore the spillover of the optical field from the mirror edges. That the transverse extent of the spontaneously emitted fields is, in important cases, finite and is of the order of the emitted wavelength multiplied by the square root of the cavity Q can be shown as follows.

One route to arrive at a finite transverse extent is given by De Martini *et al.* [12] through the uncertainty principle on the transverse position and momentum of the photon. For a planar Fabry–Perot cavity of length d and mirror reflectivities r_1 and r_2, the FWHM of the angular $\mathbf{k}$-vector distribution around the normally resonant **k**-vector is $\Delta\Theta_N = 2(fN)^{-1/2}$, where $f = \pi\sqrt{r_1 r_2}/(1 - r_1 r_2)$ is the finesse and $N = d/(\lambda/2)$ the mode number. By writing $\Delta p_x \Delta p_y = (\hbar k \Delta\Theta_N/2)^2$, one finds for cylindrical symmetry the expression for the transverse quantum correlation length $l_N = 2\lambda(fN)^{1/2} = 2\lambda\{N\pi\sqrt{r_1 r_2}/(1 - r_1 r_2)\}^{1/2}$.

Ujihara *et al.* [13] derived the mode radius through a different route. They compared the emission rate formula calculated for a symmetric planar cavity model with Purcell's enhancement factor. They used three-dimensional mode functions for the perturbation calculation. The calculated enhancement factor was $1/(2N)$ for an atom at an antinode for a normally resonant mode. Equating it to the Purcell factor $F \simeq (Q/4\pi)(\lambda^3/V)$, with $Q = N\pi/(1 - r^2)$, yields the mode radius $r_{\text{mode}} = \lambda[N/\{\pi(1 - r^2)\}]^{1/2}$.

The third simple route is to consider the photon mean free path in the transverse direction. The photon lifetime in the cavity is $t_c \approx d/\{c(1 - r_1 r_2)\}$, during which a photon propagating in the direction $\Delta\Theta_N/2$ traverses along the mirror surfaces a distance

$$l_c = ct_c\Delta\Theta_N/2 = (\lambda/2)\left[N/\{\pi\sqrt{r_1 r_2}(1 - r_1 r_2)\}\right]^{1/2}$$

For a good cavity, these three expressions agree except for numerical factors and give the transverse extent $\sim\sqrt{Q}\lambda$, which is usually smaller than an actual mirror size.

13.8
Experiments on Spontaneous Emission in a Fabry–Perot Type Cavity

Goy *et al.* [14] first observed strong shortening of the spontaneous emission lifetime of Rydberg atoms when they are made to cross a high-Q superconducting, resonant cavity. The experiment was performed with Na atoms at 340 GHz using a cavity of Q of the order of 10^6. Hulet *et al.* [15] observed inhibition of spontaneous emission from a Cs Rydberg atom at $\lambda = 0.45$mm when the atom is passed between two parallel metal plates separated by less than $\lambda/2$. The emission lifetime was increased by more than 20 times, which they attributed to the vanishing of available mode density for the radiation of relevant polarization at λ.

In the infrared region, Jhe *et al.* [16] observed similar inhibition of spontaneous emission. Using Cs atoms and passing them through a metal gap of $d = 1.1$ μm, they observed the decay of the atoms emitting at 3.49 μm. The hyperfine sublevels of the $5D_{5/2}$ state emit either σ or π polarized light corresponding, respectively, to photons polarized parallel and perpendicular to the metal surfaces. For a small gap with $d < \lambda/2$, the theoretical mode density is zero for σ polarization, while it is enhanced by a factor of $3\lambda/4d$ for π polarization. They confirmed that the emission rate of a σ emitting sublevel is reduced at most to 0.4 times the natural decay rate.

In the optical region, the observation of enhanced and inhibited spontaneous emission was made by De Martini *et al.* [17]. Using ethanol solution of tetraphenylnaphthacene dye in a 98–96% flat mirror cavity, they observed the lifetime

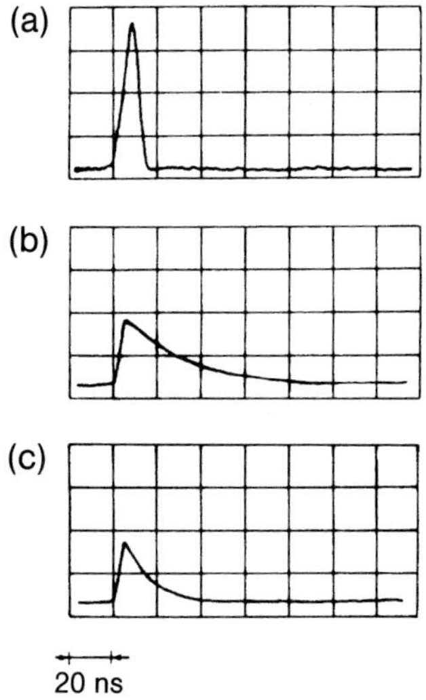

Figure 13.3 Oscilloscope traces showing (a) enhanced and (b) inhibited spontaneous emissions in a planar microcavity as compared to (c) the free-space decay.
Source: From Ref. [17]. De Martini, F., Innocenti, G., Jacobovitz, G.R., and Mataloni, P. (1987), *Phys. Rev. Lett.*, **59**, 2955, Figure 2.

of the fluorescence at 6328 Å. For cavity lengths of $\lambda/2$ and $\lambda/8$ they observed clear enhancement and suppression of the spontaneous decay. In Figure 13.3 the enhanced and inhibited spontaneous emissions as compared to the free-space decay are shown.

Heinzen *et al.* [18] observed slight changes in spontaneous emission when excited atoms of Yb crossed the focus of a 5 cm long confocal resonator. The resonator mirrors had transmissions of 2.8% and 1.8%. The fluorescence observed through one of the mirrors is enhanced by a factor of 19 when the cavity is tuned to the 556 nm fluorescence and is inhibited by a factor of 42 when it is detuned. This resulted in a fractional increase of 1.6% and decrease of 0.5% in the total emission rate.

Yamamoto *et al.* [19] observed spontaneous emission of free excitons in a GaAs quantum well that is embedded in a semiconductor microcavity composed of two distributed Bragg reflectors. When the fluorescence at 800 nm is resonant with the cavity and the quantum well is placed at the antinode of the cavity resonance, they observed an enhancement by a factor of 130, while, when the fluorescence is off-resonant and the quantum well is placed at a node, an inhibition by a factor of 30 was observed.

Norris *et al.* [20] studied the Rabi oscillation due to the two-dimensional excitons emitting in a $GaAs/Al_xGa_{1-x}As$ multiple quantum well embedded in a planar semiconductor microcavity of length $\lambda = 785$ nm and of finesse about 150. A two-dimensional exciton couples only with those field modes with the same in-plane

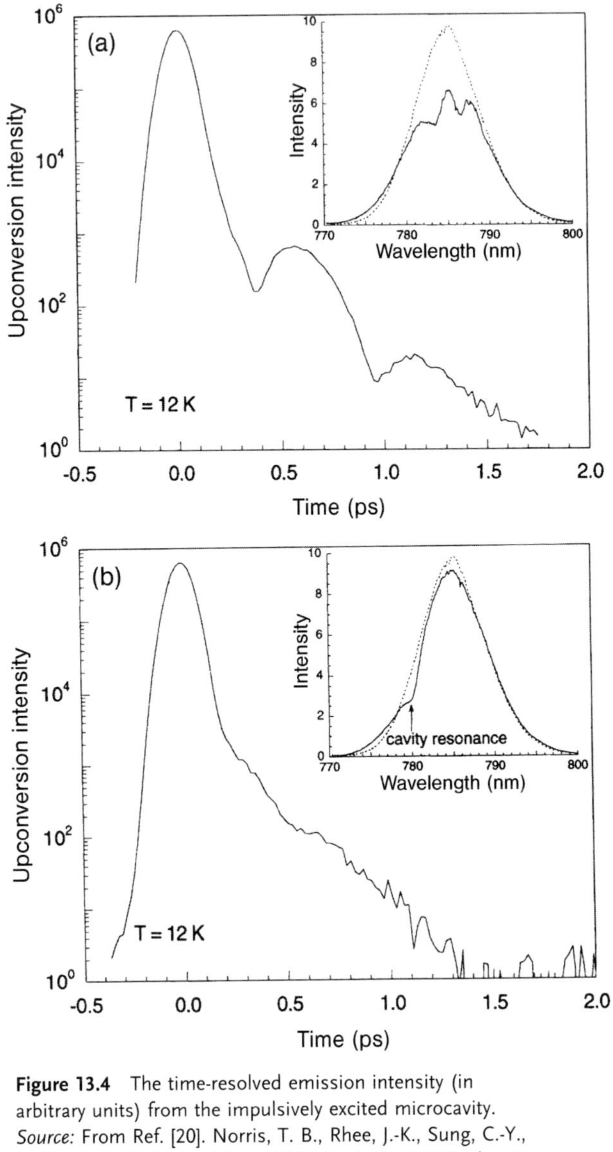

Figure 13.4 The time-resolved emission intensity (in arbitrary units) from the impulsively excited microcavity. *Source:* From Ref. [20]. Norris, T. B., Rhee, J.-K., Sung, C.-Y., Arakawa, Y., Nishioka, M., and Weisbuch, C. (1994) *Phys. Rev. B*, **50**, 14663, Figure 1.

wavevectors as its own, as contrasted to an atom, which couples with all oblique modes as long as the atom's polarization is not perpendicular to the mode polarization. They observed a damped Rabi oscillation with period of 600 fs, close to the expected value, and decay time approximately twice the cavity lifetime, $\sim 2 \times 140$ fs. In Figure 13.4 the time-resolved emission intensities from the impulsively excited cavity are shown. In Figure 13.4a, the cavity and exciton modes are near resonance. In Figure 13.4b, the cavity is detuned from the exciton resonance. In the inset, the dotted line shows the pump pulse spectrum, and the solid line shows the reflected pump spectrum, wherein the two dips correspond to the two normal modes of the system.

References

1 Feng, X.P. and Ujihara, K. (1990) *Phys. Rev., A, 41,* 2668–2676.
2 Takahashi, I. and Ujihara, K. (1997) *Phys. Rev. A, 56,* 2299–2307.
3 Heitler, W. (1954) *The Quantum Theory of Radiation,* 3rd edn, Clarendon, Oxford.
4 Milonni, P.W. and Knight, P.L. (1973) *Opt. Commun., 9,* 119–122.
5 Cook, R.J. and Milonni, P.W. (1987) *Phys. Rev. A, 35,* 5081–5087.
6 Ho, T.D. and Ujihara, K. (1999) *Phys. Rev. A, 60,* 4067–4082.
7 Sachdev, S. (1984) *Phys. Rev. A, 29,* 2627–2633.
8 Purcell, E.M. (1946) *Phys. Rev., 69,* 681.
9 Kleppner, D. (1981) *Phys. Rev. Lett., 47,* 233–236.
10 Glauber, R.J. (1963) *Phys. Rev., 130,* 2529–2539.
11 De Martini, F., Marrocco, M., Mataloni, P., Crescentini, L., and Loudon, R. (1991) *Phys. Rev. A, 43,* 2480–2497.
12 De Martini, F., Marrocco, M., and Murra, D. (1990) *Phys. Rev. Lett., 65,* 1853–1856.
13 Ujihara, K., Nakamura, A., Manba, O., and Feng, X.P. (1991) *Jpn J. Appl. Phys., 30,* 3388–3398.
14 Goy, P., Raimond, J.M., Gross, M., and Haroche, S. (1983) *Phys. Rev. Lett., 50,* 1903–1906.
15 Hulet, R.G., Hilfer, E.S., and Kleppner, D. (1985) *Phys. Rev. Lett., 55,* 2137–2140.
16 Jhe, W., Anderson, A., Hinds, E.A., Meschede, D., Moi, L., and Haroche, S. (1987) *Phys. Rev. Lett., 58,* 666–669.
17 De Martini, F., Innocenti, G., Jacobovitz, G.R., and Mataloni, P. (1987) *Phys. Rev. Lett., 59,* 2955–2958.
18 Heinzen, D.J., Childs, J.J., Thomas, J.E., and Feld, M.S. (1987) *Phys. Rev. Lett., 58,* 1320–1323.
19 Yamamoto, Y., Machida, S., Horikoshi, Y., Igeta, K., and Björk, G. (1991) *Opt. Commun., 80,* 337–342.
20 Norris, T.B., Rhee, J.-K., Sung, C.-Y., Arakawa, Y., Nishioka, M., and Weisbuch, C. (1994) *Phys. Rev. B, 50,* 14663–14666.

14
Theory of Excess Noise

A standard quantum-mechanical calculation for the light emission by an atom into a single field mode yields an emission rate proportional to the number of photons present in the mode plus one. The portion proportional to the number of photons is interpreted as the stimulated emission. The portion proportional to "one" is spontaneous emission due to vacuum fluctuation or to radiation reaction [1]. This relative rate of spontaneous emission is the basis of the laser linewidth formulas, for example the Schawlow–Townes linewidth formula [2]. This is also at the heart of the Planck radiation formula [3]. The excess noise factor discussed so far violates this notion of one extra photon for the spontaneous emission. Where do the extra photons, more than one, come from? This is the theme of the theory of excess noise.

In this chapter we review the theories of the excess noise factor. First we review the adjoint mode theory, which was developed by Siegman [4, 5] for the transverse excess noise factor instead of the longitudinal excess noise factor. For the theory of the longitudinal excess noise factor, which has been one of the main topics in the previous chapters, we follow the treatment of Champagne and McCarthy [6], adding our additional contribution. Next we review the Green's function method developed by Henry [7] and by Tromborg *et al.* [8]. Third we review the propagation method or propagation theory developed by Ujihara [9] and by Goldberg *et al.* [10] as well as by Prasad [11]. This theory has already been described and used in Chapter 11. The relation of this theory to the adjoint mode theory will be described. Finally we make reference to some sophisticated, abstract theories aimed at quantum-mechanical consistency. The transverse excess noise factor and polarization excess noise factor will be described. Some experimental results will be cited.

14.1
Adjoint Mode Theory

We recall that the excess noise factor appeared for the one-sided cavity laser and the general two-sided cavity laser but not for the quasimode cavity laser. The former laser cavities have output coupling at the end(s) of the cavity, while the quasimode cavity has no explicit output coupling but has perfect mirrors.

Output Coupling in Optical Cavities and Lasers: A Quantum Theoretical Approach
Kikuo Ujihara

ISBN: 978-3-527-40763-7

While the quasimode cavity in Chapter 4 had power orthogonal modes in the sense that

$$\int_{cavity} u_m^*(z)u_n(z)dz = \delta_{mn} \tag{14.1a}$$

the former cavities did not have power orthogonal modes of this property in one dimension: the outgoing modes defined in Chapter 1 have the property

$$\int_{cavity} u_m^*(z)u_n(z)dz \neq \delta_{mn} \tag{14.1b}$$

According to Siegman [4] this is the result of non-Hermitian boundary conditions of the cavity, and for such cavities there exist adjoint modes $v(z)$ that are bi-orthogonal to the cavity mode functions and satisfy

$$\int_{cavity} v_m(z)u_n(z)dz = \delta_{mn} \tag{14.2}$$

when properly normalized. The function $v(z)$ is the solution to the transposed equation to the original equation describing the cavity and physically corresponds to the backward-propagating wave (see Siegman [12] for details). These adjoint modes are also non-power orthogonal

$$\int_{cavity} v_m^*(z)v_n(z)dz \neq \delta_{mn} \tag{14.3}$$

Siegman [4] showed that, if, in addition to the normalization in Equation 14.2, the mode function is normalized such that

$$\int_{cavity} |u_m|^2 dz = 1 \tag{14.4}$$

the adjoint mode has the property

$$\int_{cavity} |v_m|^2 dz \geq 1 \tag{14.5}$$

We will see later that this is the mathematical origin of the excess noise factor.

In some literature [13, 14] $v'_m(z) = v_m^*(z)$ is defined as the adjoint mode function, with the inner product (integral over the specified volume) being taken by multiplying $v'_m * (z)$ $(= v_m(z))$ with the other function. The results of calculations are, of course, the same as in this book. By the way, the incoming modes discussed in Section 1.2.1 constitute adjoint modes in this sense associated with the outgoing modes.

Let the Maxwell's equation for the classical wave inside the cavity be written as

$$\nabla^2 \mathbf{E}(z,t) - \mu_0 \frac{\partial^2}{\partial t^2}\mathbf{D}(z,t) = 0 \tag{14.6a}$$

where

$$\mathbf{D}(z,t) = \varepsilon\mathbf{E}(z,t) + \mathbf{P}(z,t) \tag{14.6b}$$

Here the polarization term represents classical random noise forces. The dielectric constant ε may be complex, reflecting the presence of a gain medium. Following Champagne and McCarthy [6] we put

$$\mathbf{E}(z,t) = \sum_m 2\mathrm{Re}\left\{a_m(t) C_{Nm} \Psi_m(z) \exp(-i\omega_m t)\right\} \tag{14.7a}$$

$$\mathbf{P}(z,t) = 2\mathrm{Re}\left\{\tilde{\mathbf{P}}(z,t) \exp(-i\omega t)\right\} \tag{14.7b}$$

(We have put a factor 2 before Re in Equations 14.7a and 14.7b so as to conform with our previous definition of **E** as $\mathbf{E} = \mathbf{E}^{(+)} + \mathbf{E}^{(-)}$ rather than $\mathbf{E} = \mathrm{Re}\{\mathbf{E}^{(+)}\}$ in classical terms.) Here the summation is over the cavity mode number. We are assuming that the cavity mode functions $\Psi_m(z)$ constitute a complete set to expand the total electric field. Anticipating projection of the electric field $\mathbf{E}(z,t)$ and the polarization $\mathbf{P}(z,t)$ onto an adjoint mode function of interest, which has an angular frequency ω, we retain only one mode function in $\mathbf{E}(z,t)$ and drop the mode number m (see Equation 14.13 below)

$$\mathbf{E}(z,t) = 2\mathrm{Re}\left\{a(t) C_N \Psi(z) \exp(-i\omega t)\right\} \tag{14.8}$$

Here $a(t)$ is a variable whose squared modulus averaged over the noise reservoirs describes the total number of photons of the mode in the cavity. The constant C_N is the normalization constant for $a(t)$ to have the above property. (But note that the photon number in the mode cannot be determined independently with those of other non-orthogonal modes because of the non-orthogonality [15].) The mode function $\Psi(z)$ is the one chosen by the above-mentioned projection. It corresponds to one of the mode functions, say u_m, in Equation 14.2. It satisfies the wave equation in the cavity

$$\left(\nabla^2 + \kappa^2\right)\Psi(z) = 0 \tag{14.9}$$

and the boundary conditions. Here κ is the possibly complex wavenumber reflecting the coupling loss at the cavity end surfaces. We divide the mode function into right- and left-going waves as

$$\Psi(z) = \psi_+(z)\exp(ikz) + \psi_-(z)\exp(-ikz) \tag{14.10}$$

Here k is a real wavenumber and we are assuming that the cavity is much longer than the wavelength of the mode. This function may not be normalized with respect to the integral over the cavity length.

Substitution of Equations 14.6b, 14.7b, and 14.8 into Equation 14.6a yields

$$\begin{aligned} &-a(t)C_N\kappa^2\Psi(z) - \mu_0\Big[\varepsilon C_N\Psi(z)\{\ddot{a}(t) - 2i\omega\dot{a}(t) - \omega^2 a(t)\} \\ &\quad + \{\ddot{\tilde{\mathbf{P}}}(z,t) - 2i\omega\dot{\tilde{\mathbf{P}}}(z,t) - \omega^2\tilde{\mathbf{P}}(z,t)\}\Big] = 0 \end{aligned} \tag{14.11}$$

We assume that the variation of the noise amplitude is slow so that we have $|\omega^2\tilde{\mathbf{P}}| \gg |\omega\dot{\tilde{\mathbf{P}}}|, |\ddot{\tilde{\mathbf{P}}}|$. Ignoring the second derivative of $a(t)$ we have

$$2i\dot{a}(t)\varepsilon C_N\Psi(z) = \frac{\kappa^2 - \varepsilon\mu_0\omega^2}{\mu_0\omega}a(t)C_N\Psi(z) + \omega\tilde{\mathbf{P}}(z,t) \tag{14.12}$$

We multiply both sides of Equation 14.12 by the adjoint function $\Psi^\dagger(z)$, which corresponds to v_m in Equation 14.2, and integrate over the length of the cavity to obtain

$$\dot{a}(t) = s_0 a(t) + p(t) \tag{14.13}$$

where $s_0 = (\kappa^2 - \varepsilon\mu_0\omega^2)/2i\varepsilon\mu_0\omega$ is the net gain per unit time and

$$p(t) = \frac{-i\omega\int_{cavity}\Psi^\dagger(z)\tilde{\mathbf{P}}(z,t)dz}{2\varepsilon C_N\int_{cavity}\Psi^\dagger(z)\Psi(z)dz} \tag{14.14}$$

For simplicity we assume that we are in the linear gain regime with Re $s_0 < 0$ and evaluate the laser linewidth using Equation 14.13 and the correlation property of the noise $\tilde{\mathbf{P}}(z,t)$. Now the correlation property of the term $\mathbf{P}(z,t)$ in Equation 14.6b was given by Siegman [5] as

$$\left\langle \mathbf{P}^*(z,t)\mathbf{P}(z',t')\right\rangle = \frac{4\hbar\varepsilon}{\omega}\mathcal{g}N_2\delta(t-t')\delta(z-z') \tag{14.15}$$

where ω is the central frequency of the noise emitter. Siegman derived the coefficient on the right-hand side by equating the emitted power from a polarization in a small volume V to that from the N_2V atoms each having spontaneous emission rate 2 $\mathcal{g}$. (In the original paper by Siegman the factor 4 is written as 16, but here we have taken care of the factor 2 added before the Re sign in Equation 14.7b.) The correspondence of the noise polarization in Equation 14.15 to the noise field in Equation 11.4 or 14.110 below is shown in Appendix I.

We proceed to evaluate the correlation function just as in Equations 4.42–4.49, but noting that the central frequency of oscillation ω is truncated here. We have

$$\left\langle a^*(t+\tau)a(t)\right\rangle = \int_0^{t+\tau}\int_0^t e^{s_0^*(t+\tau-t')}e^{s_0(t-t'')}\left\langle p^*(t')p(t'')\right\rangle dt'dt'' \tag{14.16}$$

Using Equations 14.14 and 14.15 we have

$$\begin{aligned}\left\langle a^*(t+\tau)a(t)\right\rangle &= \left|\frac{\omega}{2\varepsilon C_N(\Psi^\dagger\Psi)}\right|^2\frac{4\hbar\varepsilon}{\omega}\mathcal{g}N_2\left(\Psi^{\dagger*}\Psi^\dagger\right)\\ &\quad\times\int_0^{t+\tau}\int_0^t e^{s_0^*(t+\tau-t')}e^{s_0(t-t'')}\delta(t'-t'')dt'dt''\end{aligned} \tag{14.17}$$

where the round brackets on the right-hand side indicate the integral of the quantity over the length of the cavity, for example,

$$\left(\Psi^{\dagger}\Psi\right) = \int_{cavity} \Psi^{\dagger}(z)\Psi(z)dz \tag{14.18}$$

Note that taking the complex conjugate of the first function is *not* intended in this definition. The double time integral is as in Equation 4.52

$$\int_0^{t+\tau}\int_0^{t} e^{s_0^*(t+\tau-t')}e^{s_0(t-t'')}\delta(t'-t'')dt'dt'' = \begin{cases} \dfrac{e^{s_0^*\tau}}{2|\mathrm{Re}\, s_0|}, & \tau>0 \\ \dfrac{e^{-s_0\tau}}{2|\mathrm{Re}\, s_0|}, & \tau<0 \end{cases} \tag{14.19}$$

The Fourier transform of this double integral to the angular frequency domain yields a Lorentzian line and reveals that the linewidth $\Delta\omega$ is $2|\mathrm{Re}\, s_0|$ (see Equation 4.56). Using Equation 14.19 in Equation 14.17 and setting $\tau = 0$ yields a relation between $2|\mathrm{Re}\, s_0|$ and $\langle a^*(t)a(t)\rangle$. Thus we have the linewidth in the form

$$\Delta\omega = 2|\mathrm{Re}\, s_0| = \frac{1}{\left\langle a^*(t)a(t)\right\rangle}\left|\frac{\omega}{2\varepsilon C_N\left(\Psi^{\dagger}\Psi\right)}\right|^2 \frac{4\hbar\varepsilon}{\omega} g N_2\left(\Psi^{\dagger *}\Psi^{\dagger}\right) \tag{14.20}$$

Now the normalization constant is determined by equating the stored energy to the photon energy:

$$2\varepsilon|C_N|^2\left\langle a^*(t)a(t)\right\rangle\int_{cavity}\Psi^*(z)\Psi(z)dz = \left\langle a^*(t)a(t)\right\rangle\hbar\omega \tag{14.21}$$

so that

$$|C_N|^2 = \frac{\hbar\omega}{2\varepsilon\int_{cavity}\Psi^*(z)\Psi(z)dz} \tag{14.22}$$

Thus we have

$$\Delta\omega = 2|\mathrm{Re}\, s_0| = 2gN_2\frac{\left(\Psi^*\Psi\right)\left(\Psi^{\dagger *}\Psi^{\dagger}\right)}{\left\langle a^*(t)a(t)\right\rangle\left|\left(\Psi^{\dagger}\Psi\right)\right|^2} \tag{14.23}$$

To express the linewidth in terms of the output power, we recall that in the linear gain regime the power output is the stored energy times twice the cavity decay constant: $P = 2\gamma_c\hbar\omega\langle a^*a\rangle$. Also, we have $gN\sigma_{th} = \gamma_c$ (see Equation 7.44c for example). Therefore, we have

$$\Delta\omega = \frac{4\hbar\omega\gamma_c^2}{P}\frac{N_2}{N\sigma_{th}}\frac{\left(\Psi^*\Psi\right)\left(\Psi^{\dagger *}\Psi^{\dagger}\right)}{\left|\left(\Psi^{\dagger}\Psi\right)\right|^2} \tag{14.24}$$

Comparing this result with that for the quasimode laser in Equation 4.62a, we see three different points. First, this result lacks the bad cavity and detuning effects. Second, we have no thermal noise contribution here because the correlation in Equation 14.15 takes only quantum noise into account. Third, and most important here, we have the excess noise factor

$$K_L = \frac{(\Psi^*\Psi)(\Psi^{\dagger *}\Psi^{\dagger})}{|(\Psi^{\dagger}\Psi)|^2} \tag{14.25}$$

If we compare Equation 14.24 with the Schawlow–Townes linewidth formula in Equation 4.62b, the incomplete inversion factor and the excess noise factors are added here.

By the way, we present here the method to express the output power in terms of the functional form in Equation 14.10. Referring to Figure 11.1 and Equation 14.8 the output powers $P_{1,2}$ from mirrors $M_{1,2}$ are given by

$$\begin{aligned} P_1 &= 2\varepsilon_0 c_0 |T_1 e_-(0)|^2 = 2\varepsilon_0 c_0 |T_1|^2 \langle a^* a \rangle |C_N|^2 |\psi_-(0)|^2 \\ P_2 &= 2\varepsilon_0 c_0 |T_2 e_+(d)|^2 = 2\varepsilon_0 c_0 |T_2|^2 \langle a^* a \rangle |C_N|^2 |\psi_+(d)|^2 \end{aligned} \tag{14.26}$$

The total output power is then

$$P_t = P_1 + P_2 = 2\varepsilon_0 c_0 |C_N|^2 \left\{ |T_1|^2 |\psi_-(0)|^2 + |T_2|^2 |\psi + (d)|^2 \right\} \langle a^* a \rangle \tag{14.27}$$

From Equations G.17 to G.19 in Appendix G we have

$$|T_2|^2 = \left|\frac{1}{A}\right|^2 = \frac{n(|A|^2 - |B|^2)}{|A|^2} = n(1 - |r_2|^2) \tag{14.28}$$

Using Equations 14.22 and 14.28 we have

$$P_t = \frac{c_1 \hbar\omega}{(\Psi^*\Psi)} \left\{ (1 - |r_1|^2)|\psi_-(0)|^2 + (1 - |r_2|^2)|\psi + (d)|^2 \right\} \langle a^* a \rangle \tag{14.29}$$

where we have set $\varepsilon = \varepsilon_1 = n^2\varepsilon_0$. This expression for the output power was presented by Champagne and McCarthy [6]. Now if we use the mode function for a two-sided cavity in Equation 14.52 in Example 2 below, we have $|\psi_-(0)|^2 = 1/|r_1|^2$ and $|\psi_+(d)|^2 = \exp(2\gamma d) = 1/(|r_1||r_2|)$, where we have used Equation 14.49. Further, using Equation 14.54 or $(\Psi^*\Psi)$ w recover

$$P_t = \frac{c_1}{d} \ln \frac{1}{|r_1||r_2|} \hbar\omega \langle a^* a \rangle = 2\gamma_c \hbar\omega \langle a^* a \rangle \tag{14.30}$$

as expected. Note, however, that this relation does not hold for a laser in the saturated, nonlinear gain regime, as we mentioned in Section 12.9. This is because the mode function is deformed because of the location-dependent saturated gain.

Next we examine the spontaneous emission rate. As in Section 12.12, say in Equation 12.66, we derive from Equation 14.13

$$\frac{d}{dt}\langle a^*(t)a(t)\rangle = (s_0 + s_0^*)\langle a^*(t)a(t)\rangle + R_{sp} \tag{14.31a}$$

where the last term is interpreted as the total spontaneous emission rate (see above Equation 12.70) and is given by

$$R_{sp} = \left\langle a^*(t)p(t)\right\rangle + \left\langle p^*(t)a(t)\right\rangle \tag{14.31b}$$

We have from Equation 14.13

$$a(t) = a(0)e^{s_0 t} + \int_0^t e^{s_0(t-t')}p(t')dt' \tag{14.32}$$

Since $a(0)$ is not correlated to $p(t)$ for $t > 0$ we have

$$R_{sp} = \left\langle \int_0^t e^{s_0^*(t-t')}p^*(t')\,dt' p(t)\right\rangle + \left\langle p^*(t)\int_0^t e^{s_0(t-t')}p(t')\,dt'\right\rangle \tag{14.33}$$

Then using Equations 14.14 in the first term in Equation 14.33 we have

$$\left\langle \int_0^t e^{s_0^*(t-t')}p^*(t')\,dt' p(t)\right\rangle = \frac{\omega^2 \int_0^t e^{s_0^*(t-t')} \int_{cavity}\int_{cavity} \Psi^{\dagger *}(z')\Psi^{\dagger}(z)\left\langle \tilde{\mathbf{P}}^*(z',t')\tilde{\mathbf{P}}(z,t)\right\rangle dz dz' dt'}{\left|2\varepsilon C_N \int_{cavity}\Psi^{\dagger}(z)\Psi(z)dz\right|^2} \tag{14.34}$$

The second term in Equation 14.33 can be evaluated similarly. Therefore, by use of Equation 14.15, the spontaneous emission rate becomes, using $\int_0^t \delta(t-t')\,dt' = 1/2$,

$$R_{sp} = \frac{\hbar\omega g N_2}{\varepsilon|C_N|^2}\frac{\int_{cavity}\left|\Psi^{\dagger}(z)\right|^2 dz}{\left|\int_{cavity}\Psi^{\dagger}(z)\Psi(z)dz\right|^2} \tag{14.35}$$

Substituting Equation 14.22 into Equation 14.35 we have

$$R_{sp} = 2gN_2\frac{\int_{cavity}|\Psi(z)|^2 dz \int_{cavity}\left|\Psi^{\dagger}(z)\right|^2 dz}{\left|\int_{cavity}\Psi^{\dagger}(z)\Psi(z)dz\right|^2} \tag{14.36}$$

Since $2gN_2$ is the standard spontaneous emission rate that will be obtained in a free one-dimensional space, the remaining factors give the longitudinal excess noise factor K_L (see also Equation 12.69 for the total spontaneous emission rate in the quasimode cavity). Note, however, that this rate is for the total mode, but not for individual atoms. Thus we have

$$K_L = \frac{\int_{cavity}|\Psi(z)|^2 dz \int_{cavity}\left|\Psi^{\dagger}(z)\right|^2 dz}{\left|\int_{cavity}\Psi^{\dagger}(z)\Psi(z)dz\right|^2} \tag{14.37a}$$

This is the same as the expression in Equation 14.25 obtained through evaluation of the laser linewidth. We stress again that this enhancement factor applies to the mode as a whole but is not for the individual atoms. This equation is valid for the non-normalized mode function $\Psi(z)$ and also for the non-normalized adjoint mode function $\Psi^{\dagger}(z)$. If the mode function is normalized in the sense of Equation 14.4 and the adjoint mode function is also normalized in the sense of Equation 14.2, we have, according to Equation 14.5,

$$K_L = \int_{cavity} \left|\Psi^{\dagger}(z)\right|^2 dz \geq 1 \tag{14.37b}$$

Now, what is the adjoint mode $\Psi^{\dagger}(z)$ for $\Psi(z)$ in Equation 14.10 explicitly? According to Siegman [4] the adjoint mode is the counter-propagating mode within the same structure defining the original mode. Thus the right- and left-going components of $\Psi(z)$ in Equation 14.10 are mutually adjoint. Thus we put

$$\Psi^{\dagger}(z) = C_{\Psi^{\dagger}}\{\psi_{-}(z)\exp(-ikz) + \psi_{+}(z)\exp(ikz)\} \tag{14.38}$$

where $C_{\Psi^{\dagger}}$ is the normalization constant discussed below. Therefore, for a Fabry–Perot cavity, K_L in Equation 14.37a reduces to

$$K_L = \frac{\left\{\int_{cavity} |\Psi(z)|^2 dz\right\}^2}{\left|\int_{cavity} \Psi^2(z) dz\right|^2} \tag{14.39}$$

This equation may be used for mode functions that are not normalized. This is because, if one insists on normalizing the integral in the numerator, then the normalization constant appears in the denominator, thus yielding the same result. The same result as in Equation 14.39 was reported also by Arnaud [16], and a similar result for the transverse excess noise factor (which will be discussed below) was obtained by Petermann [17]. Petermann obtained this factor as an enhancement factor for the spontaneous emission factor (the fraction in power of the emission going to the mode of interest) in a gain-guided laser.

Now we will normalize $\Psi(z)$ with an added normalization constant C_{Ψ} as in Equation 14.4:

$$\int_{cavity} |C_{\Psi}\Psi(z)|^2 dz = |C_{\Psi}|^2 \int_{cavity} \left\{|\psi_{+}(z)|^2 + |\psi_{-}(z)|^2\right\} dz = 1 \tag{14.40}$$

Also, we normalize the product as in Equation 14.2:

$$\begin{aligned}
\int_{cavity} v_m^*(z)\, u_m(z) dz \rightarrow \int_{cavity} & \Psi^{\dagger}(z)\, C_{\Psi}\Psi(z) dz \\
&= \int_{cavity} C_{\Psi^{\dagger}} C_{\Psi}\{\psi_{-}(z)\exp(-ikz) + \psi_{+}(z)\exp(ikz)\} \\
&\quad \times \{\psi_{+}(z)\exp(ikz) + \psi_{-}(z)\exp(-ikz)\} dz \\
&= \int_{cavity} 2C_{\Psi^{\dagger}} C_{\Psi}\psi_{+}(z)\psi_{-}(z)\, dz \\
&= 1
\end{aligned} \tag{14.41}$$

Next for $\Psi^{\dagger}(z)$ we have

$$\int_{cavity} |\Psi^{\dagger}(z)|^2 dz = |C_{\Psi^{\dagger}}|^2 \int_{cavity} \left\{ |\psi_+(z)|^2 + |\psi_-(z)|^2 \right\} dz = \frac{|C_{\Psi^{\dagger}}|^2}{|C_{\Psi}|^2} \tag{14.42}$$

where we have used Equation 14.40 in the last equality. Now multiplying both the numerator and the denominator of Equation 14.37a by $|C_{\Psi}|^2$ and using Equations 14.40–14.42 we obtain

$$K_L = \frac{|C_{\Psi^{\dagger}}|^2}{|C_{\Psi}|^2} \geq 1 \tag{14.43}$$

if we use Equation 14.5. Thus the evaluation of the excess noise factor reduces to that of the normalization constants [18]. The last inequality may be proved as follows. We note from Equations 14.41 and 14.42 that

$$\begin{aligned} \left| \frac{1}{C_{\Psi^{\dagger}} C_{\Psi}} \right| &= \left| \int_{cavity} 2\psi_+(z)\psi_-(z)\, dz \right| \leq \int_{cavity} 2|\psi_+(z)\psi_-(z)|\, dz \\ &\leq \int_{cavity} \left(|\psi_+(z)|^2 + |\psi_-(z)|^2 \right) dz = \frac{1}{|C_{\Psi}|^2} \end{aligned} \tag{14.44a}$$

Thus we have

$$\left| \frac{C_{\Psi^{\dagger}}}{C_{\Psi}} \right| \geq 1 \tag{14.44b}$$

Note that the equality in Equations 14.43–14.44b occurs if $\psi_+(z) = \psi_-^*(z)$ for all z. This condition expresses a flat field distribution in the cavity as in the quasimode cavity laser of Chapter 4 where we had no excess noise factor. We can say that the longitudinal excess noise factor is in a sense a barometer of the field non-uniformity. It is large for a cavity with high-transmission mirrors with high gain medium, which results in a highly non-uniform field distribution.

A different normalization scheme from the one described in this section will appear in Section 14.6.

In some of the literature, especially in works treating semiconductor lasers, the formula for the longitudinal excess noise factor in Equation 14.25 or 14.37a is modified in some respects. First, the laser medium can be dispersive. Second, the dielectric constant ε or the refractive index may be position dependent. Third, the gain or the density of inverted atoms N_2 may also be position dependent. If these matters are taken into account, the expression for the spontaneous emission rate will include all these effects, while the expression for the excess noise factor will include the dispersive effect and the position-dependent refractive index in the integrals (see, for example, Champagne and McCarthy [6]).

Up to now we have described the mode function and the adjoint mode function as defined by the "empty" cavity. This is true for below-threshold operation where the gain is linear. We saw explicitly in Chapter 9 that the cavity spatial mode is excited by the driving noise sources. For above-threshold operation where the gain is saturated, the mode function is deformed, as we saw in Chapter 10. Even for this case the adjoint mode theory can be used with the mode function properly adjusted for the saturated gain distribution. In particular, Equation 14.9 may be rewritten, for the steady state, as

$$(\nabla^2 + \varepsilon(z)\mu_0\omega^2)\Psi(z) = 0 \tag{14.45}$$

where we have set $s_0 = 0$ or $\kappa^2(z) - \varepsilon(z)\mu_0\omega^2 = 0$ (see below Equation 14.13). Here the dielectric constant $\varepsilon(z)$ contains the imaginary part corresponding to the gain, which may be saturated non-uniformly. If the field distribution consistent with this non-linear equation for the field is used, a correct result can be obtained.

The physical origin of the enhanced spontaneous emission noise is given by Siegman [4] as the correlation between the noise emissions into different cavity modes due to the non-orthogonality of the modes. We saw that the noise that drives the mode u_m is given not by the projection of the noise polarization **P** on u_m but by the projection onto the adjoint mode v_m. Thus spontaneous emission into cavity modes other than u_m can drive u_m, enhancing the spontaneous emission in the mode u_m. How this enhancement of spontaneous emission noise develops with time or with wave propagation has been discussed by many authors. Siegman [4] introduced the concept of initial wave excitation factor, which is equal to the excess noise factor and describes the total power in a selected mode just after the input plane (of an amplifier) for an input field with unit power. This factor is large when the input field is spatially mode matched not to the desired mode but to the complex conjugate of the adjoint mode. New [19] examined the time development of the injected wave in the time-reversed sense (in the form of the complex conjugate of the adjoint mode) for the case of an unstable strip resonator, and found strong confinement of the wave around the cavity axis for initial round trips, resulting in much smaller transient energy loss than expected for the self-reproducing wave. Deutsch *et al.* [15] studied the development of the photon number for a field configuration along the length of a gain-guided amplifier and found strong initial increase followed by oscillatory approach to the steady state with the correct excess noise factor. The initial strong increase in photon number resulted again from the initial field: the initial field came into the expression for the photon number in the form of the adjoint mode function rather than the mode function of interest.

We shall now look at a series of examples.

Example 14.1

In the last part of Chapter 9 we obtained the longitudinal excess noise factor (from Equation 9.114)

$$K_L = \left(\frac{\beta_c}{\gamma_c}\right)^2 = \left(\frac{1 - r^2}{2r\ln(1/r)}\right)^2$$

as the squared ratio of the integrated squared modulus of the mode function $\sin\Omega_c(z + d)/c_1$ to that of the quasimode function $\sin\{(\omega_k/c_1)(z + d)\}$.

According to Equation 14.39, this interpretation should be modified: it should be expressed as the squared ratio of the integrated squared modulus of the mode function $\sin\Omega_c(z+d)/c_1$ to the modulus of the integrated squared mode function. The complex frequency Ω_c is given in Equation 1.18b. We have

$$K_L = \frac{\left(\int_{-d}^{0} dz|\sin\{\Omega_c(z+d)/c_1\}|^2\right)^2}{\left|\int_{-d}^{0} dz(\sin\{\Omega_c(z+d)/c_1)^2\right|^2}$$
$$= \frac{(d/2)^2[(1-r^2)/\{2r\ln(1/r)\}]^2}{|(d/2)-(c_1/4\Omega_c)\sin(2\Omega_c d/c_1)|^2} \tag{14.46}$$

Here the second term in the absolute sign is of the order of the optical wavelength and can be ignored if the cavity length d is much larger than a wavelength. Then we obtain

$$K_L = \left\{\frac{1-r^2}{2r\ln(1/r)}\right\}^2 \tag{14.47}$$

Example 14.2

We want to reproduce the excess noise factor for a generalized two-sided cavity laser, Equations 11.72 and 11.107. The cavity model is depicted in Figure 11.1. We describe the right- and left-going waves inside the cavity as $a\exp(i\kappa z)$ and $b\exp(-i\kappa z)$, respectively. We require that there exist only outgoing waves outside the cavity, and write these outgoing waves as $c\exp(i\kappa z)$ and $d'\exp(-i\kappa z)$. Then we have

$$\begin{aligned} ce^{i\kappa d} &= at_2e^{i\kappa d} \\ d' &= bt_1 \\ be^{-i\kappa d} &= r_2ae^{i\kappa d} \\ a &= r_1b \end{aligned} \tag{14.48}$$

From the last two equations we have

$$r_1r_2\exp(2i\kappa d) = 1 \tag{14.49}$$

Thus we have

$$\kappa = k - i\gamma \tag{14.50}$$

where

$$\begin{aligned} k &= \frac{1}{2d}(2n\pi - \phi_1 - \phi_2) \\ \gamma &= \frac{1}{2d}\ln\frac{1}{|r_1||r_2|} \end{aligned} \tag{14.51}$$

where we have set $r_{1,2} = |r_{1,2}|e^{i\phi_{1,2}}$. Using the last equality in Equation 14.48 we may have an non-normalized mode function

$$\Psi(z) = e^{i(k-i\gamma)z} + \frac{1}{r_1}e^{-i(k-i\gamma)z} \tag{14.52}$$

Substitution of this form into the numerator of Equation 14.39 yields

$$\begin{aligned}\int_0^d |\Psi(z)|^2 dz &= \frac{1}{2\gamma}\left(e^{2\gamma d} - 1\right) - \frac{1}{2\gamma|r_1|^2}\left(e^{-2\gamma d} - 1\right)\\ &\quad + \frac{1}{2ikr_1^*}\left(e^{2ikd} - 1\right) - \frac{1}{2ikr_1}\left(e^{-2ikd} - 1\right)\end{aligned} \tag{14.53}$$

Using Equation 14.51 we obtain

$$\int_0^d |\Psi(z)|^2 dz = \frac{d}{|r_1|\ln(1/|r_1||r_2|)}\left(\frac{1}{|r_1|} - |r_1| + \frac{1}{|r_2|} - |r_2|\right) \tag{14.54}$$

where we have ignored the last two terms in Equation 14.53 assuming that $k \gg \gamma$ or $d \gg \lambda$, where λ is the wavelength of the cavity mode. Similarly, we have for the denominator in Equation 14.39

$$\int_0^d \Psi^2(z)\, dz = \frac{2d}{r_1} + \frac{e^{2i(k-i\gamma)d} - 1}{2i(k-i\gamma)} + \frac{1}{r_1^2}\frac{e^{-2i(k-i\gamma)d} - 1}{-2i(k-i\gamma)} = \frac{2d}{r_1} \tag{14.55}$$

The last equality is obtained under the same approximation as above. Using Equations 14.54 and 14.55 in Equation 14.39 we have

$$K_L = \left\{\frac{1/|r_1| - |r_1| + 1/|r_2| - |r_2|}{2\ln(1/|r_1||r_2|)}\right\}^2 \tag{14.56}$$

which is the same as in Equations 11.72 and 11.107.

Example 14.3

We consider the problem treated in Example 2 using the rule expressed in Equation 14.43. Let the normalized mode function and the normalized adjoint mode function be

$$\begin{aligned}\Psi_N(z) &= C_\Psi \Psi(z) = C_\Psi\left\{e^{i(k-i\gamma)z} + \frac{1}{r_1}e^{-i(k-i\gamma)z}\right\}\\ \Psi_N^\dagger(z) &= C_{\Psi^\dagger}\left\{\frac{1}{r_1}e^{-i(k-i\gamma)z} + e^{i(k-i\gamma)z}\right\}\end{aligned} \tag{14.57}$$

where $\Psi(z)$ is given by Equation 14.52. The subscript N stands for normalization. The first normalization condition, Equation 14.40, becomes

$$\int_0^d |C_\Psi|^2 |\Psi(z)|^2 dz = |C_\Psi|^2 \frac{d}{|r_1| \ln(1/|r_1||r_2|)} \left(\frac{1}{|r_1|} - |r_1| + \frac{1}{|r_2|} - |r_2| \right) = 1 \tag{14.58}$$

under the same approximation as in Equation 14.54. The second normalization condition corresponding to Equation 14.41 is

$$\int_0^d C_\Psi C_{\Psi^\dagger} \Psi^2(z)\, dz = C_\Psi C_{\Psi^\dagger} \frac{2d}{r_1} = 1 \tag{14.59}$$

also under the same approximation. Thus using Equation 14.39 we have

$$K_L = \frac{|C_{\Psi^\dagger} C_\Psi|^2}{|C_\Psi|^4} = \left\{ \frac{1/|r_1| - |r_1| + 1/|r_2| - |r_2|}{2 \ln(1/|r_1||r_2|)} \right\}^2 \tag{14.60}$$

Example 14.4

We cite the vector inner product method used by Hamel and Woerdman [18], applying it to the same problem as in Examples 2 and 3. The authors define a two-component vector, where the top component is the right-going wave and the bottom component is the left-going wave. The adjoint is the vector with the components interchanged:

$$\Psi(z) = \begin{pmatrix} e^{i(k-i\gamma)z} \\ \frac{1}{r_1} e^{-i(k-i\gamma)z} \end{pmatrix} \quad \text{and} \quad \Psi^\dagger(z) = \begin{pmatrix} \frac{1}{r_1} e^{-i(k-i\gamma)z} \\ e^{i(k-i\gamma)z} \end{pmatrix} \tag{14.61}$$

The inner product is defined as the integration over the cavity length of the product of the transpose of the first vector and the second vector. Thus if we use Equation 14.61 in Equation 14.37a we obtain

$$\begin{aligned}
|\Psi(z)|^2 &= \left(e^{-i(k+i\gamma)z}, \ \frac{1}{r_1^*} e^{i(k+i\gamma)z} \right) \begin{pmatrix} e^{i(k-i\gamma)z} \\ \frac{1}{r_1} e^{-i(k-i\gamma)z} \end{pmatrix} = e^{2\gamma z} + \frac{1}{|r_1|^2} e^{-2\gamma z} \\
|\Psi^\dagger(z)|^2 &= \left(\frac{1}{r_1^*} e^{i(k+i\gamma)z}, \ e^{-i(k+i\gamma)z} \right) \begin{pmatrix} \frac{1}{r_1} e^{-i(k-i\gamma)z} \\ e^{i(k-i\gamma)z} \end{pmatrix} = \frac{1}{|r_1|^2} e^{-2\gamma z} + e^{2\gamma z} \\
\Psi^\dagger(z)\Psi(z) &= \left(\frac{1}{r_1} e^{-i(k-i\gamma)z}, \ e^{i(k-i\gamma)z} \right) \begin{pmatrix} e^{i(k-i\gamma)z} \\ \frac{1}{r_1} e^{-i(k-i\gamma)z} \end{pmatrix} = \frac{2}{r_1}
\end{aligned} \tag{14.62}$$

The inner products are the respective integrations over the cavity length. It is easy to see, using Equation 14.51, that

$$K_L = \frac{\int_{cavity} |\Psi(z)|^2 dz \int_{cavity} |\Psi^\dagger(z)|^2 dz}{\left|\int_{cavity} \Psi^\dagger(z)\Psi(z)dz\right|^2}$$
$$= \frac{\{[d/\{|r_1|\ln(1/|r_1||r_2|)\}](1/|r_1| - |r_1| + 1/|r_2| - |r_2|)\}^2}{|2d/r_1|^2} \qquad (14.63)$$

which is the same as those in Equations 14.56 and 14.60. This vector product method avoids the appearance of the cross-terms that were ignored in the previous examples. Note that the rule in Equation 14.39 is not directly applicable for this formulation as opposed to the general rule in Equation 14.37a. This seeming confusion occurs because the spatial functions in Equations 14.10 and 14.38 are the same. They are different if expressed in the vector form as seen in Equation 14.61.

Example 14.5

We can consider the same problem using Equation 14.43 under the vector product formulation of the previous example. Let us set

$$\Psi_N(z) = C_\Psi \begin{pmatrix} e^{i(k-i\gamma)z} \\ \frac{1}{r_1} e^{-i(k-i\gamma)z} \end{pmatrix} \quad \text{and}$$
$$\Psi_N^\dagger(z) = C_{\Psi^\dagger} \begin{pmatrix} \frac{1}{r_1} e^{-i(k-i\gamma)z} \\ e^{i(k-i\gamma)z} \end{pmatrix} \qquad (14.64)$$

If we normalize these as in Example 3, but using the vector product concept shown in Equation 14.62, it is easy to see that the same calculations as in Equations 14.58 and 14.59 in Example 3, but without the cross-terms, will be obtained. So, we will arrive at the same excess noise factor as in Equation 14.60.

14.2 Green's Function Theory

Henry [7] and Tromborg *et al.* [8] analyzed the excess noise factor using the Green's function method. They express a frequency component of the random electric field driven by the noise sources in terms of Green's function, the Green's function being the solution to the wave equation driven by a spatial delta function noise source. The spatial distribution of the driven field component is determined by the Green's function. The result is Fourier-transformed with a resultant temporal differential equation for the amplitude of the driven field. This differential equation contains the noise source and yields the spontaneous emission rate.

Following Tromborg *et al.* [8] we consider the ω component $E_\omega(z)$ of the scalar electric field defined by

$$E(z,t) = \int_0^\infty E_\omega(z)\exp(-i\omega t)d\omega + \text{C.C.} \tag{14.65}$$

The component $E_\omega(z)$ satisfies the wave equation

$$\nabla^2 E_\omega(z) + \frac{\omega^2}{c_0^2} n^2 E_\omega(z,t) = \mu_0 \omega^2 P_\omega(z) \equiv f_\omega(z) \tag{14.66}$$

where P_ω is the ω component of the noise polarization that appeared in Equation 14.6b. We look for the Green's function $G_\omega(z,z')$ associated with Equation 14.66 that satisfies

$$\left(\nabla^2 + \frac{\omega^2}{c_0^2} n^2\right) G_\omega(z,z') = \delta(z-z') \tag{14.67}$$

Then $E_\omega(z)$ is given by

$$E_\omega(z) = \int G_\omega(z,z') f_\omega(z') dz' \tag{14.68}$$

The Green's function is given as

$$G_\omega(z,z') = \frac{Z_R(z)Z_L(z')H(z-z') + Z_R(z')Z_L(z)H(z'-z)}{W} \tag{14.69}$$

where $H(z)$ is the Heaviside unit step function. The Wronskian W is

$$W = Z_L(z)\frac{d}{dz}Z_R(z) - Z_R(z)\frac{d}{dz}Z_L(z) \tag{14.70}$$

Here $Z_L(z)$ is the solution to the homogeneous equation associated with Equation 14.66 with $f_\omega(z) = 0$, which satisfies the boundary condition at the left end of the cavity, and $Z_R(z)$ is the solution satisfying the boundary condition at the right end of the cavity:

$$\left(\nabla^2 + \frac{\omega^2}{c_0^2} n^2\right) Z_{L,R}(z) = 0 \tag{14.71}$$

Because of this relation we have

$$\frac{d}{dz} W = 0 \tag{14.72}$$

that is, W is constant over the length of the cavity.

Equations 14.68 and 14.69 suggest that, when the noise $f_\omega(z)$ is vanishingly small, W should vanish in order to have a finite electric field $E_\omega(z)$. This means that the zeros of W as a function of ω give the poles for the electric field. We assume that the system is operating in the vicinity of a pole, say the zeroth pole $\omega = \omega_0$, which may be complex. At this pole $W(\omega_0) = 0$ and Equation 14.70 shows that the functions $Z_L(z)$ and $Z_R(z)$ are proportional to each other. At this pole we can set

$$Z_L(z) = Z_R(z) = Z_0(z) \tag{14.73}$$

Then Equation 14.68 becomes

$$E_\omega(z) = \frac{Z_0(z)}{W(\omega)} \int Z_0(z')f_\omega(z')dz' \tag{14.74}$$

We assume that the functional form $Z_0(z)$ for the field is maintained even for finite values of $f_\omega(z)$. For a finite noise, the Wronskian may deviate from zero. Expanding it around the zeroth pole we have

$$W(\omega) = W(\omega_0) + \frac{\partial W}{\partial \omega}(\omega - \omega_0) = \frac{\partial W}{\partial \omega}(\omega - \omega_0) \tag{14.75}$$

Substitution of Equation 14.75 into Equation 14.74 yields

$$(\omega - \omega_0)E_\omega(z) = \frac{Z_0(z)}{\partial W/\partial \omega} \int Z_0(z')f_\omega(z')dz' \tag{14.76}$$

This form suggests that the field component is proportional to $Z_0(z)$. We set

$$E_\omega(z) = Ba_\omega Z_0(z) \tag{14.77}$$

where B is a normalization constant to be determined later and a_ω is the Fourier component of the time variation of the electric field. Substitution of Equation 14.77 into Equation 14.76 gives

$$(\omega - \omega_0)a_\omega = \frac{1}{B\,\partial W/\partial \omega} \int Z_0(z')f_\omega(z')dz' \tag{14.78}$$

We construct the field amplitude $a(t)$ by multiplying a_ω by $\exp(-i\omega t)$ and integrating over ω. The trick here is that the ω term on the left-hand side yields the time derivative $(d/dt)a(t)$. We thus have

$$\frac{d}{dt}a(t) + i\omega_0 a(t) = F_a(t) \tag{14.79}$$

where

$$F_a(t) = \frac{-i}{B\,\partial W/\partial \omega} \int Z_0(z)f(z,t)dz \tag{14.80}$$

and

$$f(z,t) = \int \exp(-i\omega t) f_\omega(z) d\omega \tag{14.81}$$

The factor $\partial W/\partial\omega$ is independent of ω and Tromborg *et al.* [8], analyzing the changes in $Z_L(z)$ and $Z_R(z)$ due to a small change in k, showed that

$$\frac{\partial W}{\partial \omega} = \frac{2k_0}{c} \int_{cavity} Z_0^2(z)\, dz \tag{14.82}$$

where k_0 corresponds to ω_0. Once we get the Langevin equation (Equation 14.79), we can obtain the total spontaneous emission rate from the diffusion coefficient for $F_a(t)$ as in Equations 14.31–14.36 above. As in Equation 14.33 the total spontaneous emission rate is

$$R_{sp} = \left\langle \int_0^t F_a^*(t')\, dt' F_a(t) \right\rangle + \left\langle F_a^*(t) \int_0^t F_a(t')\, dt' \right\rangle \tag{14.83}$$

and

$$\left\langle \int_0^t F_a^*(t')\, dt' F_a(t) \right\rangle = \frac{1}{|B|^2 |\partial W/\partial \omega|^2} \int_0^t dt \int_{cavity} dz' \int_{cavity} dz\, Z_0^*(z') Z_0(z) \langle f^*(z',t') f(z,t) \rangle \tag{14.84}$$

We use Equations 14.15 and 14.66 to obtain

$$\begin{aligned} \left\langle f^*(z',t') f(z,t) \right\rangle &= (\mu_0 \omega^2)^2 \left\langle \mathbf{P}^*(z',t') \mathbf{P}(z,t) \right\rangle \\ &= (\mu_0 \omega^2)^2 \frac{4\hbar\varepsilon}{\omega} g N_2 \delta(t'-t) \delta(z'-z) \end{aligned} \tag{14.85}$$

where we have assumed that $P_\omega(z)$ is peaked around ω. Thus we have

$$\begin{aligned} R_{sp} &= \frac{(\mu_0 \omega^2)^2}{|B|^2 |\partial W/\partial \omega|^2} \frac{4\hbar\varepsilon}{\omega} g N_2 \int_{cavity} |Z_0(z)|^2 dz \\ &= \frac{(\mu_0 \omega^2)^2}{|B|^2 \left| (2k_0/c) \int_{cavity} Z_0^2(z) dz \right|^2} \frac{4\hbar\varepsilon}{\omega} g N_2 \int_{cavity} |Z_0(z)|^2 dz \end{aligned} \tag{14.86}$$

where we have used Equation 14.82 in the second line. Finally, we determine the normalization constant B so that the stored energy in the cavity is $\langle a^*(t) a(t) \rangle$ multiplied by the photon energy $\hbar\omega$. Noting from Equations 14.65 and 14.77 that

$$E(z,t) = B a(t) Z_0(z) + \text{C.C.} \tag{14.87}$$

we have

$$2\varepsilon|B|^2\left\langle a^*(t)a(t)\right\rangle\int_{cavity}|Z_0(z)|^2dz=\hbar\omega\left\langle a^*(t)a(t)\right\rangle \tag{14.88}$$

and

$$|B|^2=\frac{\hbar\omega}{2\varepsilon\int_{cavity}|Z_0(z)|^2dz} \tag{14.89}$$

Thus we finally have

$$R_{sp}=2gN_2\frac{\left\{\int_{cavity}|Z_0(z)|^2dz\right\}^2}{\left|\int_{cavity}Z_0^2(z)dz\right|^2} \tag{14.90}$$

where we have used the relations $k_0\simeq\omega/c$ and $\mu_0\varepsilon=c^{-2}$. This yields the same form of excess noise factor as in Equation 14.39:

$$K_L=\frac{\left\{\int_{cavity}|Z_0(z)|^2dz\right\}^2}{\left|\int_{cavity}Z_0^2(z)dz\right|^2} \tag{14.91}$$

That the excess noise factor is given in this form rather than that of Equation 14.37a reflects the fact that this Green's function method assumes a Fabry–Perot cavity. The equivalence of the function Z_0 to the function $\Psi(z)$ in the previous section may be argued by comparison of Equations 14.9 and 14.71 and by comparison of the boundary conditions that these functions satisfy.

As noted in the previous section, in some of the literature, especially on semiconductor lasers, the formula for the longitudinal excess noise factor in Equation 14.91 is modified in some respects. First, the laser medium can be dispersive. Second, the dielectric constant ε or the refractive index may be position dependent. Third, the gain or the density of inverted atoms N_2 may also be position dependent. If these matters are taken into account, the expression for the spontaneous emission rate will include all these effects, while the expression for the excess noise factor will include the dispersive effect and the position-dependent refractive index in the integrals. For more details, see, for example, Henry [7] and Tromborg *et al.* [8].

We have described the function $Z_0(z)$ as defined by the "empty" cavity of index n everywhere. As in the previous section, this limitation can be relaxed if we allow the factor n^2 in Equation 14.67 to represent the index distribution inside the cavity or the (saturated) gain distribution. Then, with the mode function $Z_0(z)$ properly adjusted for the index distribution or the (saturated) gain distribution, the correct result can be obtained. In fact, for a laser with saturated gain, van Exter *et al.* [20] obtained the longitudinal excess noise factor and the power-independent part of the laser linewidth using this Green's function method.

14.3 Propagation Theory

The third theory used to derive the excess noise factor is the simplified method developed in Chapter 11 and used independently by Goldberg *et al.* [10] and by Prasad [11]. This theory may be called the propagation theory or propagation method, where the noise fields propagate to a coupling surface of the cavity, being amplified on the way, to form the total noise field at that point. The detail of this method has already been described in Chapter 11. This theory stresses the importance of the amplification of the thermal noise and the quantum noise, leading to the longitudinal excess noise factor. As mentioned at the end of Chapter 11, this theory directly shows that the excess noise factor originates from the finite mirror transmissions as well as the finite amplification of the thermal and quantum noise during one round trip in the cavity. According to New [19] this interpretation of the excess noise is in line with the view that emphasizes initial temporary enhancement resulting from the excitation of the mode in a time-reversed sense (in the complex conjugate of the adjoint mode). We stress that this propagation theory can treat thermal noise and quantum noise on an equal footing as compared to the adjoint mode theory or Green's function theory, which are not suited to treat the injected thermal noise.

In the same spirit of taking into account the amplification of spontaneous emission noise along the length of the cavity, Thompson [21] arrived at the same enhancement factor as in Equation 14.47 by considering the power aspect of the field with the longitudinal boundary conditions of a symmetric cavity taken into account.

Here, we want to show the connection of this theory to the adjoint mode theory. We do this by generalizing the propagation theory. In Chapter 11 we summed all the contributing noise fields at a coupling surface. We change this location to sum all the noise fields to a general position inside the cavity and assume that the field amplitude is the product of a time-varying amplitude operator and a cavity mode function. Then we analyze the spontaneous emission rate to derive the excess noise factor.

For simplicity, we consider the linear gain regime of a generalized two-sided cavity laser treated in Section 11.6. From Equation 11.77 the basic equation reads

$$\frac{d\hat{e}^+(d-0,t)}{dt} = s_0\hat{e}^+(d-0,t) + \frac{c_1}{2d}\frac{\gamma'}{\gamma'+\gamma'_c}\left(\hat{F}_t + \hat{F}_q\right) \tag{14.92}$$

with the thermal and quantum noise fields $\hat{F}_t$ and $\hat{F}_q$ given by Equations 11.36 and 11.41 respectively. In both of these equations, the two possible routes to the mirror at $z = d$ from the noise sources at the mirrors or at the locations of the atoms are taken into account. The noises are amplified along these routes. Now we want to change the position $z = d$ to a general position z within the cavity. This time we have four routes for both the thermal and quantum noise. For the thermal noise, the noise penetrates into the cavity at the two mirrors and goes to position z either directly or after one reflection at the opposite mirror – see Figure 14.1a as

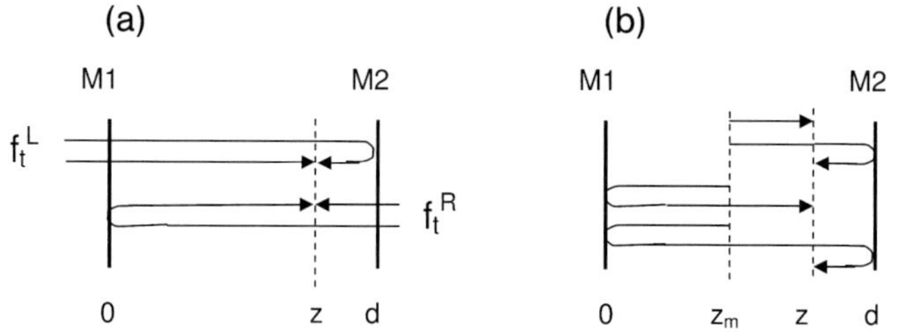

Figure 14.1 Routes for (a) the thermal noise and (b) the quantum noise to reach the position z.

well as Figure 11.1. For the quantum noise, the noise field from an atom at z_m goes to z directly, or after one reflection at either mirror, or after two successive reflections at the two mirrors – see Figure 14.1b as well as Figure 11.1.

Then Equation 11.77 may be rewritten in the form

$$\frac{d\hat{e}(z,t)}{dt} = s_0\hat{e}(z,t) + \frac{c_1}{2d}\frac{\gamma'}{\gamma' + \gamma'_c}\left\{\hat{F}_t(z,t) + \hat{F}_q(z,t)\right\} \tag{14.93}$$

where

$$\begin{aligned}\hat{F}_t(z,t) &= T'_2 e^{(ik+\alpha^0)(d-z)}\hat{f}_t^R\left(d+0,\ t-\frac{d-z}{c_1}\right)\\ &+ T'_2 r_1 e^{(ik+\alpha^0)(d+z)}\hat{f}_t^R\left(d+0,\ t-\frac{d+z}{c_1}\right)\\ &+ T'_1 e^{(ik+\alpha^0)z}\hat{f}_t^L\left(-0, t-\frac{z}{c_1}\right)\\ &+ T'_1 r_2 e^{(ik+\alpha^0)(2d-z)}\hat{f}_t^L\left(-0,\ t-\frac{2d-z}{c_1}\right)\end{aligned} \tag{14.94a}$$

and

$$\begin{aligned}\hat{F}_q(z,t) = \sum_m \Bigg\{&\hat{f}_m\left(t-\frac{|z-z_m|}{c_1}\right)e^{(ik+\alpha^0)|z-z_m|}\\ &+\hat{f}_m\left(t-\frac{z+z_m}{c_1}\right)r_1 e^{(ik+\alpha^0)(z+z_m)}\\ &+\hat{f}_m\left(t-\frac{2d-z-z_m}{c_1}\right)r_2 e^{(ik+\alpha^0)(2d-z-z_m)}\\ &+\hat{f}_m\left(t-\frac{2d-|z-z_m|}{c_1}\right)r_1 r_2 e^{(ik+\alpha^0)(2d-|z-z_m|)}\Bigg\}\end{aligned} \tag{14.94b}$$

Note that both $\hat{F}_t$ and $\hat{F}_q$ contain waves propagating to the right and the left. Thus $\hat{e}(z,t)$ also contains waves propagating in both directions. We assume that the laser is operating with a mode function in the form of Equation 14.52 expressing an outgoing mode for a generalized two-sided cavity. We also assume that the loss rate γ is slightly larger than, but nearly equal to, the gain α^0. Thus the mode function is

$$\Psi(z) = e^{i(k-i\alpha^0)z} + \frac{1}{r_1}e^{-i(k-i\alpha^0)z} \tag{14.95}$$

where k is the real wavenumber. We may write the field amplitude as

$$\hat{e}(z,t) = C_N\hat{a}(t)\Psi(z) \tag{14.96}$$

where C_N is the normalization constant enabling $\langle \hat{a}^\dagger(t)\hat{a}(t)\rangle$ to represent the number of photons in the cavity. As in Equations 14.22 and 14.89

$$|C_N|^2 = \frac{\hbar\omega}{2\varepsilon_1 \int_{cavity} |\Psi(z)|^2 dz} \tag{14.97}$$

We project Equation 14.93 onto the adjoint mode function

$$\Psi^\dagger(z) = \frac{1}{r_1}e^{-i(k-i\alpha^0)z} + e^{i(k-i\alpha^0)z} \tag{14.98}$$

Then we have formally

$$\begin{aligned} C_N\left(\Psi^\dagger\Psi\right)\frac{d}{dt}\hat{a}(t) &= s_0 C_N\left(\Psi^\dagger\Psi\right)\hat{a}(t) \\ &\quad + \frac{c_1}{2d}\frac{\gamma'}{\gamma'+\gamma'_c}\left\{\left(\Psi^\dagger\hat{F}_t\right) + \left(\Psi^\dagger\hat{F}_q\right)\right\} \end{aligned} \tag{14.99}$$

where the bracket indicates an integral over the cavity length. We have

$$\left(\Psi^\dagger\Psi\right) = \int_0^d \Psi^\dagger(z)\Psi(z)dz = \frac{2d}{r_1} \tag{14.100}$$

For simplicity, we consider the thermal part and the quantum part separately. For the thermal part, we have, by Equation 14.94a,

$$\begin{aligned} \left(\Psi^\dagger\hat{F}_t\right) &= \Big\{T'_2 de^{(ik+\alpha^0)d}\hat{f}_t^R(d+0,t) + T'_2 de^{(ik+\alpha^0)d}\hat{f}_t^R(d+0,t) \\ &\quad + T'_1\frac{d}{r_1}\hat{f}_t^L(-0,t) + T'_1 r_2 de^{(ik+\alpha^0)2d}\hat{f}_t^L(-0,t)\Big\} \\ &= 2d\left\{\frac{T'_2}{\sqrt{r_1r_2}}\hat{f}_t^R(d+0,t) + T'_1\frac{1}{r_1}\hat{f}_t^L(-0,t)\right\} \end{aligned} \tag{14.101}$$

where we have ignored the integrals of rapidly oscillating terms. Here we have replaced the time values in Equation 14.94a by t on the grounds that we are interested in the field fluctuation on a time scale larger than the cavity round-trip time. We have also used Equation 14.49 describing the steady-state condition with $\gamma \to \alpha^0$:

$$r_1r_2\exp\{2(ik+\alpha^0)d\} = 1 \tag{14.102}$$

Thus we have for the thermal noise

$$\frac{d}{dt}\hat{a}(t) = s_0\hat{a}(t) + \frac{c_1}{C_N(\Psi^\dagger\Psi)}\frac{\gamma'}{\gamma' + \gamma'_c}\frac{1}{\sqrt{r_1}} \times \left\{\frac{T_2'}{\sqrt{r_2}}\hat{f}_t^R(d+0,t) + \frac{T_1'}{\sqrt{r_1}}\hat{f}_t^L(-0,t)\right\} \tag{14.103}$$

We examine the noise photon injection rate as in Equations 12.65–12.68 or in Equations 12.72–12.80. We have

$$\begin{aligned}\frac{d}{dt}\left\langle\hat{a}^\dagger(t)\hat{a}(t)\right\rangle &= (s_0 + s_0^*)\left\langle\hat{a}^\dagger(t)\hat{a}(t)\right\rangle \\ &+ \left|\frac{c_1}{C_N(\Psi^\dagger\Psi)}\frac{\gamma'}{\gamma' + \gamma_c'}\frac{1}{\sqrt{r_1}}\right|^2\left[\left|\frac{T_2'}{\sqrt{r_2}}\right|^2\left\{\int_0^t e^{s_0^*(t-t')}\left\langle\hat{f}_t^{R\dagger}(d+0,t')\hat{f}_t^R(d+0,t)\right\rangle dt'\right.\right. \\ &\left.+\int_0^t e^{s_0(t-t')}\left\langle\hat{f}_t^{R\dagger}(d+0,t)\hat{f}_t^R(d+0,t')\right\rangle dt'\right\} \\ &\left.+\left|\frac{T_1'}{\sqrt{r_1}}\right|^2\left\{\int_0^t e^{s_0^*(t-t')}\left\langle\hat{f}_t^{L\dagger}(-0,t')\hat{f}_t^L(-0,t)\right\rangle dt' + \int_0^t e^{s_0(t-t')}\left\langle\hat{f}_t^{L\dagger}(-0,t)\hat{f}_t^L(-0,t')\right\rangle dt'\right\}\right]\end{aligned} \tag{14.104}$$

The last term in the square bracket in Equation 14.104 is the noise photon injection rate R_t. The correlation function of the thermal noise is given by the first equations of Equations 11.1b and 11.1c. We have, using Equations 14.97 and 14.100 for $|C_N|^2$ and $(\Psi^\dagger\Psi)$, respectively,

$$\begin{aligned}R_t &= \frac{2\varepsilon_1\int_{cavity}|\Psi(z)|^2 dz}{\hbar\omega}\left|\frac{\gamma'}{\gamma' + \gamma_c'}\right|^2\left(\frac{c_1}{2d}\right)^2|r_1|\left[\left|\frac{T_2'}{\sqrt{r_2}}\right|^2 + \left|\frac{T_1'}{\sqrt{r_1}}\right|^2\right]\frac{n_\omega\hbar\omega}{2\varepsilon_0 c_0} \\ &= \left|\frac{\gamma'}{\gamma' + \gamma_c'}\right|^2\frac{c_1}{4d\ln(1/|r_1||r_2|)}\frac{1}{}\left(\frac{1}{|r_1|} - |r_1| + \frac{1}{|r_2|} - |r_2|\right)^2 n_\omega\end{aligned} \tag{14.105}$$

where we have used Equations 11.52a and 14.54, and the relation $n\varepsilon_0 c_0 = \varepsilon_1 c_1$ in the second line. We have also used the integral $\int_0^t \delta(t-t')dt' = 1/2$. Using Equation 11.25 for the definition of the cavity decay rate γ_c, we have

$$\begin{aligned}R_t &= \left|\frac{\gamma'}{\gamma' + \gamma_c'}\right|^2 2\gamma_c n_\omega K_L \\ K_L &= \left\{\frac{1/|r_1| - |r_1| + 1/|r_2| - |r_2|}{2\ln(1/|r_1||r_2|)}\right\}^2\end{aligned} \tag{14.106}$$

Comparing with Equation 12.68 for the noise photon injection rate for the case of the quasimode cavity laser, we obtain the longitudinal excess noise factor that was obtained in Equation 14.56 on the basis of the adjoint mode theory.

Next we consider the quantum noise part in Equation 14.99 using Equation 14.94b:

$$
\begin{aligned}
\left(\Psi^{\dagger}\hat{F}_q\right) = & \int_0^d dz\left(\frac{1}{r_1}e^{-i(k-i\alpha^0)z} + e^{i(k-i\alpha^0)z}\right) \\
& \times \sum_m \left\{\hat{f}_m\left(t - \frac{|z-z_m|}{c_1}\right)e^{(ik+\alpha^0)|z-z_m|} + \hat{f}_m\left(t - \frac{z+z_m}{c_1}\right)r_1 e^{(ik+\alpha^0)(z+z_m)}\right. \\
& + \hat{f}_m\left(t - \frac{2d-z-z_m}{c_1}\right)r_2 e^{(ik+\alpha^0)(2d-z-z_m)} \\
& \left. + \hat{f}_m\left(t - \frac{2d-|z-z_m|}{c_1}\right)r_1 r_2 e^{(ik+\alpha^0)(2d-|z-z_m|)}\right\} \\
= & \sum_m \left[\hat{f}_m(t)\left\{z_m e^{(ik+\alpha^0)z_m} + \frac{d-z_m}{r_1}e^{-(ik+\alpha^0)z_m}\right\}\right. \\
& + \hat{f}_m(t) d e^{(ik+\alpha^0)z_m} + \hat{f}_m(t) r_2 d e^{(ik+\alpha^0)(2d-z_m)} \\
& \left. + \hat{f}_m(t) r_1 r_2 e^{(ik+\alpha^0)2d}\left\{\frac{z_m}{r_1}e^{-(ik+\alpha^0)z_m} + (d-z_m)e^{(ik+\alpha^0)z_m}\right\}\right] \\
= & 2d\sum_m \hat{f}_m(t)\left\{\frac{1}{r_1}e^{-(ik+\alpha^0)z_m} + e^{(ik+\alpha^0)z_m}\right\}
\end{aligned}
\tag{14.107}
$$

where we have ignored integrals of rapidly oscillating terms. The terms proportional to z_m cancel each other. In the last equality, we have used Equation 14.102. Here, as for the thermal noise, we have replaced the time values in Equation 14.107 by t on the grounds that we are interested in the field fluctuation on a time scale larger than the cavity round-trip time. Note that the quantity in the last curly bracket is $\Psi^{\dagger}(z_m)$. As in Equation 14.103 we have for the quantum noise

$$
\frac{d}{dt}\hat{a}(t) = s_0\hat{a}(t) + \frac{c_1}{C_N\left(\Psi^{\dagger}\Psi\right)}\frac{\gamma'}{\gamma' + \gamma'_c}\sum_m \hat{f}_m(t)\Psi^{\dagger}(z_m) \tag{14.108}
$$

Thus we have

$$
\begin{aligned}
\frac{d}{dt}\left\langle \hat{a}^{\dagger}(t)\hat{a}(t)\right\rangle = & (s_0 + s_0^*)\left\langle \hat{a}^{\dagger}(t)\hat{a}(t)\right\rangle \\
& + \left|\frac{c_1}{C_N\left(\Psi^{\dagger}\Psi\right)}\frac{\gamma'}{\gamma' + \gamma'_c}\right|^2 \sum_{m'}\sum_m \Psi^{\dagger *}(z_{m'})\Psi^{\dagger}(z_m) \\
& \times \left\{\int_0^t e^{s_0(t-t')}\left\langle \hat{f}_{m'}^{\dagger}(t)\hat{f}_m(t')\right\rangle dt' + \int_0^t e^{s_0^*(t-t')}\left\langle \hat{f}_{m'}^{\dagger}(t')\hat{f}_m(t)\right\rangle dt'\right\}
\end{aligned}
\tag{14.109}
$$

Since we have only normally ordered products of the noise operators, this time we use, instead of Equation 11.4,

$$
2\varepsilon_1 c_1\left\langle \hat{f}_m^{\dagger}(t)\hat{f}_{m'}(t')\right\rangle = \{g(1+\sigma)\hbar\omega/c_1\}\delta_{mm'}\delta(t-t') \tag{14.110}
$$

(cf. Equation 9.5c). We are assuming that half the stimulated emission rate g and the atomic inversion σ are, respectively, common for all the atoms. That is, we are assuming homogeneously broadened atoms and uniform pumping. Using Equation 14.110 in Equation 14.109 we have the total spontaneous emission rate

$$R_{sp} = \left|\frac{\gamma'}{\gamma' + \gamma_c'}\right|^2 \left|\frac{c_1}{C_N(\Psi^\dagger \Psi)}\right|^2 \frac{g(1+\sigma)\hbar\omega}{2\varepsilon_1 c_1^2} \sum_m \left|\Psi^\dagger(z_m)\right|^2 \tag{14.111}$$

Now we can write

$$\sum_m \left|\Psi^\dagger(z_m)\right|^2 = N \int_0^d \left|\Psi^\dagger(z)\right|^2 dz = N\left(\Psi^{\dagger *}\Psi^\dagger\right) \tag{14.112}$$

Using Equations 14.97 and 14.112 in Equation 14.111 we have

$$R_{sp} = \left|\frac{\gamma'}{\gamma' + \gamma_c'}\right|^2 2gN_2 \frac{(\Psi^*\Psi)\left(\Psi^{\dagger *}\Psi^\dagger\right)}{\left|\left(\Psi^\dagger\Psi\right)\right|^2} \tag{14.113}$$

Thus, by comparison with Equation 12.69 for the spontaneous emission rate in a quasimode cavity laser, we have the excess noise factor

$$K_L = \frac{(\Psi^*\Psi)\left(\Psi^{\dagger *}\Psi^\dagger\right)}{\left|\left(\Psi^\dagger\Psi\right)\right|^2} \tag{14.114}$$

which is the same as the one in Equation 14.25 derived on the basis of the adjoint mode theory. Therefore, we have shown that the formulation in the propagation theory can be converted to that in the adjoint mode theory. In this example, Equation 14.114 reduces to Equation 14.56 because of the forms of $\Psi(z)$ in Equation 14.95 and $\Psi^\dagger(z)$ in Equation 14.98. As we saw in Chapters 10 and 11, this propagation theory can be applied to a saturated gain regime, which leads not only to the longitudinal excess noise factor but also to power-independent part of the laser linewidth, which increases the linewidth. Goldberg *et al.* [10] also examined the gain saturation with the spatial hole burning taken into account. The latter effect was found to further increase the linewidth.

14.4 Three-Dimensional Cavity Modes and Transverse Effects

The adjoint mode theory described in Section 14.1 was devoted to the derivation of the longitudinal excess noise factor. However, it is easy to see that the discussion from Equation 14.6a to Equation 14.25 or to Equation 14.45 can be generalized to the three-dimensional case without much alteration. In fact, except for Equations 14.10 and 14.38 for the expression for the mode function and the adjoint mode function, respectively, all the functions of variable z can be replaced by functions of variable $\mathbf{r}$. In particular, for the functions above, we may write

$$\Psi(\mathbf{r}) = \psi_+(\mathbf{r})\exp(ikz) + \psi_-(\mathbf{r})\exp(-ikz) \tag{14.115}$$

$$\Psi^\dagger(\mathbf{r}) = \psi_-(\mathbf{r})\exp(-ikz) + \psi_+(\mathbf{r})\exp(ikz) \tag{14.116}$$

Here we assume that a set of mode functions and corresponding adjoint mode functions for a given cavity or a laser are known analytically or numerically. The other quantities $\mathbf{D}(z,t)$, $\mathbf{E}(z,t)$, and $\mathbf{P}(z,t)$ may be replaced by $\mathbf{D}(\mathbf{r},t)$, $\mathbf{E}(\mathbf{r},t)$, and $\mathbf{P}(\mathbf{r},t)$, respectively. The correlation function for noise polarization $\mathbf{P}(\mathbf{r},t)$ in Equation 14.15 may be rewritten as

$$\left\langle \mathbf{P}^*(\mathbf{r},t)\mathbf{P}(\mathbf{r}',t')\right\rangle = \frac{4\hbar\varepsilon}{\omega} g N_2 \delta(t-t')\delta(\mathbf{r}-\mathbf{r}') \tag{14.117}$$

The central equation of motion for the field amplitude reads

$$\dot{a}(t) = s_0 a(t) + p(t) \tag{14.118}$$

where

$$p(t) = \frac{-i\omega\int_{cavity}\Psi^\dagger(\mathbf{r})\tilde{\mathbf{P}}(\mathbf{r},t)d\mathbf{r}}{2\varepsilon C_N\int_{cavity}\Psi^\dagger(\mathbf{r})\Psi(\mathbf{r})d\mathbf{r}} \tag{14.119}$$

The limits of integration in the z-direction are the same as before. The integration range in the transverse direction is from the cavity axis to a certain outer surface of the mode, which may be infinity. The analysis of the laser linewidth or the spontaneous emission rate can be carried out in a similar manner as in Section 14.1. Thus the excess noise factor K corresponding to that in Equation 14.25 becomes

$$K = \frac{(\Psi^*\Psi)(\Psi^{\dagger *}\Psi^\dagger)}{|(\Psi^\dagger\Psi)|^2} = \frac{\int_{cavity}\Psi^*(\mathbf{r})\Psi(\mathbf{r})d\mathbf{r}\int_{cavity}\Psi^{\dagger *}(\mathbf{r})\Psi^\dagger(\mathbf{r})d\mathbf{r}}{\left|\int_{cavity}\Psi^\dagger(\mathbf{r})\Psi(\mathbf{r})d\mathbf{r}\right|^2} \tag{14.120}$$

Now if the mode function is expressible as the product of the longitudinal mode function $\Psi_L(z)$ and the transverse mode function $\phi_T(\mathbf{s})$, where $\mathbf{s} = (x,y)$, it can be seen that the total excess noise factor is the product of the longitudinal and transverse excess noise factors

$$K = K_L K_T, \qquad \Psi(\mathbf{r}) = \Psi_L(z)\phi_T(\mathbf{s}) \tag{14.121}$$

where

$$K_L = \frac{(\Psi_L{}^*\Psi_L)\left(\Psi_L^{\dagger *}\Psi_L^\dagger\right)}{\left|\left(\Psi_L^\dagger\Psi_L\right)\right|^2} \tag{14.122}$$

is the same as in Equation 14.25, now with the suffix L. The transverse excess noise factor K_T is given as

$$K_T = \frac{\left(\phi_T^*(\mathbf{s})\phi_T(\mathbf{s})\right)\left(\phi_T^{\dagger *}(\mathbf{s})\phi_T^{\dagger}(\mathbf{s})\right)}{\left|\left(\phi_T^{\dagger}(\mathbf{s})\phi_T(\mathbf{s})\right)\right|^2} = \frac{\int_{cs}\phi_T^*(\mathbf{s})\phi_T(\mathbf{s})d\mathbf{s}\int_{cs}\phi_T^{\dagger *}(\mathbf{s})\phi_T^{\dagger}(\mathbf{s})d\mathbf{s}}{\left|\int_{cs}\phi_T^{\dagger}(\mathbf{s})\phi_T(\mathbf{s})d\mathbf{s}\right|^2} \quad (14.123)$$

where the symbol *cs* indicates the cross-section of the cavity mode. Champagne and McCarthy [6] derived the general expression in Equation 14.120 and noted that the product form in Equation 14.121 cannot be always true.

The transverse excess noise factor equivalent to Equation 14.123 was first derived by Petermann [17] by an analysis of the spontaneous emission factor (the fraction in power of the emission going to the mode of interest) in a gain-guided semiconductor laser, and then by Siegman [4] through the adjoint mode theory for a general open optical system that has non-Hermitian boundary conditions. An ideal stable cavity laser or an index-guided laser having orthogonal transverse modes has $K_T = 1$, while a planar purely gain-guided laser has $K_T = \sqrt{2}$ (see Ref. [17]). Doumont *et al.* [22] analyzed a laser with variable reflectivity mirrors (mirrors with Gaussian reflectivity distribution along the distance from the mirror center) and gave approximate analytic expressions for the transverse excess noise factor for a stable as well as an unstable cavity. They predicted a transverse excess noise factor of 10^4–$10^5$for an unstable cavity with a large magnification and a large Fresnel number especially for higher-order transverse modes.

The transverse excess noise factor for a stable laser resonator with one or two apertures was studied by Brunel *et al.* [23]. In their study, the non-Hermitian property of the Huygens–Fresnel kernel, which determines the round-trip field development, is introduced by the presence of the apertures. They predicted a value of 100 for the observable transverse excess noise factor. The transverse excess noise factor for an unstable, confocal strip resonator was studied by New [19]. He obtained an excess noise factor in excess of 10^4 for a narrow region of the Fresnel number. He attributed such a large excess noise factor to significant difference in the shapes of the phase fronts of the mode and the adjoint mode functions. He also emphasized the importance of the initial wave excitation factor for the physical interpretation of the noise enhancement.

Firth and Yao [24] considered the transverse excess noise of misaligned cavities and predicted a value in excess of 10^{10} for a cavity that is slightly unstable in structure and has a Gaussian aperture mirror that is offset from the axis. This value was obtained for a relatively small misalignment power loss. They argued that, in view of such a large excess noise, the excess noise factor may be interpreted more physically as due to transient gain than to correlation between multiple modes.

14.5 Quantum Theory of Excess Noise Factor

In Section 14.3 the field variable was treated as an operator following the quantum-mechanical analysis in Chapter 11. However, except in Section 14.3, the field variable has been a classical variable in this chapter. How can one make the adjoint mode theory a consistent quantum theory? One way seems to be to get a quantum-mechanically correct expression for the noise polarization as was done in Section 14.3. Even in Section 14.3 the introduction of the annihilation operator in the form of Equation 14.96 and the mode normalization in the form of Equation 14.97 may need justification in view of the non-orthogonality of the relevant mode functions. It is not that the treatment in Section 14.3 is not correct quantum mechanically, but rather that the derivation of the equation used in Section 14.3 and in Chapter 11 was originally based on the field expansion in terms of the normal modes of the "universe" as was carried out in Chapters 9 and 10. The expansion of the field in normal modes of the "universe" allowed quantum-mechanically consistent analysis. In this connection, one would hope to have a general quantum-mechanical theory of the excess noise factor that extends the classical adjoint mode theory. Two of the quantization methods reported for the derivation of the excess noise factor are reviewed in this section.

14.5.1 Excess Noise Theory Based on Input–Output Commutation Rules

One method to derive the excess noise factor quantum mechanically was developed by Granjier and Poizat [25]. Here the outline of their derivation will be described. They start with a set of normalized and orthogonal mode functions. The field is expanded in terms of these normal modes and the (input and output) fields before and after one round trip in the cavity are expressed as column vectors $\{e_{in}\}$ and $\{e_{out}\}$, respectively. The character of the empty cavity is written in terms of a unitary scattering matrix as

$$\{e_{out}\} = S\{e_{in}\} \tag{14.124}$$

Since S is unitary, all the commutation relations are preserved in the input to output evolution. Then they assume that the modes can be split into "laser" modes and "loss" modes, and introduce projection operators P and Q, which project on the "laser" modes and on the "loss" modes, respectively:

$$P^2 = P, \qquad Q^2 = Q, \qquad P + Q = 1 \tag{14.125}$$

Thus we have

$$P\{e_{out}\} = PS(P+Q)\{e_{in}\} = TP\{e_{in}\} + PSQ\{e_{in}\} \tag{14.126}$$

where $T = PSP$ describes the input–output relation for the "laser" modes only. The matrix T is not unitary in general and will have eigenvectors that are non-orthogonal:

$$TU = UG \tag{14.127}$$

Here U is the matrix with columns formed by the normalized eigenvectors $\{u_n\}$ of T and G is a diagonal matrix formed by the corresponding eigenvalues γ_n. Next, one introduces the adjoint of U by

$$V = (U^{-1})^{\dagger} \tag{14.128}$$

From Equations 14.127 and 14.128 one obtains

$$V^{\dagger}T = GV^{\dagger}, \qquad T^{\dagger}V = VG^{\dagger} \tag{14.129}$$

Thus V is a matrix with columns formed by the eigenvectors $\{v_n\}$ of $T^{\dagger}$ and $G^{\dagger}$ is a diagonal matrix formed by the corresponding eigenvalues γ_n^*. It is argued that, as $\{u_n\}$ are normalized, $\{v_n\}$ cannot be normalized, but that one has the bi-orthogonality relation:

$$V^{\dagger}U = U^{\dagger}V = I \tag{14.130}$$

One may notice that the story here traces the adjoint mode theory described from Equations 14.1a–14.5. Multiplying $PV^{\dagger} = PU^{-1}$ to both sides of Equation 14.126 and using $V^{\dagger}T = GV^{\dagger}$ one has

$$PV^{\dagger}P\{e_{out}\} = G\left(PV^{\dagger}P\{e_{in}\}\right) + PV^{\dagger}PSQ\{e_{in}\} \tag{14.131}$$

This corresponds to projection onto the adjoint modes. It is assumed that one can quantize a particular "laser" mode by replacing the amplitudes $\{e_{in}\}$ and $\{e_{out}\}$ by the operators $\{\hat{a}_{in}\}$ and $\{\hat{a}_{out}\}$, respectively. As a next step, an amplifier is introduced assuming a mean-field theory. That is, the mean field along the length of the cavity is considered, ignoring the longitudinal distribution. The gain matrix is $gP + Q$ and the spontaneous emission noise is added [26, 27]. Equation 14.126 then becomes

$$P\{\hat{a}_{out}\} = g(TP\{\hat{a}_{in}\} + PSQ\{\hat{a}_{in}\}) + \sqrt{|g|^2 - 1}\, P\left\{\hat{b}_{sp}^{\dagger}\right\} \tag{14.132}$$

where $P\left\{\hat{b}_{sp}^{\dagger}\right\}$ is a column vector of spontaneous emission noise operators, each one corresponding to a mode belonging to the "laser" mode. For a completely inverted atom, the correlations read $\left\langle \hat{b}_{sp}^{\dagger}(t), \hat{b}_{sp}(t')\right\rangle = 0$ and $\left\langle \hat{b}_{sp}(t), \hat{b}_{sp}^{\dagger}(t')\right\rangle = \delta(t - t')$ (see Ref. [28]). Then, Equation 14.131 becomes

$$\begin{aligned} PV^{\dagger}P\{\hat{a}_{out}\} =& gG\left(PV^{\dagger}P\{\hat{a}_{in}\}\right) + gPV^{\dagger}PSQ\{\hat{a}_{in}\} \\ &+ \sqrt{|g|^2 - 1}\, PV^{\dagger}P\left\{\hat{b}_{sp}^{\dagger}\right\} \end{aligned} \tag{14.133}$$

We consider a lasing mode n in a steady state where $g\gamma_n = 1$. Using the identity

$$PV^{\dagger}P\{\hat{a}_{out,\,in}\} = PU^{\dagger}P\{\hat{a}_{out,\,in}\} + P(I - U^{\dagger}U)PV^{\dagger}P\{\hat{a}_{out,\,in}\} \tag{14.134}$$

one constructs $PU^\dagger P\{\hat{a}_{out}\} - gG(PU^\dagger P\{\hat{a}_{in}\})$ from Equation 14.133 and takes the nth line to get

$$\begin{aligned}&(PU^\dagger P\{\hat{a}_{out}\})_n - (PU^\dagger P\{\hat{a}_{in}\})_n \\ &\quad = +\left(gPV^\dagger PSQ\{\hat{a}_{in}\} + \sqrt{|g|^2 - 1}\, PV^\dagger P\left\{\hat{b}_{sp}^\dagger\right\}\right)_n\end{aligned} \tag{14.135}$$

where terms that are proportional to $1 - g\gamma_n$ have been omitted. The left-hand side can be interpreted as the time derivative multiplied by the cavity round-trip time of the relevant quantity for the steady state. In the language of the adjoint mode theory, this equation corresponds to Equation 14.13, where the equation of motion for the field is projected onto the adjoint mode function and the amplitude of the mode of interest is extracted. This equation also corresponds, physically, to Equation 11.35 with the first and second terms in Equation 14.135 corresponding to $\hat{F}_t$ and $\hat{F}_q$, respectively. Grangier and Poizat [25] then calculate the phase diffusion by evaluating the variance of the round-trip change in the penetrating vacuum and the spontaneous emission noise. In particular, they consider the variance of the phase quadrature of these quantities. Here the phase quadrature of $\hat{a}$ is $\hat{Y} = (\hat{a} - \hat{a}^\dagger)/(2i)$. This calculation corresponds to the evaluation of the reservoir average of the squared phase change in Equations 11.46 and 11.47, although the treatment of the gain medium is different. They define column matrices

$$\begin{aligned}P\{\delta\Gamma_{vac}\} &= (1/\gamma_n)(PV^\dagger PSQ\{\hat{Y}_{in}\}) \\ P\{\delta\Gamma_{sp}\} &= \sqrt{1/|\gamma_n|^2 - 1}(PV^\dagger P\{\hat{Y}_{sp}\})\end{aligned} \tag{14.136}$$

and derive, assuming minus zero temperature that leads to $\left\langle \hat{Y}_{sp}^2 \right\rangle = 1/4$ (see Ref. [27]) and using $\langle vac|\hat{Y}_{in}^2|vac\rangle = 1/4$, for the covariance matrices

$$\begin{aligned}P\left\langle\{\delta\Gamma_{vac}\}\{\delta\Gamma_{vac}\}^\dagger\right\rangle P &= (PVV^\dagger P - PGV^\dagger VG^\dagger P)/(4|\gamma_n|^2) \\ P\left\langle\{\delta\Gamma_{sp}\}\{\delta\Gamma_{sp}\}^\dagger\right\rangle P &= (PV^\dagger VP)(1 - |\gamma_n|^2)/(4|\gamma_n|^2)\end{aligned} \tag{14.137}$$

where the relations $Q = 1 - P$ and $PSP = T$ as well as Equation 14.129 have been used in the first line.

Finally, one compares the result with the case of an ideal single-mode laser, which is described as

$$\hat{a}_{out} = gr\hat{a}_{in} + g\sqrt{1 - |r|^2}\,\hat{b}_{vac} + \sqrt{|g|^2 - 1}\,\hat{b}_{sp}^\dagger \tag{14.138}$$

where g is the gain and r is the amplitude reflectivity of the mirror from where the vacuum noise $\hat{b}_{vac}$ comes in. Defining $\{\delta\Gamma_{vac}\}$ and $\{\delta\Gamma_{sp}\}$ as in Equation 14.136, but for the single-mode case, and setting $r = \gamma$ and $g\gamma = 1$, one obtains $\langle\delta\Gamma_{vac}^2\rangle = \left\langle\delta\Gamma_{sp}^2\right\rangle = (1 - |\gamma|^2)/(4|\gamma|^2)$. As a result, the excess noise factor for both vacuum and spontaneous emission noise becomes

Next, one considers the general quantum Langevin equation for an operator $\hat{o}$ interacting with a reservoir:

$$\frac{d}{dt}\hat{o} = -\frac{1}{2}\gamma\hat{o} + \hat{F}_{\hat{o}} \tag{14.149}$$

The relaxation constant is γ and $\hat{F}_{\hat{o}}$ is the corresponding Langevin noise operator. In the limit of short correlation time of the reservoir compared with the relaxation time of the operator, one has

$$[\hat{F}_{\hat{o}}(t), \hat{F}^{\dagger}_{\hat{o}}(t')] = \gamma[\hat{o}, \hat{o}^{\dagger}]\delta(t-t') \tag{14.150}$$

and

$$[\hat{F}_{\hat{a}_{u_n}}(t), \hat{F}^{\dagger}_{\hat{a}_{u_n}}(t')] = \gamma[\hat{a}_{u_n}, \hat{a}^{\dagger}_{u_n}]\delta(t-t') = \gamma(\phi_n|\phi_n)\delta(t-t') \tag{14.151}$$

The interaction between the field and the atoms is described by the interaction Hamiltonian

$$H_I = \int d\mathbf{x} \sum_n \hbar g_n \left\{ \hat{a}^{\dagger}_{u_n} u^*_n(\mathbf{x})\hat{\sigma}(\mathbf{x}) + \hat{\sigma}^{\dagger}(\mathbf{x})\hat{a}_{u_n} u_n(\mathbf{x}) \right\} \tag{14.152}$$

where g_n is the coupling constant and $\hat{\sigma}(\mathbf{x})$ is the atomic dipole distributed over the cavity. The equation of motion for the field mode n is

$$\frac{d}{dt}\hat{a}_{u_n} = -\frac{1}{2}\gamma_{c,n}\hat{a}_{u_n} - ig_n \int d\mathbf{x}\, \phi^*_n(\mathbf{x})\hat{\sigma}(\mathbf{x}) + F_{\hat{a}_{u_n}} \tag{14.153}$$

The adjoint mode function in the second term appears from

$$\begin{aligned} \frac{i}{\hbar}[H_I, \hat{a}_{u_n}] &= i\int d\mathbf{x} \sum_m g_m [\hat{a}^{\dagger}_{u_n} u^*_m(\mathbf{x})\hat{\sigma}(\mathbf{x}), \hat{a}_{u_n}] \\ &= -i\int d\mathbf{x} \sum_m g_m (\phi_n|\phi_m) u^*_m(\mathbf{x})\hat{\sigma}(\mathbf{x}) \end{aligned} \tag{14.154}$$

where Equation 14.148 has been used. To obtain the second term above, one uses the closure relation $\sum_m \phi_m(\mathbf{x}')u^*_m(\mathbf{x}) = \delta(\mathbf{x}'-\mathbf{x})$, which can be derived using Equation 14.146. From the equation for the atomic dipole $\hat{\sigma}(\mathbf{x})$, when the dipole relaxation is fast compared to the time variation of the population, one obtains the approximate dipole, which is roughly the sum of the population inversion $\hat{\sigma}_e - \hat{\sigma}_g$ and the Langevin force $F_{\hat{\sigma}}(\mathbf{x})$ divided by the dipolar relaxation constant γ (see Equation 4.39b, for example). Thus Equation 14.153 becomes

$$\begin{aligned} \frac{d}{dt}\hat{a}_{u_n} = &-\frac{1}{2}\gamma_{c,n}\hat{a}_{u_n} - i\frac{g_n}{\gamma}\int d\mathbf{x}\, \phi^*_n(\mathbf{x}) \sum_{n'} g_{n'}(\hat{\sigma}_e - \hat{\sigma}_g)\hat{a}_{u_{n'}} u_{n'} \\ &- i\frac{g_n}{\gamma}\int d\mathbf{x}\, \phi^*_n(\mathbf{x}) F_{\hat{\sigma}}(\mathbf{x}) + F_{\hat{a}_{u_n}} \end{aligned} \tag{14.155}$$

Assuming a well-stabilized amplitude A_0 of laser oscillation of a particular mode 0, one analyses the phase diffusion of $\hat{a}_{u_0}$. One has

$$\frac{d}{dt}\phi = -\frac{g_0}{2A_0\gamma}\left\{\left(\phi_0|F_{\hat{\sigma}}(\mathbf{x})\right) + \left(\phi_0|F_{\hat{\sigma}}(\mathbf{x})\right)^{\dagger}\right\} + \frac{1}{2iA_0}\left\{F_{\hat{a}_{un}} - F^{\dagger}_{\hat{a}_{un}}\right\} \quad (14.156)$$

Now the Einstein relation, which was mentioned at the end of Chapter 3 and which reduces to Equations 3.37 and 3.50, gives

$$\begin{aligned}\left\langle F^{\dagger}_{\hat{\sigma}}(\mathbf{x},t)F_{\hat{\sigma}}(\mathbf{x}',t')\right\rangle_R &\simeq 2\gamma\langle\hat{\sigma}_e\rangle_R\delta(t-t')\delta(\mathbf{x}-\mathbf{x}')\\ \left\langle F_{\hat{\sigma}}(\mathbf{x}',t)F^{\dagger}_{\hat{\sigma}}(\mathbf{x},t')\right\rangle_R &\simeq 2\gamma\langle\hat{\sigma}_g\rangle_R\delta(t-t')\delta(\mathbf{x}-\mathbf{x}')\end{aligned} \quad (14.157)$$

where small terms compared to γ have been omitted. The subscript R denotes the reservoir average. Also, we have

$$\begin{aligned}\left\langle F^{\dagger}_{\hat{a}_{u_0}}(t)F_{\hat{a}_{u_0}}(t')\right\rangle_R &= \gamma_{c,0}\left\langle\hat{a}^{\dagger}_{u_0}\hat{a}_{u_0}\right\rangle_R\delta(t-t')\\ \left\langle F_{\hat{a}_{u_0}}(t)F^{\dagger}_{\hat{a}_{u_0}}(t')\right\rangle_R &= \gamma_{c,0}\left\langle\hat{a}_{u_0}\hat{a}^{\dagger}_{u_0}\right\rangle_R\delta(t-t')\end{aligned} \quad (14.158)$$

The reservoir average of the products of the field amplitudes taken for the vacuum state are, by Equation 14.147,

$$\begin{aligned}\left\langle\hat{a}^{\dagger}_{u_0}\hat{a}_{u_0}\right\rangle_R &= \sum_{i,j}(\phi_0|e_i)(e_j|\phi_0)\left\langle 0|\hat{a}^{\dagger}_{e_i}\hat{a}_{e_j}|0\right\rangle_R = 0\\ \left\langle\hat{a}_{u_0}\hat{a}^{\dagger}_{u_0}\right\rangle_R &= \sum_{i,j}(\phi_0|e_i)(e_j|\phi_0)\left\langle 0|\hat{a}_{e_i}\hat{a}^{\dagger}_{e_j}|0\right\rangle_R = (\phi_0|\phi_0)\end{aligned} \quad (14.159)$$

Using Equations 14.156–14.159, the linewidth is obtained as in Chapter 10:

$$\Delta\omega = \frac{g_0^2}{2A_0^2\gamma}(\phi_0|\phi_0)(\sigma_e+\sigma_g) + \frac{1}{4A_0^2}\gamma_{c,0}(\phi_0|\phi_0) \quad (14.160a)$$

Using the steady-state condition $(g_0^2/\gamma)(\sigma_e-\sigma_g) = \frac{1}{2}\gamma_{c,0}$ and the expression for the output power $P = \gamma_{c,0}\hbar\omega_0A_0^2$, one has finally

$$\begin{aligned}\Delta\omega &= (\phi_0|\phi_0)\frac{\hbar\omega_0\gamma_{c,0}^2}{4P}\left(1+\frac{\sigma_e+\sigma_g}{\sigma_e-\sigma_g}\right)\\ &= (\phi_0|\phi_0)\frac{2\hbar\omega_0(\gamma_{c,0}/2)^2}{P}\frac{\sigma_e}{\sigma_e-\sigma_g}\end{aligned} \quad (14.160b)$$

This is $(\phi_0|\phi_0)$ times the standard linewidth (see Equation 4.82). So one obtains the excess noise factor $(\phi_0|\phi_0)$ in terms of the adjoint mode function corresponding to the laser mode function u_0 as in the previous section:

$$K = (\phi_0|\phi_0) = \int\phi_0^*\phi_0 d\mathbf{x} \quad (14.161)$$

Since u_0 is a certain sum of original basis functions e_k where the summation is taken only over $|\mathbf{k}| = \omega_0/c$, which implies zero temporal decay, it might be difficult to determine the composition of a cavity mode with finite output coupling or diffraction loss in this approach.

For another quantum-mechanical theory on the excess noise factor using the master equation (evolution equation for the system density matrix), the reader is referred to Bardroff and Stenholm [31].

14.6 Two Non-Orthogonal Modes with Nearly Equal Losses

There are cases where only two modes of the laser cavity have relatively low and nearly equal losses and other modes have higher losses. In this case, analysis with only two modes is possible and, if the two modes are non-orthogonal, the excess noise factor is determined by the interaction of the two modes. Theoretically, the excess noise factor diverges as the two modes become nearly identical.

Let us consider two orthogonal modes u_1 and u_2 of a cavity satisfying $(u_1^* u_1) = (u_2^* u_2) = 1$ and $(u_1^* u_2) = 0$. Here the bracket signifies the spatial integration over the cavity:

$$(u_i^* u_j) = \int_{cavity} u_i^* u_j d\mathbf{r} \tag{14.162}$$

Note that taking the complex conjugate of the first function is *not* intended in this definition. Let us assume that some mechanism M is introduced to couple them during the propagation in the cavity, such as

$$\frac{d}{dt}\begin{pmatrix} u_1 \\ u_2 \end{pmatrix} = M\begin{pmatrix} u_1 \\ u_2 \end{pmatrix} + \begin{pmatrix} F_1 \\ F_2 \end{pmatrix} \tag{14.163}$$

Here $F_{1,2}$ are the Langevin noises for modes $u_{1,2}$, respectively. They are mutually orthogonal and of equal magnitude. Let us assume that the matrix M has two mutually non-orthogonal eigenmodes e_1 and e_2. The adjoint mode v_1 corresponding to e_1 is orthogonal to e_2. (The adjoint mode v_2 corresponding to e_2 is orthogonal to e_1.) Now, if e_1 and e_2 are normalized to unity and the product $(v_1 e_1)$ is also normalized to unity, the excess noise factor is given by (see Equations 14.5 and 14.25)

$$K = (v_1^* v_1) = \frac{(v_1^* v_1)}{(v_1 e_1)^2} = \frac{1}{\{(v_1/|v_1|)e_1\}^2} \tag{14.164}$$

Therefore, if we abandon the normalization of $(v_1 e_1)$ and assume, instead, that v_1 is normalized (and write $v_1/|v_1| \to v_1$), we have

$$K = \frac{1}{(v_1 e_1)^2} \tag{14.165}$$

where

$$v_1 = \frac{e_1^* - e_2^*(e_1^* e_2)}{\sqrt{1 - \left|(e_1^* e_2)\right|^2}} \tag{14.166}$$

Note that it satisfies $(v_1 e_2) = 0$. Since we can show that $(v_1 e_1)^2 = 1 - \left|(e_2^* e_1)\right|^2$, we have

$$K = \frac{1}{1 - \left|(e_2^* e_1)\right|^2} \tag{14.167}$$

Therefore, we have a large excess noise factor when the two eigenmodes of the cavity have nearly identical field distributions.

As an example, let consider a case where the matrix M has the form

$$M = \begin{pmatrix} -\gamma_1 & b \\ 0 & -\gamma_2 \end{pmatrix} \tag{14.168}$$

where γ_1 and γ_2 are the damping constants of the modes u_1 and u_2, respectively, while b (assumed to be real) is the one-way coupling constant of mode u_2 to mode u_1. The eigenvalues of M are $-\gamma_1$ and $-\gamma_2$, and the corresponding normalized eigenmodes are

$$\begin{aligned} e_1 &= u_1 \\ e_2 &= \frac{u_1 + \gamma u_2}{\sqrt{1+\gamma^2}} \end{aligned} \tag{14.169}$$

respectively, where

$$\gamma = \frac{\gamma_1 - \gamma_2}{b} \tag{14.170}$$

The normalized adjoint modes to these eigenmodes are, respectively,

$$v_1 = \frac{\gamma u_1^* - u_2^*}{\sqrt{1+\gamma^2}} \quad \text{and} \quad v_2 = u_2^* \tag{14.171}$$

One can check that $(e_1^* e_1) = (e_2^* e_2) = (v_1^* v_1) = (v_2^* v_2) = 1$ and that $(e_1 v_2) = (e_2 v_1) = 0$. Writing the amplitudes of these eigenmodes as $a_1(t)$ and $a_2(t)$, respectively, one may have

$$\begin{aligned} \dot{a}_1 e_1 &= (-\gamma_1 + \alpha) a_1 e_1 + F_1' \\ \dot{a}_2 e_2 &= (-\gamma_2 + \alpha) a_2 e_2 + \frac{F_1' + \gamma F_2'}{\sqrt{1+\gamma^2}} \end{aligned} \tag{14.172}$$

where α is the gain and now the noise F_1' and F_2' should contain, in addition to F_1 and F_2, the quantum noise for u_1 and u_2, respectively, associated with the amplification. Projecting these onto respective adjoint modes and dividing by the respective integrated products, we obtain

$$\begin{aligned} \dot{a}_1 &= (-\gamma_1 + \alpha) a_1 + \frac{(v_1 F_1')}{(v_1 e_1)} \\ \dot{a}_2 &= (-\gamma_2 + \alpha) a_2 + \frac{(v_2 F_1') + \gamma (v_2 F_2')}{\sqrt{1+\gamma^2}(v_2 e_2)} \end{aligned} \tag{14.173}$$

Since v_1 and v_2 are normalized, it is easy to see that the diffusion constant of the last terms in Equation 14.173 are those of F_1' and F_2' multiplied by $(v_1 e_1)^{-2}$ and $(v_2 e_2)^{-2}$, respectively. Thus we have, using Equations 14.169–14.171,

$$\begin{aligned} K_1 &= (v_1 e_1)^{-2} = \frac{1+\gamma^2}{\gamma^2} = 1 + \left(\frac{b}{\gamma_1 - \gamma_2}\right)^2 \\ K_2 &= (v_2 e_2)^{-2} = \frac{1+\gamma^2}{\gamma^2} = 1 + \left(\frac{b}{\gamma_1 - \gamma_2}\right)^2 \end{aligned} \tag{14.174}$$

The excess noise factors for the two eigenmodes are the same. We see that both $K_{1,2}$ diverge as $(\gamma_2 - \gamma_1)/b \to 0$. In this limit, Equation 14.169 shows that $e_2 \to e_1 = u_1$. We can say that the excess noise can be very large if the difference of the decay constants of the two modes is small compared to the coupling constant of the mode u_2 to the mode u_1. In this limit, as Equation 14.169 shows, the mode e_2 becomes nearly identical to mode e_1.

Grangier and Poizat [29] gave a similar quantum-mechanical two-mode model that incorporates a loss mode instead of the noise $F_{1,2}$ in Equation 14.163 and emphasized that the laser modes are coupled by sharing common noise due to the same loss modes. Van der Lee *et al.* [32] gave an analysis of two coupled polarization modes of a laser and found a large polarization excess noise factor K_p for an induced frequency splitting close to the magnitude of dissipative coupling.

Van der Lee *et al.* [33] showed that the intensity noise of a laser with two non-orthogonal polarization modes in a gas laser is enhanced with the same polarization excess noise factor as obtained in [32] and that the noise spectrum is not white but had finite bandwidth due to the time needed for the excess noise to develop. Poizat *et al.* [34] analyzed the case of two non-orthogonal transverse modes of a semiconductor laser and showed that the intensity noise of the oscillating mode is enhanced by the presence of a non-orthogonal, second mode that is below threshold. They also showed that the excess noise factor also appears in the expression for the intensity noise.

Van Eijkelenborg *et al.* [35] analyzed the transverse excess noise in a gas laser having a hard-edged unstable cavity. They showed that, when two modes were made to have equal lowest losses by adjustment of the Fresnel number of the cavity, the excess noise factor was strongly increased. Grangier and Poizat [25] also analyzed the frequency spectrum of the excess noise and found a high excess noise factor when a second low-loss mode existed. The spectral width was shown to be the narrower the larger the excess noise factor.

14.7 Multimode Theory

A theory of a single-mode operation that takes into account the non-orthogonal, non-lasing modes was developed by Dutra *et al.* [36]. They showed that the laser spectrum is generally non-Lorentzian due to the coupling of the non-lasing modes to the oscillating mode through gain saturation.

Van Exter *et al.* [37] developed a theory of more general non-orthogonal multimode operation where the amplitudes of all the non-orthogonal modes of the system are traced and the projection onto the measured single oscillating mode inevitably picks up the contributions from other non-oscillating modes because of the non-orthogonality. They stated that the excess noise originates, for both below- and above-threshold operation, from the field fluctuations in other modes that project onto the lasing mode upon evolution. For below threshold, the dynamics of the modes (determined by the cavity geometry) determines the excess noise factor, which can be the same for the phase noise above threshold. For above threshold, the gain fluctuation, in addition to the dynamics, has a role in determining the excess intensity noise.

Van der Lee *et al.* [38] discussed the limitation in obtaining an intensity-squeezed laser light [39] exerted by the excess noise. Using essentially two-mode analysis but incorporating the effects of other non-orthogonal modes as a collected noise term, they found the upper limit of the excess noise factor to be 1.5 to obtain an intensity squeezing. They tried experimentally to reduce the intensity noise by proper mixing of a correlated non-lasing mode to the lasing mode.

14.8 Experiments on Excess Noise Factor

Here we briefly review the experimental results on the longitudinal excess noise factor, the transverse excess noise factor, and the polarization excess noise factor. We also add the experiments on intensity noise, which is related to the excess noise factor.

The longitudinal excess noise factor was observed by Hamel and Woerdman [40] using single-mode semiconductor lasers. They compared two cases of the same

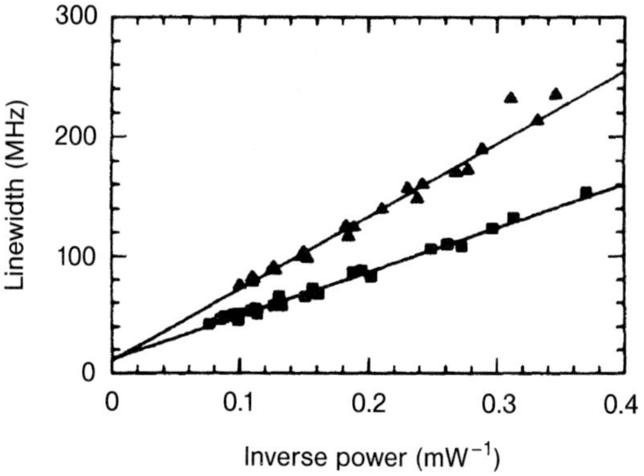

Figure 14.2 Measured laser linewidth as a function of inverse total output power. From Ref. [40]. Hamel, W.A. and Woerdman, J.P. (1990) *Phys. Rev. Lett.*, **64**, 1506, Figure 2.

overall losses. For facet reflectivity of 30%–30% they found smaller excess noise factor by a factor of 1/1.64 than for facet reflectivity of 10%–90%, in fair agreement with the prediction of 1/1.33 for the formula in Equation 14.56, showing that an asymmetry of the cavity enhances the non-orthogonality of the modes and thus excess noise. In Figure 14.2 is shown the measured laser linewidth as a function of inverse total output power. The triangles refer to the 10%–90% lasers and the squares to the 30%–30% lasers. The slopes (the linewidth $\times$ power product) of the two lines are 610 and 372 MHz mW, yielding the ratio 1.64/1 (see Equations 14.24 and 14.25 as well as Equation 14.56).

The transverse excess noise factor for a stable laser resonator with a diffracting aperture was studied by Lindberg *et al.* [41]. They used a Xe–He gas laser operating at 3.51 μm and, by inserting an aperture that is smaller in size than the mode diameter of the otherwise stable cavity, observed an excess noise factor up to 15. Emile *et al.* [42] also measured the transverse excess noise factor in a similar stable-cavity Xe–He gas laser operating at 3.51 μm. They inserted an aperture in front of one of the end mirrors and obtained an excess noise factor up to 13.4. The effect of a second aperture in front of the other mirror was studied with the longitudinal excess noise factor taken into account.

The transverse excess noise factor for an unstable laser resonator was observed by Yao *et al.* [43] using a quantum-well semiconductor laser. The round-trip magnification of the unstable cavity was 6.9. They determined the spontaneous emission factor (the ratio of the spontaneous emission power coupled to the cavity mode of interest to the total emission power) from the observed input–output curve. They deduced the excess noise factor by considering that the spontaneous emission factor for a real cavity is that of an ideal closed cavity multiplied by the transverse and longitudinal excess noise factor. They found that the excess noise factor caused by the unstable geometry was as large as 500 for pulsed operation. The numerically estimated transverse excess noise factor was 175 and the longitudinal factor was 4.

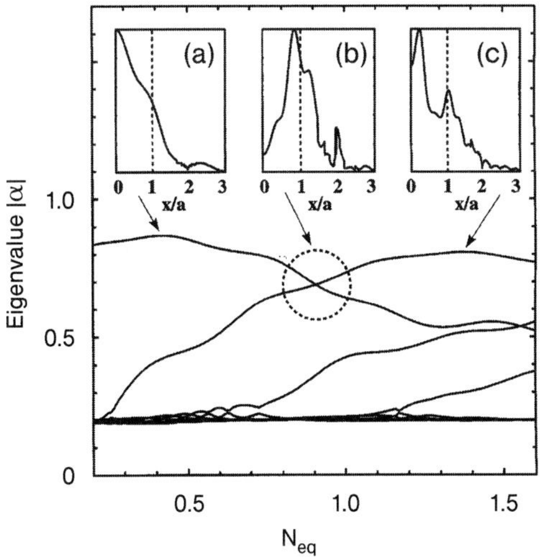

Figure 14.3 Absolute value of the eigenvalues, $|\alpha|$, of a number of transverse modes as functions of equivalent Fresnel number N_{eq} for the square resonator with $M = 1.95$. Note the mode crossing at $N_{eq} = 0.90$. The mode profiles (a), (b), and (c) are for $N_{eq} = 0.42$, 0.90, and 1.38, respectively.
Source: From Ref. [35]. van Eijkelenborg, M.A., Lindberg, Å,M., Thijssen, M.S., and Woerdman, J.P. (1996) *Phys. Rev. Lett.*, **77**, 4314, Figure 2.

Cheng *et al.* [44] observed the excess noise factor for an unstable Nd:YVO_4 laser. The cavity was composed of a convex mirror and a small flat mirror with magnification around 2. An excess noise factor as large as 330 was observed, in fair agreement with theory. Here the longitudinal excess noise factor was estimated to be only 1.1.

Van Eijkelenborg *et al.* [45] tried to express the excess noise factor of an unstable cavity laser in terms of geometrical factors concerning the diffraction loss and compared it with the non-orthogonality theory. They found reasonable agreement only for the case of magnification of 2 and for the lowest-order transverse mode. But, using a He–Xe laser, they experimentally found serious deviations for other cases, and concluded that it is unlikely that a direct relation between diffraction loss and excess noise factor exists.

Using a Xe–He gas laser with an unstable cavity operating at 3.51 μm, van Eijkelenborg *et al.* [35] observed a sharp increase in the excess noise factor when two transverse modes have a common lowest loss as the Fresnel number is varied by changing the size of the square aperture inserted. A transverse excess noise factor over 200 was observed. Figure 14.3 shows the calculated absolute value of the eigenvalue $|\alpha|$ as a function of the Fresnel number N_{eq} for a cavity with magnification M of 1.95. (The eigenvalue here is the complex multiplying factor

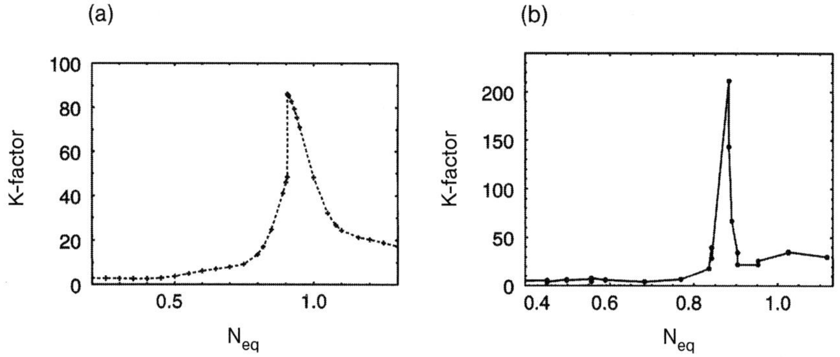

Figure 14.4 The excess noise factor K as a function of equivalent Fresnel number N_{eq}: (a) theoretical, and (b) experimental.
Source: From Ref. [35]. van Eijkelenborg, M.A., Lindberg, Å,M., Thijssen, M.S., and Woerdman, J.P. (1996), *Phys. Rev. Lett.*, **77**, 4314, Figure 3.

for the transverse field distribution associated with a round trip [4].) The insets show the calculated transverse mode profiles. Figure 14.4 shows the calculated and observed excess noise factor (K-factor) as functions of the Fresnel number N_{eq}. An abrupt increase in the K-factor is seen at the crossing point $N_{eq} = 0.9$. This is in accord with the discussions in Section 14.6.

Van der Lee *et al.* [32] studied the polarization properties of a Xe–He gas laser operating at 3.51 μm where two polarization modes are coupled by inserting a dissipative object. A polarization excess noise factor of up to 60 was observed.

The intensity noise spectrum of a He–Xe laser with two polarization modes was observed by van der Lee *et al.* [33]. They found a narrow spectrum for a large excess noise factor at zero frequency and vice versa. This shows that the excess noise factor appears only on a sufficiently long time scale and that laser dynamics is involved in determining the excess noise. The enhancement factor for the intensity noise was the same as that for the phase noise [32]. They also showed that, by proper use of a polarizer on detection to utilize the correlation between the modes, the excess noise can be greatly reduced.

Poizat *et al.* [34] studied the intensity noise of a laser diode with an oscillating TE_{00} mode and a non-oscillating TE_{10} mode. Correlation between the intensity noises of the two modes was observed to enhance the noise of the lasing mode. The intensity noise of the oscillating mode was reduced when the correlation to the side mode was decreased by adjustment of the cavity.

Van Eijkelenborg *et al.* [46] compared the intensity noises of an unstable and a stable cavity He–Xe laser. On the basis of their analysis of the linearized equations for the photon number and the inversion, they found an expression for the intensity noise spectrum that contains the excess noise factor and the spontaneous emission factor. Combining the results of a few measurements, they obtained both factors in fair agreement with theory. The theoretical excess noise factors were

82 and 1.1 for the unstable and the stable cavity, respectively, whereas the observed values were 32 and 1.1, respectively. The theoretical spontaneous emission factors were $(1.2-5.9) \times 10^{-7}$ and 3.7×10^{-6} for the unstable and the stable cavity, respectively, whereas the observed values were 0.71×10^{-7} and 2.0×10^{-6}, respectively.

References

1 Allen, L. and Eberly, J.H. (1975) *Optical Resonance and Two-Level Atoms*, John Wiley and Sons, Inc., New York.
2 Schawlow, A.L. and Townes, C.H. (1958) *Phys. Rev.*, *112*, 1940–1949.
3 Loudon, R. (1973) *The Quantum Theory of Light*, Clarendon Press, Oxford.
4 Siegman, A.E. (1989) *Phys. Rev. A*, *39*, 1253–1263.
5 Siegman, A.E. (1989) *Phys. Rev. A*, *39*, 1264–1268.
6 Champagne, Y. and McCarthy, N. (1992) *IEEE J. Quantum Electron.*, *28*, 128–135.
7 Henry, C.H. (1986) *J. Lightwave Technol.*, *LT-4*, 288–297.
8 Tromborg, B., Olesen, H., and Pan, X. (1991) *IEEE J. Quantum Electron.*, *27*, 178–192.
9 Ujihara, K. (1984) *IEEE J. Quantum Electron.*, *QE-20*, 814–818.
10 Goldberg, P., Milonni, P.W., and Sundaram, B. (1991) *Phys. Rev. A*, *44*, 4556–4563.
11 Prasad, S. (1992) *Phys. Rev. A*, *46*, 1540–1559.
12 Siegman, A.E. (1986) *Lasers*, University Science Books, Mill Valley, CA.
13 Marani, R. and Lax, M. (1995) *Phys. Rev. A*, *52*, 2376–2387.
14 Dutra, S.M. and Nienhuis, G. (2000) *Phys. Rev. A*, *62*, 063805.
15 Deutsch, I.H., Garrison, J.C., and Wright, E.M. (1991) *J. Opt. Soc. Am. B*, *8*, 1244–1251.
16 Arnuad, J. (1986) *Opt. Quantum Electron.*, *18*, 335–343.
17 Petermann, K. (1979) *IEEE J. Quantum Electron.*, *QE-15*, 566–570.
18 Hamel, W.A. and Woerdman, J.P. (1989) *Phys. Rev. A*, *40*, 2785–2787.
19 New, G.C.H. (1995) *J. Mod. Opt.*, *42*, 799–810.
20 van Exter, M.P., Kuppens, S.J.M., and Woerdman, J.P. (1995) *Phys. Rev. A*, *51*, 809–816.
21 Thompson, G.H.B. (1980) *Physics of Semiconductor Laser Devices*, John Wiley and Sons, Ltd, Chichester.
22 Doumont, J.L., Mussche, P.L., and Siegman, A.E. (1989) *IEEE J. Quantum Electron.*, *25*, 1960–1967.
23 Brunel, M., Ropars, G., Floch, A.L., and Bretenaker, F. (1997) *Phys. Rev. A*, *55*, 4563–4567.
24 Firth, W.J. and Yao, A.M. (2005) *Phys. Rev. Lett.*, *95*, 07903.
25 Grangier, P. and Poizat, J.-P. (1999) *Eur. Phys. J. D*, *7*, 99–105.
26 Caves, C.M. (1982) *Phys. Rev. D*, *26*, 1817–1839.
27 Gardiner, C.W. (1991) *Quantum Noise*, Springer, Berlin.
28 Yamamoto, Y. and Imoto, N. (1986) *IEEE J. Quantum Electron.*, *QE-22*, 2032–2042.
29 Grangier, P. and Poizat, J.-P. (1998) *Eur. Phys. J. D*, *1*, 97–104.
30 Cheng, Y.J. and Siegman, A.E. (2003) *Phys. Rev. A*, *68*, 043808.
31 Bardroff, P.J. and Stenholm, S. (1999) *Phys. Rev. A*, *60*, 2529–2533.
32 van der Lee, A.M., van Druten, N.J., Mieremet, A.L., van Eijkelenborg, M.A., Lindberg, Å.M., van Exter, M.P., and Woerdman, J.P. (1997) *Phys. Rev. Lett.*, *79*, 4357–4360.
33 van der Lee, A.M., van Exter, M.P., Mieremet, A.L., van Druten, N.J., and Woerdman, J.P. (1998) *Phys. Rev. Lett.*, *81*, 5121–5124.
34 Poizat, J.-Ph., Chang, T., and Grangier, P. (2000) *Phys. Rev. A*, *61*, 043807.
35 van Eijkelenborg, M.A., Lindberg, Å.M., Thijssen, M.S., and Woerdman, J.P. (1996) *Phys. Rev. Lett.*, *77*, 4314–4317.

36 Dutra, S.M., Joosten, K., Nienhuis, G., van Druten, N.J., van der Lee, A.M., van Exter, M.P., and Woerdman, J.P. (**1999**) *Phys. Rev. A*, *59*, 4699–4702.

37 van Exter, M.P., van Druten, N.J., van der Lee, A.M., Dutra, S.M., Nienhuis, G., and Woerdman, J.P. (**2001**) *Phys. Rev. A*, *63*, 043801.

38 van der Lee, A.M., van Druten, N.J., van Exter, M.P., Woerdman, J.P., Poizat, J.-P., and Grangier, P. (**2000**) *Phys. Rev. Lett.*, *85*, 4711–4714.

39 Machida, S., Yamamoto, Y., and Itaya, Y. (**1987**) *Phys. Rev. Lett.*, *58*, 1000–1003.

40 Hamel, W.A. and Woerdman, J.P. (**1990**) *Phys. Rev. Lett.*, *64*, 1506–1509.

41 Lindberg, Å.M., van Eijkelenborg, M.A., Joosten, K., Nienhuis, G., and Woerdman, J.P. (**1998**) *Phys. Rev. A*, *57*, 3036–3039.

42 Emile, O., Brunel, M., Bretenaker, F., and Floch, A.L. (**1998**) *Phys. Rev. A*, *57*, 4889–4893.

43 Yao, G., Cheng, Y.C., Harding, C.M., Sherrick, S.M., Waters, R.G., and Largent, C. (**1992**) *Opt. Lett.*, *17*, 1207–1209.

44 Cheng, Y.-J., Fanning, C.G., and Siegman, A.E. (**1996**) *Phys. Rev. Lett.*, *77*, 627–630.

45 van Eijkelenborg, M.A., Lindberg, Å.M., Thijssen, M.S., and Woerdman, J.P. (**1997**) *Phys. Rev. A*, *55*, 4556–4562.

46 van Eijkelenborg, M.A., van Exter, M.P., and Woerdman, J.P. (**1998**) *Phys. Rev. A*, *57*, 571–579.

15
Quantum Theory of the Output Coupling of an Optical Cavity

An optical system often includes an optical cavity or an optical resonator. A laser is a typical example. An ideal optical cavity has perfect boundaries and consequently has a set of discrete eigenmodes that are mutually orthogonal. Chapter 4 dealt with such a cavity. A real optical cavity has output coupling so as to allow waves to go into and to come out of it. This coupling inevitably introduces cavity loss and an imperfect boundary. The modes of the cavity become lossy and they are no longer orthogonal to each other. The consequences of the presence of a real cavity in a laser have been described in depth up to now in this book. As we have seen, one aspect of the consequences is the appearance of thermal noise associated with the output coupling (and with other unwanted losses). Another aspect is the appearance of the excess noise factor associated with the laser linewidth that is due to local output coupling at the mirrors. This second aspect has been interpreted in terms of the modes and the adjoint modes associated with the non-Hermiticity of the system. We saw that the thermal noise was also enhanced by the excess noise factor.

The quantum-theoretical treatment of cavity loss or output coupling is a hard task, as we have seen. This belongs to the common topic of the quantum theory of an open system where a system is coupled to its reservoir(s). The accuracy of the description of the coupling determines the extent to which the theory is applicable.

The treatment of output coupling in this book has been based on the expansion of the field in terms of the continuous, normal modes of the "universe," which were defined in a large box including the optical cavity, and led to cavity decay as well as thermal Langevin noise [1, 2]. The analyses of laser linewidth revealed the existence of the excess noise factor [3, 4].

Other methods of field expansion were developed to derive the cavity decay and the Langevin forces due to the output coupling and to obtain the expression for the output field [5, 6, 7–8]. These methods, which are based on field expansion in terms of some continuous field modes, are called quantum field theories.

Since the 1980s, the consequences of the output coupling of an optical cavity have been extensively discussed in relation to the nature of the squeezed state of an optical field that is generated in an optical parametric oscillator. (A "squeezed state" here is a state where the fluctuation of one quadrature is suppressed below the level that both quadratures preserve when the field is not squeezed and the fluctuations of both

Output Coupling in Optical Cavities and Lasers: A Quantum Theoretical Approach
Kikuo Ujihara

ISBN: 978-3-527-40763-7

quadratures are the same. For this state, the other quadrature has an increased fluctuation.) There was an apparent discrepancy between theory and measurements on the degree of squeezing. The theory was first developed for the field inside the cavity [9], while the measurements were of course done outside. Surprisingly, the measurements revealed a larger degree of squeezing than the theoretical prediction [10]. So, a theory of optical squeezing was needed that could calculate the degree of squeezing outside the cavity. This led to the development of the so-called input–output theory for an optical cavity. This theory, also termed quantum noise theory, was based on a system–reservoir model that is similar to the one described in Appendix C, and gave the relation between the input to and the output from the cavity in addition to the relation between the cavity decay constant and the fluctuation of the input noise field. The standard reservoir theory was developed by Haken [11] and Lax [12]. The input–output theory and related theories were developed by Collet and Gardiner [13], Gardiner and Collett [14], Carmichael [15], and Yamamoto and Imoto [16].

More recently, a generation of various non-classical quantum states of the light field in a cavity and transfer to another cavity have been studied for use in quantum information technology, such as quantum computation and quantum communication. In these systems, the maintenance of the quantum state is of crucial importance. Degradation of a quantum state due to unwanted contact with other systems or reservoirs is called "decoherence," which should be avoided as much as possible. In this context, the effect of unwanted noise (noise other than that associated with finite transmission of the mirrors) on the performance, or the input–output characteristics, of the cavity is of great importance. Semenov *et al.* [17] approached this problem by input–output theory, and Khanbekyan *et al.* [18] used the Green's function method. The Green's function method is often used to treat distributed losses in an absorptive dielectric that exists within a cavity or in the output mirror(s) [19].

In this chapter we review some of these quantum theories on output coupling of an optical cavity.

15.1 Quantum Field Theory

There are two schemes of field quantization. One scheme quantizes the field by giving the total field vector potential $\mathbf{A}(\mathbf{r},t)$ and the canonical momentum field $\Pi(\mathbf{r},t)$ a suitable commutation relation and expresses the total Hamiltonian in terms of them. The second scheme is based on the field expansion in terms of the normal modes of the whole space including the cavity, the "universe," which satisfy the orthonormality condition. One quantizes each mode by imposing suitable commutation relations on the expansion coefficients and their time derivatives. Most of the literature follows the latter scheme.

15.1.1 Normal Mode Expansion

15.1.1.1 The One-Sided Cavity Discussed in Chapters 1, 2, and 5–10

Here we summarize the results on the one-sided cavity discussed in Chapters 1, 2, and 5–10 using the second scheme. The cavity consists of a lossless, non-dispersive dielectric of dielectric constant ε_1 that is bounded by a perfect conductor at $z=-d$ and by a vacuum at $z=0$.

For a single cavity mode at ω_c, we have derived, using the approximate power spectrum for a single cavity mode in Equation 2.56, the Langevin equation (Equation 2.63)

$$\frac{d}{dt}\hat{E}^{(+)}(z,t) = -(\gamma_c + i\omega_c)\hat{E}^{(+)}(z,t) + \hat{f}(z,t)$$

where (Equation 2.70b)

$$\left\langle \hat{f}^{\dagger}(z',t')\hat{f}(z,t)\right\rangle = \frac{2\gamma_c\hbar\omega_c\langle n_{\omega_c}\rangle}{\varepsilon_1 d}u^{*}_{\omega_c}(z')u_{\omega_c}(z)\delta(t-t')$$

So, we have a Markovian noise for the field mode ω_c. Here γ_c is the cavity decay constant due to output coupling, $u_{\omega_c}(z)$ is the universal mode function at the resonance frequency ω_c, and $\langle n_{\omega_c}\rangle$ is the expectation value of the number of thermal photons per universal mode at ω_c.

This Langevin force $\hat{f}(z,t)$ should correspond to the Langevin force $\hat{\Gamma}_f(t)$ for the quasimode introduced in Equation 3.35 in which the equation was written for the annihilation operator $\hat{a}$ of the quasimode. For comparison we write

$$\begin{aligned}\hat{E}^{(+)}(z,t) &= B\hat{a}_c(t)u_{\omega_c}(z)\\ \hat{f}(z,t) &= B\hat{\Gamma}_c(t)u_{\omega_c}(z)\end{aligned} \tag{15.1}$$

where the constant B is for normalization so that $\hat{a}^{\dagger}_c(t)\hat{a}_c(t)$ expresses the number of photons in the cavity. As in Equations 14.88 and 14.89 we have

$$|B|^2 = \frac{\hbar\omega}{2\varepsilon_1\int_{cavity}|u_{\omega_c}(z)|^2 dz} = \frac{\hbar\omega}{\varepsilon_1 d} \tag{15.2}$$

where we have used the expression $u_{\omega_c}(z) = \sin\{\omega_c(z+d)/c_1\}$ (see Equation 1.41b). Therefore, Equation 2.70b cited above becomes

$$\left\langle \hat{\Gamma}^{\dagger}_c(t')\hat{\Gamma}_c(t)\right\rangle = 2\gamma_c\langle n_{\omega_c}\rangle\delta(t-t') \tag{15.3}$$

We see that this corresponds to the property in Equations 3.36 and 3.37 for the Langevin force introduced for the quasimode. In Equation 15.6a below we will give another Langevin equation applicable to the one-sided cavity that is more rigorous in the sense that the output coupling at the mirror, as well as the cavity field distribution, is taken into account exactly, thus leading to the correct excess noise factor.

In Equation 10.71 we have shown that the thermal noise that affects the field inside the cavity is the thermal noise that penetrated into the cavity from outside. This relation can be rewritten as

$$e^{ikd}\hat{f}^{+}_t(-0,t) - (e^{-ikd}/r)\hat{f}^{-}_t(-0,t) = -(T'/r)\hat{f}^{-}_{ot}(+0,t) \tag{15.4}$$

Except for a phase factor, the terms on the left-hand side appear in Equation 10.71 or 10.69 as the effective thermal noise at $z=-0$, the inner surface of the coupling mirror. The first term is the right-going thermal noise at $z=-0$ and the second term is the left-going noise amplified during one round trip by $1/r$ with π phase shift at the perfect mirror at $z=-d$. Equation 15.4 says that the sum is equal to the ambient noise that penetrated into the cavity with transmission coefficient T' and was amplified by $1/r$ with a π phase shift. The correlation properties of the ambient noise $\hat{f}_{ot}^{-}(+0,t)$ are the same as those given in Equation 11.1 for the ambient thermal noise $f_t^{R,L}$:

$$
\begin{aligned}
2\varepsilon_0 c_0\left\langle \hat{f}_{ot}^{-\dagger}(+0,t)\hat{f}_{ot}^{-}(+0,t')\right\rangle &= n_\omega \hbar\omega\delta(t-t') \\
2\varepsilon_0 c_0\left\langle \hat{f}_{ot}^{-}(+0,t)\hat{f}_{ot}^{-\dagger}(+0,t')\right\rangle &= (n_\omega+1)\hbar\omega\delta(t-t')
\end{aligned}
\tag{15.5}
$$

We note that the noise $\hat{f}_{ot}^{-}(+0,t)$, which was originally a superposition of the initial values of the modes of the "universe" as in Equation 10.70, has simple correlation functions as a "collective" thermal noise operator. However, it is difficult, if we follow the calculations in Chapter 10 or in Chapter 9, to relate this thermal noise with the cavity decay constant as in Equation 15.3 or in Equations 3.36 and 3.37.

In order to relate the thermal noise to the cavity decay constant, we need to rewrite the field equation of motion as in Equation 2.63 cited above using an equation like Equation 15.1, where the optical field and the thermal field are cast on the same spatial functions and see the relation between the remaining temporal factors. One method to do this is to go from, for example, Equation 10.69 for $\hat{e}^{+}(-0,t)$ to the corresponding equation for $\hat{e}(z,t)$ as we did in Section 14.3 and cast the equations on to the appropriate adjoint mode (in this case $\sin\{\Omega_c^*(z+d)/c_1\}$). Then, through a procedure similar to Equations 14.103–14.106 one can obtain the normally ordered correlation function for the thermal noise and get its relation to the cavity decay constant. The anti-normally ordered correlation function may be obtained similarly. Here, instead of deriving a Langevin equation for the one-sided cavity considered in Chapters 9 and 10, we derive a Langevin equation for a general two-sided cavity according to the directions mentioned above. We have Equation 14.103, which can be rewritten, after using Equations 14.97, 14.100, and 14.54, as well as Equation 11.52a, as

$$
\frac{d}{dt}\hat{a}(t) = -\gamma_c\hat{a}(t) + K_L^{1/4}\left\{K_2^{1/4}\hat{b}^R(d+0,t) + K_1^{1/4}\hat{b}^L(-0,t)\right\}
\tag{15.6a}
$$

where

$$
\hat{b}^{R,L} = \sqrt{\frac{2\varepsilon_0 c_0}{\hbar\omega}}\sqrt{2\gamma_c}\hat{f}^{R,L}, \qquad \left\langle \hat{b}^{R,L\dagger}(t)\hat{b}^{R,L}(t')\right\rangle = 2\gamma_c n_\omega\delta(t-t')
\tag{15.6b}
$$

and

$$
K_L = \left(\frac{\beta_{c1}+\beta_{c2}}{\gamma_c}\right)^2, \qquad K_{1,2} = \left(\frac{\beta_{c1,2}}{\gamma_c}\right)^2, \qquad \beta_{c1,2} = \frac{1-|r_{1,2}|^2}{2|r_{1,2}|}
\tag{15.6c}
$$

Here we have used Equation 11.31 with $\langle\sigma_z\rangle = 0$ and set $\omega = \omega_c$ to treat an empty cavity. The factor $\gamma'/(\gamma'+\gamma_c')$ coming from the dispersion of the medium has also been omitted. The Langevin noise forces defined in Equation 15.6b have standard forms of correlation, but these are multiplied by factors related to the excess noise factor in the above equation. Equation 15.6a shows clearly the separate contributions of the noise from the two mirrors. The K_L in Equation 15.6c is the same as those in Equations 11.72 and 11.107. By setting $r_1 \to -1$ and $r_2 \to r$ we have the Langevin equation for the one-sided cavity treated in Chapters 9 and 10. In this case K_L reduces to those in Equations 9.106 and 10.112. Note that this derivation of the Langevin force is more rigorous than that in Equation 15.3 above in the sense that the output coupling at the mirrors, as well as the cavity field distribution, is taken into account exactly. If applied to the one-sided cavity, we have

$$K_L^{1/2} K_2^{1/2} \left\langle \hat{b}^{R\dagger}(t) \hat{b}^R(t') \right\rangle = \left(\frac{\beta_c}{\gamma_c} \right)^2 2\gamma_c n_\omega \delta(t - t') \tag{15.6d}$$

which is larger than Equation 15.3 by the factor $(\beta_c/\gamma_c)^2$, which appeared in Chapters 9 and 10 as the excess noise factor.

In Chapter 10 we have examined in detail the relation between the field inside the cavity, the field coupled out from the cavity, and the ambient thermal field. We have from Equations 10.137 and 10.127, respectively

$$\hat{e}_o^+(z,t) = Te^{ikd}\hat{e}^+\left(-0, t - \frac{z}{c_0}\right) + r'\hat{f}_{ot}^-\left(+0, t - \frac{z}{c_0}\right)$$

$$\hat{e}_o^-(z,t) = \hat{f}_o^-(z,t) = \hat{f}_{ot}^-(z,t)$$

Here the suffix o signifies a wave existing outside the cavity, and the $+$ $(-)$ sign designates a right-going (left-going) wave. We see that, outside the cavity, the right-going waves are the waves transmitted from inside and the thermal wave reflected at the mirror. The left-going wave is only the thermal field, as expected. Note that, as shown in Equations 10.26 and 10.130 and meant by $\hat{f}_{ot}^-(z,t)$ above, the quantum noise does not appear outside the cavity in its raw form. Its effect is contained in $\hat{e}^+(-0, t - z/c_0)$ in the first term in Equation 10.137.

In Chapter 10 we have ignored the second term in Equation 10.137 in evaluating the linewidth. The consequence of taking this term into account was discussed by Yamamoto and Imoto [16] (using an input–output theory similar to that of Carmichael [15]), who found a constant term in the phase noise spectrum in addition to that obtained for inside the cavity. (They also found changes in the photon statistics.) The thermal noise contained in the first term in Equation 10.137 is, by Equation 10.69 together with Equation 10.71, proportional to the time integral of $\hat{f}_{ot}^-(z,t)$, which is roughly in phase quadrature with the second term. Thus it can be shown that the inclusion of the second term will bring no interference and add a constant (white) term in the power spectrum.

15.1.1.2 A One-Sided Cavity with a Dielectric-Slab Mirror

Knöll *et al.* [6] considered an empty one-sided cavity of length L with a perfect reflector at $z = 0$. The coupling mirror is composed of a dielectric slab of thickness d with refractive index n, and the inside and outside of the cavity are vacuum. The refractive index distribution is $n(x) = 1$, for $0 \le x \le L$, $L + d \le x$, and $n(x) = n$, for $L \leq x \leq L + d$. Eventually, the thickness d will be made to be very small. The vector potential $A(k, x)$ is given by

$$\frac{d^2}{dx^2} A(k, x) + n^2(x) k^2 A(k, x) = 0, \qquad k^2 = \omega^2 / c^2 \tag{15.7}$$

giving

$$A(k, x) = \left[\frac{\hbar}{4\pi F \varepsilon_0 \omega}\right]^{1/2} \begin{cases} \bar{T}(\omega)(e^{ikx} - e^{-ikx}), & 0 \leq x \leq L \\ [\bar{T}(\omega)/\bar{T}^*(\omega)] e^{ikx} - e^{-ikx}, & L \leq x \end{cases} \tag{15.8a}$$

where F is the mirror area. Here the function $\bar{T}(\omega)$ is the spectral response function of the cavity, which reads

$$\bar{T}(\omega) = \frac{\bar{t}(\omega)}{1 + \bar{r}(\omega) e^{2iL\omega/c}} \tag{15.8b}$$

where c is the velocity of light in vacuum and $\bar{t}(\omega)$ and $\bar{r}(\omega)$ are the transmission and reflection coefficients, respectively, of the mirror, which should satisfy

$$\begin{aligned} |\bar{t}(\omega)|^2 + |\bar{r}(\omega)|^2 &= 1 \\ \bar{t}^*(\omega)\bar{r}(\omega) + \bar{t}(\omega)\bar{r}^*(\omega) &= 0 \end{aligned} \tag{15.8c}$$

Assuming that the frequency dependences of $\bar{t}(\omega)$ and $\bar{r}(\omega)$ are small, we write $\bar{t}(\omega) = \bar{t} = |\bar{t}| e^{i\phi}$ and $\bar{r}(\omega) = \bar{r} = |\bar{r}| e^{i\psi}$. In this case, the poles of $\bar{T}(\omega)$ giving the cavity resonance are given, assuming a good cavity ($|\bar{t}|^2 \ll 1$), as

$$\begin{aligned} \Omega_m &= \omega_m - i\Gamma/2 \\ \omega_m &= m\frac{\pi c}{L} + \frac{c}{2L}(\pi - \psi) \\ \Gamma &= -\frac{c}{L}\ln(1 - |\bar{t}|^2)^{1/2} \approx \frac{c}{2L}|\bar{t}|^2 \end{aligned} \tag{15.9}$$

where m is an integer. In the vicinity of a cavity mode m, Equation 15.8b reads

$$\bar{T}(\omega) \approx \left(\frac{c}{2L}\Gamma\right)^{1/2} \frac{e^{i\phi}}{\Gamma/2 - i(\omega - \omega_m)} \tag{15.10}$$

The positive frequency part of the electric field operator is given as

$$\hat{E}^{(+)}(x, t) = i \int_0^\infty dk\, \omega A(k, x) \hat{a}(k, t) \tag{15.11}$$

Defining the propagation function

$$K^{(+)}(x_1, t_1; x_2, t_2) = -\frac{1}{\hbar} \int_0^\infty dk\, \omega A(k, x_1) A^*(k, x_2) e^{-i\omega(t_1 - t_2)} \tag{15.12}$$

we have, using the orthogonal property of the mode functions,

$$\hat{E}^{(+)}(x, t) = -2\varepsilon_0 F \int_0^\infty dx' K^{(+)}(x, t; x', t') \hat{E}^{(+)}(x', t') \tag{15.13}$$

Here we ignore the source terms that drive the electric field (although these are included in the authors' paper) but concentrate on the thermal field. The propagation function for the case where both x_1 and x_2 are inside the cavity reads

$$\begin{aligned} K^{(+)}(x_1, t_1; x_2, t_2) = -\frac{1}{4\pi F\varepsilon_0 c}\Big[& G_1\{t_1 - t_2 - (x_1 - x_2)/c\} \\ & + G_1^*\{t_2 - t_1 - (x_1 - x_2)/c\} \\ & - G_1\{t_1 - t_2 - (x_1 + x_2)/c\} \\ & - G_1^*\{t_2 - t_1 - (x_1 + x_2)/c\}\Big] \end{aligned} \tag{15.14}$$

The propagation function for the case $x_1 > L$ and $0 < x_2 < L$ is

$$\begin{aligned} K^{(+)}(x_1, t_1; x_2, t_2) = -\frac{1}{4\pi F\varepsilon_0 c}\Big[& G_2\{t_1 - t_2 - (x_1 - x_2)/c\} \\ & + G_2^*\{t_2 - t_1 - (x_1 - x_2)/c\} \\ & - G_2\{t_1 - t_2 - (x_1 + x_2)/c\} \\ & - G_2^*\{t_2 - t_1 - (x_1 + x_2)/c\}\Big] \end{aligned} \tag{15.15}$$

The functions $G_1(t)$ and $G_2(t)$ are given as

$$\begin{aligned} G_1(t) &= \int_0^\infty d\omega |\bar{T}(\omega)|^2 e^{-i\omega t} \simeq \sum_m \frac{\pi c}{L} e^{-i\omega_m t - (\Gamma/2)|t|} \\ G_2(t) &= \int_0^\infty d\omega\, \bar{T}(\omega) e^{-i\omega t} \simeq \sum_m 2\pi \left[\frac{c}{2L}\Gamma\right]^{1/2} e^{i\phi}\Theta(t) e^{-i\{\omega_m - i(\Gamma/2)\}t} \end{aligned} \tag{15.16}$$

Here $\Theta(t)$ is the unit step function. The second expression in each of $G_1(t)$ and $G_2(t)$ have been obtained under the approximation that we have a good cavity ($|\bar{t}|^2 \ll 1$) and the assumption that we are interested in the field variation that is slower than ω. The latter assumption has allowed us to extend the lower limit of the frequency integral to $-\infty$.

We further assume that the cavity field can be (approximately) expressed in terms of the standing waves and that we are concerned with a time scale that is larger than the cavity round-trip time but smaller than the cavity decay time ($2L/c \ll \Delta\tau \ll \Gamma^{-1}$). In this case, the propagation function in Equation 15.14 for both x and x' inside the cavity may be rewritten as

$$\begin{aligned} K^{(+)}(x, t; x', t') = & -\frac{1}{F\varepsilon_0 L}\sum_m \exp\left\{-i\left(\omega_m - i\frac{\Gamma}{2}\right)(t - t')\right\} \\ & \times \sin(\omega_m x/c)\sin(\omega_m x'/c) \end{aligned} \tag{15.17}$$

The propagation function in Equation 15.15 for $x' > L$ and $0 < x < L$ may be rewritten as

$$K^{(+)}(x,t;x',t') = \frac{i}{F\varepsilon_0 L}\sum_m \left[\frac{c}{2L}\Gamma\right]^{1/2} e^{i\phi} \times \exp\left\{-i\left(\omega_m - i\frac{\Gamma}{2}\right)(t-t'-x'/c)\right\}\sin(\omega_m x/c) \tag{15.18}$$

Now we introduce the normalized cavity mode function as

$$A_m(x) = \left[\frac{\hbar}{LF\varepsilon_0\omega_m}\right]^{1/2}\sin(\omega_m x/c), \qquad 0<x<L \tag{15.19}$$

and write the field operator in terms of the mode creation operators $\hat{a}_m(t)$:

$$\hat{E}^{(+)}(x,t) = i\sum_m \omega_m A_m(x)\hat{a}_m(t) \tag{15.20}$$

Then, using the property of $G_2(t)$ in Equation 15.16, one can show that Equations 15.13, 15.17, and 15.18 yield

$$\hat{a}_m(t) = \hat{a}_m(t')\exp\left\{-i\left(\omega_m - i\frac{\Gamma}{2}\right)(t-t')\right\} + \Gamma^{1/2}e^{i\phi}\int_{t'}^{t}\exp\left\{-i\left(\omega_m - i\frac{\Gamma}{2}\right)(t-\tau)\right\}\hat{b}_m(\tau)d\tau \tag{15.21}$$

where

$$\hat{b}_m(\tau) = -\left[\frac{2\varepsilon_0 Fc}{\hbar\omega_m}\right]^{1/2}\hat{E}_{in}^{(+)}(\tau) \tag{15.22}$$

Here $\hat{E}_{in}^{(+)}(\tau)$ stems from the incoming part (e^{-ikx}) of the outer field in Equation 15.8a. (The outgoing part (e^{ikx}) does not contribute due to destructive interference with the preceding factor.) In Equation 15.21, τ is defined as $\tau = t' + x'/c$.

Differentiation of Equation 15.21 with respect to time yields the Langevin equation

$$\dot{\hat{a}}_m(t) = -i\left(\omega_m - i\frac{\Gamma}{2}\right)\hat{a}_m(t) + \Gamma^{1/2}e^{i\phi}\hat{b}_m(t) \tag{15.23}$$

Here the coefficient $\Gamma/2$ is the damping rate and the incoming field of $\hat{b}_m(t)$ provides the thermal Langevin noise. The output field $\hat{E}_{out}^{(+)}(t)$ comes from the outgoing part (e^{ikx}) in Equation 15.8a. Following a similar procedure that led to Equation 15.21, the authors show that the total input–output relation is

$$\hat{E}_{out}^{(+)}(t-x/c) = \bar{r}\hat{E}_{in}^{(+)}(t-x/c) + \sum_m \Gamma^{1/2}e^{i\phi}\left[\frac{\hbar\omega_m}{2\varepsilon_0 Fc}\right]^{1/2}\hat{a}_m(t-x/c) \tag{15.24}$$

The authors further examine the commutation relations for the cavity field mode operator $\hat{a}_m(t)$, and the incoming and outgoing field operators, $\hat{b}_m(t)$ and $\hat{c}_m(t)$. Here $\hat{c}_m(t)$ is given by $\hat{E}^{(+)}_{out}(\tau)$ as in Equation 15.22. Under the good cavity approximation and the assumption on the time scale mentioned above, they find the proper commutation relations:

$$\begin{aligned}
&\left[\hat{a}_m, \hat{a}^{\dagger}_{m'}\right] = \delta_{mm'}, && [\hat{a}_m, \hat{a}_{m'}] = 0 = \left[\hat{a}^{\dagger}_m, \hat{a}^{\dagger}_{m'}\right] \\
&\left[\hat{b}_m(t), \hat{b}^{\dagger}_{m'}(t')\right] = \delta_{mm'}\delta(t-t'), && \left[\hat{b}_m(t), \hat{b}_{m'}(t')\right] = 0 = \left[\hat{b}^{\dagger}_m(t), \hat{b}^{\dagger}_{m'}(t')\right] \\
&\left[\hat{c}_m(t), \hat{c}^{\dagger}_{m'}(t')\right] = \delta_{mm'}\delta(t-t'), && [\hat{c}_m(t), \hat{c}_{m'}(t')] = 0 = \left[\hat{c}^{\dagger}_m(t), \hat{c}^{\dagger}_{m'}(t')\right]
\end{aligned} \tag{15.25}$$

Also, the causality is derived as

$$\begin{aligned}
&\left[\hat{a}_m(t), \hat{b}_{m'}(t')\right] = 0 = \left[\hat{a}_m(t), \hat{b}^{\dagger}_{m'}(t')\right], && t<t' \\
&[\hat{a}_m(t), \hat{c}_{m'}(t')] = 0 = \left[\hat{a}_m(t), \hat{c}^{\dagger}_{m'}(t')\right], && t>t'
\end{aligned} \tag{15.26}$$

The first equation says that the future input does not affect the present field inside the cavity, and the second says that the past outgoing wave does not affect the present field inside the cavity. The authors also outline the way to construct the (multi-space-time) correlation functions of the outgoing field in terms of those of the internal and incoming fields.

15.1.1.3 **Other Works on Normal Mode Expansion**

Historically, the first paper on the thermal Langevin force on the cavity field as the superposition of initial values of the modes of the "universe" was published by Lang and Scully [20] using the one-sided cavity model similar to that used by Knöll *et al.* [6] in the previous subsection. They showed that the correlation function of the force is related to the cavity damping rate so as to fulfill the proper fluctuation–dissipation theorem. Gea-Banacloche *et al.* [21] used a similar cavity model to describe the input–output relation for application to the problem of squeezing in a cavity mode relative to the squeezing of the field coupled out of the cavity. Baseia *et al.* [22, 23] also used a similar cavity model to analyze the laser operation using the modes of the "universe" but going to the "collective" mode amplitude, and examined the relation of the internal and external fields through essentially a semiclassical approach. Glauber and Lewenstein [5] considered a general scattering and transmission problem in the presence of a non-uniform, linear dielectric. They used the normal mode expansion of the field and quantized the field by treating the expansion coefficients as the operators. They also expanded the field in terms of the plane waves and found a Hamiltonian in a non-diagonal form. The two quantization schemes are discussed in relation to the scattering theory and field fluctuation due to the dielectric. The results are applied to the problem of spontaneous emission and emission by a charged particle.

15.1.2 Natural Mode Quantization

A natural mode of a cavity with transmission loss is, as we saw in Chapter 1, a decaying mode. Quantization schemes of such modes were described in Chapter 14 in relation to the theory of the excess noise factor. One was due to Grangier and Poizat [24], who divided the "universal modes" into laser modes and loss modes by use of projection operators, and introduced the cavity mode as those corresponding to the laser. The other was presented by Cheng and Siegman [25], who also divided the "universal modes" into system eigenmodes and the remaining modes. Both of these two papers utilized the concept of adjoint mode to derive the excess noise in a laser. These two works were formal and gave no connection with realistic cavity decay.

Dutra and Nienhuis [7] gave a direct approach to quantize the natural decaying modes of a leaky cavity. They also depended on the concept of adjoint mode in quantizing the decaying modes. Here we show excerpts of the last.

The cavity model considered is a one-dimensional cavity with a perfect mirror at $x=-L$ and a coupling surface at $x=0$. The interior of the cavity is a vacuum and the outside region $0<x<\infty$ is filled with a dielectric of refractive index n_d. The amplitude reflection coefficient of the coupling surface for a wave incident from inside the cavity is r. The electric field is polarized in the y-direction. The natural mode of the cavity, which has only an outgoing wave outside, is

$$g(c\kappa_n, x) = \begin{cases} e^{i\kappa_n x} + re^{-i\kappa_n x}, & -L \le x<0 \\ (1+r)e^{i\kappa_n n_d x}, & 0<x \end{cases} \tag{15.27}$$

where $\kappa_n = k_n - i\gamma$ with $k_n = (\pi/L)n$ (integer n) and $\gamma = (1/2L)\ln(1/|r|)$. The adjoint mode, which has only an incoming wave outside, is

$$\tilde{g}(c\kappa_m^*, x) = \begin{cases} e^{i\kappa_m^* x} + \dfrac{1}{r}e^{-i\kappa_m^* x}, & -L \le x<0 \\ \dfrac{1+r}{r}e^{-i\kappa_m^* n_d x}, & 0<x \end{cases} \tag{15.28}$$

Note that the adjoint mode here is the complex conjugate of that defined in Chapter 14 (see below Equation 14.5). The natural mode diverges at large x and is not suitable for quantization in this form. One looks for other functions suitable for outside the cavity, which the authors claim is the natural mode for the outside region that satisfies the boundary condition at the interface $x=0$ and at $x=+\infty$. It reads

$$G(ck, x) = \begin{cases} (1-r)e^{-ikx}, & x<0 \\ e^{-ikn_d x} - re^{ikn_d x}, & 0<x \end{cases} \tag{15.29}$$

where k is real. The adjoint mode for this is

$$\tilde{G}(ck,x) = \begin{cases} \dfrac{r-1}{r} e^{ikx}, & x<0 \\ e^{-ikn_d x} - \dfrac{1}{r} e^{ikn_d x}, & 0<x \end{cases} \tag{15.30}$$

Because, for inside the cavity, the mode function and the adjoint mode function are not strictly orthogonal, the authors seek exact orthogonality. For this purpose, they introduce spinor notation, as in Example 4 in Chapter 14 (see Equation 14.61), where the upper and lower members signify the right- and left-going wave, respectively. These members are given formally as

$$\widehat{g}(c\kappa_n, x) = \frac{1}{\sqrt{8L}} \begin{bmatrix} g(c\kappa_n, x) - \frac{i}{n(x)\kappa_n} \frac{\partial}{\partial x} g(c\kappa_n, x) \\ g(c\kappa_n, x) + \frac{i}{n(x)\kappa_n} \frac{\partial}{\partial x} g(c\kappa_n, x) \end{bmatrix} \tag{15.31}$$

and the adjoint $\tilde{\widehat{g}}(c\kappa_m^*, x)$ can be obtained by replacing $g(c\kappa_n, x)$ by $\tilde{g}(c\kappa_m^*, x)$ and κ_n by κ_m^*. The factor $\sqrt{8L}$ is for normalization in Equation 15.32 below. Physically, the upper and lower members in the spinor are shown to be equal to $E(x) + cB(x)/n(x)$ and $E(x) - cB(x)/n(x)$, which are, respectively, right- and left-going waves in the cavity. It can be shown that

$$\int_{-L}^{0} dx\, \tilde{\widehat{g}}^{\dagger}(c\kappa_m^*, x) \widehat{g}(c\kappa_n, x) n^2(x) = \delta_{nm} \tag{15.32}$$

Note that the dagger sign here denotes the transpose and the transposed quantity should be complex conjugated. Thus $\widehat{g}(c\kappa_n, x)$ and its adjoint $\tilde{\widehat{g}}(c\kappa_m^*, x)$ constitute the cavity modes and their adjoint modes as discussed in Chapter 14.

The spinor form of the outside mode and its adjoint are constructed in just the same manner as for inside. The resultant spinor $\widehat{G}(ck, x)$ and its adjoint $\tilde{\widehat{G}}(ck, x)$ satisfy

$$\int_{0}^{\infty} dx\, \tilde{\widehat{G}}^{\dagger}(ck', x) \widehat{G}(ck, x) n^2(x) = \delta(k - k') \tag{15.33}$$

The field can be written in the spinor form as

$$\mathbf{F}(x) = \frac{1}{2} \begin{bmatrix} E(x) + \dfrac{c}{n(x)} B(x) \\ E(x) - \dfrac{c}{n(x)} B(x) \end{bmatrix} \tag{15.34}$$

The field inside the cavity can be expanded in terms of the spinors $\widehat{g}(c\kappa_n, x)$ and $\tilde{\widehat{g}}(c\kappa_m^*, x)$. Similarly, the field outside can be expanded in terms of $\widehat{G}(ck, x)$ and $\tilde{\widehat{G}}(ck, x)$. The expansion coefficient for the mode is given by the projection of the field onto the adjoint mode and vice versa:

$$\begin{aligned}
\mathbf{F}_{cav}(x) &= \sum_{n=-\infty}^{\infty} \bar{g}(c\kappa_n, x) \int_{-L}^{0} dx' \,\tilde{\bar{g}}^{\dagger}(c\kappa_n^*, x')\mathbf{F}(x')n^2(x') \\
\tilde{\mathbf{F}}_{cav}(x) &= \sum_{n=-\infty}^{\infty} \tilde{\bar{g}}(c\kappa_n^*, x) \int_{-L}^{0} dx' \,\bar{g}^{\dagger}(c\kappa_n, x')\mathbf{F}(x')n^2(x') \\
\mathbf{F}_{out}(x) &= \int_{-\infty}^{\infty} dk \,\bar{G}(ck, x) \int_{0}^{\infty} dx' \,\tilde{\bar{G}}^{\dagger}(ck, x')\mathbf{F}(x')n^2(x') \\
\tilde{\mathbf{F}}_{out}(x) &= \int_{-\infty}^{\infty} dk \,\tilde{\bar{G}}(ck, x) \int_{0}^{\infty} dx' \,\bar{G}^{\dagger}(ck, x')\mathbf{F}(x')n^2(x')
\end{aligned} \tag{15.35}$$

For treating the field at the coupling surface $x=0$, we represent the field as

$$\begin{aligned}
\mathbf{F}(x) = \tfrac{1}{2}\lim_{\varepsilon\to 0^+} \Big[&\left\{\mathbf{F}_{cav}(x) + \tilde{\mathbf{F}}_{cav}(x)\right\}\Theta(\varepsilon - x) \\
&+ \left\{\mathbf{F}_{out}(x) + \tilde{\mathbf{F}}_{out}(x)\right\}\Theta(\varepsilon + x)\Big]
\end{aligned} \tag{15.36}$$

This expression can be shown to allow for the correct value of $\mathbf{F}(0)$ regardless of the actual value of the refractive index at $x=0$.

The next task is to quantize the field. The quantization is carried out by regarding the expansion coefficients in Equation 15.35 as operators. For example, we set

$$\begin{aligned}
\hat{\mathbf{F}}_{cav}(x) &= \sum_{n=-\infty}^{\infty} \sqrt{\frac{\hbar c\kappa_n}{2\varepsilon_0}}\hat{a}_n \bar{g}(c\kappa_n, x) \\
\hat{a}_n &= \sqrt{\frac{2\varepsilon_0}{\hbar c\kappa_n}} \int_{-L}^{0} dx' \,\tilde{\bar{g}}^{\dagger}(c\kappa_n^*, x')\hat{\mathbf{F}}(x')n^2(x')
\end{aligned} \tag{15.37}$$

where the field $\hat{\mathbf{F}}(x)$ is now an operator (see below). Other operators, $\hat{b}_n$, $\hat{a}_{out}(k)$, and $\hat{b}_{out}(k)$ are defined similarly. The $\hat{b}$ operators are quantized versions of the expansion coefficients for adjoint modes. Then using Equation 15.36 in

$$H = \varepsilon_0 \int_{-L}^{\infty} dx \,\hat{\mathbf{F}}^{\dagger}(x)\hat{\mathbf{F}}(x)n^2(x)$$

one obtains the expression for the Hamiltonian in terms of the mode operators. It contains terms of $\hat{b}_n^{\dagger}\hat{a}_n$, $\hat{a}_n^{\dagger}\hat{b}_n$, $\hat{a}_n^{\dagger}\hat{a}_{n'}$, and $\hat{b}_n^{\dagger}\hat{b}_{n'}$, as well as of $\hat{b}_{out}^{\dagger}(k)\hat{a}_{out}(k)$, $\hat{a}_{out}^{\dagger}(k)\hat{b}_{out(k)}$, $\hat{a}_{out}^{\dagger}(k)\hat{a}_{out}(k')$, and $\hat{b}_{out}^{\dagger}(k)\hat{b}_{out}(k')$.

As is expected from the form of Equation 15.36, where the operators for the cavity modes and the outside modes appear only in separate spatial regions, the Hamiltonian has no cross-terms between the cavity mode and the outside mode operators. This is in contrast to, for example, Equation C.1 in Appendix C for the reservoir model or Equation 15.45 below for the projection operator method, where the system (cavity) and the reservoir (channel) modes are quantized independently of each other and their interaction is expressed by the cross-terms of the operators for the system and for the reservoir. How can the interaction between the inner cavity modes and the outer modes occur without cross-terms here? The answer is that in this formalism the commutator between the cavity operators and outside field operators do not vanish in general (while the corresponding operators commute in the case of the system–reservoir model or in the case of the projection operator method cited below due to their mutual independence).

Let us examine the commutators. For this purpose, we need the commutators concerning the field $\hat{\mathbf{F}}(x)$. These commutators are obtained through the modes of the "universe" approach, as follows. We write the mode of the universe as $U(\omega\ x)$, where ω is a continuous variable, which comes from the universe model that extends to $x \to \infty$ (cf. Section 1.4). In a similar fashion to that leading to Equation 1.75, the authors derive the closure relation

$$\int_0^\infty d\omega\, U^*(\omega, x)U(\omega, x') = \frac{\delta(x-x') - \delta(x+x'+2L)}{n^2(x)} \tag{15.38}$$

where the second term appears because of the presence of a perfect mirror at $x = -L$ (see below Equation 1.78). Then, introducing the continuous creation and annihilation operators $\hat{a}(\omega)$ and $\hat{a}^\dagger(\omega)$, the electric field and the magnetic flux can be written as

$$\begin{aligned} \hat{E}(x) &= \int_0^\infty d\omega \sqrt{\frac{\hbar\omega}{\varepsilon_0}}\, U(\omega, x)\hat{a}(\omega) + \text{H.C.} \\ \hat{B}(x) &= -i\int_0^\infty d\omega \sqrt{\frac{\hbar}{\varepsilon_0\omega}}\frac{\partial}{\partial x}\, U(\omega, x)\hat{a}(\omega) + \text{H.C.} \end{aligned} \tag{15.39}$$

Using Equation 15.38 together with the commutation relation $[\hat{a}(\omega), \hat{a}^\dagger(\omega')] = \delta(\omega - \omega')$ we have

$$[\hat{D}(x), \hat{B}(x')] = i\hbar\frac{\partial}{\partial x'}\{\delta(x-x') - \delta(x+x'+2L)\} \tag{15.40}$$

where $\hat{D}(x) = \varepsilon_0 n^2(x)\hat{E}(x)$ is the electric displacement operator.

Using Equation 15.40 with the quantized form of Equation 15.34, we can evaluate the commutators involving the operator in Equation 15.37 and similar expressions. The cavity modes $\hat{a}_n$ and $\hat{a}^\dagger_{n'}$ do not commute, expressing the non-orthogonality, and similarly for the adjoint modes:

$$[\hat{a}_n, \hat{a}^\dagger_{n'}] = \frac{1}{r^2}[\hat{b}_n, \hat{b}^\dagger_{n'}]^* = \frac{i}{4L\sqrt{\kappa_n\kappa^*_{n'}}}\frac{\kappa_n + \kappa^*_{n'}}{\kappa_n - \kappa^*_{n'}}\frac{r^2-1}{r^2} \tag{15.41}$$

The operators of the outside region have essentially delta-correlated commutators:

$$\begin{aligned} [\hat{a}_{out}(k), \hat{a}^\dagger_{out}(k')] &= \frac{1}{r^2}[\hat{b}_{out}(k), \hat{b}^\dagger_{out}(k')]^* \\ &= \frac{1}{\sqrt{kk'}}\left\{\frac{1+r^2}{2r^2}k\delta(k-k') - i\frac{1-r^2}{4\pi r^2}(k+k')P\frac{1}{k-k'}\right\} \end{aligned} \tag{15.42}$$

The mode operator $\hat{a}_n$ and the adjoint mode operator $\hat{b}^\dagger_{n'}$ have the familiar form of the commutator expressing the bi-orthogonality and similarly for the outside region:

$$\left[\hat{a}_n, \hat{b}^\dagger_{n'}\right] = \delta_{nn'}, \qquad \left[\hat{a}_{out}(k), \hat{b}^\dagger_{out}(k')\right] = \delta(k - k') \tag{15.43}$$

The important inside–outside relations are obtained as

$$\begin{aligned}\left[\hat{a}_n, \hat{b}^\dagger_{out}(k)\right] &= r^2\left[\hat{a}_n, \hat{a}_{out}(k)\right] = \left[\hat{b}^\dagger_n, \hat{a}_{out}(k)\right] \\ &= \left[\hat{b}^\dagger_n, \hat{b}_{out}(k), \right] = \frac{i}{2}\sqrt{\frac{1-r^2}{L\pi\kappa_n k}}\end{aligned} \tag{15.44}$$

The formulation of this theory is applicable to a cavity of arbitrary transmission loss, because no assumptions or approximations concerning the reflectivity r have been made. The authors further discuss the motion of the spinor field $\widehat{g}(c\kappa_n, x)$ and speculate that the laser excess noise factor K may be given by $K = \int_{-L}^{0} dx\, \tilde{\widehat{g}}^\dagger(c\kappa_n, x)\widehat{g}(c\kappa_n, x)n^2(x)$. (The value of K thus obtained is $(1-|r|^2)/\{2|r|^2\ln(1/|r|)\}$, which is different from Equation 14.47 for the one-sided cavity. Since the spinor for the mode $\widehat{g}(c\kappa_n, x)$ is not normalized here, a factor

$$\int_{-L}^{0} dx\, \widehat{g}^\dagger(c\kappa_n, x)\widehat{g}(c\kappa_n, x)n^2(x) = (1-|r|^2)/\{2\ln(1/|r|)\}$$

should be multiplied on the right-hand side of the proposed formula for correct evaluation of K – see Equation 14.25.)

15.1.3 Projection Operator Method

Viviescas and Hackenbroich [8] considered the quantization of the field in the presence of a spatially non-uniform dielectric and optical cavities defined by mirrors of arbitrary shape. They introduced the projection operators for the inside and outside regions of the cavity. After projection operations on the field equations, they obtained the working Hamiltonian

$$\begin{aligned}H &= \sum_\lambda \hbar\omega_\lambda a^\dagger_\lambda a_\lambda + \sum_m \int d\omega\, \hbar\omega b^\dagger_m(\omega)b_m(\omega) \\ &\quad + \hbar\sum_\lambda\sum_m \int d\omega \left\{ W_{\lambda m}(\omega)a^\dagger_\lambda b_m(\omega) + \text{H.C.}\right\}\end{aligned} \tag{15.45}$$

where λ stands for the cavity modes and m stands for the "channels" representing the outside region. The coefficient $W_{\lambda m}(\omega)$ comes from the boundary conditions. We note that the Hamiltonian in Equation 15.45 is in a similar form as the system–reservoir Hamiltonian discussed in Appendix C. The Langevin equation for the cavity mode can be derived as in Appendix C. It can be shown that different cavity modes are coupled via the damping forces, and the noise forces for different cavity modes are correlated because the cavity modes couple to the same external channels. (This latter point makes contact with the assertion by Grangier and Poizat [24] on the "loss-induced coupling.")

15.2 Quantum Noise Theory

The thermal noise force associated with the cavity decay was traditionally treated by the system–reservoir model as in Appendix C, where the cavity modes are discrete and each cavity mode interacts with the reservoir modes of a fairly broad spectrum. The latter modes are independent of the cavity modes. The coupling strength is assumed to be constant over a wide frequency range. A fluctuating force was derived, which assured the preservation of the commutation relation for the cavity mode on the reservoir average. The output to the reservoir (outside region) was not considered seriously.

15.2.1 The Input–Output Theory by Time Reversal

To treat the output from a cavity with output coupling, Gardiner and Collett [14] developed a theory, called input–output theory, that paid attention not only to incoming noise but also to the outgoing field, which is the main quantity to be measured. They considered a system interacting with a heat bath (reservoir) described by the Hamiltonian

$$\begin{aligned} H &= H_{sys} + H_B + H_{int} \\ H_B &= \hbar \int_{-\infty}^{\infty} d\omega\, \omega b^\dagger(\omega) b(\omega) \\ H_{int} &= i\hbar \int_{-\infty}^{\infty} d\omega\, \kappa(\omega) \left\{ b^\dagger(\omega) c - c^\dagger b(\omega) \right\} \end{aligned} \tag{15.46}$$

where $b(\omega)$ ($b^\dagger(\omega)$) are boson annihilation (creation) operators for the bath, which satisfy $[b(\omega), b^\dagger(\omega')] = \delta(\omega - \omega')$, and c is one of the system operators. The factor $k(\omega)$ is the coupling constant (here assumed to be real). The equations of motion for the bath operator and a system operator a read

$$\begin{aligned} \dot{b}(\omega) &= -i\omega b(\omega) + \kappa(\omega) c \\ \dot{a} &= -\frac{i}{\hbar}[a, H_{sys}] + \int_{-\infty}^{\infty} d\omega\, \kappa(\omega) \left\{ b^\dagger(\omega)[a, c] - [a, c^\dagger] b(\omega) \right\} \end{aligned} \tag{15.47}$$

Solving for $b(\omega)$ with the initial value $b_0(\omega)$ at $t = t_0$ and substituting the result into the second equation, we have

$$\begin{aligned} \dot{a} = &-\frac{i}{\hbar}[a, H_{sys}] + \int_{-\infty}^{\infty} d\omega\, \kappa(\omega) \left\{ e^{i\omega(t-t_0)} b_0^\dagger(\omega)[a, c] - [a, c^\dagger] e^{-i\omega(t-t_0)} b_0(\omega) \right\} \\ &+ \int_{-\infty}^{\infty} d\omega\, \kappa^2(\omega) \int_{t_0}^{t} dt' \left\{ e^{i\omega(t-t')} c^\dagger(t')[a, c] - [a, c^\dagger] e^{-i\omega(t-t')} c(t') \right\} \end{aligned} \tag{15.48}$$

where the time variables for a and c are omitted for brevity.

To proceed further, we assume that the coupling coefficient $\kappa(\omega)$ is a constant and write $\kappa(\omega) = \sqrt{\gamma/2\pi}$. We define the input field by

$$b_{in}(t) = \frac{1}{\sqrt{2\pi}}\int_{-\infty}^{\infty} d\omega\, e^{-i\omega(t-t_0)} b_0(\omega) \tag{15.49}$$

Using the commutator $[b_0(\omega), b_0^\dagger(\omega')] = \delta(\omega-\omega')$, we can show the commutation relation $[b_{in}(t), b_{in}^\dagger(t')] = \delta(t-t')$. Equation 15.48 is now rewritten as a Langevin equation:

$$\begin{aligned}\dot{a} = &-\frac{i}{\hbar}[a, H_{sys}] - \Bigg[[a, c^\dagger]\left\{\frac{1}{2}\gamma c + \sqrt{\gamma}\, b_{in}(t)\right\} \\ &- \left\{\frac{1}{2}\gamma c^\dagger + \sqrt{\gamma}\, b_{in}^\dagger(t)\right\}[a, c]\Bigg]\end{aligned} \tag{15.50}$$

If the operator that couples with the bath is a and $H_{sys} = \hbar\omega_0(a^\dagger a + \frac{1}{2})$, we have

$$\dot{a} = -i\omega a - \tfrac{1}{2}\gamma a - \sqrt{\gamma}\, b_{in}(t) \tag{15.51}$$

This reproduces the damping term and the Langevin force term as obtained in Appendix C. Note, however, that the appearance of the damping does not need the incoherence of $b_{in}(t)$. The latter can be thermal or coherent or a mixture of them. Assume that we consider a future time t_1 ($>t$), integrate the first of Equations 15.47 with the temporal boundary condition $b(\omega)_{t=t_1} = b_1(\omega)$, and define

$$b_{out}(t) = \frac{1}{\sqrt{2\pi}}\int_{-\infty}^{\infty} d\omega\, e^{-i\omega(t-t_1)} b_1(\omega) \tag{15.52}$$

Then, we obtain an alternative equation to Equation 15.50, a time-reversed Langevin equation:

$$\begin{aligned}\dot{a} = &-\frac{i}{\hbar}[a, H_{sys}] - \Bigg[[a, c^\dagger]\left\{-\frac{1}{2}\gamma c + \sqrt{\gamma}\, b_{out}(t)\right\} \\ &- \left\{-\tfrac{1}{2}\gamma c^\dagger + \sqrt{\gamma}\, b_{out}^\dagger(t)\right\}[a, c]\Bigg]\end{aligned} \tag{15.53}$$

For $c = a$ and $H_{sys} = \hbar\omega_0(a^\dagger a + \frac{1}{2})$ we have

$$\dot{a} = -i\omega a + \tfrac{1}{2}\gamma a - \sqrt{\gamma}\, b_{out}(t) \tag{15.54}$$

Comparing Equations 15.50 and 15.53, or Equations 15.51 and 15.54 for $c = a$, yields the input–output relation

$$\begin{aligned} b_{out}(t) - b_{in}(t) &= \sqrt{\gamma} c(t) \\ b_{out}(t) - b_{in}(t) &= \sqrt{\gamma} a(t) \end{aligned} \tag{15.55}$$

To show the first line more directly, one can use the first line in Equation 15.47 with Equations 15.49 and 15.53. The quantities $b_{in}(t)$ and $b_{out}(t)$ can be interpreted as the input to and output from the system, and Equation 15.55 is the boundary condition relating the input, the output, and the internal modes. Assuming causality, the authors deduce the commutators

$$\begin{aligned} [a(t), b_{in}(t')] &= 0, \qquad t<t' \\ [a(t), b_{out}(t')] &= 0, \qquad t>t' \end{aligned} \tag{15.56}$$

Combining these with Equation 15.55 we have

$$\begin{aligned} [a(t), b_{in}(t')] &= -u(t-t')\sqrt{\gamma}[a(t), c(t')] \\ [a(t), b_{out}(t')] &= u(t'-t)\sqrt{\gamma}[a(t), c(t')] \end{aligned} \tag{15.57}$$

where $u(t)$ is the unit step function.

The authors further develop the theory of the quantum stochastic differential equation and the master equation (equation of motion for the density matrix) for the system and the bath to calculate the correlation functions of the output field in terms of those for the input and the internal fields, which are beyond the scope of this book. The same authors [13] also consider the results of having a second coupling mirror, which introduces an additional noise source.

15.2.2
The Input–Output Theory by the Boundary Condition

Another method of deriving the input–output relation was given by Carmichael [15]. The author uses a reservoir model, which is composed of quantized outer traveling modes incident on a semitransparent mirror of a ring cavity. These modes are partially reflected. The reflected reservoir modes are superimposed with the output from the cavity.

A periodic boundary condition is imposed on the reservoir modes, where the period is from $z=-L/2$ to $z=L/2$ with the mirror at $z=0$, and the paths of the modes are deflected by 90° at the mirror – see Figure 15.1.

The Hamiltonians of the cavity mode, the reservoir, and their interaction are written, respectively, as

$$\begin{aligned} H_S &= \hbar\omega_c a^\dagger a \\ H_R &= \sum_j \hbar\omega_j r_j^\dagger r_j \\ H_{SR} &= \sum_j \hbar\left(\kappa_j r_j a^\dagger + k_j^* r_j^\dagger a\right) \end{aligned} \tag{15.58}$$

where a and $a^\dagger$ are the annihilation and creation operators for the cavity mode, respectively, while r_j and $r_j^\dagger$ are the annihilation and creation operators, respectively, for the jth reservoir mode of frequency ω_j. The constant κ_j is the coupling constant, which will be determined later. (The author considers also intracavity interaction, a second coupling mirror, as well as reservoirs that

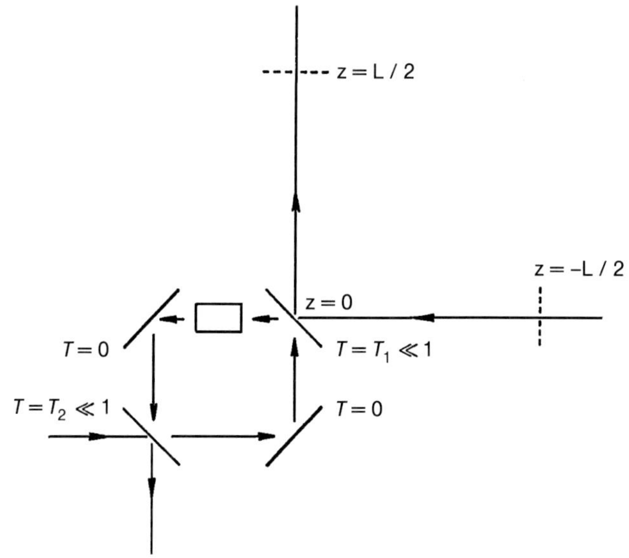

Figure 15.1 Schematic representation of ring–cavity system.
Source: From Ref. [15]. Carmichael, H.J. (1987) *J. Opt. Soc. Am. B*, *4*, 1588, Figure 1.

are not directly coupled to the cavity mode. These are omitted here for simplicity.)

The positive frequency part of the external (reservoir) field is written as

$$\hat{E}^{(+)}(z,t) = i\sum_j \left(\frac{\hbar\omega_j}{2\varepsilon_0 AL}\right)^{1/2} r_j(t)\exp(ik_j z) \tag{15.59}$$

Here A is the cross-sectional area of the reservoir field. As in Appendix C (see Equation C.12), the damping constant of the cavity, and consequently the coupling constant, are given by

$$\begin{aligned} \gamma_c &= \pi\rho|\kappa_{\omega_c}|^2 = \pi(L/2\pi c)|\kappa_{\omega_c}|^2 \\ |\kappa_{\omega_c}| &= \sqrt{2\gamma_c}\sqrt{c/L} \end{aligned} \tag{15.60}$$

where $\rho = L/2\pi c$ is the density of modes of the reservoir modes. From Equation 15.58 we have

$$\begin{aligned} \dot{a} &= -i\omega_c a - i\sum_j \kappa_j r_j \\ \dot{r}_j &= -i\omega_j r - i\kappa_j^* a \end{aligned} \tag{15.61}$$

Integrating the second equation and using Equation 15.59 we obtain

$$\hat{E}^{(+)}(z,t) = \sum_j \left(\frac{\hbar\omega_j}{2\varepsilon_0 AL}\right)^{1/2} r_j(0)\exp\{-i\omega_j(t-z/c)\}$$
$$+\exp\{-i\omega_c(t-z/c)\}\frac{1}{2\pi}\int_0^\infty d\omega\left(\frac{\hbar\omega}{2\varepsilon_0 cA}\right)^{1/2}\sqrt{\frac{L}{c}}\kappa^*(\omega) \qquad (15.62)$$
$$\times\int_0^t dt'\,\tilde{a}(t')\exp\{-i(\omega_c-\omega)(t'-t+z/c)\}$$

where $\tilde{a}(t) = a(t)e^{i\omega_c t}$. The first term is the free reservoir field. We assume that the time variation of $\tilde{a}(t)$ is slow compared to the optical frequency, or the bandwidth of $\tilde{a}(t)$ is much narrower than ω_c, so that $\sqrt{\omega}\kappa^*(\omega)$ can be replaced by $\sqrt{\omega_c}\kappa^*(\omega_c)$. Then the lower limit of the frequency integral can be extended to $-\infty$. Defining the second term in Equation 15.62 as the source term $\hat{E}_s^{(+)}(z,t)$ and using Equation 15.60, we have

$$\hat{E}_s^{(+)}(z,t) = \exp\{-i\omega_c(t-z/c)\}\left(\frac{\hbar\omega_c}{2\varepsilon_0 cA}\right)^{1/2}$$
$$\times e^{-i\phi}\sqrt{2\gamma_c}\int_0^t dt'\,\tilde{a}(t')\delta(t'-t+z/c) \qquad (15.63)$$

where ϕ is the phase of $\kappa^*(\omega_c)$. Performing the time integral we obtain

$$\hat{E}^{(+)}(z,t) = \sum_j \left(\frac{\hbar\omega_j}{2\varepsilon_0 AL}\right)^{1/2} r_j(0)\exp\{-i\omega_j(t-z/c)\}$$
$$+\left(\frac{\hbar\omega_c}{2\varepsilon_0 cA}\right)^{1/2} e^{-i\phi}\sqrt{2\gamma_c}\begin{cases} a(t-z/c), & ct > z > 0 \\ (1/2)a(t), & z = 0 \end{cases} \qquad (15.64)$$

It is easy to see that the photon flux due to $\hat{E}_s^{(+)}(z,t)$ collected over the area A is equal to $2\gamma_c\langle a^\dagger(t-z/c)a(t-z/c)\rangle$, the mean number of photons inside the cavity multiplied by the power damping rate of the cavity. We now define field operators $r(z,t)$ and $r^\dagger(z,t)$ for the reservoir in photon flux units by

$$r(z,t) = e^{i\phi}\left(\frac{2\varepsilon_0 cA}{\hbar\omega_c}\right)^{1/2}\hat{E}^{(+)}(z,t) \qquad (15.65)$$

Then Equation 15.64 reads

$$r(z,t) = r_f(z,t) + \begin{cases} \sqrt{2\gamma_c}a(t-z/c), & ct>z>0 \\ \frac{1}{2}\sqrt{2\gamma_c}a(t), & z=0 \end{cases} \qquad (15.66)$$

where $r_f(z,t)$ corresponds to the first term in Equation 15.64 and exists for the region $-\infty < z < \infty$ with $L \to \infty$. We have for $z = +0$ and for $z = 0$

$$\begin{aligned} r(+0,t) &= r_f(+0,t) + \sqrt{2\gamma_c}a(t-0) \\ r(0,t) &= r_f(0,t) + \tfrac{1}{2}\sqrt{2\gamma_c}a(t) \\ &= r(+0,t) - \tfrac{1}{2}\sqrt{2\gamma_c}a(t) \end{aligned} \tag{15.67}$$

The last two lines yields the input–output relation

$$r(+0,t) = r_f(0,t) + \sqrt{2\gamma_c}a(t) \tag{15.68}$$

This shows that the output field is the sum of the reservoir field and the internal field coupled out of the cavity. (This corresponds to Equation 15.55 where $b_{in}(t)$ and $b_{out}(t)$ correspond to $r_f(0,t)$ and $r(+0,t)$, and γ to $2\gamma_c$, respectively.) Now using Equations 15.59, 15.60, and 15.65, we have for $z=0$

$$\begin{aligned} r(0,t) &= e^{i\phi}\left(\frac{2\varepsilon_0 cA}{\hbar\omega_c}\right)^{1/2}\hat{E}^{(+)}(0,t) = e^{i\phi}\left(\frac{2\varepsilon_0 cA}{\hbar\omega_c}\right)^{1/2} i\sum_j\left(\frac{\hbar\omega_j}{2\varepsilon_0 AL}\right)^{1/2} r_j(t) \\ &\simeq ie^{i\phi}\sqrt{\frac{c}{L}}i\sum_j r_j(t) \simeq i\frac{1}{\sqrt{2\gamma_c}}\sum_j \kappa_j r_j(t) \end{aligned} \tag{15.69}$$

Using this result and the second line of Equation 15.67 in Equation 15.61, we have the Langevin equation

$$\dot{a} = -i\omega_c a - \gamma_c a - \sqrt{2\gamma_c}r_f(t) \tag{15.70}$$

(This corresponds to Equation 15.51 with $2\gamma_c \to \gamma$ and $r_f(t) \to b_{in}(t)$.) If the reservoir is initially in the vacuum state, we have

$$\langle r_f(t)r_f(t')\rangle = \left\langle r_f^\dagger(t)r_f^\dagger(t')\right\rangle = \left\langle r_f^\dagger(t)r_f(t')\right\rangle = 0$$

Also, it can be shown that $\left\langle r_f(t)r_f^\dagger(t')\right\rangle = \delta(t-t')$. The author further calculates correlation functions between a system operator $\hat{s}(t)$ and $r_f(t)$ or $r(t)$ using the Langevin equation (Equation 15.70) and the master equation for the cavity–reservoir system (the latter is beyond the scope of this book).

15.2.3
Another Quantum Noise Theory

Semenov *et al.* [17] considered the absorption or scattering loss in the coupling mirror. These affect the input wave before entering the cavity and the output wave after leaving the cavity. It also introduces a feedback of the output back to the cavity. The authors modeled these effects by setting three half-mirrors outside a lossless semitransparent mirror. This theory extends the input–output theories considered so far by introducing two new noise sources with added feedback and explores new forms of Langevin equations and input–output relations.

15.3
Green's Function Theory

Gruner and Welsch [19] considered the input–output relation for dispersive and absorptive linear dielectric layers. Taking into account the fact that an absorptive medium is associated with distributed noise sources, the authors solve the wave equation for a dielectric layer incorporating the noise sources as a distributed driving term. The solution is a convolution of the noise term and the Green's function that satisfies an equation of the form in Equation 14.67. Here n^2 is replaced by the complex permittivity $\varepsilon(x, \omega)$. The dielectrics should satisfy the Kramers–Kronig relation. The authors construct a spatial quantum Langevin equation associated with wave propagation and investigate the spatial input–output relation for the multilayered structure.

Khanbekyan *et al.* [18] considered the absorption loss in the coupling mirror of a cavity that Semenov *et al.* [17] considered using the input–output formalism. The authors, using the Green's function method and assuming a slow amplitude variation, show that the cavity mode obeys the quantum Langevin equation and investigate the input–output relations. The mirror loss introduces additional noise terms into the input–output relation. The problem of extracting the quantum state of the cavity mode was studied.

15.4
Quasimode Theory

Dalton *et al.* [26] considered the quantization of linear optical devices including radiating atoms. The quasimode functions for the device are obtained by solving the Helmholtz equation for a spatially dependent electrical permittivity that is specially designed to have an ideal quasimode. The Hamiltonian for the electromagnetic field is found to be equivalent to those of a set of harmonic oscillators, but these oscillators are coupled. The atoms are shown to be coupled only to certain types of quasimodes. The emission from an atom inside the cavity is described as a two-step process: de-excitation of an atom with the creation of a cavity quasimode photon; and annihilation of the quasimode photon with the creation of an external quasimode photon.

15.5
Summary

Here we briefly compare the results of the theories of output coupling of an optical cavity in terms of input–output relation and of the Langevin equation.

First, the input–output relations given by Equations 10.137, 15.24, 15.55, and 15.68 all relate the output field to the input and internal field. In the case of our treatment (Equation 10.137) and of the approximate universal modes (Equation 15.24), the input field is associated with the reflection coefficient of the coupling mirror. Other treatments lack this factor. In the case of quantum noise theories (Equations 15.55 and 15.68), the mirror transmission is assumed to be small. The contribution of the inner field is associated with the accurate transmission coefficient in our case (Equation 10.137), while all other cases have expressions in terms of the square root of the power damping rate. This reflects the neglect of the inner field distribution or the use of the approximate mode function in these other theories. Note that our formulation lead to different constants of output coupling for below- and above-threshold operation of the laser, as shown in Section 12.9.

The natural mode theory gives the non-commuting operator relations between the inside modes and the outside modes, but does not give a concrete input–output relation. In the cases of the approximate universal modes (Equation 15.24), the input–output relation involving multiple cavity modes is given. The definitions of the input and output fields in the cases of the quantum noise theory (Equation 15.55) do not fit with our intuition. This is because the fields in these theories are not instantaneous values but are associated with fictitious past or future times coming from the time-reversal concept.

Second, the Langevin equation for the field inside the cavity is given in Equations 15.6a, 15.23, 15.51, and 15.70, where the noise forces are interpreted as coming from the outer free fields. The noise correlations are associated with the power damping rate of the cavity, which may be reinterpreted as the penetration rate of the outer noise power. The case of our relation (Equation 15.6a) is associated with the excess noise factor (see Equation 15.6d) in addition to the above penetration rate. The natural mode theory does not give any Langevin equation for the cavity mode.

We may conclude that, except for the capability of treating several cavity modes at the same time, our methods of normal mode expansion yield the most natural input–output relation as well as the most general Langevin equation despite the complexity of the calculations. The problem of multiple cavity modes would not be difficult to solve if we had tackled it from the outset. The quantum field theories in general need complicated calculations associated with a cavity structure, although the results are sometimes in simple forms. On the contrary, the quantum noise theories are simple in calculations but are not applicable to a cavity with large transmission loss. Most theories, quantum field theories or quantum noise theories, lack some delicate information such as mirror transmission coefficient or excess noise factor.

15.6
Equations for the Output Coupling and Input–Output Relation

For ease of comparison, we enumerate, in this section, the equations for the output coupling and input–output relation given by several authors.

Ujihara (this book, [4]):

$$\frac{d}{dt}\hat{a}(t) = -\gamma_c \hat{a}(t) + K_L^{1/4}\left\{K_2^{1/4}\hat{b}^R(d+0,t) + K_1^{1/4}\hat{b}^L(-0,t)\right\}$$

$$\hat{e}_o^+(z,t) = Te^{ikd}\hat{e}^+\left(-0, t-\frac{z}{c_0}\right) + r'\hat{f}_{ot}^-\left(+0, t-\frac{z}{c_0}\right)$$

$$\left\langle \hat{b}^{R\dagger}(t)\hat{b}^R(t')\right\rangle = 2\gamma_c n_\omega \delta(t-t')$$

Knöll *et al.* [6]:

$$\dot{\hat{a}}_m(t) = -i\left(\omega_m - i\frac{\Gamma}{2}\right)\hat{a}_m(t) + \Gamma^{1/2}e^{i\phi}\hat{b}_m(t)$$

$$\dot{\hat{a}}_m(t) = -i\left(\omega_m + i\frac{\Gamma}{2}\right)\hat{a}_m(t) + \Gamma^{1/2}e^{-i\phi}\hat{c}_m(t)$$

$$\hat{E}_{out}^{(+)}(t-x/c) = \bar{r}\hat{E}_{in}^{(+)}(t-x/c) + \sum_m \Gamma^{1/2}e^{i\phi}\left[\frac{\hbar\omega_m}{2\varepsilon_0 Fc}\right]^{1/2}\hat{a}_m(t-x/c)$$

$$\left[\hat{a}_m, \hat{a}_{m'}^\dagger\right] = \delta_{mm'}, \qquad [\hat{a}_m, \hat{a}_{m'}] = 0 = \left[\hat{a}_m^\dagger, \hat{a}_{m'}^\dagger\right]$$

$$\left[\hat{b}_m(t), \hat{b}_{m'}^\dagger(t')\right] = \delta_{mm'}\delta(t-t'), \qquad \left[\hat{b}_m(t), \hat{b}_{m'}(t')\right] = 0 = \left[\hat{b}_m^\dagger(t), \hat{b}_{m'}^\dagger(t')\right]$$

$$\left[\hat{c}_m(t), \hat{c}_{m'}^\dagger(t')\right] = \delta_{mm'}\delta(t-t'), \qquad [\hat{c}_m(t), \hat{c}_{m'}(t')] = 0 = \left[\hat{c}_m^\dagger(t), \hat{c}_{m'}^\dagger(t')\right]$$

Dutra and Nienhuis [7]:

$$\mathbf{F}(x) = \tfrac{1}{2}\lim_{\varepsilon\to 0^+}\left[\left\{\mathbf{F}_{cav}(x) + \tilde{\mathbf{F}}_{cav}(x)\right\}\Theta(\varepsilon - x) + \left\{\mathbf{F}_{out}(x) + \tilde{\mathbf{F}}_{out}(x)\right\}\Theta(\varepsilon + x)\right]$$

$$[\hat{a}_n, \hat{a}_{n'}^\dagger] = \frac{1}{r^2}[\hat{b}_n, \hat{b}_{n'}^\dagger]^* = \frac{i}{4L\sqrt{\kappa_n\kappa_{n'}^*}}\frac{\kappa_n + \kappa_{n'}^*}{\kappa_n - \kappa_{n'}^*}\frac{r^2-1}{r^2}$$

$$[\hat{a}_n, \hat{b}_{n'}^\dagger] = \delta_{nn'}, \qquad [\hat{a}_{out}(k), \hat{b}_{out'}^\dagger(k')] = \delta(k-k')$$

$$[\hat{a}_n, \hat{b}_{out'}^\dagger(k)] = r^2[\hat{a}_n(k), \hat{a}_{out}(k)] = [\hat{b}_n^\dagger, \hat{a}_{out}(k)]$$

$$= [\hat{b}_n^\dagger, \hat{b}_{out}(k)] = \frac{i}{2}\sqrt{\frac{1-r^2}{L\pi\kappa_n k}}$$

Gardiner and Collett [14]:

$$\dot{a} = -i\omega a - \tfrac{1}{2}\gamma a - \sqrt{\gamma}\, b_{in}(t)$$

$$\dot{a} = -i\omega a + \tfrac{1}{2}\gamma a - \sqrt{\gamma}\, b_{out}(t)$$

$$b_{out}(t) - b_{in}(t) = \sqrt{\gamma}c$$
$$b_{out}(t) - b_{in}(t) = \sqrt{\gamma}a \qquad (c = a)$$

$$[b_{in}(t), b_{in}^{\dagger}(t')] = \delta(t - t')$$

Carmichael [15]:

$$r(+0, t) = r_f(0, t) + \sqrt{2\gamma_c}a(t)$$

$$\dot{a} = -i\omega_c a - \gamma_c a - \sqrt{2\gamma_c}r_f(t)$$

$$\left\langle r_f(t) r_f^{\dagger}(t') \right\rangle = \delta(t - t')$$

References

1 Ujihara, K. (1975) *Phys. Rev. A, 12,* 148–158.
2 Ujihara, K. (1978) *Phys. Rev. A, 18,* 659–670.
3 Ujihara, K. (1977) *Phys. Rev. A, 16,* 652–658.
4 Ujihara, K. (1984) *Phys. Rev. A, 29,* 3253 –3263.
5 Glauber, R.J. and Lewenstein, M. (1991) *Phys. Rev. A, 43,* 467–491.
6 Knöll, L., Vogel, W., and Welsch, D.-G. (1991) *Phys. Rev. A, 43,* 543–553.
7 Dutra, S.M. and Nienhuis, G. (2000) *Phys. Rev. A, 62,* 063805.
8 Viviescas, C. and Hackenbroich, G. (2003) *Phys. Rev. A, 67,* 013805.
9 Milburn, G. and Walls, D.F. (1981) *Opt. Commun., 39,* 401–404.
10 Wu, L.-A., Kimble, H.J., Hall, J.L., and Wu, H. (1986) *Phys. Rev. Lett., 57,* 2520–2523.
11 Haken, H. (1970) *Laser theory*, in *Licht und Materie, IC,* Handbuch der Physik, vol. XXV/2c (eds. S. Flügge and L. Genzel), Springer, Berlin.
12 Lax, M. (1966) *Phys. Rev., 145,* 110–129.
13 Collett, M.J. and Gardiner, C.W. (1984) *Phys. Rev. A, 30,* 1386–1391.
14 Gardiner, C.W. and Collett, M.J. (1985) *Phys. Rev. A, 31,* 3761–3774.
15 Carmichael, H.J. (1987) *J. Opt. Soc. Am. B, 4,* 1588–1603.
16 Yamamoto, Y. and Imoto, N. (1986) *IEEE J. Quantum Electron., QE-22,* 2032–2042.
17 Semenov, A.A., Vasylyev, D. Yu., Vogel, W., Khanbekyan, M., and Welsch, D.-G. (2006) *Phys. Rev. A, 74,* 033803.
18 Khanbekyan, M., Knöll, L., Welsch, D.-G., Semenov, A.A., and Vogel, W. (2005) *Phys. Rev. A, 72,* 053813.
19 Gruner, T. and Welsch, D.-G. (1996) *Phys. Rev. A, 54,* 1661–1677.
20 Lang, R. and Scully, M.O. (1973) *Opt. Commun., 9,* 331–335.
21 Gea-Banacloche, J., Lu, N., Pedrotti, L.M., Prasad, S., Scully, M.O., and Wódkiewicz, K. (1990) *Phys. Rev. A, 41,* 369–380.
22 Penaforte, J.C. and Baseia, B. (1984) *Phys. Rev. A, 30,* 1401–1406.
23 Guedes, I., Penaforte, J.C., and Baseia, B. (1989) *Phys. Rev. A, 40,* 2463–2470.
24 Grangier, P. and Poizat, J.-P. (1999) *Eur. Phys. J., 7,* 99–105.
25 Cheng, Y.J. and Siegman, A.E. (2003) *Phys. Rev. A, 68,* 043808.
26 Dalton, B.J., Barnett, S.M., and Knight, P.L. (1999) *J. Mod. Opt., 46,* 1315–1341.

Appendices

Appendix A
Integration for the Field Hamiltonian

Here we derive Equation 1.47 of the text. We are assuming that

$$\varepsilon(z) = \begin{cases} \varepsilon_1, & -d<z<0 \\ \varepsilon_0, & 0<z<L \end{cases}$$
$$\mu(z) = \mu_0, \qquad -d<z<L \tag{A.1}$$

We start with Equations 1.44 and 1.46. We substitute these into Equation 1.45 using Equation 1.41 for the mode function. Now we have

$$\begin{aligned} H &= \int_{-d}^{L} \left[\frac{\varepsilon(z)}{2} \left(\frac{\partial}{\partial t} A(z,t) \right)^2 + \frac{1}{2\mu(z)} \left(\frac{\partial}{\partial z} A(z,t) \right)^2 \right] dz \\ &= \int_{-d}^{L} \left[\frac{\varepsilon(z)}{2} \left(\sum_k P_k U_k(z) \right)^2 + \frac{1}{2\mu_0} \left(\sum_k Q_k \frac{\partial}{\partial z} U(z) \right)^2 \right] dz \end{aligned} \tag{A.2}$$

Because of the orthonormality in Equation 1.42, the first term in the integration becomes

$$\int_{-d}^{L} \frac{\varepsilon(z)}{2} \left(\sum_k P_k U_k(z) \right)^2 dz = \frac{1}{2} \sum_k P_k^2 \int_{-d}^{L} \varepsilon(z) U_k^2(z) dz = \frac{1}{2} \sum_k P_k^2 \tag{A.3}$$

Output Coupling in Optical Cavities and Lasers: A Quantum Theoretical Approach
Kikuo Ujihara

ISBN: 978-3-527-40763-7

The second term is

$$\int_{-d}^{L}\left[\frac{1}{2\mu_0}\left(\sum_k Q_k\frac{\partial}{\partial z}U_k(z)\right)^2\right]dz$$
$$=\frac{1}{2\mu_0}\sum_{k,k'}Q_kQ_{k'}\int_{-d}^{L}\frac{\partial}{\partial z}U_k(z)\frac{\partial}{\partial z}U_{k'}(z)dz$$
$$=\frac{1}{2\mu_0}\sum_{k,k'}Q_kQ_{k'}U_k(z)\frac{\partial}{\partial z}U_{k'}(z)\Big|_{-d}^{0}+\frac{1}{2\mu_0}\sum_{k,k'}Q_kQ_{k'}U_k(z)\frac{\partial}{\partial z}U_{k'}(z)\Big|_{0}^{L} \tag{A.4}$$
$$-\frac{1}{2\mu_0}\sum_{k,k'}Q_kQ_{k'}\int_{-d}^{0}U_k(z)\left(\frac{\partial}{\partial z}\right)^2U_{k'}(z)dz$$
$$-\frac{1}{2\mu_0}\sum_{k,k'}Q_kQ_{k'}\int_{0}^{L}U_k(z)\left(\frac{\partial}{\partial z}\right)^2U_{k'}(z)dz$$

Since the mode function vanishes at both ends of the space, at $z=-d$ and $z=L$, and the mode function and its derivative are continuous at the interface, $z=0$, the first and second terms on the right-hand side vanish. Also, since the mode functions satisfy the Helmholtz equations (Equations 1.32b), the second derivatives in the third and fourth terms can be replaced by $-(k'_1)^2$ and $-(k'_0)^2$, respectively. From Equation 1.33 we have

$$\frac{(k'_{1,0})^2}{\mu_0}=\frac{(\omega_{k'})^2\varepsilon_{1,0}\mu_0}{\mu_0}=(\omega_{k'})^2\varepsilon_{1,0} \tag{A.5}$$

for inside and outside the cavity, respectively. Therefore, the third and fourth terms in Equation A.4 reduce to

$$-\frac{1}{2\mu_0}\sum_{k,k'}Q_kQ_{k'}\int_{-d}^{0}U_k(z)\left(\frac{\partial}{\partial z}\right)^2U_{k'}(z)dz$$
$$-\frac{1}{2\mu_0}\sum_{k,k'}Q_kQ_{k'}\int_{0}^{L}U_k(z)\left(\frac{\partial}{\partial z}\right)^2U_{k'}(z)dz$$
$$=\frac{1}{2}\sum_{k,k'}Q_kQ_{k'}\int_{-d}^{0}(\omega_{k'})^2\varepsilon_1U_k(z)U_{k'}(z)dz$$
$$+\frac{1}{2}\sum_{k,k'}Q_kQ_{k'}\int_{0}^{L}(\omega_{k'})^2\varepsilon_0U_k(z)U_{k'}(z)dz \tag{A.6}$$
$$=\frac{1}{2}\sum_{k,k'}Q_kQ_{k'}(\omega_{k'})^2\int_{-d}^{L}\varepsilon(z)U_k(z)U_{k'}(z)dz$$
$$=\frac{1}{2}\sum_{k,k'}Q_kQ_{k'}(\omega_{k'})^2\delta_{kk'}$$
$$=\frac{1}{2}\sum_k\omega_k^2Q_k^2$$

where we have used the orthonormality relation in Equation 1.42 again. Thus adding the results in Equations A.3 and A.6, we arrive at Equation 1.47.

Appendix B
Energy Eigenstates of a Single Field Mode

Here we show that the energy eigenstate of a field mode k satisfies the eigenvalue equation

$$\hat{H}_k|n_k\rangle = E_{k,n}|n_k\rangle = \left(n_k + \frac{1}{2}\right)\hbar\omega_k|n_k\rangle, \qquad n_k = 0, 1, 2, 3, \ldots \tag{B.1}$$

where, from Equation 2.9, the Hamiltonian is

$$\hat{H}_k = \hbar\omega_k\left(\hat{a}_k^\dagger\hat{a}_k + \frac{1}{2}\right) \tag{B.2}$$

We have to show that

$$\hbar\omega_k\left(\hat{a}_k^\dagger\hat{a}_k + \frac{1}{2}\right)|n_k\rangle = \hbar\omega_k\left(n_k + \frac{1}{2}\right)|n_k\rangle, \qquad n_k = 0, 1, 2, 3, \ldots \tag{B.3}$$

Let us write the eigenvalue equation as

$$\hbar\omega_k\left(\hat{a}_k^\dagger\hat{a}_k + \frac{1}{2}\right)|u_{k,j}\rangle = E_{k,j}|u_{k,j}\rangle \tag{B.4}$$

where the suffix j denotes the jth eigenvalue and the corresponding jth eigenstate. Let us multiply $\hat{a}_k^\dagger$ from the left on both sides:

$$\hbar\omega_k\hat{a}_k^\dagger\left(\hat{a}_k^\dagger\hat{a}_k + \frac{1}{2}\right)|u_{k,j}\rangle = E_{k,j}\hat{a}_k^\dagger|u_{k,j}\rangle \tag{B.5}$$

By the commutation rule in Equation 2.8, the left-hand side is

$$\begin{aligned}\hbar\omega_k\hat{a}_k^\dagger\left(\hat{a}_k^\dagger\hat{a}_k + \frac{1}{2}\right)|u_{k,j}\rangle &= \hbar\omega_k\left\{\hat{a}_k^\dagger(\hat{a}_k^\dagger\hat{a}_k) + \frac{1}{2}\hat{a}_k^\dagger\right\}|u_{k,j}\rangle \\ &= \hbar\omega_k\left\{\hat{a}_k^\dagger(\hat{a}_k\hat{a}_k^\dagger - 1) + \frac{1}{2}\hat{a}_k^\dagger\right\}|u_{k,j}\rangle \\ &= \hbar\omega_k\left\{\left(\hat{a}_k^\dagger\hat{a}_k + \frac{1}{2}\right)\hat{a}_k^\dagger - \hat{a}_k^\dagger\right\}|u_{k,j}\rangle\end{aligned} \tag{B.6}$$

Adding $\hbar\omega_k\hat{a}_k^\dagger|u_{k,j}\rangle$ to both sides of Equation B.5 then yields

$$\hbar\omega_k\left(\hat{a}_k^\dagger\hat{a}_k + \frac{1}{2}\right)\hat{a}_k^\dagger|u_{k,j}\rangle = (E_{k,j} + \hbar\omega_k)\hat{a}_k^\dagger|u_{k,j}\rangle \tag{B.7a}$$

We add parentheses below merely for clarity:

$$\hbar\omega_k\left(\hat{a}_k^\dagger\hat{a}_k+\frac{1}{2}\right)\left(\hat{a}_k^\dagger|u_{k,j}\rangle\right)=\left(E_{k,j}+\hbar\omega_k\right)\left(\hat{a}_k^\dagger|u_{k,j}\rangle\right) \tag{B.7b}$$

This equation says that the state $\hat{a}_k^\dagger|u_{k,j}\rangle$ is a new eigenstate of $\hat{H}_k$ with the eigenvalue $E_{k,j}+\hbar\omega_k$. We write the state as $c_{j+1}|u_{k,j+1}\rangle$. If we repeat the above procedure, we see that $(\hat{a}_k^\dagger)^n|u_{k,j}\rangle$ is an eigenstate with the eigenvalue $E_{k,j}+n\hbar\omega_k$, where n is a non-negative integer. We see that the operator $\hat{a}_k^\dagger$ acts to increase the energy by $\hbar\omega_k$.

Next, let us multiply $\hat{a}_k$ from the left on both sides of Equation B.4:

$$\hbar\omega_k\hat{a}_k\left(\hat{a}_k^\dagger\hat{a}_k+\frac{1}{2}\right)|u_{k,j}\rangle=E_{k,j}\hat{a}_k|u_{k,j}\rangle \tag{B.8}$$

By the commutation rule in Equation 2.8, the left-hand side is

$$\begin{aligned}\hbar\omega_k\hat{a}_k\left(\hat{a}_k^\dagger\hat{a}_k+\frac{1}{2}\right)|u_{k,j}\rangle&=\hbar\omega_k\left\{(\hat{a}_k\hat{a}_k^\dagger)\hat{a}_k+\frac{1}{2}\hat{a}_k\right\}|u_{k,j}\rangle\\&=\hbar\omega_k\left\{(\hat{a}_k^\dagger\hat{a}_k+1)\hat{a}_k+\frac{1}{2}\hat{a}_k\right\}|u_{k,j}\rangle\\&=\hbar\omega_k\left\{\left(\hat{a}_k^\dagger\hat{a}_k+\frac{1}{2}\right)\hat{a}_k+\hat{a}_k\right\}|u_{k,j}\rangle\end{aligned} \tag{B.9}$$

Subtracting $\hbar\omega_k\hat{a}_k|u_{k,j}\rangle$ from both sides of Equation B.8 then yields

$$\hbar\omega_k\left(\hat{a}_k^\dagger\hat{a}_k+\frac{1}{2}\right)\hat{a}_k|u_{k,j}\rangle=\left(E_{k,j}-\hbar\omega_k\right)\hat{a}_k|u_{k,j}\rangle \tag{B.10a}$$

We again add parentheses below merely for clarity:

$$\hbar\omega_k\left(\hat{a}_k^\dagger\hat{a}_k+\frac{1}{2}\right)\left(\hat{a}_k|u_{k,j}\rangle\right)=\left(E_{k,j}-\hbar\omega_k\right)\left(\hat{a}_k|u_{k,j}\rangle\right) \tag{B.10b}$$

This equation says that the state $\hat{a}_k|u_{k,j}\rangle$ is a new eigenstate of $\hat{H}_k$ with the eigenvalue $E_{k,j}-\hbar\omega_k$. We write the state as $c_{j-1}|u_{k,j-1}\rangle$. If we repeat the above procedure, we see that $(\hat{a}_k)^m|u_{k,j}\rangle$ is an eigenstate with the eigenvalue $E_{k,j}-m\hbar\omega_k$, where m is a non-negative integer. We see that the operator $\hat{a}_k$ acts to reduce the energy by $\hbar\omega_k$.

From the above results, the operator $\hat{a}_k^\dagger$ is now interpreted as the photon creation operator and $\hat{a}_k$ as the photon annihilation operator.

Since the energy eigenvalue should be non-negative, we should have in general

$$E_{k,j}-m\hbar\omega_k\geq 0 \tag{B.11}$$

If we write the state of the smallest eigenvalue $E_{k,min}$ as $|min\rangle$, then we have $E_{k,min}-\hbar\omega_k<0$, and $\hat{a}_k|min\rangle$ should vanish to avoid eigenstates with negative energies. Thus we have

$$\hat{a}_k|min\rangle = 0 \tag{B.12}$$

Applying the Hamiltonian on $|min\rangle$ we have

$$\hat{H}_k|min\rangle = \hbar\omega_k\left(\hat{a}_k^\dagger\hat{a}_k + \frac{1}{2}\right)|min\rangle = \frac{1}{2}\hbar\omega_k|min\rangle \tag{B.13}$$

Therefore, the minimum energy eigenvalue is $\frac{1}{2}\hbar\omega_k$. The corresponding eigenstate is usually written as

$$|min\rangle = |0\rangle \tag{B.14}$$

Thus, applying the operator $\hat{a}_k^\dagger$ sequentially, we can generate the states $|n_k\rangle$ with energies $(1+\frac{1}{2})\hbar\omega_k$, $(2+\frac{1}{2})\hbar\omega_k, \ldots,$ $(n+\frac{1}{2})\hbar\omega_k, \ldots$. Thus we have the energy eigenstates and the energy levels

$$\begin{aligned}\hat{H}_k|n_k\rangle &= E_{k,n}|n_k\rangle \\ E_{k,n} &= (n_k + \frac{1}{2})\hbar\omega_k, \qquad n_k = 0, 1, 2, 3, \ldots\end{aligned} \tag{B.15}$$

The integer n_k is the eigenvalue for the photon number operator $\hat{a}_k^\dagger\hat{a}_k$,

$$\hat{a}_k^\dagger\hat{a}_k|n_k\rangle = n_k|n_k\rangle, \qquad n_k = 0, 1, 2, 3, \ldots \tag{B.16}$$

and represents the photon number in the mode.

Finally, we determine the coefficients $c_{n\pm1}$ that appear on operating $\hat{a}_k^\dagger$ or $\hat{a}_k$ on a state $|n_k\rangle$. We assume that the eigenstates are normalized so that

$$\langle n_k|n_k\rangle = 1, \qquad n_k = 0, 1, 2, 3, \ldots \tag{B.17}$$

According to the statement below Equation B.7b, we write

$$\hat{a}_k^\dagger\left|u_{k,n}\right\rangle = c_{n+1}\left|u_{k,n+1}\right\rangle \tag{B.18}$$

Taking the squared modulus of both sides we have

$$\left\langle u_{k,n}\middle|\hat{a}_k\hat{a}_k^\dagger\middle|u_{k,n}\right\rangle = \left\langle u_{k,n+1}\middle|c_{n+1}^*c_{n+1}\middle|u_{k,n+1}\right\rangle \tag{B.19}$$

Using the commutation relation in Equation 2.8 and Equation B.16 as well as Equation B.17, we have

$$|c_{n+1}|^2 = n_k + 1 \tag{B.20}$$

We can arbitrarily choose the phase of c_{n+1} to be zero, so that

$$c_{n+1} = \sqrt{n_k + 1} \tag{B.21}$$

Similarly, according to the statement below Equation B.10b, we write

$$\hat{a}_k\left|u_{k,n}\right\rangle = c_{n-1}\left|u_{k,n-1}\right\rangle \tag{B.22}$$

Taking the squared modulus of both sides we have

$$\left\langle u_{k,n}\middle|\hat{a}_k^\dagger \hat{a}_k\middle|u_{k,n}\right\rangle = \left\langle u_{k,n-1}\middle|c_{n-1}^* c_{n-1}\middle|u_{k,n-1}\right\rangle \tag{B.23}$$

Using Equations B.16 and B.17 we have

$$|c_{n-1}|^2 = n_k \tag{B.24}$$

Again choosing the phase to be zero, we have

$$c_{n-1} = \sqrt{n_k} \tag{B.25}$$

Therefore, from Equations B.18 and B.21 and from Equations B.22 and B.25, respectively, we can write

$$\begin{aligned} \hat{a}_k^\dagger|n_k\rangle &= \sqrt{n_k+1}|n_k+1\rangle, & n_k &= 0,1,2,3,\ldots \\ \hat{a}_k|n_k\rangle &= \sqrt{n_k}|n_k-1\rangle, & n_k &= 1,2,3,\ldots \end{aligned} \tag{B.26}$$

and from Equations B.12 and B.14

$$\hat{a}_k|0_k\rangle = 0 \tag{B.27}$$

In the above, we have derived the energy eigenstates on the basis of the commutation relation for the creation and annihilation operators for the mode using the Hamiltonian in the form of the photon number operator plus one-half, multiplied by the photon energy. If, instead, we use the original Hamiltonian in Equation 2.3, the time-independent Schrödinger equation will formally read

$$\frac{1}{2}(\hat{P}_k^2 + \omega_k^2 \hat{Q}_k^2)\Psi(Q_k) = E_k\Psi(Q_k) \tag{B.28}$$

and the corresponding differential equation will read

$$\left(-\frac{\hbar^2}{2}\frac{\partial^2}{\partial Q_k^2} + \frac{\omega_k^2 Q_k^2}{2}\right)\Psi(Q_k) = E_k\Psi(Q_k) \tag{B.29}$$

It is known that the solutions to this equation are the Hermite–Gaussian functions with the same eigenvalues as obtained above. Therefore, each eigenstate in the form of a ket has a corresponding expression in the form of a Hermite–Gaussian function of the coordinate Q_k. See, for example, Schiff [1].

Appendix C
The Reservoir Model for the Cavity Loss

Here we describe a reservoir model for the cavity loss and derive the correlation function for the associated random noise force following Haken [2].

We assume that the cavity loss is caused by a large number of loss oscillators that are coupled to the cavity mode in question. The total Hamiltonian for the cavity mode and the loss oscillators may be written in the form

$$\hat{H} = \hbar\omega_c \hat{a}^\dagger \hat{a} + \sum_\omega \hbar\omega \hat{b}^\dagger_\omega \hat{b}_\omega + \hbar \sum_\omega \left(\kappa_\omega \hat{b}^\dagger_\omega \hat{a} + \kappa^*_\omega \hat{b}_\omega \hat{a}^\dagger \right) \tag{C.1}$$

The first term is the Hamiltonian for the free cavity field. The vacuum energy is subtracted, since it does not affect the interaction with the reservoir, as can be easily verified (see Equation 13.20). The second term is the Hamiltonian for the loss oscillators, ω being the oscillation frequency of the oscillator. The third term is the interaction Hamiltonian under the rotating-wave approximation. The coupling constant κ_ω is assumed to be slowly varying with frequency.

By the Heisenberg equation, the equation of motion for the field and the loss oscillators becomes

$$\frac{d}{dt}\hat{a} = -i\omega_c \hat{a} - i \sum_\omega \kappa^*_\omega \hat{b}_\omega \tag{C.2}$$

$$\frac{d}{dt}\hat{b}_\omega = -i\omega \hat{b}_\omega - i\kappa_\omega \hat{a} \tag{C.3}$$

Integrating Equation C.3 we have

$$\hat{b}_\omega = -i\kappa_\omega \int_0^t e^{-i\omega(t-t')} \hat{a}(t')dt' + \hat{b}_\omega(0)e^{-i\omega t} \tag{C.4}$$

Substituting this into Equation C.2 we have

$$\frac{d}{dt}\hat{a} = -i\omega_c \hat{a} - \sum_\omega |\kappa_\omega|^2 \int_0^t e^{-i\omega(t-t')} \hat{a}(t')dt' - i \sum_\omega \kappa^*_\omega \hat{b}_\omega(0) e^{-i\omega t} \tag{C.5}$$

Truncating the cavity resonance frequency, we set

$$\hat{a}(t) = \tilde{a}(t)e^{-i\omega_c t} \tag{C.6}$$

Then we have

$$\frac{d}{dt}\tilde{a} = -\sum_\omega |\kappa_\omega|^2 \int_0^t e^{-i(\omega-\omega_c)(t-t')} \tilde{a}(t')dt' - i \sum_\omega \kappa^*_\omega \hat{b}_\omega(0) e^{i(\omega_c-\omega)t} \tag{C.7}$$

Here we assume that the density of the loss oscillators per unit angular frequency is $\rho(\omega)$ and the density is a slowly varying function of the frequency. We have

$$\sum_\omega |\kappa_\omega|^2 e^{-i(\omega-\omega_c)(t-t')} = \int_{-\omega_c}^{\infty} \rho(x+\omega_c)|\kappa_{x+\omega_c}|^2 e^{-ix(t-t')}dx \tag{C.8}$$

where we have set $\omega - \omega_c = x$. Since the integrand is important only around $x = 0$, we may take $\rho|\kappa|^2$ outside the integral sign and replace the lower limit of integration by $-\infty$ with minimal error. Thus we may write

$$\sum_\omega |\kappa_\omega|^2 e^{-i(\omega-\omega_c)(t-t')} \approx 2\pi\rho(\omega_c)|\kappa_{\omega_c}|^2 \delta(t-t') \tag{C.9}$$

Thus Equation C.7 reduces to

$$\frac{d}{dt}\tilde{a} = -\pi\rho(\omega_c)|\kappa_{\omega_c}|^2\tilde{a}(t) - i\sum_{\omega}\kappa^*_{\omega}\hat{b}_{\omega}(0)e^{i(\omega_c-\omega)t} \tag{C.10}$$

where we have used the property of the delta function that

$$\int_0^t f(t')\delta(t-t')dt' = \frac{1}{2}f(t) \tag{C.11a}$$

Going back to $\hat{a}(t)$ we have

$$\frac{d}{dt}\hat{a} = -i\omega_c\hat{a} - \gamma_c\hat{a}(t) - i\sum_{\omega}\kappa^*_{\omega}\hat{b}_{\omega}(0)e^{-i\omega t} \tag{C.11b}$$

where

$$\gamma_c = \pi\rho(\omega_c)|\kappa_{\omega_c}|^2 \tag{C.12}$$

Now the last term in Equation C.11b gives a fluctuating force $\hat{\Gamma}_f(t) = \tilde{\Gamma}_f(t)e^{-i\omega_c t}$. Then $\tilde{\Gamma}_f(t)$ gives the fluctuating force for the cavity mode amplitude, the correlation function of which is

$$\begin{aligned}\left\langle \tilde{\Gamma}^\dagger_f(t')\tilde{\Gamma}_f(t)\right\rangle &= \left\langle \sum_{\omega'}\kappa_{\omega'}\hat{b}^\dagger_{\omega'}(0)e^{i(\omega'-\omega_c)t'}\sum_{\omega}\kappa^*_{\omega}\hat{b}_{\omega}(0)e^{-i(\omega-\omega_c)t}\right\rangle \\ &= \sum_{\omega}|\kappa_{\omega}|^2\left\langle \hat{b}^\dagger_{\omega}\hat{b}_{\omega}(0)\right\rangle e^{-i(\omega-\omega_c)(t-t')}\end{aligned} \tag{C.13}$$

$$\begin{aligned}\left\langle \tilde{\Gamma}_f(t')\tilde{\Gamma}^\dagger_f(t)\right\rangle &= \left\langle \sum_{\omega'}\kappa^*_{\omega'}\hat{b}_{\omega'}(0)e^{-i(\omega'-\omega_c)t'}\sum_{\omega}\kappa_{\omega}\hat{b}^\dagger_{\omega}(0)e^{i(\omega-\omega_c)t}\right\rangle \\ &= \sum_{\omega}|\kappa_{\omega}|^2\left\langle \hat{b}_{\omega}\hat{b}^\dagger_{\omega}(0)\right\rangle e^{i(\omega-\omega_c)(t-t')}\end{aligned} \tag{C.14}$$

where we have assumed the independence of different loss oscillators: $\left\langle \hat{b}_{\omega'}(0)\hat{b}^\dagger_{\omega}(0)\right\rangle = 0$ for $\omega \neq \omega'$. Here the bracket signifies an ensemble average with respect to the reservoir of the quantum-mechanical expectation value. If we write the average number of oscillator bosons as

$$\left\langle \hat{b}^\dagger_{\omega}\hat{b}_{\omega}(0)\right\rangle = n_{\omega} \tag{C.15}$$

the corresponding average is

$$\left\langle \hat{b}_{\omega}\hat{b}^\dagger_{\omega}(0)\right\rangle = n_{\omega} + 1 \tag{C.16}$$

Now, assuming that n_{ω} is also a slowly varying function of ω, the summations in Equations C.13 and C.14 may be performed as in Equation C.9, with the result that

$$\left\langle \tilde{\Gamma}_f^\dagger(t')\tilde{\Gamma}_f(t) \right\rangle = 2\pi|\kappa_{\omega_c}|^2\rho(\omega_c)n_{\omega_c}\delta(t-t') \\ = 2\gamma_c n_{\omega_c}\delta(t-t') \tag{C.17}$$

$$\left\langle \tilde{\Gamma}_f(t')\tilde{\Gamma}_f^\dagger(t) \right\rangle = 2\pi|\kappa_{\omega_c}|^2\rho(\omega_c)(n_{\omega_c}+1)\delta(t-t') \\ = 2\gamma_c(n_{\omega_c}+1)\delta(t-t') \tag{C.18}$$

If the approximation made in obtaining Equation C.9, namely that we replace $-\omega_c$ by $-\infty$, is relaxed to the one where we replace the lower limit of integration 0 by $-\infty$, the above correlations may be rewritten as

$$\left\langle \hat{\Gamma}_f^\dagger(t')\hat{\Gamma}_f(t) \right\rangle = 2\pi|\kappa_{\omega_c}|^2\rho(\omega_c)n_{\omega_c}\delta(t-t') \\ = 2\gamma_c n_{\omega_c}\delta(t-t') \tag{C.19}$$

$$\left\langle \hat{\Gamma}_f(t')\hat{\Gamma}_f^\dagger(t) \right\rangle = 2\pi|\kappa_{\omega_c}|^2\rho(\omega_c)(n_{\omega_c}+1)\delta(t-t') \\ = 2\gamma_c(n_{\omega_c}+1)\delta(t-t') \tag{C.20}$$

Thus Equation C.11 may be rewritten as

$$\frac{d}{dt}\hat{a} = -i\omega_c\hat{a} - \gamma_c\hat{a}(t) + \hat{\Gamma}_f(t) \tag{C.21}$$

with Equations C.19 and C.20 taken as valid. Thus the presence of the reservoir made up of loss oscillators results in a damping of the cavity mode plus a fluctuating force. Then, just as in Equations 3.35–3.41, we can show the preservation of the commutation relation:

$$\left\langle \hat{a}(t)\hat{a}^\dagger(t) - \hat{a}^\dagger(t)\hat{a}(t) \right\rangle = 1 \tag{C.22}$$

Appendix D
Derivation of Equation 7.29: The Laplace-Transformed Solution

Here we derive Equation 7.29. We first set $z = -d$ in Equations 7.24a and 7.24b to give

$$(s+\gamma')\{L^+(-d,s) - V^+(-d,s)\} \\ = GN\left[-\frac{1}{1-r''(s)}\int_{-d}^{0}\exp\{-(z_m+d)s/c_1\}L^-(z_m,s)dz_m \right. \\ \left. + \frac{r''(s)}{1-r''(s)}\int_{-d}^{0}\exp\{(z_m+d)s/c_1\}L^+(z_m,s)dz_m\right] \tag{D.1}$$

and

$$
\begin{aligned}
&(s+\gamma')\{L^-(-d,s)-V^-(-d,s)\}\\
&\quad = GN\left[\int_{-d}^{0}\exp\{(z_m+d)s/c_1\}L^-(z_m,s)dz_m\right.\\
&\qquad -\frac{r''(s)}{1-r''(s)}\int_{-d}^{0}\exp\{(z_m+d)s/c_1\}L^+(z_m,s)dz_m\\
&\qquad \left.+\frac{r''(s)}{1-r''(s)}\int_{-d}^{0}\exp\{(z_m+d)s/c_1\}L^-(z_m,s)dz_m\right]
\end{aligned}
\tag{D.2}
$$

Comparing these two equations we obtain

$$
L^+(-d,s)-V^+(-d,s)=-\{L^-(-d,s)-V^-(-d,s)\}
\tag{D.3}
$$

Next we set $z = 0$ in Equations 7.24a and 7.24b to give

$$
\begin{aligned}
&(s+\gamma')\{L^+(0,s)-V^+(0,s)\}\\
&\quad = GN\left[-\frac{1}{1-r''(s)}\int_{-d}^{0}\exp\{-(z_m+2d)s/c_1\}L^-(z_m,s)dz_m\right.\\
&\qquad \left.+\frac{1}{1-r''(s)}\int_{-d}^{0}\exp\{z_m s/c_1\}L^+(z_m,s)dz_m\right]
\end{aligned}
\tag{D.4}
$$

and

$$
\begin{aligned}
&(s+\gamma')\{L^-(0,s)-V^-(0,s)\}\\
&\quad = GN\left[-\frac{r''(s)}{1-r''(s)}\int_{-d}^{0}\exp\{(z_m+2d)s/c_1\}L^+(z_m,s)dz_m\right.\\
&\qquad \left.+\frac{r''(s)}{1-r''(s)}\int_{-d}^{0}\exp(-z_m s/c_1)L^-(z_m,s)dz_m\right]
\end{aligned}
\tag{D.5}
$$

Comparing these two equations we have

$$
\begin{aligned}
L^-(0,s)-V^-(0,s) &= -r''(s)\exp(2ds/c_1)\{L^+(0,s)-V^+(0,s)\}\\
&= -r'\{L^+(0,s)-V^+(0,s)\}
\end{aligned}
\tag{D.6}
$$

where we have used Equation 7.25.

Then we set $z = 0$ in Equations 7.28a and 7.28b to obtain

$$\begin{aligned} L^+(0,s) - V^+(0,s) = &\int_{-d}^{0} \exp\left[\left\{\frac{s}{c_1} - \frac{GN}{(s+\gamma')}\right\} z_m\right] \frac{GN}{(s+\gamma')} V^+(z_m,s)dz_m \\ &+ \exp\left[\left\{-\frac{s}{c_1} + \frac{GN}{(s+\gamma')}\right\} d\right] \{L^+(-d,s) - V^+(-d,s)\} \end{aligned} \tag{D.7}$$

$$\begin{aligned} L^-(0,s) - V^-(0,s) = &-\int_{-d}^{0} \exp\left[-\left\{\frac{s}{c_1} - \frac{GN}{(s+\gamma')}\right\} z_m\right] \frac{GN}{(s+\gamma')} V^-(z_m,s)dz_m \\ &+ \exp\left[\left\{\frac{s}{c_1} - \frac{GN}{(s+\gamma')}\right\} d\right] \{L^-(-d,s) - V^-(-d,s)\} \end{aligned} \tag{D.8}$$

Addition of Equation D.7 multiplied by r' and Equation D.8 yields a null sum for the left-hand sides because of Equation D.6. Thus the similar sum of the right-hand sides is also zero. Therefore, eliminating $\{L^-(-d,s) - V^-(-d,s)\}$ using Equation D.3, we obtain

$$\begin{aligned} &r' \int_{-d}^{0} \exp\left[\left\{\frac{s}{c_1} - \frac{GN}{(s+\gamma')}\right\} z_m\right] \frac{GN}{(s+\gamma')} V^+(z_m,s)dz_m \\ &+ r' \exp\left[\left\{-\frac{s}{c_1} + \frac{GN}{(s+\gamma')}\right\} d\right] \{L^+(-d,s) - V^+(-d,s)\} \\ &- \int_{-d}^{0} \exp\left[-\left\{\frac{s}{c_1} - \frac{GN}{(s+\gamma')}\right\} z_m\right] \frac{GN}{(s+\gamma')} V^-(z_m,s)dz_m \\ &- \exp\left[\left\{\frac{s}{c_1} - \frac{GN}{(s+\gamma')}\right\} d\right] \{L^+(-d,s) - V^+(-d,s)\} = 0 \end{aligned} \tag{D.9}$$

We finally have

$$\begin{aligned} \{L^+(-d,s) - V^+(-d,s)\} = &\frac{GN}{(s+\gamma')} \left[\int_{-d}^{0} \exp\left\{-\left(\frac{s}{c_1} - \frac{GN}{s+\gamma'}\right) z_m\right\}\right. \\ &\times V^-(z_m,s)dz_m - r' \int_{-d}^{0} \exp\left\{\left(\frac{s}{c_1} - \frac{GN}{s+\gamma'}\right) z_m\right\} \\ &\left.\times V^+(z_m,s)dz_m\right] \left[r' \exp\left\{-\left(\frac{s}{c_1} - \frac{GN}{s+\gamma'}\right) d\right\}\right. \\ &\left.- \exp\left\{\left(\frac{s}{c_1} - \frac{GN}{s+\gamma'}\right) d\right\}\right]^{-1} \end{aligned} \tag{D.10}$$

The solution for $\{L^-(-d,s) - V^-(-d,s)\}$ is obtained as the right-hand side of Equation D.10 multiplied by (-1) because of Equation D.3. Multiplying both the numerator and the denominator by $\exp(-ds/c_1)$ and writing $V^+(z,s) = \theta^+(z)$ and $V^-(z,s) = \theta^-(z)$ (because the initial driving field is assumed to be a delta function of time), we arrive at Equation 7.29.

Appendix E
Integrated Absolute Squared Field Strength of the Cavity Resonant Mode

In order to prove Equations 10.101 and 10.102, we evaluate

$$I = \int_{-d}^{0} dz_m \left(\frac{|e^-(z_m)|^2 + |e^+(z_m)|^2}{|e^+(-d)|^2} \right) \tag{E.1}$$

Let us set

$$x(z) = |e^-(z)|^2 + |e^+(z)|^2 \tag{E.2}$$

From Equations 8.35a and 8.35b cited above Equation 10.94a, we have

$$\begin{aligned} \frac{d}{dz} x(z) &= \frac{2\mathrm{Re}\,\alpha^0}{1 + |E_{z/s}|^2} \left(|e^+(z)|^2 - |e^-(z)|^2 \right) \\ &= \frac{2\mathrm{Re}\,\alpha^0}{1 + \{x(z)/E_s^2\}} \left(\pm\sqrt{x^2(z) - 4C} \right) \end{aligned} \tag{E.3}$$

where the constant C is defined in Equations 10.98a and 10.98c. Since 2 Re α^0 is positive and because of Equation 10.98b, Equations 10.94a and 10.94b show that $|e^+(z)|^2 - |e^-(z)|^2$ is never negative. Thus we can choose the positive sign in Equation E.3. We have

$$\int_{-d}^{0} 2\mathrm{Re}\,\alpha^0 dz = \int_{x(-d)}^{x(0)} \frac{\left\{ 1 + \left(x/|E_s|^2 \right) \right\}}{\sqrt{x^2 - 4C}} dx \tag{E.4}$$

Integrating both sides

$$\begin{aligned} 2\mathrm{Re}\,\alpha^0 d &= \ln\left| x + \sqrt{x^2 - 4C} \right| \Bigg|_{x(-d)}^{x(0)} + \frac{\sqrt{x^2 - 4C}}{|E_s|^2} \Bigg|_{x(-d)}^{x(0)} \\ &= \ln\left| x(0) + \sqrt{x^2(0) - 4C} \right| - \ln\left| x(-d) + \sqrt{x^2(-d) - 4C} \right| \\ &\quad + \frac{\sqrt{x^2(0) - 4C}}{|E_s|^2} - \frac{\sqrt{x^2(-d) - 4C}}{|E_s|^2} \end{aligned} \tag{E.5}$$

Since

$$\begin{aligned} \sqrt{x^2(0) - 4C} &= |e^+(0)|^2 - |e^-(0)|^2 \\ \sqrt{x^2(-d) - 4C} &= |e^+(-d)|^2 - |e^-(-d)|^2 = 0 \end{aligned} \tag{E.6}$$

where we have used Equation 10.98b, we have

$$\begin{aligned} 2\mathrm{Re}\,\alpha^0 d &= \ln\left|\,2|e^+(0)|^2\,\right| - \ln\left|\,2|e^+(-d)|^2\,\right| + \frac{|e^+(0)|^2 - |e^-(0)|^2}{E_s^2} \\ &= \ln\left(\frac{|e^+(0)|^2}{|e^+(-d)|^2}\right) + \frac{|e^+(0)|^2 - |e^-(0)|^2}{|E_s|^2} \end{aligned} \tag{E.7}$$

Now from Equations 8.27a and 8.27b we have

$$\begin{aligned} |e^+(0)|^2 &= |e^+(-d)|^2 \exp\{I(0) + I^*(0)\} \\ |e^-(0)|^2 &= |e^-(-d)|^2 \exp\{-I(0) - I^*(0)\} \end{aligned} \tag{E.8}$$

and from Equations 8.28a and 8.29

$$I(0) + I^*(0) = 2\mathrm{Re}\left(\alpha^0 I\right) = \mathrm{Re}\left[\ln\left\{\frac{-1}{r\exp(2ikd)}\right\}\right] = \ln\left(\frac{1}{r}\right) \tag{E.9}$$

(The I in Equation E.9 should not be confused with that in Equation E.1.) Therefore we have

$$\begin{aligned} |e^+(0)|^2 &= |e^+(-d)|^2/r \\ |e^-(0)|^2 &= |e^-(-d)|^2 r \end{aligned} \tag{E.10}$$

Thus Equation E.7 reads

$$2\mathrm{Re}\,\alpha^0 d = \ln\left(\frac{1}{r}\right) + \left(\frac{1}{r} - r\right)\frac{|e^+(-d)|^2}{|E_s|^2} \tag{E.11}$$

This gives

$$|e^+(-d)|^2 = \sqrt{C} = |E_s|^2\left(\frac{1}{r} - r\right)^{-1}\left\{2\mathrm{Re}\,\alpha^0 d - \ln\left(\frac{1}{r}\right)\right\} \tag{E.12}$$

Next we multiply both sides of Equation E.3 by x and rewrite the result as in Equation E.4 to have

$$2\mathrm{Re}\,\alpha^0 \int_{-d}^{0} x(z)dz = \int_{x(-d)}^{x(0)} \frac{\left\{x + \left(x^2/|E_s|^2\right)\right\}}{\sqrt{x^2 - 4C}}dx \tag{E.13}$$

The right-hand side is evaluated as

$$
\begin{aligned}
&\int_{x(-d)}^{x(0)} \frac{x+(x^2-4C+4C)/|E_s|^2}{\sqrt{x^2-4C}}dx \\
&= \sqrt{x^2-4C}\Big|_{x(-d)}^{x(0)} + \frac{1}{2|E_s|^2}\left(x\sqrt{x^2-4C} - 4C\ln\left|x+\sqrt{x^2-4C}\right|\right)\Big|_{x(-d)}^{x(0)} \\
&+ \frac{4C}{|E_s|^2}\left(\ln\left|x+\sqrt{x^2-4C}\right|\right)\Big|_{x(-d)}^{x(0)} \\
&= |e^+(0)|^2 - |e^-(0)|^2 + \frac{1}{2|E_s|^2}\left(|e^+(0)|^2+|e^-(0)|^2\right)\left(|e^+(0)|^2-|e^-(0)|^2\right) \\
&+ \frac{2C}{|E_s|^2}\ln\left(\frac{|e^+(0)|^2}{|e^+(-d)|^2}\right)
\end{aligned}
\tag{E.14}
$$

where we have used Equation E.6. Using Equation E.10 again we have

$$
\begin{aligned}
&\int_{x(-d)}^{x(0)} \frac{x+(x^2-4C+4C)/|E_s|^2}{\sqrt{x^2-4C}}dx \\
&= \left(\frac{1}{r}-r\right)|e^+(-d)|^2 + \left(\frac{1}{r^2}-r^2\right)\frac{|e^+(-d)|^4}{2|E_s|^2} + \frac{2|e^+(-d)|^4}{|E_s|^2}\ln\left(\frac{1}{r}\right)
\end{aligned}
\tag{E.15}
$$

where we have used Equation 10.98c. Thus the original integral in Equation E.1,

$$
I = \int_{-d}^{0} dz_m\left(\frac{|e^-(z_m)|^2+|e^+(z_m)|^2}{|e^+(-d)|^2}\right) = \frac{\int_{-d}^{0} x(z_m)dz_m}{|e^+(-d)|^2}
\tag{E.16}
$$

is Equation E.15 divided by $|e^+(-d)|^2$ and by $2\mathrm{Re}\,\alpha^0$. Thus

$$
I = \frac{1}{2\mathrm{Re}\,\alpha^0}\left\{\left(\frac{1}{r}-r\right) + \left(\frac{1}{r^2}-r^2\right)\frac{|e^+(-d)|^2}{2|E_s|^2} + \frac{2|e^+(-d)|^2}{|E_s|^2}\ln\left(\frac{1}{r}\right)\right\}
\tag{E.17}
$$

Here we can use Equation E.12 for $|e^+(-d)|^2/E_s^2$. Thus

$$
\begin{aligned}
I = \frac{1}{2\mathrm{Re}\,\alpha^0}&\left[\left(\frac{1}{r}-r\right) + \left\{\frac{1}{2}\left(\frac{1}{r^2}-r^2\right) + 2\ln\left(\frac{1}{r}\right)\right\}\left(\frac{1}{r}-r\right)^{-1}\right. \\
&\left.\times\left\{2\mathrm{Re}\,\alpha^0 d - \ln\left(\frac{1}{r}\right)\right\}\right]
\end{aligned}
\tag{E.18}
$$

Now from Equation 10.94c we have

$$
2\mathrm{Re}\,\alpha^0 = \frac{2\gamma g N\sigma^0}{(\nu_0-\omega)^2+\gamma^2} \equiv D\sigma^0
\tag{E.19}
$$

where we have defined the constant D, which is equal to $2\wp N/c_1$ by Equation 10.94c. The threshold condition is obtained from Equation E.12 by setting $|e^+(-d)|^2 = 0$. We have

$$2\mathrm{Re}\,\alpha_{th}^0 d - \ln\left(\frac{1}{r}\right) = 0 \tag{E.20a}$$

or

$$Dd\sigma_{th}^0 - \ln\left(\frac{1}{r}\right) = 0 \tag{E.20b}$$

Thus by Equations E.19 and E.20b

$$\begin{aligned} I &= \frac{1}{D\sigma^0}\left[\left(\frac{1}{r}-r\right) + \left\{\frac{1}{2}\left(\frac{1}{r^2}-r^2\right) + 2\ln\left(\frac{1}{r}\right)\right\}\left(\frac{1}{r}-r\right)^{-1}\left\{D\sigma^0 d - D\sigma_{th}^0 d\right\}\right] \\ &= \left[\left(\frac{1}{r}-r\right)\frac{1}{D\sigma^0} + \left\{\frac{1}{2}\left(\frac{1}{r^2}-r^2\right) + 2\ln\left(\frac{1}{r}\right)\right\}\left(\frac{1}{r}-r\right)^{-1} d\left\{1-\frac{\sigma_{th}^0}{\sigma^0}\right\}\right] \\ &= d\left[\left(\frac{1}{r}-r\right)\frac{\sigma_{th}^0}{\ln(1/r)\sigma^0} + \left\{\frac{1}{2}\left(\frac{1}{r^2}-r^2\right) + 2\ln\left(\frac{1}{r}\right)\right\}\left(\frac{1}{r}-r\right)^{-1}\left\{1-\frac{\sigma_{th}^0}{\sigma^0}\right\}\right] \end{aligned} \tag{E.21}$$

Using the definitions $\gamma_c = (c_1/2d)\ln(1/r)$ (Equation 1.18) and $\beta_c = (c_1/2d)(1-r^2)/(2r)$ (Equation 6.35), we can rearrange the terms as

$$I = 2d\frac{\beta_c}{\gamma_c}\left\{1 + \left(1-\frac{\sigma_{th}^0}{\sigma^0}\right)g(r)\right\} = 2d\frac{\beta_c}{\gamma_c}\left\{1 + \frac{\Delta}{1+\Delta}g(r)\right\} \tag{E.22}$$

where Δ is the fractional excess atomic inversion

$$\Delta = \frac{\sigma^0 - \sigma_{th}^0}{\sigma_{th}^0} \tag{E.23}$$

and the function of the reflection coefficient r,

$$\begin{aligned} g(r) &= 2\left(\frac{\ln(1/r)}{(1-r^2)/r}\right)^2 + \frac{\frac{1}{2}\ln(1/r)\,(1-r^4)/r^2}{[(1-r^2)/r]^2} - 1 \\ &= \frac{1}{2}\left(\frac{\gamma_c}{\beta_c}\right)^2 + \frac{1+r^2}{4r}\frac{\gamma_c}{\beta_c} - 1 \end{aligned} \tag{E.24}$$

is monotonically decreasing from $+\infty$ to 0 as r goes from 0 to 1.

Appendix F
Some Rules on the Absolute Squared Amplitudes and Evaluation of the Integrated Intensity

From Equation 11.5 we have in general that

$$\begin{aligned} |e^{+}(z')|^2 &= |e^{+}(z)|^2 \exp\left\{\int_z^{z'} 2\mathrm{Re}\,\alpha(z)dz\right\} \\ |e^{-}(z')|^2 &= |e^{-}(z)|^2 \exp\left\{-\int_z^{z'} 2\mathrm{Re}\,\alpha(z)dz\right\} \end{aligned} \tag{F.1a}$$

In particular we have

$$\begin{aligned} |e^{+}(d)|^2 &= |e^{+}(0)|^2 \exp\left\{\int_0^{d} 2\mathrm{Re}\,\alpha(z)dz\right\} \\ |e^{-}(d)|^2 &= |e^{-}(0)|^2 \exp\left\{-\int_0^{d} 2\mathrm{Re}\,\alpha(z)dz\right\} \end{aligned} \tag{F.1b}$$

We define a neutral point z_c, where

$$|e^{+}(z_c)|^2 = |e^{-}(z_c)|^2 = \sqrt{C} \tag{F.2}$$

Then we have

$$\begin{aligned} |e^{+}(z_c)|^2 &= |e^{+}(0)|^2 \exp\left\{\int_0^{z_c} 2\mathrm{Re}\,\alpha(z)dz\right\} \\ |e^{-}(z_c)|^2 &= |e^{-}(0)|^2 \exp\left\{-\int_0^{z_c} 2\mathrm{Re}\,\alpha(z)dz\right\} \end{aligned} \tag{F.3}$$

and

$$\begin{aligned} |e^{+}(z_c)|^2 &= |e^{+}(d)|^2 \exp\left\{-\int_{z_c}^{d} 2\mathrm{Re}\,\alpha(z)dz\right\} \\ |e^{-}(z_c)|^2 &= |e^{-}(d)|^2 \exp\left\{\int_{z_c}^{d} 2\mathrm{Re}\,\alpha(z)dz\right\} \end{aligned} \tag{F.4}$$

From Equations 11.10 and F.1b we obtain

$$|e^{+}(d)|^2 = |e^{+}(0)|^2 \frac{1}{|r_1||r_2|} \quad \text{and} \quad |e^{-}(d)|^2 = |e^{-}(0)|^2 |r_1||r_2| \tag{F.5}$$

Combining Equations 11.6 and 11.9 we have

$$|e^{+}(0)|^2 = |r_1|\sqrt{C} \quad \text{and} \quad |e^{-}(d)|^2 = |r_2|\sqrt{C} \tag{F.6}$$

Using these in Equation F.5 we have

$$|e^{+}(d)|^2 = \frac{\sqrt{C}}{|r_2|} \quad \text{and} \quad |e^{-}(0)|^2 = \frac{\sqrt{C}}{|r_1|} \tag{F.7}$$

Next we evaluate the integral I in Equation 11.12. Since we know the value of the constant C (see Equation 11.22), Equation 11.17 gives the value of $X(z)$ completely. Thus the integral I in Equation 11.12 can be evaluated as follows. Multiplying both sides of Equation 11.17 by $X = |e^{+}(z)|^2 + |e^{-}(z)|^2$, we have

$$\begin{aligned} 2\mathrm{Re}\left\{\alpha^0\right\}X \quad dz &= \pm\frac{\{X+(X^2/|E_s|^2)\}dX}{\sqrt{X^2-4C}} \\ &= \pm\left[\frac{X}{\sqrt{X^2-4C}}+\frac{\sqrt{X^2-4C}}{|E_s|^2}+\frac{4C}{|E_s|^2\sqrt{X^2-4C}}\right]dX \end{aligned} \tag{F.8}$$

Here the plus sign applies for $z > z_c$ and the minus sign for $z < z_c$. Integrating, we obtain

$$\begin{aligned} &2\,\mathrm{Re}\left\{\alpha^0\right\}I \\ &= -\int_{X(0)}^{X(z_c)}\frac{\{X+(X^2/|E_s|^2)\}dX}{\sqrt{X^2-4C}}+\int_{X(z_c)}^{X(d)}\frac{\{X+(X^2/|E_s|^2)\}dX}{\sqrt{X^2-4C}} \\ &= -\sqrt{X^2-4C}\Big|_{X(0)}^{X(z_c)}-\frac{1}{2|E_s|^2}\left(X\sqrt{X^2-4C}-4C\ln\left|X+\sqrt{X^2-4C}\right|\right)\Big|_{X(0)}^{X(z_c)} \\ &-\frac{4C}{|E_s|^2}\ln\left|X+\sqrt{X^2-4C}\right|\Big|_{X(0)}^{X(z_c)}+\sqrt{X^2-4C}\Big|_{X(z_c)}^{X(d)} \\ &+\frac{1}{2|E_s|^2}\left(X\sqrt{X^2-4C}-4C\ln\left|X+\sqrt{X^2-4C}\right|\right)\Big|_{X(z_c)}^{X(d)} \\ &+\frac{4C}{|E_s|^2}\ln\left|X+\sqrt{X^2-4C}\right|\Big|_{X(z_c)}^{X(d)} \\ &= |e^-(0)|^2-|e^+(0)|^2-\frac{1}{2|E_s|^2}\Big\{-4C\ln\left(|e^+(z_c)|^2+|e^-(z_c)|^2\right) \\ &-\left(|e^-(0)|^2+|e^+(0)|^2\right)\left(|e^-(0)|^2-|e^+(0)|^2\right)+4C\ln\left(2|e^-(0)|^2\right)\Big\} \\ &-\frac{4C}{|E_s|^2}\left\{\ln\left(|e^+(z_c)|^2+|e^-(z_c)|^2\right)-\ln\left(2|e^-(0)|^2\right)\right\} \\ &+|e^+(d)|^2-|e^-(d)|^2+\frac{1}{2|E_s|^2}\Big\{\left(|e^+(d)|^2+|e^-(d)|^2\right)\left(|e^+(d)|^2-|e^-(d)|^2\right) \\ &-4C\ln\left(2|e^+(d)|^2\right)+4C\ln\left(|e^+(z_c)|^2+|e^-(z_c)|^2\right)\Big\} \\ &+\frac{4C}{|E_s|^2}\left\{\ln\left(2|e^+(d)|^2\right)-\ln\left(|e^+(z_c)|^2+|e^-(z_c)|^2\right)\right\} \\ &= |e^-(0)|^2-|e^+(0)|^2-\frac{2C}{|E_s|^2}\left\{\ln\left(|e^+(z_c)|^2+|e^-(z_c)|^2\right)-\ln\left(2|e^-(0)|^2\right)\right\} \\ &+\frac{1}{2|E_s|^2}\left(|e^-(0)|^2+|e^+(0)|^2\right)\left(|e^-(0)|^2-|e^+(0)|^2\right) \\ &+|e^+(d)|^2-|e^-(d)|^2+\frac{2C}{|E_s|^2}\left\{\ln\left(2|e^+(d)|^2\right)-\ln\left(|e^+(z_c)|^2+|e^-(z_c)|^2\right)\right\} \\ &+\frac{1}{2|E_s|^2}\left\{\left(|e^+(d)|^2+|e^-(d)|^2\right)\left(|e^+(d)|^2-|e^-(d)|^2\right)\right. \end{aligned} \tag{F.9}$$

Substituting Equations F.2, F.6, and F.7 into Equation F.9, we obtain

$$\begin{aligned}2\mathrm{Re}\{\alpha^0\}I=&\sqrt{C}\left(\frac{1}{|r_1|}-|r_1|\right)-\frac{2C}{|E_s|^2}\ln|r_1|+\frac{C}{2|E_s|^2}\left(\frac{1}{|r_1|}+|r_1|\right)\left(\frac{1}{|r_1|}-|r_1|\right)\\&+\sqrt{C}\left(\frac{1}{|r_2|}-|r_2|\right)-\frac{2C}{|E_s|^2}\ln|r_2|+\frac{C}{2|E_s|^2}\left(\frac{1}{|r_2|}+|r_2|\right)\left(\frac{1}{|r_2|}-|r_2|\right)\end{aligned} \tag{F.10}$$

We consider the integrated intensity scaled to $\sqrt{C}=|e^{\pm}(z_c)|^2$:

$$\begin{aligned}\frac{I}{\sqrt{C}}=\frac{1}{2\mathrm{Re}\{\alpha^0\}}&\left\{\left(\frac{1}{|r_1|}-|r_1|+\frac{1}{|r_2|}-|r_2|\right)+\frac{2\sqrt{C}}{|E_s|^2}\ln\left(\frac{1}{|r_1||r_2|}\right)\right.\\&\left.+\frac{\sqrt{C}}{2|E_s|^2}\left(\frac{1}{|r_1|^2}-|r_1|^2+\frac{1}{|r_2|^2}-|r_2|^2\right)\right\}\end{aligned} \tag{F.11}$$

For the remaining $\sqrt{C}$ on the right-hand side, we substitute Equation 11.22 to obtain

$$\begin{aligned}\frac{I}{\sqrt{C}}=\frac{1}{2\mathrm{Re}\{\alpha^0\}}&\left[\frac{(|r_1|+|r_2|)(1-|r_1||r_2|)}{|r_1||r_2|}\right.\\&+\left\{2\mathrm{Re}\{\alpha^0\}d-\ln\left(\frac{1}{|r_1||r_2|}\right)\right\}\frac{|r_1||r_2|}{(|r_1|+|r_2|)(1-|r_1||r_2|)}\\&\left.\times\left\{2\ln\left(\frac{1}{|r_1||r_2|}\right)+\frac{1}{2}\frac{(|r_1|^2+|r_2|^2)(1-|r_1|^2|r_2|^2)}{|r_1|^2|r_2|^2}\right\}\right]\end{aligned} \tag{F.12}$$

We use Equation 11.23 or

$$2d\mathrm{Re}\{\alpha_{th}^0\}=\ln\left(\frac{1}{|r_1||r_2|}\right) \tag{F.13}$$

and

$$\frac{\mathrm{Re}\{\alpha_{th}^0\}}{\mathrm{Re}\{\alpha^0\}}=\frac{\sigma_{th}^0}{\sigma^0} \tag{F.14}$$

Thus

$$\frac{1}{\mathrm{Re}\{\alpha^0\}}=\frac{\sigma_{th}^0}{\sigma^0}\frac{1}{\mathrm{Re}\{\alpha_{th}^0\}}=\frac{\sigma_{th}^0}{\sigma^0}\frac{2d}{\ln(1/|r_1||r_2|)} \tag{F.15}$$

Then, we have

$$\begin{aligned}\frac{I}{\sqrt{C}}=&\,d\frac{\sigma_{th}^0}{\sigma^0}\frac{(|r_1|+|r_2|)(1-|r_1||r_2|)}{|r_1||r_2|}\Big/\ln\left(\frac{1}{|r_1||r_2|}\right)\\&+d\left(1-\frac{\sigma_{th}^0}{\sigma^0}\right)\frac{|r_1||r_2|}{(|r_1|+|r_2|)(1-|r_1||r_2|)}\\&\times\left\{2\ln\left(\frac{1}{|r_1||r_2|}\right)+\frac{1}{2}\frac{(|r_1|^2+|r_2|^2)(1-|r_1|^2|r_2|^2)}{|r_1|^2|r_2|^2}\right\}\end{aligned} \tag{F.16}$$

Finally, we introduce the fractional excess atomic inversion Δ by

$$\Delta = \frac{\sigma^0 - \sigma^0_{th}}{\sigma^0_{th}} \tag{F.17}$$

Then noting that $\sigma^0_{th}/\sigma^0 = 1/(1+\Delta) = 1 - \Delta/(1+\Delta)$ and that $1 - (\sigma^0_{th}/\sigma^0) = \Delta/(1+\Delta)$, we rearrange the terms to obtain

$$\begin{aligned} \frac{I}{\sqrt{C}} &= d\frac{(|r_1|+|r_2|)(1-|r_1||r_2|)/|r_1||r_2|}{\ln(1/|r_1||r_2|)} \\ &\times \left[1+\frac{\Delta}{1+\Delta}\right. \\ &\times \left.\left\{\frac{2(\ln(1/|r_1||r_2|))^2+\frac{1}{2}\ln(1/|r_1||r_2|)\left(|r_1|^2+|r_2|^2\right)\left(1-|r_1|^2|r_2|^2\right)/|r_1|^2|r_2|^2}{((|r_1|+|r_2|)(1-|r_1||r_2|)/|r_1||r_2|)^2}-1\right\}\right] \end{aligned} \tag{F.18}$$

Appendix G
Derivation of Equations 11.52a and 11.52b: Treatment of a Multilayered Dielectric Mirror

Here we derive Equations 11.52a and 11.52b for mirrors M1 and M2. As shown in Figure 11.1, the cavity comprises a lossless and non-dispersive dielectric of length d for which the dielectric constant is ε_1, the light velocity is c_1, and the refractive index is $n = \sqrt{\varepsilon_1/\varepsilon_0}$. The outside regions to the right and left of the cavity are both half free spaces (vacuum) for which the dielectric constants are ε_0 and the light velocities are c_0. The mirror M1 is at the interface of the cavity dielectric and the left half free space, and the mirror M2 is at the interface of the cavity dielectric and the right half free space. The mirrors are assumed to be made up of several thin layers of lossless and non-dispersive dielectrics. We assume that the thickness of either mirror can be ignored as compared to the cavity length d. Figure G.1 depicts the structure of M2, for example.

Let us assume that layers a, b, c, etc. are coated successively on the cavity dielectric. The first layer of dielectric constant ε_a and light velocity c_a extends from $z = z_1$ to $z = z_a$ with thickness $l_a = z_a - z_1$; the second layer of ε_b and c_b extends from $z = z_a$ to $z = z_b$ with thickness $l_b = z_b - z_a$; and so on. The magnetic

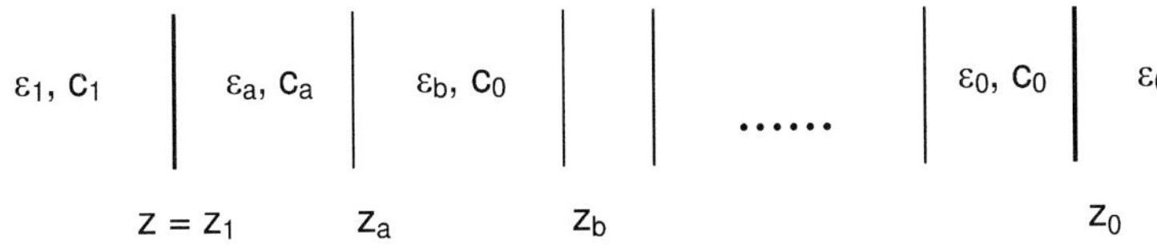

Figure G.1 The multi-dielectric-layer mirror.

permeability is assumed to be μ_0 everywhere. We write the vector potential of the nth region as

$$A_n(z,t) = u_n(z)e^{-i\omega t}, \qquad n = 1, a, b, ..., o, 0 \tag{G.1}$$

$$u_n(z) = \alpha_n e^{ik_n z} + \beta_n e^{-ik_n z} \tag{G.2}$$

where region 1 is the region inside the cavity; the layered region o is that facing the outer free space; and the region 0 is the outer free space. The first term in Equation G.2 represents the right-going wave and the second term the left-going wave in the nth region. First we want to show that the quantities in regions 1 and 0 are related in the form

$$\begin{pmatrix} \alpha_1 e^{ik_1 z_1} \\ \beta_1 e^{-ik_1 z_1} \end{pmatrix} = \begin{pmatrix} A & B \\ B^* & A^* \end{pmatrix} \begin{pmatrix} \alpha_0 e^{ik_0 z_o} \\ \beta_0 e^{-ik_0 z_o} \end{pmatrix} \tag{G.3}$$

where the quantities $A = A(\omega)$ and $B = B(\omega)$ are some definite functions of ω determined by the mirror structure; and A^* and B^* are their respective complex conjugates.

The electric and magnetic fields are given as

$$E_n(z) = -i\omega(\alpha_n e^{ik_n z} + \beta_n e^{-ik_n z}) \tag{G.4}$$

$$\mu_0 H_n(z) = ik_n\alpha_n e^{ik_n z} - ik_n\beta_n e^{-ik_n z} = (i\omega/c_n)\left(\alpha_n e^{ik_n z} - \beta_n e^{-ik_n z}\right) \tag{G.5}$$

At any boundary, both the electric and magnetic fields should be continuous. Thus at the right interface z_n of the nth region we have

$$\alpha_n e^{ik_n z_n} + \beta_n e^{-ik_n z_n} = \alpha_{n+1} e^{ik_{n+1} z_n} + \beta_{n+1} e^{-ik_{n+1} z_n} \tag{G.6}$$

$$(1/c_n)\left(\alpha_n e^{ik_n z_n} - \beta_n e^{-ik_n z_n}\right) = (1/c_{n+1})\left(\alpha_{n+1} e^{ik_{n+1} z_n} - \beta_{n+1} e^{-ik_{n+1} z_n}\right) \tag{G.7}$$

Therefore for the boundary at $z = z_n$ we have

$$\begin{pmatrix} \alpha_n e^{ik_n z_n} \\ \beta_n e^{-ik_n z_n} \end{pmatrix} = M_{n,n+1}(z_n) \begin{pmatrix} \alpha_{n+1} e^{ik_{n+1} z_n} \\ \beta_{n+1} e^{-ik_{n+1} z_n} \end{pmatrix} \tag{G.8}$$

where

$$M_{n,n+1}(z_n) = \begin{pmatrix} \frac{1}{2}(1 + c_n/c_{n+1}) & \frac{1}{2}(1 - c_n/c_{n+1}) \\ \frac{1}{2}(1 - c_n/c_{n+1}) & \frac{1}{2}(1 + c_n/c_{n+1}) \end{pmatrix} \tag{G.9}$$

The propagation in the $(n + 1)$th region is described as

$$\begin{pmatrix} \alpha_{n+1}e^{ik_{n+1}z_n} \\ \beta_{n+1}e^{-ik_{n+1}z_n} \end{pmatrix} = M_{n+1}(z_n, z_{n+1}) \begin{pmatrix} \alpha_{n+1}e^{ik_{n+1}z_{n+1}} \\ \beta_{n+1}e^{-ik_{n+1}z_{n+1}} \end{pmatrix} \tag{G.10}$$

where

$$M_{n+1}(z_n, z_{n+1}) = \begin{pmatrix} e^{-ik_{n+1}l_{n+1}} & 0 \\ 0 & e^{ik_{n+1}l_{n+1}} \end{pmatrix} \tag{G.11}$$

It is clear that the matrix that should appear in Equation G.3 is

$$M_{1,0}(z_1, z_o) = M_{1,a}(z_1)M_a(z_{1,}z_a)M_{a,b}(z_a)M_b(z_a, z_b)\cdots M_{o,0}(z_o) \tag{G.12}$$

As shown by Equations G.9 and G.11, all the component matrices have the property of the matrix in Equation G.3, that is, the lower left element and the lower right element are, respectively, the complex conjugate of the upper right element and upper left element. But it is easy to show that a product of two such matrices has the same property:

$$\begin{aligned} \begin{pmatrix} \alpha & \beta \\ \beta^* & \alpha^* \end{pmatrix} \begin{pmatrix} \gamma & \delta \\ \delta^* & \gamma^* \end{pmatrix} &= \begin{pmatrix} \alpha\gamma + \beta\delta^* & \alpha\delta + \beta\gamma^* \\ \beta^*\gamma + \alpha^*\delta^* & \beta^*\delta + \alpha^*\gamma^* \end{pmatrix} \\ &= \begin{Bmatrix} \alpha\gamma + \beta\delta^* & \alpha\delta + \beta\gamma^* \\ (\alpha\delta + \beta\gamma^*)^* & (\alpha\gamma + \beta\delta^*)^* \end{Bmatrix} \end{aligned} \tag{G.13}$$

Therefore, the product of an arbitrary number of matrices having this property also has this same property. Thus the matrix in Equation G.12 and, consequently, that in Equation G.3 should have this property. Thus we have proved the form of the matrix in Equation G.3.

Now we prove Equations 11.52a and 11.52b using Equation G.3. First consider the energy conservation among the input and output waves to the mirror M2. Because the layers are all lossless, the sum of the input powers should be equal to the sum of the output powers:

$$\varepsilon_1 c_1 \omega^2 |\alpha_1|^2 + \varepsilon_0 c_0 \omega^2 |\beta_0|^2 = \varepsilon_1 c_1 \omega^2 |\beta_1|^2 + \varepsilon_0 c_0 \omega^2 |\alpha_0|^2 \tag{G.14}$$

Since $\varepsilon_1 c_1/(\varepsilon_0 c_0) = n$ we have

$$n(|\alpha_1|^2 - |\beta_1|^2) = |\alpha_0|^2 - |\beta_0|^2 \tag{G.15}$$

Using Equation G.3

$$\begin{aligned} |\alpha_1|^2 - |\beta_1|^2 &= \left|A\alpha_0 e^{ik_0 z_o} + B\beta_0 e^{-ik_0 z_o}\right|^2 - \left|B^*\alpha_0 e^{ik_0 z_o} + A^*\beta_0 e^{-ik_0 z_o}\right|^2 \\ &= (|A|^2 - |B|^2)(|\alpha_0|^2 - |\beta_0|^2) \end{aligned} \tag{G.16}$$

The above two equations give

$$n(|A|^2 - |B|^2) = 1 \tag{G.17}$$

Next, from Figure 11.1 and Equation G.3 we have

$$r_2 = \left.\frac{\beta_1 e^{-ik_1 z_1}}{\alpha_1 e^{ik_1 z_1}}\right|_{\beta_0=0} = \frac{B^*}{A}, \qquad r_2' = \left.\frac{\alpha_0 e^{ik_0 z_0}}{\beta_0 e^{-ik_0 z_0}}\right|_{\alpha_1=0} = -\frac{B}{A} \tag{G.18}$$

$$T_2 = \left.\frac{\alpha_0 e^{ik_0 z_o}}{\alpha_1 e^{ik_1 z_1}}\right|_{\beta_0=0} = \frac{1}{A} \tag{G.19}$$

where the expression for r_2', the reflectivity for the wave incident from outside, has been given for completeness. When $\alpha_1 = 0$, Equation G.3 yields

$$\begin{aligned} 0 &= A\alpha_0 e^{ik_0 z_o} + B\beta_0 e^{-ik_0 z_o} \\ \beta_1 e^{-ik_1 z_i} &= B^* \alpha_0 e^{ik_0 z_o} + A^* \beta_0 e^{-ik_0 z_o} \end{aligned} \tag{G.20}$$

Thus

$$T_2' = \left.\frac{\beta_1 e^{-ik_1 z_1}}{\beta_0 e^{-ik_0 z_o}}\right|_{\alpha_1=0} = -B^* \frac{B}{A} + A^* = \frac{|A|^2 - |B|^2}{A} \tag{G.21}$$

From Equations G.18 and G.19 we have

$$\frac{|T_2|^2}{|r_2|} = \frac{|1/A|^2}{|B/A|} = \frac{1}{|AB|} \tag{G.22}$$

Using Equation G.17

$$\frac{|T_2|^2}{|r_2|} = \frac{n(|A|^2 - |B|^2)}{|AB|} = n\left(\left|\frac{A}{B}\right| - \left|\frac{B}{A}\right|\right) = n\left(\frac{1}{|r_2|} - |r_2|\right) \tag{G.23}$$

Also, from Equations G.18 and G.21 we have, using Equation G.17 again,

$$\begin{aligned} \frac{|T_2'|^2}{|r_2|} &= \frac{(|A|^2 - |B|^2)^2/|A|^2}{|B/A|} = (|A|^2 - |B|^2)\left(\left|\frac{A}{B}\right| - \left|\frac{B}{A}\right|\right) \\ &= \frac{1}{n}\left(\frac{1}{|r_2|} - |r_2|\right) \end{aligned} \tag{G.24}$$

Thus we have proved Equations 11.52a and 11.52b for mirror M2. These calculations leading to Equations G.23 and G.24 can be repeated for mirror M1, with results where r_2 is replaced by r_1. To do this, we may rotate Figure 11.1 by 180° so that the mirror M1 comes to the right-hand side of the cavity. Then, except for the particular forms of the constants A and B in Equation G.3, all the procedure from Equations G.1 to G.24 are applicable with replacement of the suffix 2 by suffix 1 for r, r', T, and T''.

Appendix H
Spontaneous Emission Spectrum Observed Outside the Cavity

Here we show that the spontaneous emission spectrum observed outside the cavity is given in general by Equation 13.65. We start with Equation 13.58. We first use the definition in Equation 13.61 of $D(t)$. Thus Equation 13.58 reads

$$
\begin{aligned}
I(z_B,t) &= \left|\sum_j \mu_A^* \frac{1}{2}\omega_j U_j(z_B)U_j(z_A)e^{-i(\omega_j-\omega_A)t}\int_0^t dt'\, e^{i\omega_j t'}e^{-i\omega_A t'}C_u(t')\right|^2 \\
&= |\mu_A|^2|\omega_A|^2\left|\sum_j \frac{1}{2}U_j(z_B)U_j(z_A)e^{-i(\omega_j-\omega_A)t}\int_0^t dt'\, e^{i\omega_j t'}D(t')\right|^2
\end{aligned}
\tag{H.1}
$$

where we have taken ω_j out of the summation sign and replaced it by ω_A because the spectrum of $D(t)$ is sharply peaked at ω_A. We formally rewrite $D(t')$ in the form of the inverse Fourier transform:

$$
\begin{aligned}
&I(z_B,t) \\
&= |\mu_A|^2|\omega_A|^2\left|\sum_j \frac{1}{2}U_j(z_B)U_j(z_A)e^{-i(\omega_j-\omega_A)t}\int_0^t dt'\, e^{i\omega_j t'}\int_{-\infty}^{\infty} D(\omega)e^{-i\omega t'}d\omega\right|^2 \\
&= |\mu_A|^2|\omega_A|^2\left|\int_{-\infty}^{\infty}\sum_j \frac{1}{2}U_j(z_B)U_j(z_A)e^{-i(\omega_j-\omega_A)t}\int_0^t dt'\, e^{i\omega_j t'}e^{-i\omega t'}D(\omega)d\omega\right|^2
\end{aligned}
\tag{H.2}
$$

where we have moved the integration sign for frequency to the top. As we are now considering the spectrum, we may take the time t to be large, so that the time integral becomes

$$
\int_0^t dt'\, e^{i\omega_j t'}e^{-i\omega t'} \to \int_0^{\infty} dt'\, e^{i(\omega_j-\omega)t'} = i\zeta(\omega_j-\omega) \tag{H.3}
$$

where the zeta function was defined in Equation 2.53b and is here meaningful only for $\omega_j = \omega$. Thus we have

$$
\begin{aligned}
I(z_B,t) &= |\mu_A|^2|\omega_A|^2 \\
&\times\left|\int_{-\infty}^{\infty}\left\{\sum_j \frac{1}{2}U_j(z_B)U_j(z_A)\zeta(\omega_j-\omega)\right\}e^{-i(\omega-\omega_A)t}D(\omega)d\omega\right|^2
\end{aligned}
\tag{H.4}
$$

Now the quantity in the curly bracket is i times the response function $Y(z_B,z_A,\omega)$ as given by Equation 2.53a. Thus we have

$$I(z_B, t) = |\mu_A|^2|\omega_A|^2 \left| \int_{-\infty}^{\infty} Y(z_B, z_{A,}\omega)e^{-i(\omega-\omega_A)t} D(\omega)d\omega \right|^2$$
$$= |\mu_A|^2|\omega_A|^2 \left| \int_{-\infty}^{\infty} Y(z_B, z_{A,}\omega)D(\omega)e^{-i\omega t} d\omega \right|^2 \quad \text{(H.5)}$$

Thus the field amplitude at the observation point z_B is proportional to the integral inside the absolute value sign, which is in the form of the inverse Fourier transform of $Y(z_B, z_{A,}\omega)D(\omega)$. Thus the Fourier spectrum of the field amplitude is proportional to $Y(z_B, z_{A,}\omega)D(\omega)$ and for the power spectrum we have

$$S(z_A, z_B, \omega) = \omega_A^2|\mu_A|^2|D(\omega)|^2|Y(z_B, z_A, \omega)|^2 \quad \text{(H.6)}$$

Appendix I
Correspondence of the Noise Polarization to the Noise Field

Here we show the correspondence of the noise polarization in Equation 14.15 and the noise field in Equation 14.110. Both of these represent the quantum noise associated with spontaneous emission. The temporal and spatial correlation for the noise polarization in the classical sense is written from Equation 14.15 as

$$\langle \mathbf{P}^*(z, t)\mathbf{P}(z', t')\rangle = \frac{4\hbar\varepsilon}{\omega} g N_2 \delta(t - t')\delta(z - z')$$

and from Equation 14.110 the reservoir average of the quantum-mechanical expectation value for the product of the noise electric field is

$$2\varepsilon_1 c_1 \left\langle \hat{f}_m^\dagger(t)\hat{f}_{m'}(t')\right\rangle = \{g(1+\sigma)\hbar\omega/c_1\}\delta_{mm'}\delta(t - t')$$

These noise terms appear in Equations 14.12 and 14.93, respectively, as

$$2i\dot{a}(t)\varepsilon C_N \Psi(z) = \frac{\kappa^2 - \varepsilon\mu_0\omega^2}{\mu_0\omega} a(t)C_N\Psi(z) + \omega\tilde{\mathbf{P}}(z, t)$$

and

$$\frac{de(z, t)}{dt} = s_0 e(z, t) + \frac{c_1}{2d} F_q(z, t)$$

In Equation 14.93 cited above, we have omitted the factor $\gamma'/(\gamma' + \gamma_c')$ describing the bad cavity and detuning effects, which are not considered in Equation 14.15. Also, in Equation 14.93 above, the thermal noise term has been omitted. $F_q(z,t)$ and $\hat{f}_m(t)$ are related as in Equation 14.94b. In view of the expression for the electric field in Equation 14.8, Equation 14.12 above may be rewritten formally as

$$\frac{d}{dt}e(z,t) = s_0 e(z,t) + \frac{\omega}{2i\varepsilon}\tilde{\mathbf{P}}(z,t) \tag{I.1}$$

where e (z,t) is the slowly varying envelope of the positive frequency part of the electric field. In view of Equation 14.107, where the noise source projected onto the adjoint mode function reproduces the adjoint mode function, the noise F_q (z,t) may be written effectively as

$$F_q(z,t) = 2d\sum_m \hat{f}_m \delta(z - z_m) \tag{I.2}$$

We may convert the discrete noise source in Equation I.2 into a spatially continuous noise field

$$F_q(z,t) = 2d\sum_m \hat{f}_m \delta(z - z_m) \to 2d\hat{f}(z,t) \tag{I.3}$$

with the property

$$\begin{aligned}\left\langle \hat{f}^\dagger(z',t')\hat{f}(z,t)\right\rangle &= \sum_{m'}\sum_m \delta(z'-z_{m'})\delta(z-z_m)\left\langle \hat{f}^\dagger_{m'}(t')\hat{f}_m(t)\right\rangle \\ &= \sum_m \delta(z'-z)\left\langle \hat{f}^\dagger_m(t')\hat{f}_m(t)\right\rangle \\ &= \frac{g(1+\sigma)\hbar\omega/c_1}{2\varepsilon_1 c_1} N\delta(z'-z)\delta(t-t')\end{aligned} \tag{I.4}$$

where Equation 14.110 has been used in the third line. The spatial density of atoms N per unit length has been taken into account. Now, by Equations I.2 and I.3, Equation 14.93 is rewritten as

$$\frac{de(z,t)}{dt} = s_0 e(z,t) + c_1 \hat{f}(z,t) \tag{I.5}$$

and the last terms of Equations I.1 and I.5 correspond to each other. We take the correlation functions of these two and compare them. For Equation I.1 we have

$$\begin{aligned}\left\langle \left\{\frac{\omega}{2i\varepsilon}\tilde{\mathbf{P}}(z,t)\right\}^* \frac{\omega}{2i\varepsilon}\tilde{\mathbf{P}}(z,t)\right\rangle &= \frac{\omega^2}{4\varepsilon^2}\frac{4\hbar\varepsilon}{\omega} g N_2 \delta(t-t')\delta(z-z') \\ &= \frac{\hbar\omega}{\varepsilon} g N_2 \delta(t-t')\delta(z-z')\end{aligned} \tag{I.6}$$

where Equation 14.15 has been used. For Equation I.5 we have

$$\begin{aligned}\left\langle \left\{c_1\hat{f}(z,t)\right\}^\dagger c_1\hat{f}(z,t)\right\rangle &= c_1^2 \frac{g(1+\sigma)\hbar\omega/c_1}{2\varepsilon_1 c_1} N\delta(z'-z)\delta(t-t') \\ &= \frac{g(1+\sigma)\hbar\omega}{2\varepsilon_1} N\delta(z'-z)\delta(t-t')\end{aligned} \tag{I.7}$$

where Equation I.4 has been used. Since $N(1+\sigma) = 2N_2$ and the ε in Equation I.6 is ε_1 if applied to the case of the two-sided cavity in Figure 11.1, the correlation function in Equation I.6 for the noise "polarization" has an exact correspondence to that for the noise "field" in Equation I.7. Note that Equation I.6 is a classical reservoir average, whereas Equation I.7 is a reservoir average of a quantum-mechanical expectation value.

References

1 Schiff, L.I. (**1955**) *Quantum Mechanics*, 2nd edn, McGraw-Hill, New York.

2 Haken, H. (**1970**) *Laser theory*, in *Licht und Materie, IC*, Handbuch der Physik, vol. XXV/2c (eds. S. Flügge and L. Genzel), Springer, Berlin.

Index

Output Coupling in Optical Cavities and Lasers: A Quantum Theoretical Approach
Kikuo Ujihara

ISBN: 978-3-527-40763-7